AF737794

Applied Elasticity and Plasticity

Applied Elasticity and Plasticity

Mumtaz Kassir

CRC Press
Taylor & Francis Group
Boca Raton London New York

CRC Press is an imprint of the
Taylor & Francis Group, an **informa** business

CRC Press
Taylor & Francis Group
6000 Broken Sound Parkway NW, Suite 300
Boca Raton, FL 33487-2742

© 2018 by Taylor & Francis Group, LLC
CRC Press is an imprint of Taylor & Francis Group, an Informa business

No claim to original U.S. Government works

Printed on acid-free paper

International Standard Book Number-13: 978-1-138-06191-0 (Hardback)

This book contains information obtained from authentic and highly regarded sources. Reasonable efforts have been made to publish reliable data and information, but the author and publisher cannot assume responsibility for the validity of all materials or the consequences of their use. The authors and publishers have attempted to trace the copyright holders of all material reproduced in this publication and apologize to copyright holders if permission to publish in this form has not been obtained. If any copyright material has not been acknowledged please write and let us know so we may rectify in any future reprint.

Except as permitted under U.S. Copyright Law, no part of this book may be reprinted, reproduced, transmitted, or utilized in any form by any electronic, mechanical, or other means, now known or hereafter invented, including photocopying, microfilming, and recording, or in any information storage or retrieval system, without written permission from the publishers.

For permission to photocopy or use material electronically from this work, please access www.copyright.com (http://www.copyright.com/) or contact the Copyright Clearance Center, Inc. (CCC), 222 Rosewood Drive, Danvers, MA 01923, 978-750-8400. CCC is a not-for-profit organization that provides licenses and registration for a variety of users. For organizations that have been granted a photocopy license by the CCC, a separate system of payment has been arranged.

Trademark Notice: Product or corporate names may be trademarks or registered trademarks, and are used only for identification and explanation without intent to infringe.

Visit the Taylor & Francis Web site at
http://www.taylorandfrancis.com

and the CRC Press Web site at
http://www.crcpress.com

Printed and bound in the United States of America by
Edwards Brothers Malloy on sustainably sourced paper

Contents

Preface

The purpose of this book is to introduce senior level, graduate students, and practicing engineers in civil, mechanical, aeronautical, and metallurgical engineering to the theories of elasticity and plasticity. It is based on lectures given by the author over many years at City College of New York–City University of New York (CCNY–CUNY).

After a brief introduction to elementary mechanics of materials in Chapter 1, Chapter 2 deals with the analysis of stress, including all analytical and graphical properties of conventional, deviatoric, and true stresses in two- and three-dimensional coordinate systems.

Engineering, mathematical, and deviatoric strains are discussed in Chapter 3. Graphical methods of determining strains as well as strain gauges are covered.

Data obtained from simple static tests of materials are used in Chapter 4 to derive the various elastic and plastic stress–strain relationships. Basic yield criteria and hardening rules are also discussed. The Haigh–Westergaard geometrical representation of the state of stress is used to represent any elastic and plastic states of stress in a body.

Chapter 5 covers torsion of structural members. Elastic, plastic, and work-hardening materials are used as examples to illustrate the basic principles. Torsion of thin-walled members and restrained structural members are included.

Chapter 6 presents two-dimensional problems (plane stress and plane strain) using rectangular and polar coordinate systems. Elastic and plastic solutions are provided for beam loaded by concentrated force, uniform and hydrostatic loads, and moments; wedge loaded by line loads and moments; semi-infinite solid subjected to vertical and horizontal loads and moments; rotating rings; and solid and annular disks (thin, variable thickness, and uniform strength).

Chapter 7 completes the elastic and plastic analyses of two-dimensional problems by discussing thick-walled cylindrical tubes with various end conditions (closed, open, and plane-strain condition) subjected to internal and external pressures and combined with external loads and moments. Compound and multilayered cylinders are discussed. Influence of strain hardening is included. Thick-walled spherical shell is discussed in the elastic, plastic, and strain hardening ranges. Influence of change in geometry of the sphere is studied. Chapter 7 also includes study of stress concentration factors. Analytical elastic, plastic, and work-hardening solutions of infinite sheet containing circular hole subjected to various loads are discussed. Methods of determining the stress-intensity factors in structural members containing cracks are examined.

For problems where analytical solutions do not exist, the need for computer codes such as SAP, ANSYS, ABACUS, and others is discussed.

Chapter 8 deals with the induced deformation and thermal stresses in structural components subjected to nonuniform changes in temperature. Examples include beams, circular thin disks (solid and annular), and thick-walled cylindrical and spherical vessels. Initiation of yield, elastic–plastic expansion as well as influences of internal pressure and rotational stresses are examined for vessels with various wall thickness ratios.

Three-dimensional analytical solutions are developed in Chapter 9 for a number of structural components, such as rotation of a circular disk, deformation of a circular cylinder subjected to forces acting on the surface, contact between two spherical bodies, arbitrary loads acting on the boundary of semi-infinite solid, and loads acting on the surfaces of a semi-infinite cut or crack in a large body. Wherever possible the results are compared to the corresponding two-dimensional problem considered in Chapters 6 and 7.

Chapter 10 deals with three-dimensional thermal stresses induced in structures experiencing nonuniform change in temperature. Solutions are obtained for long circular cylinder (solid and annular), sphere (solid and with central hole), and large bodies containing flaws or cracks where the crack surface is subjected to temperature, temperature gradient, or it is insulated and subjected to the path of heat flows.

Every chapter contains a number of solved examples, problems to be solved, and a list of references. Solutions to selected problems from the chapters are given in Chapter 11.

I wish to thank Professors Michel Ghosn and Feridun Delale of CCNY–CUNY for reviewing some sections of the manuscript and the many students who took my course. All contributed to various aspects of improving the manuscript.

Author

Mumtaz Kassir earned a BSc (Hons.) degree in building and construction engineering at the University of Manchester (UK) in 1957. In 1963 and 1966, he earned an MS in structural engineering at Stanford University and a PhD in applied mechanics at Lehigh University, respectively.

Dr. Kassir has worked in civil engineering and construction works in Baghdad, Iraq, during 1957–1958. He also worked as an instructor at the University of Baghdad during 1959–1961 and at CCNY (City College of New York): CUNY (City University of New York) Department of Civil Engineering as an assistant professor during 1966–1970, associate professor during 1971–1975, and professor during 1976–1998.

He was associate dean for Graduate Studies and Executive Officer of the PhD Program in Engineering at the Graduate Center and CCNY of CUNY during 1999–2013 and retired in 2014.

Dr. Kassir and G. C. Sih wrote a book titled *Three-Dimensional Crack Problems*. It is 452 pages and was published by Noord–Hoff International Publishing, The Netherlands in 1975.

Dr. Kassir's research interests have resulted in a total of 52 journals papers and numerous technical reports and conference presentations in the United States, Greece, Holland, China, and Japan.

Dr. Kassir has received several grants from the following agencies as principal investigator and co-principal investigator.

He was principal investigator at National Science Foundation (NSF) and Brookhaven National Laboratory (Long Island, New York) during 1968–1995. He was co-principal investigator with Lehigh University faculty–Naval Research, AFOSR (Air Force Office of Scientific Research), NASA (National Aeronautics and Space Administration), during 1970–1980 and also co-principal investigator with Alfano and Ho at CCNY–AFOSR.

1

Introduction

When external loads are applied to structural components, internal forces and deformations are induced. The objective of this book is to determine the induced stresses and strains, instead of focusing on internal forces and deformations (changes in lengths, angles, shapes, etc.). This is because the measured effect of a load on a specimen of the material depends on the cross-sectional area. In addition, changes in lengths and angles are measured relative to their initial or instantaneous values. It is the purpose of this book to determine the response of structural components to external loads as manifested by the induced stresses and strains. Elastic and plastic behaviors are considered using continuum mechanics theories. When loads are removed, the behavior is elastic if the component returns to its original shape. The behavior is called plastic if permanent deformations remain. Whether the behavior is elastic or plastic, the response is determined by studying three distinct aspects of the behavior as follows:

1. *Equilibrium*: This ensures that external and internal forces (or stresses) are in equilibrium. Basically, this requires the stresses in any structural component to satisfy six equations of statics in space (three equations of force equilibrium and three equations of moment equilibrium).
2. *Geometry of deformation*: Under the action of applied loads, structures deform as a result of changes in shape and volume. It is essential to consider the nature of the deformation in determining the response to external loads. This involves studying the geometry of deformation and determining the governing equations. In continuum mechanics, the relations between strains and displacements form the governing equations of deformation.
3. *Mechanical behavior of the material*: The third aspect of the analysis involves the relation between stress and strain as determined by standard tests in the laboratory. For elastic bodies, Hooke's law governs the mathematical stress–strain relations up to and including the yield point indicating maximum elastic behavior. Beyond the yield point, the behavior of the material is plastic, strain hardening occurs during the deformation until the ultimate stress and material failure. Several mathematical models are available to predict the plastic behavior.

In order to illustrate the procedure, the strength-of-materials approach to solving few simple problems is briefly reviewed in the following sections, References 1.1, 1.2.

1.1 Axial Extension

Consider a prismatic member subjected to two equal forces applied in opposite directions along the longitudinal axis (Figure 1.1). The origin of coordinates is located at the center

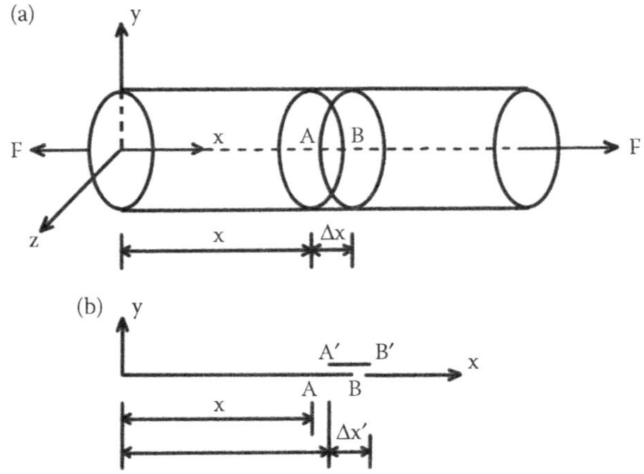

FIGURE 1.1
Elongation of axially loaded prismatic member.

of the cross section in the x-direction. Because of the nature of loading, the member remains straight after deformation and is expected to stretch or shorten along the x-axis.

Deformation: The features of the deformation are assumed as follows: (i) the axis of the member remains straight, (ii) plane cross sections remain plane parallel to the y–z plane, and displace along the x-axis.

Consider two plane cross sections with distance Δx apart as shown in Figure 1.1a. In the deformed state, these planes are displaced and length Δx becomes Δx′. The conventional axial strain is defined as

$$\varepsilon_x = \mathrm{Lim}_{\Delta x \to 0} \frac{\Delta x' - \Delta x}{\Delta x} \tag{1.1}$$

The axial strain is constant on any cross section and may vary with x, but it is independent of y and z. The deformation of structural members subjected to axial loading does not depend on the material, it is applicable to all materials.

Equilibrium: The overall equilibrium is satisfied because the external forces are equal and opposite. Since, on any cross section, the internal force has magnitude F and it is along the horizontal axis, the normal stress is given by $\sigma_x = F/A$, where A is the area of the cross section. It can be assumed that the remaining normal stresses, σ_y and σ_z, vanish throughout the member. Consider an element of area $\Delta A = (\Delta y)(\Delta z)$ at a typical cross section of the member (Figure 1.2), then $\Delta F = \sigma_x \Delta A$. Integrating over the cross-sectional area, it follows that

$$F = \iint_A \sigma_x dA = \sigma_x A \tag{a}$$

and

$$\sigma_x = \frac{F}{A} \tag{1.2}$$

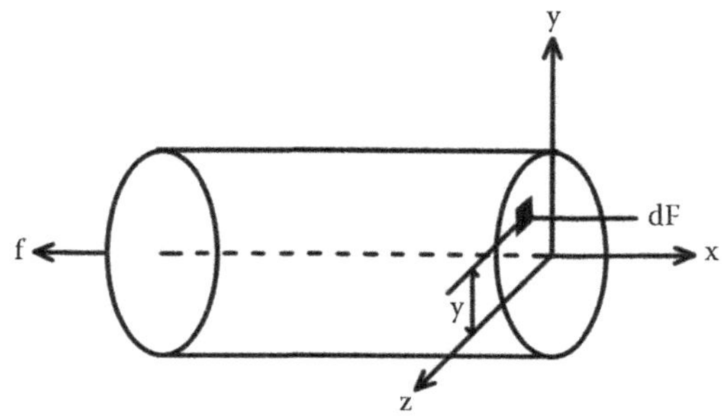

FIGURE 1.2
Internal forces of axially loaded member.

The moments of ΔF about the y and z axes are $\Delta M_y = z\Delta F$ and $\Delta M_z = y\Delta F$, respectively. Hence,

$$M_y = \iint_A z\sigma_x dA = \sigma_x \iint_A zdA$$
$$M_z = \iint_A y\sigma_x dA = \sigma_x \iint_A ydA \tag{b}$$

Because the centroid of the cross section is the origin of the coordinates, it follows that $\iint_A zdA = \iint_A ydA = 0$. Hence, $M_y = M_z = 0$, and the axial force is the only internal force in the member.

The elongation (or contraction) of the member is given by

$$e = \int_0^L (dx' - dx) = \int_0^L \varepsilon_x dx \tag{1.3}$$

Here, L is the original length of the member.

Mechanical behavior of material is determined by performing standard tensile test on a specimen of the material to find the axial stress and strain until failure. For a typical metal, the experimental data yield the curve shown in Figure 1.3a.

The initial experimental data are represented by the straight line 0A, where A represents the yield point of the material. The stress and strain at yield are represented by σ_0 and ε_0, respectively. This behavior is commonly referred to as elastic or reversible behavior because loading and unloading traces the same line 0A and the material recovers its initial position after removal of the load. In the elastic range, the relationship between stress and stain is represented by the one-dimensional (1D) Hooke's law

$$\varepsilon_x = \frac{\sigma_x}{E} \tag{1.4}$$

where E is the modulus of elasticity of the material (sometimes referred to as Young's modulus). Since E is constant and ε_x is a function of x only, it follows that σ_x is independent

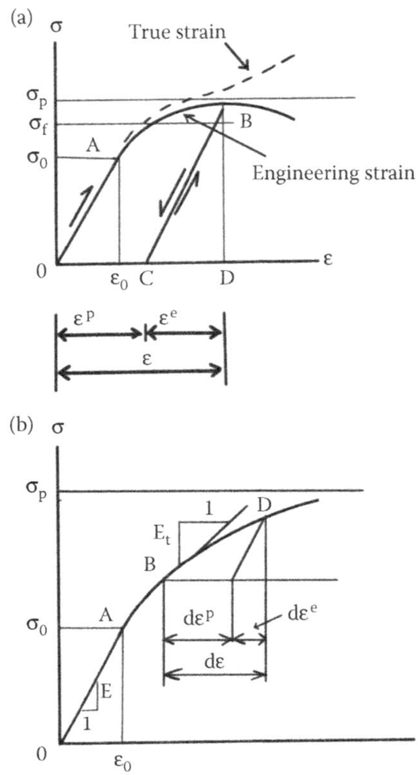

FIGURE 1.3
Stress–strain diagram of a ductile material, (a) engineering and true stress versus strain and (b) plastic stress–strain incruent.

of y and z but can vary with x. Beyond point A, the material behaves plastically, plastic deformation is irreversible, that is, permanent. Also, there is no volume change of the material during plastic deformation, that is, the material is incompressible and Poisson's ratio is equal to 1/2. In the plastic range, the slope of the stress–strain curve decreases progressively and eventually the material fails at the rupture stress shown in Figure 1.3a. The solid line in Figure 1.3 represents the engineering stress while the dotted line is obtained by plotting the "true" stress, which is defined as the load divided by the instantaneous cross-sectional area of the member. The ultimate or plastic stress is denoted by σ_p. If the load is removed from a point in the plastic range, say point B, in Figure 1.3a, a straight line BC parallel to the initial line 0A is traced indicating that a permanent strain, $\varepsilon_p = 0C$, is developed when the load is completely removed. If the load is applied again, the stress–strain relation follows line CB. The total strain at point B consists of elastic and plastic components, $\varepsilon = \varepsilon^e + \varepsilon^p$ (skipping the x-subscript).

In the plastic range, because there is no unique relation between stress and strain, except in dealing with the *"Deformation theory of plasticity,"* an incremental approach is used to determine the response of structural components to applied loads. A small increment BD in Figure 1.3b is considered, it consists of elastic and plastic parts,

$$d\varepsilon = d\varepsilon^e + d\varepsilon^p \tag{1.5a}$$

The corresponding stress increment is

$$d\sigma = E_t d\varepsilon = E d\varepsilon^e = E_p d\varepsilon^p \tag{1.5b}$$

Here, E is the modulus of elasticity of the material, E_t is the tangent modulus, which is slope of the tangent to the stress–strain curve at point B in Figure 1.3b, and E_p is the plastic modulus. The tangent modulus is obtained from experimental stress–strain data obtained under monotonic loading condition using a given hardening rule. It is typically a fraction of Young's modulus, that is, $E_t = k\,E$, $k \leq 1$. The plastic modulus E_p is discussed further in Chapter 4, Section 4.2.

Note that Equations 1.5a and 1.5b give

$$\frac{1}{E_t} = \frac{1}{E} + \frac{1}{E_p} \tag{1.6}$$

Some typical models of stress–strain relationships for elastic–plastic materials include the following.

Elastic–Perfectly Plastic: Here, hardening of the material is neglected after reaching the yield stress.

$$\begin{aligned} \varepsilon_x &= \frac{\sigma_x}{E}, \quad \sigma_x < \sigma_0 \\ \varepsilon_x &= \frac{\sigma_x}{E} + C, \quad \sigma_x = \sigma_0 \end{aligned} \tag{1.7}$$

where C is a positive constant. In this model, once the yield stress is reached, the strain increases without limit.

Linear-Hardening Material: In this model, the material behavior is modeled by two straight line segments as shown in Figure 1.4. The initial segment represents elastic behavior up to the yield stress, σ_0, and the plastic part is represented by another straight line having a slope

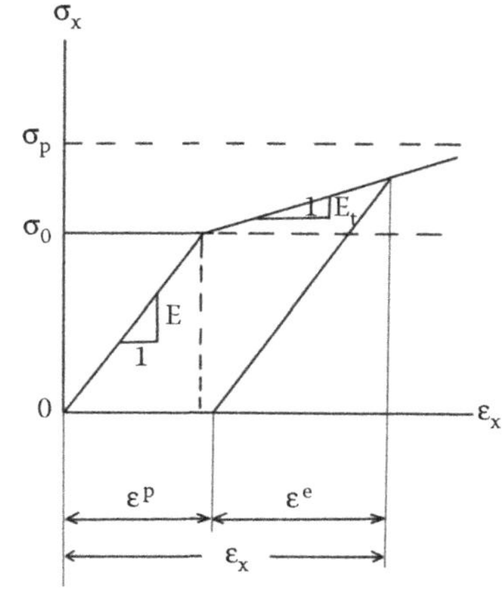

FIGURE 1.4
Stress–strain diagram of a linear-hardening material.

equal to the tangent modulus, E_t. The stress–strain relationship is expressed as

$$\begin{aligned} \varepsilon_x &= \frac{\sigma_x}{E}, \quad \sigma_x \le \sigma_0 \\ \varepsilon_x &= \frac{\sigma_x}{E} + \frac{\sigma_x - \sigma_0}{E_t}, \quad \sigma_x > \sigma_0 \end{aligned} \tag{1.8}$$

The tangent modulus is typically a fraction of Young's modulus, $E_t = k\,E$, $k \le 1$. When $k = 0$, the upper straight line in Figure 1.4 becomes horizontal (elastic–perfectly plastic material) and the elongation increases without bound. Additional models are given in Chapter 4.

EXAMPLE 1.1

A rigid bar AB of negligible weight supports a load, W, at the free end B as shown in Figure 1.5. The cylindrical wire, CD, is made of a linear-hardening material with yield stress $\sigma_0 = 36 \times 10^3$ psi, modulus of elasticity $E = 29 \times 10^6$ psi, and $E_t = 0.1\,E$. Find the relation between W and the angle of rotation, θ, of the bar.

Solution

Denote the tension in wire CD by F. The moments of forces, W and F, about point A give $F = 4\,W$. Denote the angle of rotation of the bar by θ and the elongation of the wire by e. Because the angle of rotation is small, it follows that

$$e = 6 \sin\theta \approx 6\theta \tag{a}$$

The elongation in the elastic range is computed by inserting the first equation in relations (1.8) into Equation 1.4 to yield

$$e = \frac{FL}{EA} = \frac{4W(12)}{29 \times 10^6 (\pi)(1/2)^2} = 2.107 \times 10^{-6}\,W$$

provided that W is less than $\sigma_0 A/4$.

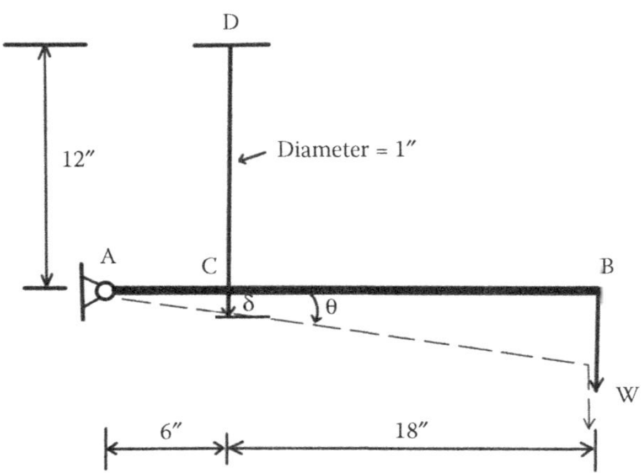

FIGURE 1.5
Example 1.1.

Thus,

$$\theta \leq 0.351 \times 10^{-6}\,W, \quad W \leq 7069\,\text{lbs} \tag{b}$$

When W is greater than $(\sigma_0 A/4)$, the second equation in Equation 1.8 is inserted into Equation 1.4 giving $e = (FL/EA) + (L/E_t A)(F - F_0)$, $F > F_0$, where F_0 is the yield value of F. Carrying out the calculation, it is found that $e = 23.181 \times 10^{-6}\,W - 0.149$.

Thus,

$$\theta = 3.864 \times 10^{-6}\,W - 0.0248, \quad W > 7069\,\text{lbs} \tag{c}$$

EXAMPLE 1.2

A fixed ended member consisting of length L is subjected to an axial load at section AA at distance L/3 from the left support as shown in Figure 1.6. The material's behavior is modeled as elastic–perfectly plastic with yield stress $\sigma_0 = 250$ MPa and modulus of elasticity $E = 200$ GPa. If the cross-sectional area is rectangular (2 cm × 4 cm) and L = 0.9 m, determine:

a. The load that initiates yield, F_e, the ultimate load, F_u, and the corresponding displacement, u, of section AA.
b. The residual stresses if the ultimate load is removed.

Solution

The free body diagram of the member consists of two segments as shown in Figure 1.6b. R_a and R_b denote the reactions at the ends. Equilibrium of the axial forces gives $R_a + R_b = F$. In terms of stresses $(\sigma_a = R_a/A,\ \sigma_b = -R_b/A)$, equilibrium is satisfied by

$$\sigma_a - \sigma_b = \frac{F}{A} \tag{a}$$

Since the ends are restrained, the deformation of the member implies that total extension of the two segments is zero, that is, $e_a + e_b = 0$. In term of strains, the equation of compatibility of deformation is

$$a\varepsilon_a + b\varepsilon_b = 0 \tag{b}$$

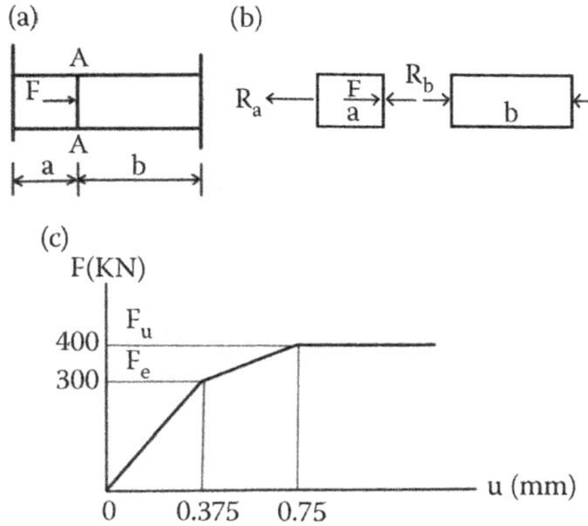

FIGURE 1.6
Example 1.2. (a), (b) FBD and (c) F versus u (mm).

a. Considering the elastic behavior, Equation (b) in term of stresses is

$$a\sigma_a + b\sigma_b = 0 \tag{c}$$

Solving for the stresses from Equations (a) and (c) gives $\sigma_a = bF/LA$ and $\sigma_b = -aF/LA$. Since $b = 0.6$ and $a = 0.3$ m, the magnitude of σ_a is greater than that of σ_b. Hence, segment a will yield first while segment b remains elastic. The magnitude of the force that initiates yield is

$$F_e = \left(\frac{L}{b}\right)\sigma_0 A = 300\ \text{KN} \tag{d}$$

The strains are $\varepsilon_a = \varepsilon_0$ and $\varepsilon_b = (a/b)\varepsilon_0$. The displacement of section AA is $u = a\ \varepsilon_0 = 0.375$ mm.

When the applied force increases further, the yield stress in segment a remains constant while the load is totally carried by segment b till it yields at stress $(-\sigma_0)$. Equation (a) gives the ultimate or plastic load as

$$F_u = 2\sigma_0 A\ = 400\ \text{KN} \tag{e}$$

The corresponding strains are $\varepsilon_b = -\ \varepsilon_0$, $\varepsilon_a = (b/a)\varepsilon_0$, and the displacement of section AA is $u = b\ \varepsilon_0 = 0.75$ mm.

The variation of the applied force, F, with the displacement, u, is shown in Figure 1.6c.

b. Unloading from the ultimate force follows elastic behavior. Thus, adding an increment—$F_u = -2\sigma_0 A$ to the current stresses—the residual stresses are

$$\begin{aligned} \sigma_a^R &= \sigma_0 - \frac{2b}{L}\sigma_0 \\ \sigma_b^R &= -\sigma_0 + \frac{2a}{L}\sigma_0 \end{aligned} \tag{f}$$

The residual stresses obtained from Equation (f) are

$$\sigma_a^R = \sigma_b^R = -0.333 \times 250 = -83.25\ \text{MPa}$$

1.2 Bending of Symmetric Beams

Consider a prismatic beam bent by two equal and opposite moments M, applied at the ends as shown in Figure 1.7. During bending the top fibers contract while the bottom ones elongate, and in between, there is a neutral plane with zero deformation. The origin of rectangular coordinates, 0, is chosen on the neutral plane at the centroid of the cross section, 0x is the longitudinal axis, 0y is the axis of symmetry, and 0z is the bending axis. The non-vanishing stress is σ_x.

Equilibrium: The nontrivial equations of equilibrium are the sums of forces along the axis of the beam and moments about the z-axis, namely, $\Sigma F_x = 0$, $\Sigma M_z = 0$, which give rise to the equations

$$\iint_A \sigma_x dA = 0 \tag{1.9}$$

$$\iint_A \sigma_x y dA = -M \tag{1.10}$$

where A is the cross-sectional area of the beam.

Deformation: Under the action of the end moments, the upper part of the beam is in compression and the lower part is in tension. This implies that the upper longitudinal fibers shorten and the lower lines elongate, and somewhere in between there is a neutral surface, which experiences zero strain. The x-axis is taken along the longitudinal axis of the beam in the neutral surface and the intersection with the y-axis gives rise to the neutral axis of the cross section. It can be shown experimentally by using strain gages to measure the strains that the longitudinal axis of the beam is bent into a circular arc and any cross sections of the beam, such as AA or BB in Figure 1.7, remain plane and normal to the longitudinal lines. Now, we use the geometry of the deformed beam to compute the longitudinal strain, ε_x, of a typical longitudinal line of length dx at distance y above the x-axis. The undeformed length is $dx = \rho d\theta$, where ρ is the radius of curvature of the longitudinal axis. The deformed length is $(\rho - y)\, d\theta$. It follows that the elongation of the line element is $-y\,(dx/\rho)$ and the longitudinal strain is

$$\varepsilon_x = -y/\rho \tag{1.11}$$

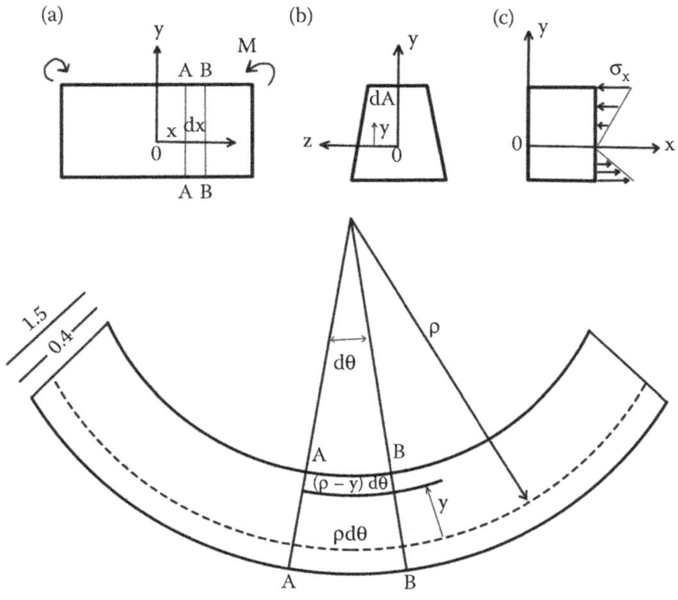

FIGURE 1.7
Deformations of a straight beam in pure bending, (a) beam section, (b) cross section, and (c) stress distribution.

which shows that the longitudinal strain varies linearly with the distance y from the neutral surface. The accompanying transverse strains in the y and z directions are

$$\varepsilon_y = \varepsilon_z = -\upsilon \frac{y}{\rho} \tag{a}$$

Here, υ denotes Poisson's ratio of the material.

Note that up to now the material of the beam did not enter the analysis and hence the above equations apply to beams of all materials and at every stage of the loading. To complete the analysis, the mechanical properties of the material are invoked.

Linear Elastic Material: Hooke's law for 1D stress is

$$\sigma_x = E\varepsilon_x = -E\frac{y}{\rho} \tag{1.12}$$

where E is Young's modulus of elasticity.

Inserting the axial stress from Equation 1.12 into Equation 1.9 and noting that E and ρ are constants, Equation 1.9 yields

$$\iint_A y\,dA = 0 \tag{1.13}$$

Equation 1.13 implies that the first moment of the cross-sectional area with respect to the z-axis is zero. Hence for elastic beams subjected to pure bending, *the neutral axis of the beam passes through the centroid of the cross section.* The neutral axis experiences no deformation.

Inserting Equation 1.12 into Equation 1.10, and noting that $\iint_A y^2 dA = I$, where I is the moment of inertia of the cross section of the beam, it follows that the moment–curvature relation in the elastic range is

$$\frac{1}{\rho} = \frac{M}{EI} \tag{1.14}$$

Substituting the expression for the radius of curvature from Equation 1.12 gives

$$\sigma_x = -\frac{My}{I} \tag{1.15}$$

If the beam is bent by transverse loads acting in the plane of symmetry, the bending moment varies along the length of the beam and the corresponding variation in the curvature causes deflection of the bent axis commonly known as the deflection curve. The deflections due to axial forces and transverse shears are small compared to that due to bending and may be neglected.

If M_e denotes the moment that initiates yield then, $1/\rho_e = M_e/EI$, ρ_e is the radius of curvature at yield initiation. And

$$\frac{M}{M_e} = \frac{\rho_e}{\rho}, \quad M \le M_e \tag{1.16}$$

For a beam with rectangular cross section, width b and height 2 h, the moment that initiates yield at the top and bottom fibers, M_e, has magnitude

$$M_e = \frac{\sigma_0 I}{(h/2)} = \frac{bh^2}{6}\sigma_0 \tag{1.17}$$

where σ_0 denotes the yield stress of the material.

The strain at any point is given as

$$\varepsilon_x = \frac{\sigma_x}{E} = -\frac{My}{EI} \tag{b}$$

Other Materials: For nonelastic behavior, it is assumed that there is a neutral axis normal to the axis of symmetry of the cross section but not necessarily coinciding with the centroidal axis. During bending of straight beams, cross sections remain plane and normal to the neutral axis. The equations of equilibrium, Equations 1.9, 1.10, and 1.11, which describe the geometry of deformation of the beam, are used to locate the neutral axis and the bending moment. The procedure is illustrated by considering commonly used cross sections.

Consider a beam with rectangular cross section, width b and height h, whose material is elastic–perfectly plastic with yield stress σ_0. The beam is subjected to bending moments M at the ends. As the bending moment increases beyond the moment that initiates yield, M_e, the strains at the top and bottom fibers of the beam exceed the yield strain ε_0 and plastic zones are formed as shown in the shaded regions of Figure 1.8. If c denotes the distance of the plastic zone above and below the neutral axis, the bending stress σ_x across the cross section is given by

$$\begin{aligned} \sigma_x &= \mp\sigma_0, \quad -\frac{h}{2} \le y \le -c \\ \sigma_x &= \frac{y}{c}\sigma_0, \quad |y| \le c \end{aligned} \tag{c}$$

The bending moment is obtained from Equation 1.10. Because of symmetry of the cross section, the integration can be carried out using an element dA = bdy. Hence,

$$M = -\iint_A y\sigma_x \, dA = -2b \int_0^{h/2} \sigma_x \, dy$$

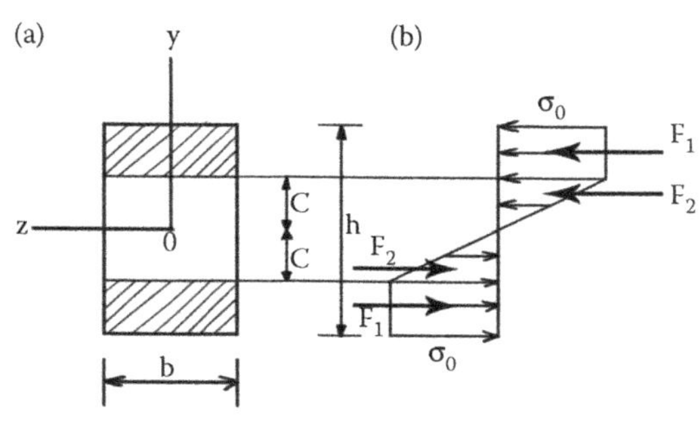

FIGURE 1.8
Elastic–plastic stresses in a beam of rectangular cross section subjected to pure bending (a) plastic zones (shaded) and (b) stress–strain diagram.

Substituting the values of the normal stress from Equation (d), the moment M acting on the cross section is evaluated as follows:

$$M = 2\frac{b\sigma_0}{c}\int_0^c y^2 dy + 2b\sigma_0 \int_c^{h/2} y dy = \sigma_0 b\left[\frac{h^2}{4} - \frac{c^2}{3}\right] \tag{1.18}$$

It should be noted that formula (1.18) may be obtained by taking moments of the stress resultants F_1 and F_2 shown in Figure 1.8b about the neutral axis. The magnitude of F_1 is equal to $b(h/2-c)\sigma_0$ and its line of action is located at distance $1/2(c + h/2)$ from the neutral axis. Similarly, $F_2 = 1/2\sigma_0 bc$ acting at distance $= (2/3)c$. Hence,

$$M = 2\left[\sigma_0 b\left(\frac{h^2}{4} - c^2\right) + \frac{2}{3}\sigma_0 b\left(\frac{h^2}{4} - \frac{c^2}{3}\right)\right]$$

as in Equation 1.18. The equilibrium of axial forces acting on the beam can be easily confirmed by substituting the stress distribution, Equation (c), into Equation 1.9 and performing the integration. This fact implies that for this beam the neutral axis passes through the centroid of the cross section at all stages of the loading, elastic, elastic–plastic, and completely plastic.

When $c = h/2$, Equation 1.18 gives the value of the moment that initiates yield, $M_e = (bh^2/6)\sigma_0$, as in Equation 1.17. When $c = 0$, the cross section becomes completely plastic, and the maximum moment carried by the beam is

$$M_p = \frac{bh^2}{4}\sigma_0 \tag{1.19}$$

Formulas of the maximum plastic moment M_p are useful in plastic analysis and design of structural components. For this reason, a *plastic section modulus*, Z, is introduced where

$$M_p = \sigma_0 Z \tag{1.20}$$

A table containing the values of the plastic section moduli of common structural shapes is available in Reference 1.3.

The ratio of the plastic to the maximum elastic moments is commonly referred to as the shape factor of the cross section

$$\text{Shape Factor} = \frac{M_p}{M_e} \tag{1.21}$$

For a rectangular cross section, the shape factor $= 3/2$.

Solving for c from Equation 1.18 and using the value of M_p, it is found that

$$c = \frac{h}{2}\left[3\left(1 - \frac{M}{M_p}\right)\right]^{1/2}, \quad M_e \le M \le M_p \tag{1.22}$$

Since $\varepsilon_0 = c/\rho = \sigma_0/E$, it follows from Equations 1.14 and 1.15 that the moment–curvature relation in the plastic range is

$$\frac{1}{\rho} = \frac{2\sigma_0}{Eh[3(1 - M/M_p)]^{1/2}}, \quad M_e \le M \le M_p \tag{1.23}$$

When M approaches M_p, the radius of curvature increases without limit and a plastic hinge is formed.

EXAMPLE 1.3

A beam with rectangular cross section, width $b = 6$ in. and depth $d = 12$ in., is bent so that the axial strain at the top is 0.003 (Figure 1.9). The material is assumed to be ideally plastic with yield stress 30×10^3 psi and $E = 30 \times 10^6$ psi. Find the depth of the yield zones in the beam and the bending moment causing the deformation.

Solution

The axial strain at $y = 6$ in. (top fiber) is related to the radius of curvature, R, by the relation in Equation 1.11, that is,

$$-0.003 = -6/R, \; R = 2000 \text{ in.}$$

The strain that initiates yield is obtained by dividing the yield stress by Young's modulus, that is, $30 \times 10^3/30 \times 10^6 = 0.001$. It follows that the top and bottom fibers of the beam yield first and then the yield zone spreads from the outside boundaries to a distance c above and below the neutral axis. The distance *c* is found from relation (1.11) as $c = (0.001)\,(2000) = 2$ in.

Hence, the depth of the plastic zones is 4 in. from the top and bottom fibers.

The moment causing the deformation is computed from Equation 1.10 as follows:

$$M = -2b\int_0^6 \sigma_x y\, dy = 2b\frac{E}{R}\int_0^2 y^2\, dy - 2b(-30 \times 10^3)\int_2^6 y\, dy$$

Evaluating the integrals, the moment is $M = 6,240$ Kip-in.

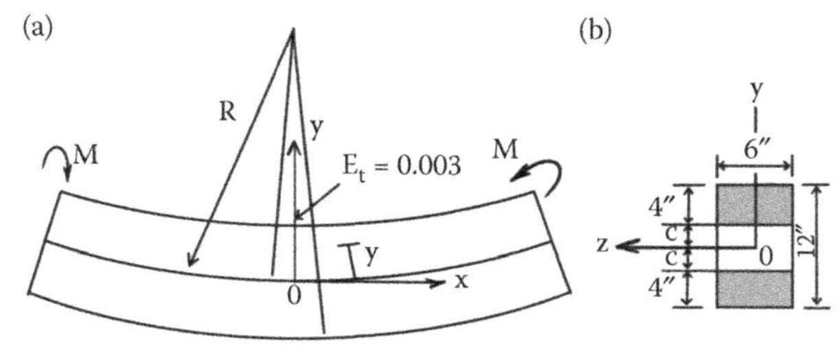

FIGURE 1.9
Example 1.3.

1.3 Torsion of Circular Members

Consider a solid circular bar of radius, a, length, L, fixed at one end and twisted by a torque T applied at the other end (free torsion). The origin of polar coordinates is located at the center of the cross section with the z-axis located along the axis of the bar as shown in Figure 1.10a.

The applied torque causes rotation about the central axis and induces shearing stresses and strains across any cross section of the bar normal to the z-axis. Because of polar symmetry, it is reasonable to assume that plane sections normal to the central axis remain plane after twisting. Also, for small angles of twist the length and radii of the bar will not change. These assumptions are no longer true for bars with noncircular cross sections as evidenced by known experimental results. Since one end of the bar is held firmly, the other end will rotate in the direction of the applied torque (Figure 1.10a), line AB moves to AB′. The angle γ, measured in radians, between lines AB and AB′ is the shear strain at the surface of the bar. Also, point B moves to B′ and the angle, ϕ, between lines CB and CB′ on the cross section is the angle of twist of the bar. The unit angle of twist, θ (should not be confused with polar coordinates), is obtained by dividing the angle of twist by the length of the bar

$$\theta = \frac{\phi}{L} \tag{a}$$

For small angles, $\tan\gamma = \gamma = BB'/L = a\,\phi$, it follows that by using relation (a), for pure torsion the shear strain at the surface of the bar is $\gamma = a\theta$.

In general, the shear strain at any point of the bar, distance r from the central axis, is given by the relation

$$\gamma = r\theta \tag{b}$$

For elastic material, using Hooke's law, the shear stress is

$$\tau = G\gamma = Gr\theta \tag{1.24}$$

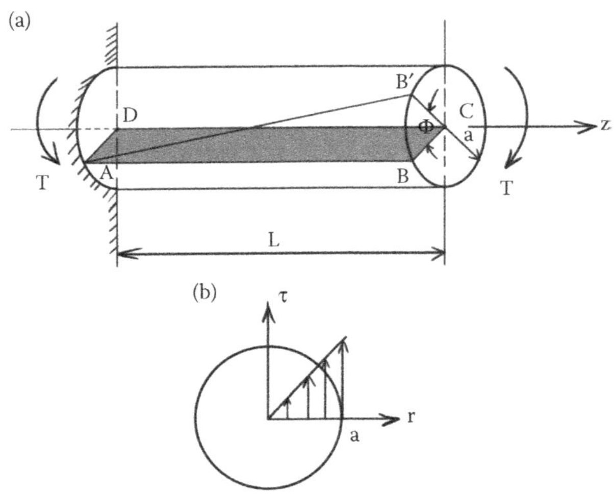

FIGURE 1.10
Circular bar subject to torque T, (a) variations of surface strains and (b) variation of shear stress.

where G is the shear modulus of elasticity. Equations (a) and 1.24 show that the shear stress and strain vary linearly with the radial distance from the center.

Equilibrium of the bar implies that the resultant moment of the shear stresses acting on the cross section must equal to the applied torques, that is,

$$T = \iint_A \tau(r) r dA = G\theta \iint_A r^2 dA$$
$$= G\theta J \qquad (1.25)$$

where J is the polar moment of inertia of the cross section. Hence, $\theta = TL/G$, and the angle of twist of the bar is

$$\phi = \frac{TL}{GJ} \qquad (1.26)$$

The distribution of the shear stress across the cross section is obtained from Equation 1.24.

$$\tau = \frac{Tr}{J} \qquad (1.27)$$

The shear stress varies linearly with the distance from the axis of the bar (Figure 1.10b) and the maximum shear stress occurs at $r = a$, $\tau_{max} = 2T/(\pi a^3)$.

The maximum elastic torque, which initiates yield in the bar, is given by

$$T_e = \frac{\pi \tau_0 a^3}{2} \qquad (1.28)$$

where τ_0 is the yield stress in shear. The corresponding unit angle of twist is

$$\theta_e = \frac{\tau_0}{Ga} \qquad (1.29)$$

When the shear stress exceeds the yield strength, the torsion problem becomes nonlinear and is discussed in Chapter 5.

EXAMPLE 1.4

A solid tapered bar is twisted by torques applied at the end, Figure 1.11.

The radius of the cross section varies linearly, from a, at the left-hand end to b ($b > a$), at the other end. For elastic behavior of the material, derive a formula for the angle of twist ϕ of the bar.

Solution

Consider an element of the bar of length dz at distance z from the left-hand end. The diameter of this section is given by

$$d = 2a + (2b - 2a)z/L \qquad (a)$$

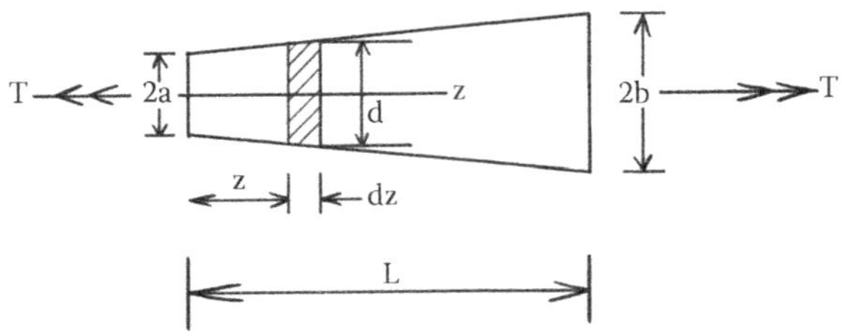

FIGURE 1.11
Example 1.4.

The polar moment of inertia of the section is

$$J(z) = \frac{\pi}{2}\left(a + \frac{b-a}{L}z\right)^4 \tag{b}$$

Hence, if the taper of the bar is very small, we have $d\phi = T\,dz/GJ$. Writing $b - a = c\,L$, c is constant, in relation (b) it follows that

$$\phi = \frac{2T}{\pi G}\int_0^L \frac{dz}{(a+cz)^4} = \frac{2T}{\pi G}\left[-\frac{1}{3c(a+cz)^3}\right]_0^L$$

Simplifying the algebra and introducing the dimensionless parameter $\eta = b/a$, the angle of twist of the bar is found to be

$$\phi = \frac{2TL}{\pi G a^4}\left(\frac{\eta^2 + \eta + 1}{3\eta^2}\right) \tag{c}$$

The angle of twist of a tapered bar with $\eta = 1.2$ is 0.7 times that of the prismatic bar where $\eta = 1$.

Problems

1.1 A steel member is initially 1.5 m long with a cross-sectional area of 100 mm^2. The member is stretched uniformly to an axial strain of 0.015 by a uniform load distributed on the ends. Find (a) the load needed to cause the strain; (b) the length of the member if the load is removed.
Assume linear-hardening material with $\sigma_0 = 250$ MPa, $E = 200$ GPa, and $E_t = E/12$.
Ans: 46.15 kN, 1.513 mm.

1.2 A stepped up axial member consisting of two segments AB and BC is subjected to an axial load F(kip) at the free end, Figure P1.2. Section AB has a cross-sectional area of 1 in.^{2}, yield stress 40 ksi and section BC has area of 0.5 in.^{2} and yield stress of

60 ksi. Assuming the elastic–perfectly plastic material with $E = 30 \times 10^6$ psi, find the relation between F and the axial extension e (in.).

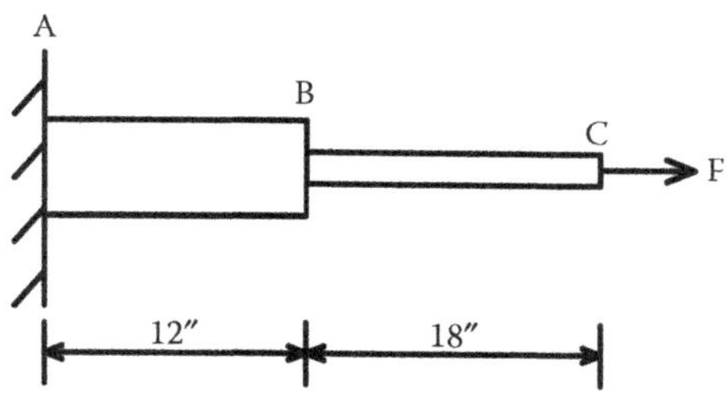

FIGURE P1.2

Ans: $e = 0.0016\,F$, $F < 25\,K$.

1.3 A structural member of length L, cross-sectional area A, and weight W is fixed at the top end and hangs vertically under its own weight. Assuming the elastic behavior, find the total axial elongation. The modulus of elasticity is E and the yield stress is σ_0.

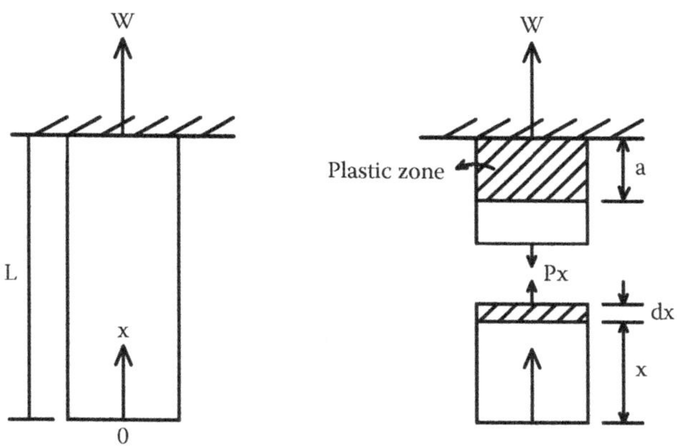

FIGURE P1.3

Ans: $e = \dfrac{WL}{2EA}$, $W \le \sigma_0 A$.

1.4 Repeat problem (1.3) by assuming linear-hardening material with yield stress, σ_0, Young's modulus, E, and tangent modulus, E_t. Show that the axial elongation of the member is

$$e = \frac{WL}{2EA} + \frac{WL}{E_t A}\left(1 - \frac{\sigma_0 A}{W}\right)^2, \quad W > \sigma_0 A$$

What is the value of e when $E_t = 0$? Explain.

1.5 A tapered conical member made of steel, length $L = 1.0$ m and radius at the base, $a = 25$ mm, hangs vertically under its own weight (Figure P1.5). Find the total elongation of the member. Assume elastic behavior, $E = 200$ GPa, and mass density $= 7800\ \text{kg/m}^3$.

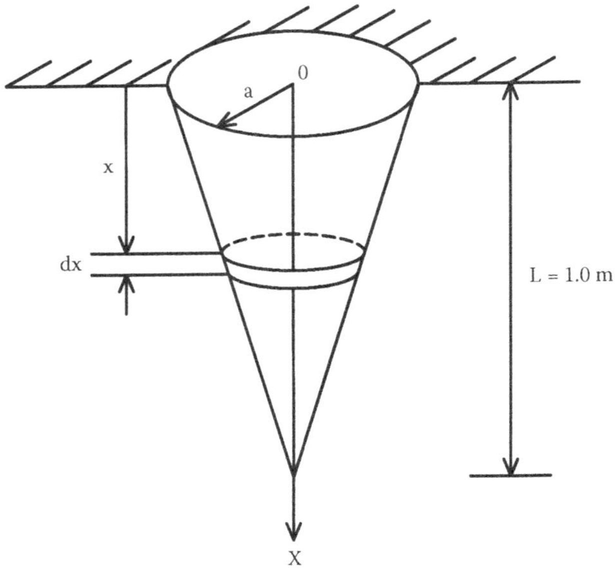

FIGURE P1.5

Ans: $e = 0.064$ mm.

1.6 A uniform rod AB of length, L, cross-sectional area, A, and mass density, ρ, has a mass m attached at end B as shown in Figure P1.6. The rod rotates with a constant angular velocity ω about a vertical axis through A. Assuming the elastic behavior, find the elongation of the rod.

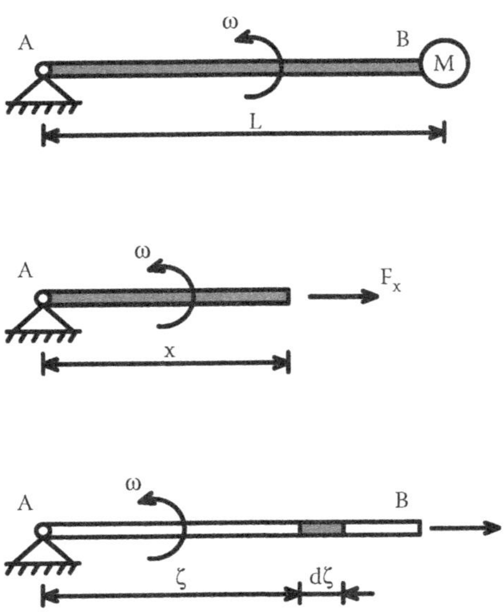

FIGURE P1.6

Ans: $e = \omega^2\left[\frac{mL^2}{EA} + \frac{\rho L^3}{3E}\right].$

1.7 Two beams, one with a circular cross section, radius = a, and the other with a rectangular cross section, width = b, depth h = 2b, are made of the same ideally elastic–plastic material. If both beams have the same cross-sectional area, which one can carry the greater moment before yielding is initiated.

Ans: $M_{rect}/M_{circ} = 1.67$.

1.8 Find the shape factor of a prismatic beam of circular cross section of radius = a when subjected to pure bending.

Ans: 1.698.

1.9 A straight beam has the T—cross section shown in Figure P1.9. Find the location of the neutral axis during pure bending in the vertical plane. Assume (a) elastic behavior, (b) perfectly plastic behavior. The dimensions given in the figure are in millimeter.

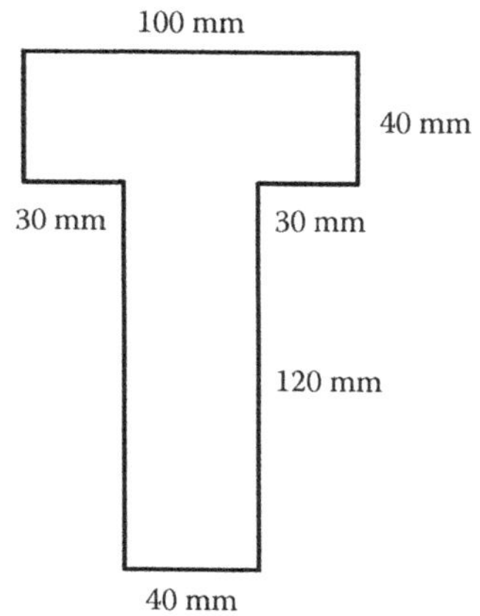

FIGURE P1.9

Ans: 49.4 mm, 30 mm.

1.10 A beam of rectangular cross section, width b and depth d, is constructed from a material having a stress–strain diagram consisting of two linear portions producing modulus of elasticity E_1 in tension and E_2 in compression. Assuming that the beam is subjected to positive-bending moment M and cross sections remain plane, show that (a) the normal strain on the outer surfaces of the beam can be written in the form

$$\varepsilon = \frac{d}{R}\left(\frac{\sqrt{E_2}}{\sqrt{E_1} + \sqrt{E_2}}\right)$$

(b) The bending moment M is given by

$$M = \frac{4E_1E_2I}{R[\sqrt{E_1} + \sqrt{E_2}]^2}$$

where R is the radius of curvature in bending and I is the moment of inertia of the cross section about the neutral axis.

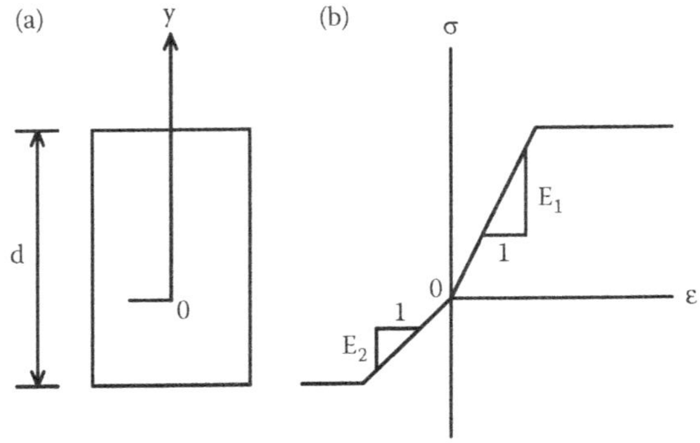

FIGURE P1.10

1.11 A prismatic shaft, 1.0 m long, is fixed at both ends A and B. The shaft is subjected to torques of 500 and 1000 N m at points C and D, respectively (Figure P1.11). Assuming the elastic behavior, find the reactive torques at the ends.

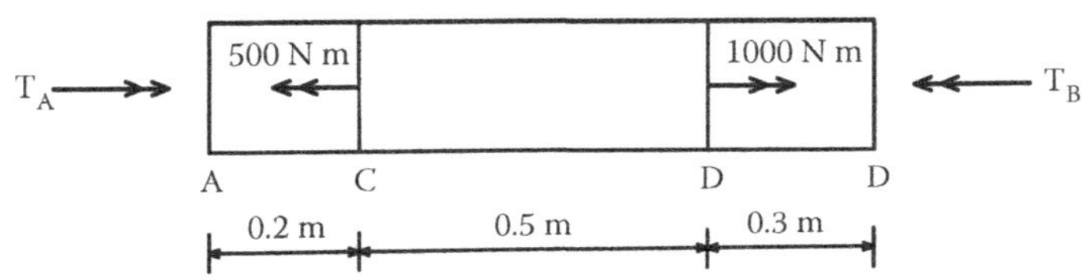

FIGURE P1.11

Ans: 100 N m, 60 N m.

1.12 (a) A cantilever circular shaft of radius 0.05 m and length 1.2 m is loaded by a continuously distributed torque of intensity $t_0 = 1000$ N m/m as shown in Figure P1.12. The shear modulus G = 80 GPa. Find the angular rotation at the free end. (b) If both ends are fixed, find the reactive torques.

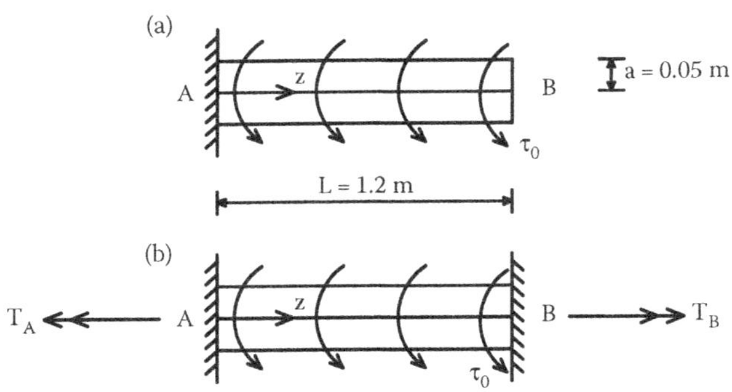

FIGURE P1.12

Ans: 0.33°, $T_A = T_B = 500$ N m.

1.13 Repeat the previous problem if the shaft is subjected to linearly varying torque with intensity zero at end A and intensity $t_0 = 1000$ N/m/m at B.

Ans: 0.11°, $T_A = 200$ N m, $T_B = 400$ N m.

List of Elementary Formulas

1. Prismatic member of linearly elastic material:
   a. Axial loading

$$\sigma = \frac{P}{A}$$

   b. Bending: normal stress

$$\sigma = -\frac{My}{I}; \quad \sigma_{max} = -\frac{Mc}{I}$$

   Shear stress

$$\tau = \frac{VQ}{Ib}$$

   c. Torsion

$$\tau = \frac{T\rho}{J}; \quad \tau_{max} = \frac{Tr}{J}$$

   where σ = normal axial and bending stress, τ = shearing stress due to vertical shear force V and torque T. M is the bending moment, I = moment of inertial about neutral axis of the cross section, Q = the moment of the area beyond the level at which the shear is calculated about the neutral axis, J = polar moment of inertia of circular cross section, and b = width of the member at which the shear stress is calculated.

2. Thin-walled pressure vessels:
   a. Cylinder

$$\sigma_\theta = \frac{pr}{t}$$

$$\sigma_{axial} = \frac{pr}{2t}$$

   b. Sphere

$$\sigma = \frac{pr}{2t}$$

   Here, σ_θ = tangential stress in cylinder wall, σ_{axial} = axial stress in cylinder wall, σ = membrane stress in sphere wall, p = internal pressure, r = mean radius, and t = wall thickness of pressure vessel.

References

1.1. F. P. Beer and R. Russel, Jr., *Mechanics of Materials*, McGraw-Hill Book Co., New York, 1992.
1.2. J. M. Gere and S. P. Timoshenko, *Strength of Materials*, 3rd edition, PWS-Kent Publishing Co., Boston, MA, 1990.
1.3. *Manual of Steel Construction*, American Institute of Steel Construction (AISC), latest edition.

2

Analysis of Stress

When a structural component is deformed by the action of external forces, the arrangement of molecules in the material is changed and internal forces are developed to satisfy equilibrium. The internal forces give rise to internal stresses. Stress is measured in units of force divided by units of area. In the SI system of units, stress is measured in Newton per square meter, abbreviated as N/m^2. A Newton per square meter is a Pascal (Pa), and because the Pascal is a small quantity, the mega Pascal (MPa) is commonly used to measure stress where $MPa = 10^6$ Pa. In the customary U.S. units, stress is measured in pound per square inch (psi) or kip per square inch (ksi). One kip is a kilo pound or 1000 pound (lb), and one $ksi = 10^3$ psi. To examine the state of stress at an arbitrary point, P, in a structure, we use an imaginary cut and expose a plane containing the point P as shown in Figure 2.1a. There are infinite such planes passing through the point P in the structure, each plane is identified by its normal. Introduce an x, y, z system of coordinates with the origin located at the point P, the x-axis is oriented normal to the exposed plane, and the y- and z-axes are located in the plane as shown in Figure 2.1b. The internal forces and moments acting on this plane, which are required to maintain equilibrium of the free-body diagram shown in Figure 2.1b, consist of forces $\mathbf{F} = (F_x, F_{xy}, F_{xz})$ and moments $\mathbf{M} = (M_x, M_y, M_z)$. The forces consist of a normal force F_x and two tangential or shearing forces F_{xy} and F_{xz} along the y- and z-direction, respectively, while the moments are oriented along the axes.

Consider an element of area dA surrounding the point P. On this area element, there are element forces, $d\mathbf{F} = (dF_x, dF_{xy}, dF_{xz})$, and element moments, $d\mathbf{M} = (dM_x, dM_y, dM_z)$. In the limit as $dA \rightarrow 0$, the normal stress σ_x and the shearing stresses τ_{xz} and τ_{yz} are defined as follows:

$$\sigma_x = dF_x/dA \tag{2.1a}$$

$$\tau_{xy} = dF_{xy}/dA \tag{2.1b}$$

$$\tau_{xz} = dF_{xz}/dA \tag{2.1c}$$

The ratios $d\mathbf{M}/dA$ are assumed zero. If the moment ratios, $d\mathbf{M}/dA$, do not vanish as dA approaches zero, a "moment-stress" theory in continuum mechanics can be developed. However, it has limited practical applications and is not discussed in the book.

The normal and shearing stresses measure the intensities of the internal forces acting on the element through point P. A normal stress is commonly denoted by the Greek letter, *sigma,* and requires one subscript designating the normal to the exposed plane. Thus, σ_x denotes the normal stress on a plane whose normal is the x-axis. A shearing stress is denoted by the Greek letter, *tau,* and requires two subscripts where the first subscript denotes the normal to the plane and the second subscript designates the direction of the shearing stress. Accordingly, τ_{xy} denotes a shearing stress on a plane normal to the x-axis and points along an axis parallel to the y-direction.

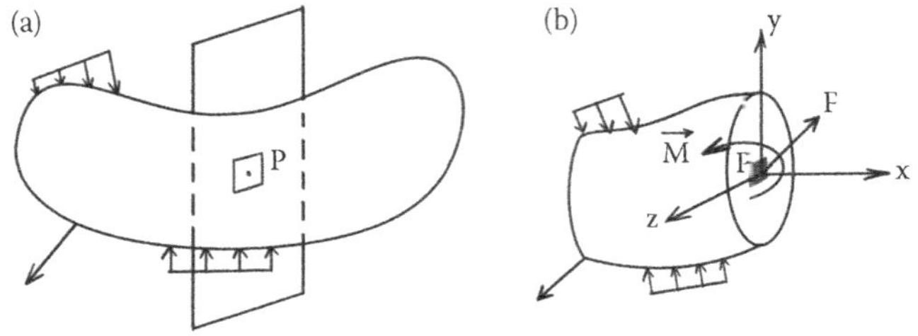

FIGURE 2.1
Internal forces and moments, (a) section through point P and (b) internal force F and moment $\vec{M}$.

The corresponding forces acting on the surface of the element are obtained by summing up the stress times the surface area. Thus, the normal force (F_x) and tangential forces (F_{xy}, F_{xz}) are given by

$$F_x = \iint_A \sigma_x dA, \quad F_{xy} = \iint_A \tau_{xy} dA, \quad F_{xz} = \iint_A \tau_{xz} dA$$

The moments of the forces about the reference axes are obtained as follows:

$$M_x = \iint_A (y\tau_{xz} - z\tau_{xy}) dA$$

$$M_y = \iint_A z\sigma_x dA$$

$$M_z = \iint_A y\sigma_x dA$$

The sign convention of the stress components is discussed in the next section.

2.1 The Stress at a Point

In order to define the stress at point P, we need to determine the normal and the shearing stresses on all planes passing through P. However, as we shall prove later, it is sufficient to pass three mutually perpendicular planes through point P as shown in Figure 2.2, and a total of nine stress components are required to define the stress at point P in the material.

With reference to rectangular coordinates, the nine stress components are arranged in a 3×3 matrix form, where each row represents the stresses acting on one of the three orthogonal planes through the point P, namely, the plane whose normal is x, y, or z.

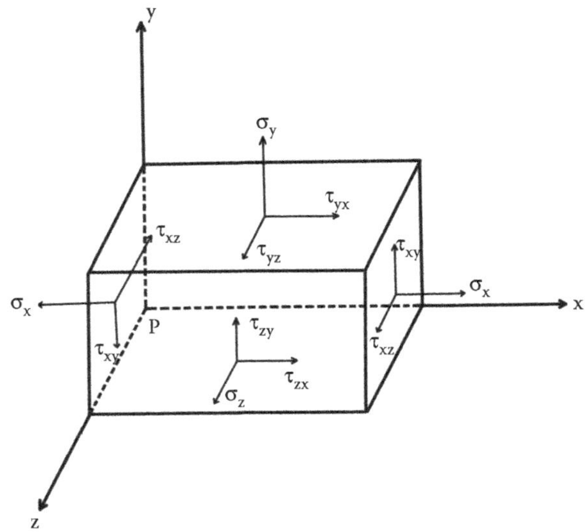

FIGURE 2.2
Rectangular stress components.

Thus, the stress at the point P, σ_{ij}, i, j = x, y, z, i

$$\sigma_{ij} = \begin{pmatrix} \sigma_x & \tau_{xy} & \tau_{xz} \\ \tau_{yx} & \sigma_y & \tau_{yz} \\ \tau_{zx} & \tau_{zy} & \sigma_z \end{pmatrix} \tag{2.2}$$

Mathematically, the stress is a *tensor* of the second kind and is governed by the specific rules and operations of such tensor quantities.

The stress components shown in Figure 2.2 are all positive because the sign convention we elect to use follows the coordinate directions. For normal stress, this sign convention implies that a tensile stress is always positive and a compressive stress is always negative. To confirm this conclusion, it is observed that the normal stresses σ_x in Figure 2.2 point along the positive outer normal on the positive side and the negative outer normal on the negative side of the x-axis. Hence, both quantities are positive stresses. A compressive normal stress is always negative because the direction of the stress opposes the direction of the outward normal to the plane. Similarly, when the outer normal and the direction of the shearing stress are both positive and negative, the resulting shear stress is positive. And when the outer normal and the direction of the shearing stress are positive in one and negative in the other, the resulting shear stress is always negative.

The nine stress components mentioned in Equation 2.2 are functions of the external loads and the geometry of the member. They vary from point to point in the member and can be determined subject to satisfying the governing equations and boundary conditions. As we shall explain later, the governing equations are derived based on the requirements of statics equilibrium, the geometry of deformation, and the stress–strain relations of the material.

Now we show that the stress matrix in Equation 2.2 is symmetric, that is, $\sigma_{ij} = \sigma_{ji}$, i, j = x, y, z. Consider a plane element in the xy-plane of sides dx and dy, as shown in Figure 2.3. The origin of the coordinates is chosen at point P. On the left side of the element, the stresses

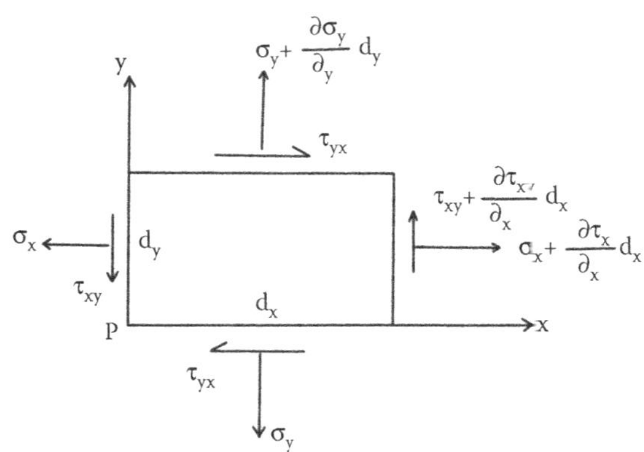

FIGURE 2.3
Variation of stress in a plane element.

are σ_x and τ_{xy}, and since the stresses vary with x and y, they can be expanded in Taylor's series, and to the first order of approximation, the stresses on the right side of the element are $\sigma_x + (\partial\sigma_x/\partial x)dx$ and $\tau_{xy} + (\partial\tau_{xy}/\partial x)dx$, respectively.

Using the moment about the z-axis equation of equilibrium, $\sum M_z = 0$, for the forces acting on the element results in

$$\left(\tau_{xy} + \frac{\partial\tau_{xy}}{\partial x}dx\right)(dy)dx - \left(\tau_{yx} + \frac{\partial\tau_{yx}}{\partial y}dy\right)(dx)dy - \left(\frac{\partial\sigma_x}{\partial x}dx\right)(dy)\frac{dy}{2} + \left(\frac{\partial\sigma_y}{\partial y}dy\right)(dx)\frac{dx}{2} = 0 \tag{2.3}$$

Dividing each term by dxdy and letting the element shrink to the point P, that is, letting $dx \to 0$, $dy \to 0$, it follows that $\tau_{yx} = \tau_{xy}$. Similarly, by taking moments about the x- and y-axes, it can be shown that $\tau_{zx} = \tau_{xz}$ and $\tau_{yz} = \tau_{zy}$. In other words, the stress matrix is symmetric and the state of stress at a point requires six independent components, three normal stresses (σ_j, $j = x, y, z$), and three shear stresses (τ_{ij}, $i, j = x, y, z$). With such characteristics, the stress at a point in a loaded component is a *symmetric tensor of the second kind.*

True stress: The conventional or engineering stress discussed above is the load per unit original area of the cross section of the member. It should be noted that true stress is defined as the load per the current cross section of the area of the member and can be significantly different from the conventional stress (see Section 3.2 in Chapter 3 and Reference 2.1).

2.2 Equations of Equilibrium

Consider the element at point P of a structural component (Figure 2.4). In the previous section, the three moment equilibrium equations were used to show that the 3 × 3 stress matrix is symmetric. In this section, the three force-equilibrium equations are applied to derive the differential equations governing the distribution of stress throughout the component. Applying the force–equilibrium equation along the x-axis, namely, $\sum F_x = 0$, and noting

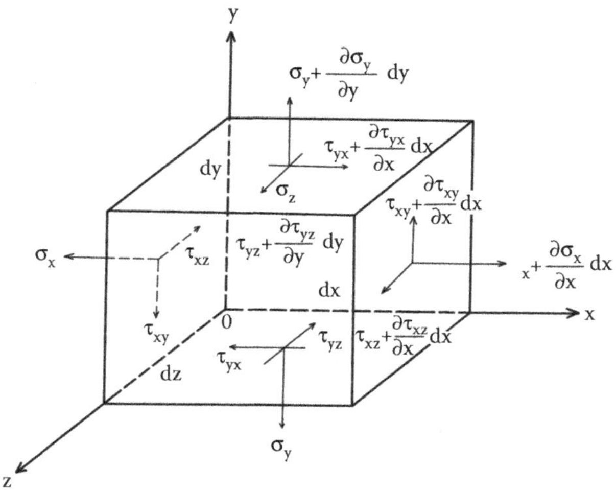

FIGURE 2.4
3D stress system.

that force = stress × area, results in

$$\left(\frac{\partial \sigma_x}{\partial x} dx\right) dydz + \left(\frac{\partial \tau_{yx}}{\partial y} dy\right) dxdz + \left(\frac{\partial \tau_{zx}}{\partial z} dz\right) dydx = 0 \tag{2.4}$$

Dividing by dxdydz and taking the limit as the element shrinks to the point P, the equilibrium equation in the x-direction is obtained. Similarly, the equilibrium equations in the y- and z-directions yield two more equations so that overall equilibrium is obtained by satisfying the equations

$$\begin{aligned} \frac{\partial \sigma_x}{\partial x} + \frac{\partial \tau_{xy}}{\partial y} + \frac{\partial \tau_{xz}}{\partial z} &= 0 \\ \frac{\partial \tau_{yx}}{\partial x} + \frac{\partial \sigma_y}{\partial y} + \frac{\partial \tau_{yz}}{\partial z} &= 0 \\ \frac{\partial \tau_{zx}}{\partial x} + \frac{\partial \tau_{zy}}{\partial y} + \frac{\partial \sigma_z}{\partial z} &= 0 \end{aligned} \tag{2.5}$$

For two-dimensional (2D) problems modeled in the xy-plane, the equilibrium equations reduce to

$$\begin{aligned} \frac{\partial \sigma_x}{\partial x} + \frac{\partial \tau_{xy}}{\partial y} &= 0 \\ \frac{\partial \tau_{yx}}{\partial x} + \frac{\partial \sigma_y}{\partial y} &= 0 \end{aligned} \tag{2.6}$$

In the presence of body forces B_x, B_y, and B_z along the reference axes, Equations 2.5 and 2.6 are modified by adding the body forces to the left-hand side of each equation. Thus, in the presence of body forces, the equilibrium equations governing the three-dimensional (3D) state of stress are

$$\begin{aligned}
&\frac{\partial \sigma_x}{\partial x} + \frac{\partial \tau_{xy}}{\partial y} + \frac{\partial \tau_{xz}}{\partial z} + B_x = 0 \\
&\frac{\partial \tau_{yx}}{\partial x} + \frac{\partial \sigma_y}{\partial y} + \frac{\partial \tau_{yz}}{\partial z} + B_y = 0 \\
&\frac{\partial \tau_{zx}}{\partial x} + \frac{\partial \tau_{zy}}{\partial y} + \frac{\partial \sigma_z}{\partial z} + B_z = 0
\end{aligned} \tag{2.7}$$

Examples of body forces include gravitational force due to the weight of the component, centrifugal force due to rotation of the component, and magnetic field force. The body forces are expressed in units of force divided by units of volume.

2.3 Stress Vector (Traction)

A stress vector or traction is a convenient vector parameter, introduced to simplify the mathematical analysis of the state of stress. Making use of unit vectors, **i, j,** and **k**, along the x-, y- and z-reference axes, the tractions associated with the three orthogonal faces of the element shown in Figure 2.2 are

$$\begin{aligned}
\mathbf{T}_x &= \sigma_x \mathbf{i} + \tau_{xy} \mathbf{j} + \tau_{xz} \mathbf{k} \\
\mathbf{T}_y &= \tau_{yx} \mathbf{i} + \sigma_y \mathbf{j} + \tau_{yz} \mathbf{k} \\
\mathbf{T}_z &= \tau_{zx} \mathbf{i} + \tau_{zy} \mathbf{j} + \sigma_z \mathbf{k}
\end{aligned} \tag{2.8}$$

It follows that in this notation, the stress, σ_x, is readily obtained by evaluating the dot product $\mathbf{i} \cdot \mathbf{T}_x$. Similarly, τ_{xy} is obtained from the product $\mathbf{j} \cdot \mathbf{T}_x$, and so on.

2.4 Stresses on an Inclined Plane

Consider the equilibrium of a small tetrahedron element, PABC, whose edges PA, PB, and PC coincide with the coordinate axes as shown in Figure 2.5. The inclined plane ABC has an outward unit normal $\mathbf{n} = l\,\mathbf{i} + m\,\mathbf{j} + n\,\mathbf{k}$, with (l, m, n) denoting the direction cosines of the normal to plane ABC, that is, $l = \cos\alpha$, $m = \cos\beta$, and $n = \cos\gamma$ where α, β, and γ are the angles made by the normal and the coordinate axes as indicated in Figure 2.5b. Because **n** is a unit vector, its direction cosines satisfy the geometrical identity

$$l^2 + m^2 + n^2 = 1 \tag{2.9}$$

Across surfaces PBC, PCA, and PAB, the given stresses, σ_{ij}, i, j = x, y, z, are specified, and in order to enforce equilibrium, the internal forces induced on the plane ABC must balance

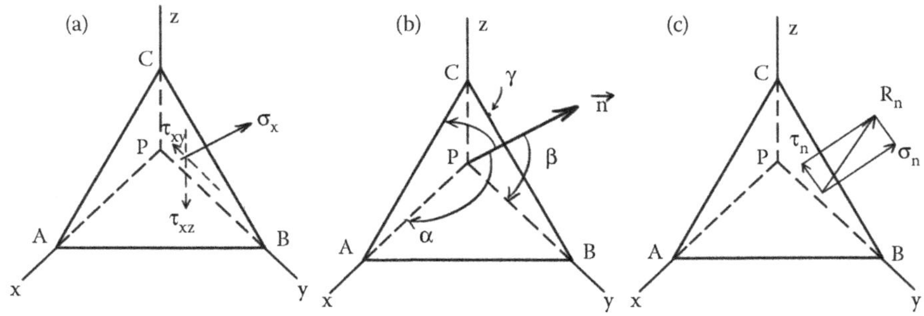

FIGURE 2.5
Stresses on an inclined plane; (a) tetrahedron element PABC; (b) $\vec{n}$ is normal to plane PABC, and (c) resultant stress R_n on plane ABC.

the specified forces on the sides of the element. The traction on plane ABC is expressed as

$$\mathbf{T}_n = \sigma_n \mathbf{n} + \tau_{ns}\mathbf{s} + \tau_{nt}\mathbf{t} \tag{2.10}$$

where **n, s,** and **t** are orthogonal unit vectors emanating from a point on the inclined plane. It is also convenient to express the traction $\mathbf{T_n}$ in term of stress components p_x, p_y, and p_z acting on the plane ABC in the x-, y-, and z-directions, thus

$$\mathbf{T_n} = p_x\mathbf{i} + p_y\mathbf{j} + p_z\mathbf{k} \tag{2.11}$$

Let δA denote the area of the inclined plane. It follows from solid geometry that the surfaces PBC, PAC, and PAB have areas lδA, mδA, and nδA, respectively. Thus, the forces on the four surfaces of the tetrahedron element are as follows:

$$\begin{aligned} \mathbf{F}_x &= (\sigma_x\mathbf{i} + \tau_{xy}\mathbf{j} + \tau_{xz}\mathbf{k})l\delta A \\ \mathbf{F}_y &= (\tau_{yx}\mathbf{i} + \sigma_y\mathbf{j} + \tau_{yz}\mathbf{k})m\delta A \\ \mathbf{F}_z &= (\tau_{zx}\mathbf{i} + \tau_{zy}\mathbf{j} + \sigma_z\mathbf{k})n\delta A \\ \mathbf{F}_n &= (p_x\mathbf{i} + p_y\mathbf{j} + p_z\mathbf{k})\delta A \end{aligned} \tag{2.12}$$

Force equilibrium of the element requires $\mathbf{F}_n = \mathbf{F_x} + \mathbf{F}_y + \mathbf{F_z}$, that is,

$$(p_x\mathbf{i} + p_y\mathbf{j} + p_z\mathbf{k})\delta A = (\sigma_x l + \tau_{xy}m + \tau_{xz}n)\delta A\mathbf{i} + (\tau_{yx}l + \sigma_y m + \tau_{yz}n)\delta A\mathbf{j}$$
$$+ (\tau_{zx}l + \tau_{zy}m + \sigma_z n)\delta A\mathbf{k}$$

Equating the coefficients of unit vectors **i, j, k,** on both sides of the equation, one obtains

$$\begin{aligned} p_x &= \sigma_x l + \tau_{xy}m + \tau_{xz}n \\ p_y &= \tau_{yx}l + \sigma_y m + \tau_{yz}n \\ p_z &= \tau_{zx}l + \tau_{zy}m + \sigma_z n \end{aligned} \tag{2.13}$$

Now $\sigma_n = \mathbf{n} \cdot \mathbf{T}_n = (l\mathbf{i} + m\mathbf{j} + n\mathbf{k}) \cdot (p_x\mathbf{i} + p_y\mathbf{j} + p_z\mathbf{k}) = l\,p_x + m\,p_y + n\,p_z$.

Making use of the expressions in Equations 2.13, it follows that

$$\sigma_n = \sigma_x l^2 + \sigma_y m^2 + \sigma_z n^2 + 2(\tau_{xy} lm + \tau_{yz} mn + \tau_{zx} nl) \tag{2.14}$$

The components of the shear stress acting on the inclined plane are given by the dot products: $\tau_{ns} = \mathbf{s} \cdot \mathbf{T_n}$ and $\tau_{nt} = \mathbf{t} \cdot \mathbf{T_n}$. Suppose that the direction cosines of the unit vectors **s** and **t** relative to the x-, y-, and z-axes are (l_1, m_1, n_1) and (l_2, m_2, n_2), respectively, it follows that

$$\tau_{ns} = (l_1\mathbf{i} + m_1\mathbf{j} + n_1\mathbf{k}) \cdot (p_x\mathbf{i} + p_y\mathbf{j} + p_z\mathbf{k})$$

Using Equations 2.13 and performing the dot product, it is found that

$$\tau_{ns} = ll_1\sigma_x + mm_1\sigma_y + nn_1\sigma_z + (lm_1 + l_1m)\tau_{xy} + (mn_1 + nm_1)\tau_{yz} + (nl_1 + ln_1)\tau_{zx} \tag{2.15a}$$

$$\tau_{nt} = ll_2\sigma_x + mm_2\sigma_y + nn_2\sigma_z + (lm_2 + l_2m)\tau_{xy} + (mn_2 + nm_2)\tau_{yz} + (nl_2 + n_2l)\tau_{zx} \tag{2.15b}$$

Here, the direction cosines of **s** and **t**, l_i, m_i, n_i, $i = 1, 2$, satisfy the geometrical relations

$$l_i^2 + m_i^2 + n_i^2 = 1, \; i = 1, \; 2$$

and since **s** and **t** are orthogonal unit vectors, we have the additional relation

$$l_1l_2 + m_1m_2 + n_1n_2 = 0$$

The resultant shear stress across the inclined plane is given by Pythagoras' theorem as $\tau_n = (\tau_{ns}^2 + \tau_{nt}^2)^{1/2}$. Observing that

$$\mathbf{T_n} \cdot \mathbf{T_n} = T_n^2 = \sigma_n^2 + \tau_{ns}^2 + \tau_{nt}^2 = p_x^2 + p_y^2 + p_z^2$$

Hence, the magnitude of the resultant shear stress is

$$\tau_n = (T_n^2 - \sigma_n^2)^{1/2} = (p_x^2 + p_y^2 + p_z^2 - \sigma_n^2)^{1/2} \tag{2.16}$$

where p_x, p_y, and p_z are the stress components given in Equations 2.13 and σ_n is given by Equation 2.14.

The direction cosines of the resultant shear stress relative to rectangular axes, say, l_s, m_s, and n_s, are found as follows: The stress components of T_n, as defined in Equation 2.11, along the x-, y-, and z-axes as defined in Equation 2.11 are expressed in terms of the normal and resultant shear stresses, σ_n and τ_n, by the relations

$$p_x = \sigma_n l + \tau_n l_s$$

$$p_y = \sigma_n m + \tau_n m_s$$

$$p_z = \sigma_n n + \tau_n n_s$$

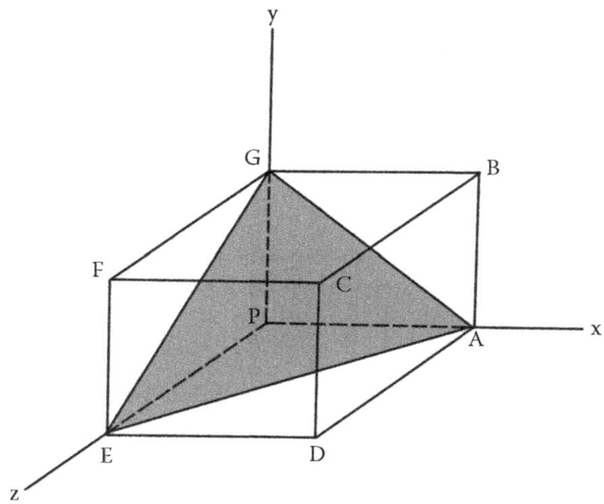

FIGURE 2.6
Example 2.2.

Hence,

$$l_s = \frac{p_x - l\sigma_n}{\tau_n}, \quad m_s = \frac{p_y - m\sigma_n}{\tau_n}, \quad n_s = \frac{p_z - n\sigma_n}{\tau_n} \tag{2.17}$$

which shows that the direction cosines of the resultant shear stress are proportional to its Cartesian components.

The magnitude of the resultant stress, R_n, acting on the inclined plane ABC (Figures 2.5c and 2.6) is calculated from

$$R_n^2 = \sigma_n^2 + \tau_n^2$$

and using Equation 2.16, it is found that

$$R_n^2 = p_x^2 + p_y^2 + p_z^2 \tag{2.18}$$

Suppose that the resultant stress, R_n, makes angles θ_x, θ_y, θ_z with the x-, y-, z-axes, respectively. Since p_x, p_y, and p_z are the components of the resultant stress in the x-, y-, and z-directions, it follows that the direction cosines of R_n are

$$\begin{aligned} \cos\theta_x &= \frac{p_x}{R_n} \\ \cos\theta_y &= \frac{p_y}{R_n} \\ \cos\theta_z &= \frac{p_z}{R_n} \end{aligned} \tag{2.19}$$

EXAMPLE 2.1

The state of stress at a point 0 in a loaded member relative to the x-, y-, and z-axes is given by

$$\begin{bmatrix} 20 & 10 & 5 \\ 10 & -30 & -15 \\ 5 & -15 & 5 \end{bmatrix} \text{MPa}$$

Determine the normal, shear, and resultant stresses on a plane through 0 whose normal makes 65° with the x-axis and 35° with the y-axis.

Solution

First, we calculate the direction cosines of the normal to the plane through point 0.

$l = \cos 65° = 0.423$, $m = \cos 35° = 0.819$. Since $l^2 + m^2 + n^2 = 1$, it follows that $n = 0.388$.

Substituting in Equation 2.14, the normal stress is found as

$$\begin{aligned}\sigma_n &= 20(0.423)^2 + (-30)(0.819)^2 + 5(0.388)^2 + 2[10(0.423)(0.819) + (-15)(0.819)(0.388) \\ &\quad + 5(0.388)(0.423)] \\ &= -16.78\,\text{MPa}.\end{aligned}$$

Next, in order to compute the shear stress, τ_n, the stress, components p_x, p_y, and p_z are first computed from Equation 2.13

$$p_x = 20(0.423) + 10(0.819) + 5(0.388) = 18.59\,\text{MPa}$$

$$p_y = 10(0.423) + (-30)(0.819) + 5(0.388) = -26.16\,\text{MPa}$$

$$p_z = 5(0.423) + (-15)(0.819) + 5(0.388) = -8.23\,\text{MPa}$$

Substituting in Equation 2.16 leads to $\tau_n = 28.59\,\text{MPa}$.

Hence, the resultant stress $R_n = \sqrt{\sigma_n{}^2 + \tau_n{}^2} = 33.13\,\text{MPa}$.

EXAMPLE 2.2

The rectangular components of the state of stress at a point, P, in a loaded member are given by the array

$$\begin{bmatrix} 10 & 5 & 3 \\ 5 & -5 & 4 \\ 3 & 4 & 15 \end{bmatrix} \text{MPa}$$

Find the normal stress σ_n and shear stress τ_n acting on a plane through P parallel to the surface AEG in the parallelepiped shown in Figure 2.6. Find also the direction of the shear stress.

Solution

Assume that the normal to the plane through point P, parallel to the surface AEG, is $\mathbf{n} = l\mathbf{i} + m\mathbf{j} + n\mathbf{k}$. To find the direction cosines of this normal, we observe that unit vectors along directions AE and AG are given by

$$\mathbf{e}_{AE} = \frac{-4i + 3k}{5} \quad \text{and} \quad \mathbf{e}_{AG} = \frac{-4i + 3j}{5}$$

Since these unit vectors are located in the plane AEG, they make 90° with the unit normal **n**, and hence the dot products of these unit vectors with **n** are equal to zero.

$$\mathbf{e}_{AE} \cdot \mathbf{n} = -\frac{4}{5}l + \frac{3}{5}n = 0 \quad \text{giving } n = \frac{4}{3}l;$$

$$\text{similarly, } \mathbf{e}_{AG} \cdot \mathbf{n} = -\frac{4}{5}l + \frac{3}{5}m = 0 \quad \text{giving } m = \frac{4}{3}l$$

Substituting in $l^2 + m^2 + n^2 = 1$, it follows that $l = \frac{3}{\sqrt{41}}$ and $m = n = \frac{4}{\sqrt{41}}$.

Next, p_x, p_y, and p_z are computed from Equation 2.13 as $p_x = 9.683$ MPa, $p_y = 1.718$ MPa, and $p_z = 13.275$ MPa.

The normal stress follows from $\sigma_n = lp_x + mp_y + np_z$ (or from Equation 2.14) as $\sigma_n = 13.903$ MPa.

The resultant shear is computed from Equation 2.13 as

$$\tau_n = [(9.683)^2 + (1.718)^2 + (13.275)^2 - (13.903)^2]^{\frac{1}{2}} = 8.924 \text{ MPa}$$

The direction cosines of the shear stress are found from Equations 2.17 as

$$l_s = 0.355; \quad m_s = -0.781; \quad n_s = 0.514$$

EXAMPLE 2.3

The rectangular components of the stress at a point 0 in a loaded member is given by

$$\begin{bmatrix} \sigma_x & 20 & 30 \\ 20 & 50 & 15 \\ 30 & 15 & -40 \end{bmatrix} \text{MPa}$$

Find the value of σ_x for which there will be a traction-free plane passing through the point 0 and compute the direction cosines of the normal to the plane.

Solution

Let (l, m, n) represents the direction cosines of the normal to the required plane through the point 0. Since the plane is traction free, the stress components p_x, p_y, and p_z acting on the plane must vanish. This implies the following (using Equation 2.13):

$$\begin{aligned} \sigma_x l + 20m + 30n &= 0 \\ 20l + 50m + 15n &= 0 \\ 30l + 15m - 40n &= 0 \end{aligned} \qquad \text{(a)}$$

A nontrivial solution exists if the determinant of the coefficients of l, m, and n in Equation (a) vanishes, that is,

$$\begin{vmatrix} \sigma_x & 20 & 30 \\ 20 & 50 & 15 \\ 30 & 15 & -40 \end{vmatrix} = 0$$

which gives $\sigma_x = -4.94$ MPa. Substituting the value of σ_x in the first two equations of the set (a) and eliminating n gives $m = -0.562\,l$. The third equation in Equation (a) gives $n = 0.539\,l$.

Hence

$$l^2 + m^2 + n^2 = 1$$

It follows that the direction cosines of the normal to the plane are given by

$$l = 0.789, \quad m = -0.443, \quad \text{and} \quad n = 0.425$$

2.5 Transformation of Stress

Let σ_{ij} represent the state of stress at a point, 0, in a loaded member relative to rectangular axes 0x, 0y, and 0z. Suppose that 0x′, 0y′, and 0z′ represent another set of rectangular axes emanating from the same origin as shown in Figure 2.7. The axes 0x′, 0y′, and 0z′ make angles α_j, β_j, and γ_j, $j = 1, 2, 3$ with the original axes 0x, 0y, and 0z, respectively. Denote the direction cosines of the rotated axes with respect to the original axes by l_j, m_j, n_j ($j = 1, 2, 3$), that is, $l_j = \cos\alpha_j$, $m_j = \cos\beta_j$, and $n_j = \cos\gamma_j$.

Table 2.1 gives a summary of the direction cosines of the new axes.

Note that if the coordinates of an arbitrary point, P, of the element in the original coordinate system are x, y, z, the coordinates of P in the rotated system are obtained by finding the projections of the line 0P on axes 0x′, 0y′, and 0z′. Since $\overrightarrow{0P} = x\mathbf{i} + y\mathbf{j} + z\mathbf{k}$, the

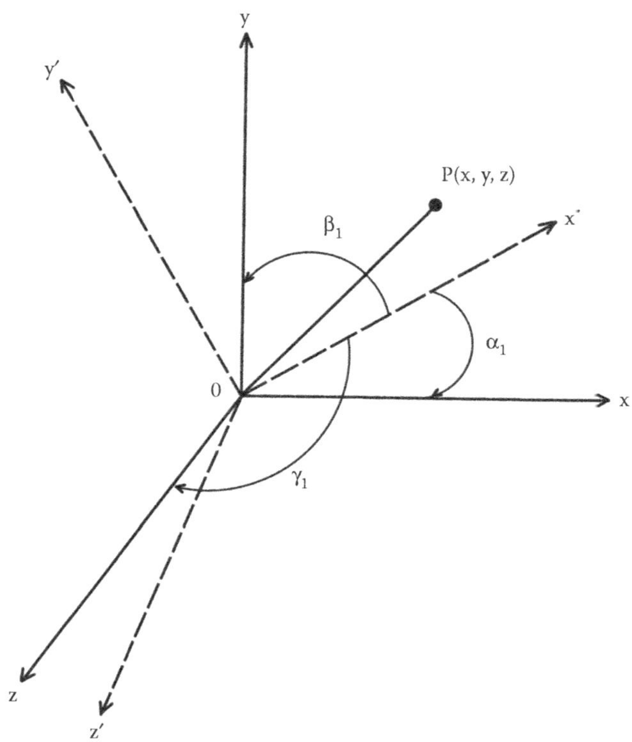

FIGURE 2.7
Rotation of coordinate axes ($xyz \rightarrow x'y'z'$).

TABLE 2.1

Direction Cosines of $0x'$, $0y'$, and $0z'$

	x	y	z
x'	l_1	m_1	n_1
y'	l_2	m_2	n_2
z'	l_3	m_3	n_3

coordinates of P along the x'-axis are thus obtained as

$$\vec{i'} \cdot 0P = (l_1\mathbf{i} + m_1\mathbf{j} + n\mathbf{k}) \cdot (x\mathbf{i} + y\mathbf{j} + z\mathbf{k}) = (l_1x + m_1y + n_1z)$$

Similar expressions are obtained for the projections of 0P along the $0y'$ and $0z'$ axes. It follows that the coordinates of P in the rotated system are $[(l_1x+m_1y+n_1z), (l_2x + m_2y + n_2z), \text{and } (l_3x + m_3y + n_3z)]$.

In order to find the state of stress with reference to the rotated system of coordinates, the normal stress acting on a plane whose normal is the $0x'$ axis is obtained from Equation 2.14 as

$$\sigma_{x'} = l_1^2\sigma_x + m_1^2\sigma_y + n_1^2\sigma_z + 2(l_1m_1\tau_{xy} + m_1n_1\tau_{yz} + n_1l_1\tau_{zx}) \tag{2.20a}$$

Similarly, the two components of the shearing stress on the same plane are obtained from Equations 2.15a and b as

$$\tau_{x'y'} = l_1l_2\sigma_x + m_1m_2\sigma_y + n_1n_2\sigma_z + (l_1m_2 + l_2m_1)\tau_{xy} + (m_1n_2 + m_2n_1)\tau_{yz} + (n_1l_2 + n_2l_1)\tau_{zx} \tag{2.20b}$$

$$\tau_{x'z'} = l_1l_3\sigma_x + m_1m_3\sigma_y + n_1n_3\sigma_z + (l_1m_3 + l_3m_1)\tau_{xy} + (m_1n_3 + m_3n_1)\tau_{yz} + (n_1l_3 + n_3l_1)\tau_{zx} \tag{2.20c}$$

The remaining stress components relative to the y'- and z'-axes are determined in the same manner and they are

$$\sigma_{y'} = l_2^2\sigma_x + m_2^2\sigma_y + n_2^2\sigma_z + 2(l_2m_2\tau_{xy} + m_2n_2\tau_{yz} + n_2l_2\tau_{zx}) \tag{2.20d}$$

$$\sigma_{z'} = l_3^2\sigma_x + m_3^2\sigma_y + n_3^2\sigma_z + 2(l_3m_3\tau_{xy} + m_3n_3\tau_{yz} + n_3l_3\tau_{zx}) \tag{2.20e}$$

$$\tau_{y'z'} = l_2l_3\sigma_x + m_2m_3\sigma_y + n_2n_3\sigma_z + (l_2n_3 + l_3n_2)\tau_{xy} + (m_2n_3 + m_3n_2)\tau_{yz} + (n_2l_3 + n_3l_2)\tau_{zx} \tag{2.20f}$$

In matrix notation, the transformed stress tensor is written as

$$\sigma'_{ij} = N\sigma_{ij}\,N^T, \quad i, j = x, y, \text{ and } z \tag{2.21}$$

where N is the 3×3 matrix of the direction cosines, σ_{ij} is the matrix of the stress tensor, and N^T is the transpose of N, that is,

$$\sigma'_{ij} = \begin{pmatrix} l_1 & m_1 & n_1 \\ l_2 & m_2 & n_2 \\ l_3 & m_3 & n_3 \end{pmatrix} \begin{pmatrix} \sigma_x & \tau_{xy} & \tau_{xz} \\ \tau_{yx} & \sigma_y & \tau_{yz} \\ \tau_{zx} & \tau_{zy} & \sigma_z \end{pmatrix} \begin{pmatrix} l_1 & l_2 & l_3 \\ m_1 & m_2 & m_3 \\ n_1 & n_2 & n_3 \end{pmatrix}$$

Based on the orthogonality of the unit vectors along the original and rotated coordinate systems, the direction cosines in Table 2.1 satisfy the following geometrical relations:

$$l_j^2 + m_j^2 + n_j^2 = 1; \quad i,j = 1, 2, 3 \tag{2.22a}$$

$$\begin{aligned} l_1 l_2 + m_1 m_2 + n_1 n_2 &= 0 \\ l_1 l_3 + m_1 m_3 + n_1 n_3 &= 0 \\ l_2 l_3 + m_2 m_3 + n_2 n_3 &= 0 \end{aligned} \tag{2.22b}$$

For a 2D state of stress in the xy-plane in which the x′-axis makes angle θ with the positive x-axis, it follows that

$$\begin{aligned} &l_1 = \cos\theta, \quad m_1 = \sin\theta, \quad n_1 = 0 \\ &l_2 = -\sin\theta, \quad m_2 = \cos\theta, \quad n_2 = 0 \\ &l_3 = m_3 = 0, \quad n_3 = 1 \end{aligned}$$

and Equations 2.19 and 2.20 yield

$$\begin{aligned} \sigma_{x'} &= \sigma_x \cos^2\theta + \sigma_y \sin^2\theta + \tau_{xy} 2\sin\theta\cos\theta \\ &= \frac{1}{2}(\sigma_x + \sigma_y) + \frac{1}{2}(\sigma_x - \sigma_y)\cos 2\theta + \tau_{xy}\sin 2\theta \end{aligned} \tag{2.23a}$$

$$\begin{aligned} \sigma_{y'} &= \sigma_x \sin^2\theta + \sigma_y \cos^2\theta - 2\tau_{xy}\sin\theta\cos\theta \\ &= \frac{1}{2}(\sigma_x + \sigma_y) - \frac{1}{2}(\sigma_x - \sigma_y)\cos 2\theta - \tau_{xy}\sin 2\theta \end{aligned} \tag{2.23b}$$

$$\tau_{x'y'} = (\sigma_y - \sigma_x)\sin\theta\cos\theta + (\cos^2\theta - \sin^2\theta)\tau_{xy} = -\frac{1}{2}(\sigma_x - \sigma_y)\sin 2\theta + \tau_{xy}\cos 2\theta$$

A convenient way of writing Equations 2.23 is as follows:

$$\begin{aligned} \sigma_{x'} + \sigma_{y'} &= \sigma_x + \sigma_y \\ \sigma_{y'} - \sigma_{x'} + 2i\tau_{x'y'} &= (\sigma_y - \sigma_x + 2i\tau_{xy})e^{2i\theta} \end{aligned} \tag{2.24}$$

where i stands for the imaginary number $i = \sqrt{-1}$.

EXAMPLE 2.4

The state of stress at a point in a loaded member relative to an x, y, and z coordinate system is given as

$$\begin{bmatrix} 50 & 80 & -40 \\ 80 & 100 & 40 \\ -40 & 40 & 20 \end{bmatrix} \text{MPa}$$

Determine the stresses relative to the x′, y′, and z′ coordinate system obtained by rotating the yz-plane through an angle of 25° counterclockwise about the x-axis (Figure 2.8).

Solution

The direction cosines of the x′, y′, z′ system are

$$l_1 = \cos 0^\circ = 1, \quad m_1 = n_1 = \cos 90^\circ = 0$$

$$l_2 = 0, \quad m_2 = \cos 25^\circ = 0.906, \quad n_2 = \cos 65^\circ = 0.423$$

$$l_3 = 0, \quad m_3 = \cos 115^\circ = -0.423, \quad n_3 = \cos 25^\circ = 0.906$$

The transformed stresses are computed using Equations 2.20 or 2.21. Using Equation 2.21, it is found that

$$\sigma'_{ij} = \begin{pmatrix} 1 & 0 & 0 \\ 0 & 0.906 & 0.423 \\ 0 & -0.423 & 0.906 \end{pmatrix} \begin{pmatrix} 50 & 80 & -40 \\ 80 & 100 & 40 \\ -40 & 40 & 20 \end{pmatrix} \begin{pmatrix} 1 & 0 & 0 \\ 0 & 0.906 & -0.423 \\ 0 & 0.423 & 0.906 \end{pmatrix}$$

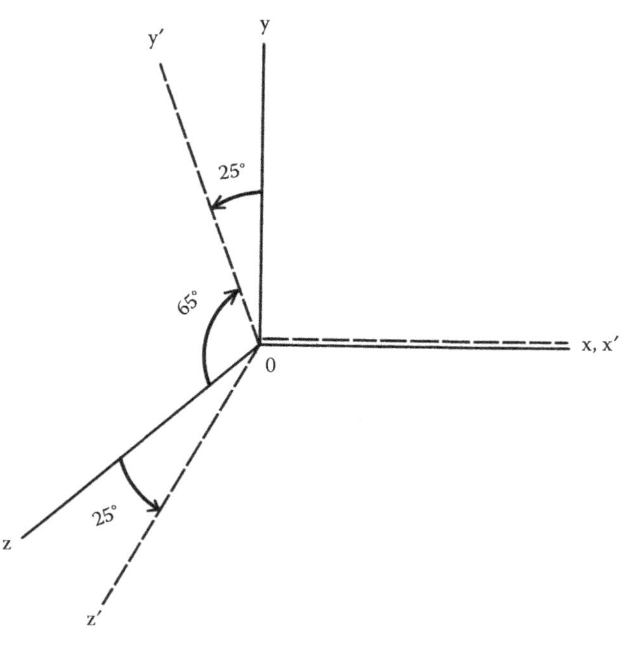

FIGURE 2.8
Rotating of y_3-plane through 25° counterclockwise.

Performing the matrix multiplication, it is found that

$$\sigma'_{ij} = \begin{pmatrix} 50 & 55.56 & -70.08 \\ 55.56 & 116.32 & -4.98 \\ -70.08 & -4.98 & 3.65 \end{pmatrix}$$

2.6 Principal Stresses

At any point in a stressed component, the normal and shear stresses have maximum and minimum values for varying orientations of the inclined plane through that point. These extreme stresses are known as *principal stresses* and their associated directions as *principal directions.* In this section, we will discuss methods of finding the principal stresses and the direction cosines of the normals to the planes on which they act. The cases of normal and shear stresses are considered separately.

2.6.1 Principal Normal Stresses

Let us assume that the state of stress at a point P in a loaded member is given with reference to the rectangular axes. Consider a plane containing the point P whose normal has variable direction cosines, l, m, n, relative to the same axes. The normal stress σ_n is computed by Equation 2.14, namely

$$\sigma_n = \sigma_x l^2 + \sigma_y m^2 + \sigma_z n^2 + 2(\tau_{xy} lm + \tau_{yz} mn + \tau_{zx} nl) \tag{a}$$

Since two of those variables are independent, the extreme values of the normal stress are obtained by differentiation with respect to the independent variables. If l and m are the independent variables, then we need to obtain $(\partial\sigma_n/\partial l) = 0$ and $(\partial\sigma_n/\partial m) = 0$.

Differentiating σ_n with respect to l and m, and noting from the geometric relation, $l^2 + m^2 + n^2 = 1$, that is,

$$\frac{\partial n}{\partial l} = -\frac{l}{n}, \quad \frac{\partial n}{\partial m} = -\frac{m}{n} \tag{2.25}$$

It follows that

$$\frac{\partial\sigma_n}{\partial l} = 2[l\sigma_x + m\tau_{xy} + n\tau_{xz} - \frac{l}{n}(l\tau_{zx} + m\tau_{zy} + n\sigma_z)]$$
$$= 2\left(p_x - \frac{l}{n}p_z\right) = 0$$

The last equation gives $\frac{p_x}{l} = \frac{p_z}{n}$.
Similarly,

$$\frac{\partial\sigma_n}{\partial m} = 2[l\tau_{yx} + m\sigma_y + n\tau_{yz} - \frac{m}{n}(l\tau_{zx} + m\tau_{zy} + n\sigma_z)] = 2\left(p_y - \frac{m}{n}p_z\right) = 0$$

which yields, $\frac{p_y}{m} = \frac{p_z}{n}$.

Hence,

$$\frac{p_x}{l} = \frac{p_y}{m} = \frac{p_z}{n} = \sigma \tag{2.26}$$

where σ is an arbitrary constant.

Equation 2.26 implies that $p_x = l\sigma$, $p_y = ml$, and $p_z = nl$. Since $\sigma_n = lp_x + mp_y + np_z$ and $\tau_n = (p_x^2 + p_y^2 + p_z^2 - \sigma_n^2)^{1/2}$, it is readily concluded that $\sigma_n = \sigma$ and $\tau_n = 0$. This proves that normal stresses with stationary values act on planes free from shearing stresses and the resultant stress, $R_n = (\sigma_n^2 + \tau_n^2)^{1/2} = \sigma_n$, acts in the direction of the normal to the plane. Such stresses are known as *principal stresses* and the corresponding planes as *principal planes*.

Inserting the relations into the set (2.26) into (2.13) provides the equations

$$l(\sigma_x - \sigma) + m\tau_{xy} + n\tau_{xz} = 0 \tag{2.27a}$$

$$l\tau_{xy} + m(\sigma_y - \sigma) + n\tau_{yz} = 0 \tag{2.27b}$$

$$l\tau_{xz} + m\tau_{yz} + n(\sigma_z - \sigma) = 0 \tag{2.27c}$$

The set of homogeneous Equations 2.27 yields a nontrivial solution for l, m, and n if the determinant of the coefficients vanish, that is,

$$\begin{vmatrix} \sigma_x - \sigma & \tau_{xy} & \tau_{xz} \\ \tau_{yx} & \sigma_y - \sigma & \tau_{yz} \\ \tau_{zx} & \tau_{zy} & \sigma_z - \sigma \end{vmatrix} = 0 \tag{2.28}$$

Expanding the determinant leads to a cubic equation in σ

$$\sigma^3 - I_1\sigma^2 + I_2\sigma - I_3 = 0 \tag{2.29}$$

where the coefficients, I_1, I_2, and I_3, known as stress invariants, are given by

$$\begin{aligned} I_1 &= \text{sum of the diagonal terms} \\ &= \sigma_x + \sigma_y + \sigma_z \end{aligned} \tag{2.30a}$$

$$\begin{aligned} I_2 &= \text{sum of cofactors of the diagonal terms} \\ &= \begin{vmatrix} \sigma_y & \tau_{yz} \\ \tau_{zy} & \sigma_z \end{vmatrix} + \begin{vmatrix} \sigma_x & \tau_{xz} \\ \tau_{zx} & \sigma_z \end{vmatrix} + \begin{vmatrix} \sigma_x & \tau_{xy} \\ \tau_{yx} & \sigma_y \end{vmatrix} \\ &= \sigma_x\sigma_y + \sigma_y\sigma_z + \sigma_z\sigma_x - \tau_{xy}^2 - \tau_{yz}^2 - \tau_{zx}^2 \end{aligned} \tag{2.30b}$$

$$I_3 = \begin{vmatrix} \sigma_x & \tau_{xy} & \tau_{xz} \\ \tau_{yx} & \sigma_y & \tau_{yz} \\ \tau_{zx} & \tau_{zy} & \sigma_z \end{vmatrix} = \sigma_x\sigma_y\sigma_z + 2\tau_{xy}\tau_{yz}\tau_{zx} - \sigma_x\tau_{yz}^2 - \sigma_y\tau_{zx}^2 - \sigma_z\tau_{xy}^2 \tag{2.30c}$$

The principal stresses, σ_1, σ_2, σ_3, are the roots of Equation 2.29. Because of the principal stresses are physical quantities, the roots must be real numbers and the coefficients of the cubic equation must be invariants, that is, they are identical regardless of the reference stress coordinates.

The roots are determined either algebraically or numerically by standard methods. In some cases, it is convenient to obtain the roots by first finding one root, $\sigma = \sigma_a$, either by trial and error or by plotting Equation 2.29 with σ as abscissa. Dividing the cubic equation by $(\sigma - \sigma_a)$, a quadratic equation is obtained with simple solution to provide the other two roots.

A direct algebraic method to solve the cubic equation involves introducing a new variable, x, such that

$$\sigma = Ax + (1/3)I_1, \quad A = \text{constant} \tag{a}$$

Inserting Equation (a) into the cubic equation and simplifying the algebra results in

$$x^3 - (R/A^2)x + (Q/A^3) = 0 \tag{b}$$

where

$$R = \frac{1}{3}I_1^2 - I_2$$

$$Q = \frac{1}{3}I_1I_2 - \frac{2}{27}I_1^3 - I_3$$

Next, consider the standard trigonometric identity: $\text{Cos}\, 3\theta = 4\, \text{Cos}^3\, \theta - 3\, \text{Cos}\, \theta$.
If $x = \text{Cos}\, \theta$, then the trigonometric identity assumes the form

$$x^3 - (3/4)x - (1/4)\cos 3\theta = 0 \tag{c}$$

It is clear that Equations (b) and (c) are identical provided that $(R/A^2) = (3/4)$ or $A = 2(R/3)^{1/2}$ and $\text{Cos}\, 3\theta = -4\,(Q/A^3) = -Q/(2T)$ with $T = A^3/8 = (R^3/27)^{1/2}$. Since the equation

$$\cos 3\theta = -Q/(2T) \quad \text{or} \quad 3\theta = \cos^{-1}(-Q/2T) \tag{d}$$

provides three values for the angle θ, namely, (i) θ, (ii) $\theta + 120°$, (iii) $\theta - 120°$, $0 < \theta < 60°$.
It follows from Equation (a) that the three roots of the cubic equation are

$$\begin{aligned}
\sigma_a &= 2\left(\frac{R}{3}\right)^{1/2} \cos\theta + \frac{I_1}{3} \\
\sigma_b &= 2\left(\frac{R}{3}\right)^{1/2} \cos(\theta + 120°) + \frac{I_1}{3} \\
\sigma_c &= 2\left(\frac{R}{3}\right)^{1/2} \cos(\theta + 120°) + \frac{I_1}{3}
\end{aligned} \tag{2.31}$$

The third root in Equations 2.31 may be written as $\sigma_c = 2\left(\frac{R}{3}\right)^{1/2} (\cos\theta + 240°) + \frac{I_1}{3}$.
Once numerical values of the principal stresses are obtained from Equations 2.31, the principal stresses can be arranged in the conventional order of magnitude, $\sigma_1 > \sigma_2 > \sigma_3$.

In terms of the principal stresses, Equation 2.29 becomes

$$(\sigma - \sigma_1)(\sigma - \sigma_2)(\sigma - \sigma_3) = 0 \tag{e}$$

and the stress invariants reduce to

$$\begin{aligned} I_1 &= \sigma_1 + \sigma_2 + \sigma_3 \\ I_2 &= \sigma_1\sigma_2 + \sigma_2\sigma_3 + \sigma_3\sigma_1 \\ I_3 &= \sigma_1\sigma_2\sigma_3 \end{aligned} \tag{f}$$

The principal stresses are the characteristic values or eigen vectors of the stress tensor, σ_{ij}, i, j = x, y, z and the associated direction cosines are the eigen vectors of each principal stress.

Direction cosines: The direction cosines (eigenvectors) of the principal stresses, σ_j, j = 1,2,3, are found by determining the directions cosines l_j, m_j, n_j, in Equations 2.27 when σ is replaced by σ_j, j = 1, 2, 3. Because only two of Equations 2.27 are independent equations, an additional equation, namely, the identity, $l_j^2 + m_j^2 + n_j^2 = 1$, is needed to determine the direction cosines. Thus, the direction cosines of the principal stresses may be obtained by solving the following set of equations:

$$\begin{aligned} \frac{l_j}{n_j} &= \frac{\tau_{xy}\tau_{yz}\tau_{xz} - (\sigma_y - \sigma_j)}{(\sigma_x - \sigma_j)(\sigma_y - \sigma_j) - \tau_{xy}^2} \\ \frac{m_j}{n_j} &= \frac{\tau_{xy}\tau_{xy} - \tau_{yz}(\sigma_x - \sigma_j)}{(\sigma_x - \sigma_j)(\sigma_y - \sigma_j) - \tau_{xy}^2} \\ l_j^2 &+ m_j^2 + n_j^2 = 1 \end{aligned} \tag{2.32a}$$

Here, j = 1, 2, 3.

Instead of solving two linear equations and one second-order equation simultaneously, the following method provides simpler approach to determine the directions of the principal stresses. Equations 2.27 are written in matrix form as

$$\begin{pmatrix} (\sigma_x - \sigma_j) & \tau_{xy} & \tau_{xz} \\ \tau_{yx} & (\sigma_y - \sigma_j) & \tau_{yz} \\ \tau_{zx} & \tau_{zy} & (\sigma_z - \sigma_j) \end{pmatrix} \begin{pmatrix} l_j \\ m_j \\ n_j \end{pmatrix} = 0 \tag{g}$$

The cofactors of the determinant of this matrix using the elements of the first row are

$$\begin{aligned} a_j &= \begin{vmatrix} (\sigma_y - \sigma_j) & \tau_{yz} \\ \tau_{zy} & (\sigma_z - \sigma_j) \end{vmatrix} \\ b_j &= -\begin{vmatrix} \tau_{yx} & \tau_{yz} \\ \tau_{zx} & (\sigma_z - \sigma_j) \end{vmatrix} \\ c_j &= \begin{vmatrix} \tau_{yx} & (\sigma_y - \sigma_j) \\ \tau_{zx} & \tau_{zy} \end{vmatrix} \end{aligned} \tag{h}$$

Hence, the direction cosines (or eigen vectors) of the principal stresses are then given by

$$l_j = a_j k_j, \quad m_j = b_j k_j, \quad n_j = c_j k_j, \quad j = 1, 2, 3 \tag{2.32b}$$

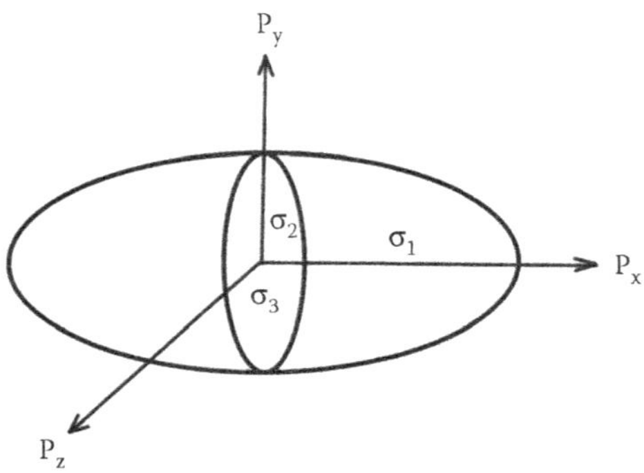

FIGURE 2.9
Stress ellipsoid.

where

$$k_j = \frac{1}{\sqrt{a_j^2 + b_j^2 + c_j^2}}$$

It is clear that the solution in Equation 2.32b satisfies the identity $l_j^2 + m_j^2 + n_j^2 = 1$.

Stress ellipsoid: It is clear from Equation 2.26 that the principal stresses satisfy the relations

$$l = \frac{p_x}{\sigma_1}, \quad m = \frac{p_y}{\sigma_2}, \quad n = \frac{p_z}{\sigma_3}$$

Hence, it follows that

$$\left(\frac{p_x}{\sigma_1}\right)^2 + \left(\frac{p_y}{\sigma_2}\right)^2 + \left(\frac{p_z}{\sigma_3}\right)^2 = 1 \tag{2.33}$$

which represents an ellipsoidal surface with the semiaxes as the principal stresses at the point (Figure 2.9). Note that Equation 2.33 represents a geometrical interpretation of the fact that the principal stresses are extreme values of the normal stress. In case of equal principal stresses, the ellipsoid reduces to a sphere. Any three mutually perpendicular axes can be used as the principal axes.

2.6.2 Principal Shear Stresses

In order to determine the maximum and minimum shear stresses and the planes on which they act, it is convenient to consider the principal axes of the state of stress as the reference axes. Suppose that 0x, 0y, 0z denote the coordinate axes associated with the principal stresses σ_1, σ_2, σ_3, respectively, the components of the stress vector across an arbitrary plane whose normal has direction cosines l, m, n are (Equation 2.13) $p_x = l\,\sigma_1$, $p_y = m\,\sigma_2$, $p_z = n\,\sigma_3$. The normal stress on the same plane is given by Equation 2.14.

$$\sigma_n = l^2\sigma_1 + m^2\sigma_2 + n^2\sigma_3 \tag{2.34}$$

The resultant shear stress, τ_n, is obtained as follows:

$$\tau_n^2 = T_n^2 - \sigma_n^2 = l^2\sigma_1^2 + m^2\sigma_2^2 + n^2\sigma_3^2 - (l^2\sigma_1 + m^2\sigma_2 + n^2\sigma_3)^2 \quad (2.35)$$

The maximum and minimum values of the resultant shear stress τ_n are found by treating the variables l and m in Equation 2.35 as the independent variables, differentiating with respect to l and m, and observing that $\frac{\partial n}{\partial l} = -\frac{l}{n}$ and $\frac{\partial n}{\partial m} = -\frac{m}{n}$, it follows that stationary values of the resultant shear stress are obtained from the relations

$$2\tau_n \frac{\partial \tau_n}{\partial l} = 2l(\sigma_1 - \sigma_2)[(1 - 2l^2)(\sigma_1 - \sigma_3) - 2m^2(\sigma_2 - \sigma_3)] = 0 \quad (2.36a)$$

$$2\tau_n \frac{\partial \tau_n}{\partial m} = 2m(\sigma_2 - \sigma_3)[(1 - 2m^2)(\sigma_2 - \sigma_3) - 2l^2(\sigma_1 - \sigma_2)] = 0 \quad (2.36b)$$

Equations 2.36 are satisfied by setting $l = m = 0$, which implies that $n = \pm 1'$. Substituting these values of l, m, and n into Equation 2.35, it is found that the stationary shear stress, τ_n, has a minimum value equal to zero and acts on a principal plane whose normal is the z-axis. Equations 2.36 are also satisfied by $l = 0$, $m = n = \pm 1/\sqrt{2}$, and by $m = 0$, $l = n = \pm 1/\sqrt{2}$, which provide the principal or maximum values of τ_n. Thus, by inserting $m = 0, l = n = \pm 1/\sqrt{2}$ in Equation 2.35, one obtains the first principal shear stress $\tau_1 = 1/2\,|\sigma_1 - \sigma_3|$. In this case, $m = \cos 90°$ and $l = n = \cos 45°$. Hence, τ_1 acts on a plane whose normal makes equal angles with the directions of the principal stresses σ_1 and σ_3 (see the shaded plane in Figure 2.10). The associated normal stress acting on the same plane is obtained from Equation 2.14 and has the value $1/2(\sigma_1 + \sigma_3)$. Similarly, by setting $l = 0$, $m = n = \pm 1/\sqrt{2}$ in Equation 2.35, a second principal shear stress is obtained as $\tau_2 = 1/2\,|\sigma_2 - \sigma_3|$, and it acts on a plane that makes equal angles with the directions of σ_2

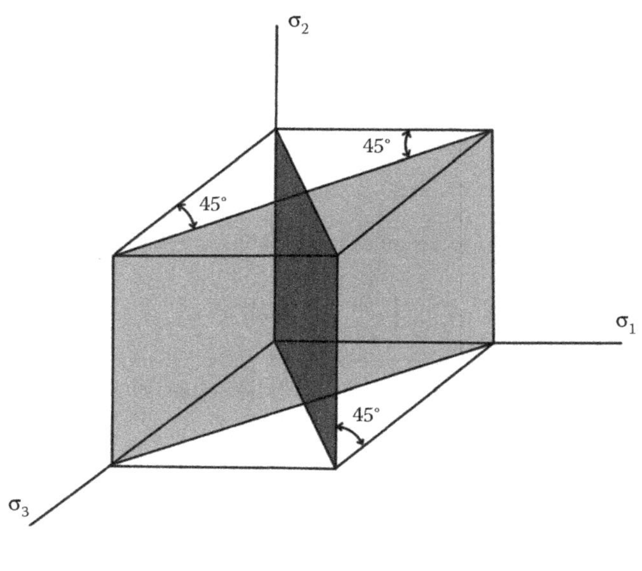

FIGURE 2.10
Planes of principal sheer stress.

and σ_3. The associated normal stress on the same plane is $1/2(\sigma_2 + \sigma_3)$. Moreover, by treating l and n as independent variables, new equations similar to Equations 2.41 are obtained for the stationary values of the shear stress. Upon setting $n = 0$, $l = m = \pm 1/\sqrt{2}$ in the new equations and in Equation 2.35, it can be readily shown that a third principal stress is obtained as $\tau_3 = 1/2\,|\sigma_2 - \sigma_3|$ and acts on a plane that bisects the angles between the directions of σ_1 and σ_2. The value of the associated normal stress is $(1/2)(\sigma_1 + \sigma_2)$.

In sum, the magnitudes of the three principal shear stresses are

$$\tau_1 = \frac{1}{2}(\sigma_1 - \sigma_3), \quad \tau_2 = \frac{1}{2}(\sigma_2 - \sigma_3), \quad \tau_3 = \frac{1}{2}(\sigma_1 - \sigma_2) \tag{2.37}$$

And they act on planes that bisect the planes of the principal stresses. Thus, if the principal normal stresses are arranged such that $\sigma_1 > \sigma_2 > \sigma_3$, the absolute maximum shear stress acts on a plane whose normal bisects the directions of σ_1 and σ_3 and its value is given by

$$\tau_{max} = \tau_1 = \frac{1}{2}(\sigma_1 - \sigma_3) \tag{2.38}$$

It is interesting to observe that when $\sigma_1 = \sigma_2 = \sigma_3$, that is, hydrostatic loading, τ_n vanishes (see Equation 2.35).

Direction cosines: The approach used to determine the direction cosines of the principal shearing stresses, τ_j, $j = 1,2,3$, is similar to that used in deriving the direction cosines of the resultant shear stress acting on an inclined given in Equation 2.17.

The components of the normal stress along the coordinate axes are $l\sigma_1$, $m\sigma_2$, $n\sigma_3$. It follows that the direction cosines, l_{sj}, m_{sj}, n_{sj}, of τ_j are as follows:

$$\begin{aligned} l_{sj} &= l\left(\frac{\sigma_1 - \sigma_n}{\tau_j}\right) \\ m_{sj} &= m\left(\frac{\sigma_2 - \sigma_n}{\tau_j}\right) \\ n_{sj} &= n\left(\frac{\sigma_3 - \sigma_n}{\tau_j}\right) \end{aligned} \tag{2.39}$$

Here, $j = 1, 2, 3$.

EXAMPLE 2.5

The state of stress relative to rectangular coordinates is given by

$$\begin{pmatrix} 3 & 2 & 4 \\ 2 & 1 & 5 \\ 4 & 5 & 2 \end{pmatrix} \text{MPa}$$

Find the principal stresses and the associated direction cosines.

Solution

The stress invariants are determined by inserting the given stresses into Equations 2.30a, b, and c. The results are

$$I_1 = 6\,\text{MPa}, \quad I_2 = -34\,(\text{MPa})^2, \quad I_3 = -13\,(\text{MPa})^3$$

Next, the parameters R, Q, T, and θ are calculated using Equations (b) and (d) in Section 2.6.1.

$R = 46\,(\text{MPa})^2,\quad Q = -71\,(\text{MPa})^3,\quad T = 60.042\,(\text{MPa})^3,\quad 3\theta = \cos^{-1}(0.591) = 53.75°$
$\theta = 17.017°$

Introducing the above values into Equations 2.31 gives $\sigma_a = 9.452$ MPa, $\sigma_b = -3.813$ MPa, $\sigma_c = 0.361$ MPa.

Hence, the principal stresses arranged in the conventional manner are

$$\sigma_1 = 9.452$$
$$\sigma_2 = 0.361\,\text{MPa}$$
$$\sigma_3 = -3.813\,\text{MPa}$$

The associated direction cosines are found using Equations 2.32b

$$l_1, m_1, n_1 = (0.561,\ 0.561,\ 0.647)$$
$$l_2, m_2, n_2 = (-0.795,\ 0.550,\ 0.247)$$
$$l, m_3, n_3 = (0.230,\ 0.653,\ -0.721)$$

It is readily confirmed that the direction cosines satisfy Equation 2.9 and the three principal axes are perpendicular to each other (see solved Example 2.8 for general proof of such property).

EXAMPLE 2.6

The state of stress relative to the x, y, z coordinates is given by

$$\begin{pmatrix} 4 & 2 & 3 \\ 2 & 6 & 1 \\ 3 & 1 & 5 \end{pmatrix} \text{ksi}$$

Given that the magnitude of one of the principal stresses is 9 ksi, determine

1. The magnitudes of the other principal stresses
2. The direction cosines of the major principal stress

Solution

The stress invariants from Equations 2.30 are $I_1 = 15$ ksi, $I_2 = 60\,(\text{ksi})^2$, $I_3 = 54\,(\text{ksi})^3$.

It follows that Equation 2.29 is

$$\sigma^3 - 15\sigma^2 + 60\sigma - 54 = 0$$

Since $\sigma = 9$ is one root of this equation, dividing the equation by the factor $(\sigma - 9)$, the equation is reduced to

$$(\sigma - 9)(\sigma^2 - 6\sigma + 6) = 0$$

The other principal stresses are obtained by solving the quadratic part of the previous equation:

1. Hence, the principal stresses are $\sigma_1 = 9$ ksi, $\sigma_2 = 4.73$ ksi, $\sigma_3 = 1.27$ ksi.
2. The direction cosines of the major principal stress may be obtained by applying Equations 2.32b or in this example the use of Equations 2.32a is illustrated. With

$\sigma_1 = 9$ ksi, the first two equations in Equation 2.32a give

$$-5l + 2m + 3n = 0$$
$$2l - 3m + n = 0$$

Eliminating l_1 yields $m_1 = n_1$, and the third equation in Equation 2.32a gives

$$l = m = n\frac{1}{\sqrt{3}}$$

(The axis of the major principal stress makes equal angles with the reference axes.)

EXAMPLE 2.7

The stresses at a point across a given plane are $\sigma_n = 10$ ksi and $\tau_n = 6$ ksi. If two angles of the direction cosines of σ_n and τ_n relative to the principal axes are (55°, 45°) and (130°, 86°), respectively, determine the principal stresses at the point.

Solution

We start by finding the third direction cosine of the given stresses relative to the principal directions. If the direction cosines of the normal and shear stresses relative to the principal axes are (l, m, n) and (l_s, m_s, n_s), respectively, we have

$$\text{for } \sigma_n\text{: } l = \cos 55° = 0.574;\ m = \cos 45° = 0.707;\ n = (1 - l^2 - m^2)^{1/2} = 0.413$$
$$\tau_n\text{: } l_s = \cos 130° = -0.643;\ m_s = \cos 86° = 0.0698;\ n_s = (1 - l_s^2 - m_s^2)^{1/2} = 0.763$$

The proper values of n and n_s are those that satisfy the identity $ll_s + mm_s + nn_s = 0$. It is readily confirmed that $n = 0.413$ and $n_s = 0.763$.

Relative to the principal axes, the direction cosines of the resultant shear are obtained using Equation 2.17 provided that the substitutions $p_x = \sigma_1 l$, $p_y = \sigma_2 m$, and $p_z = \sigma_3 n$ are used in those equations (see also Equations 2.39). Thus, we have

$$l_s = l\left(\frac{\sigma_1 - \sigma_n}{\tau_n}\right)$$
$$m_s = m\left(\frac{\sigma_2 - \sigma_n}{\tau_n}\right)$$
$$n_s = n\left(\frac{\sigma_3 - \sigma_n}{\tau_n}\right)$$

Substituting for the direction cosines of the given normal and shear stresses and solving for the principal stresses from the above equations, we obtain the principal stresses at the point

$$\sigma_1 = \sigma_n + \left(\frac{l_s}{l}\right)\tau_n = 3.28\ \text{ksi}$$
$$\sigma_2 = \sigma_n + \left(\frac{m_s}{m}\right)\tau_n = 10.59\ \text{ksi}$$
$$\sigma_3 = \sigma_n + \left(\frac{n_s}{n}\right)\tau_n = 21.08\ \text{ksi}$$

EXAMPLE 2.8

Show that the principal axes are normal to one another.

Solution

The principal directions are found from the solution of any two equations of Equations 2.27 and the geometrical identity (2.9). Let (l_1, m_1, n_1) and (l_2, m_2, n_2) denote the direction cosines of σ_1 and σ_2, respectively. It follows from Equations 2.27 that the direction cosines of σ_1 satisfy the equations

$$\begin{aligned} l_1(\sigma_x - \sigma_1) + m_1\tau_{xy} + n_1\tau_{xz} &= 0 \\ l_1\tau_{yx} + m_1(\sigma_y - \sigma_1) + n_1\tau_{yz} &= 0 \\ l_1\tau_{zx} + m_1\tau_{zy} + n_1(\sigma_z - \sigma_1) &= 0 \end{aligned} \qquad \text{(a)}$$

Multiplying Equations (a) by l_2, m_2, n_2, respectively, and adding the results gives

$$l_1l_2(\sigma_x - \sigma_1) + m_1m_2(\sigma_y - \sigma_1) + n_1n_2(\sigma_z - \sigma_1) + (l_2m_1 + l_1m_2)\tau_{xy} + (l_2n_1 + l_1n_2)\tau_{xz} + (m_2n_1 + m_1n_2)\tau_{yz} \qquad \text{(b)}$$

The direction cosines of the principal stress σ_2 satisfy three equations identical to the set (a) provided that the subscript 1 is replaced by the subscript 2. Upon multiplying the new set by l_1, m_1, n_1, respectively, and adding one obtains

$$l_2l_1(\sigma_x - \sigma_2) + m_2m_1(\sigma_y - \sigma_2) + n_2n_1(\sigma_z - \sigma_2) + (l_1m_2 + l_2m_1)\tau_{xy} + (l_1n_2 + l_2n_1)\tau_{xz} + (m_1n_2 + m_2n_1)\tau_{yz} \qquad \text{(c)}$$

Note that Equation (c) can be obtained from Equation (b) by interchanging the subscripts 1 and 2. Subtracting Equation (b) from (c) yields

$$(\sigma_1 - \sigma_2)(l_1l_2 + m_1m_2 + n_1n_2) = 0 \qquad \text{(d)}$$

Hence, if $\sigma_1 \neq \sigma_2$, it follows that $l_1l_2 + m_1m_2 + n_1n_2 = 0$, confirming that the principal directions are orthogonal.

In the degenerate case when $\sigma_1 = \sigma_2$, the direction cosines of σ_1 and σ_2 are not unique. However, the third principal axis is uniquely determined from any two of Equations 2.27 and 2.9, and all directions normal to this axis are principal directions. When all three principal stresses are identical (hydrostatic state of stress), any direction through the point is a principal direction.

2.7 Stress Deviator Tensor

A stress deviator tensor, s_{ij}, is obtained by subtracting the average (mean) of three normal stresses, $\sigma_m = (1/3)\ (\sigma_x + \sigma_y + \sigma_z)$, from the stress tensor σ_{ij}. The shear stresses remain unchanged. Thus, a stress deviator tensor (also known as reduced tensor) is given by the following array:

$$s_{ij}\begin{pmatrix} s_x & \tau_{xy} & \tau_{xz} \\ \tau_{yx} & s_y & \tau_{yz} \\ \tau_{zx} & \tau_{zy} & s_z \end{pmatrix} = \begin{pmatrix} \sigma_x - \sigma_m & \tau_{xy} & \tau_{xz} \\ \tau_{yx} & \sigma_y - \sigma_m & \tau_{yz} \\ \tau_{zx} & \tau_{zy} & \sigma_z - \sigma_m \end{pmatrix} \qquad (2.40)$$

where $\sigma_x - \sigma_m = (1/3)\ (2\sigma_x - \sigma_y - \sigma_z)$ and similar expressions for the other two diagonal terms. Note that $s_x + s_y + s_z = 0$, indicating that the deviatoric normal stresses cannot all have the same sign. Also, the difference between any two normal stress components of the stress deviator tensor is the same as the difference between the corresponding components of the original stress tensor, for example, $s_x - s_y = \sigma_x - \sigma_y$, etc. Since $\sigma_m = (1/3)\ I_1$, which is the same for all coordinate axes, it is called a *hydrostatic or spherical* stress. The hydrostatic stress is also known as volumetric or dilatational stress because it produces changes in the body dimensions without distortion. The deviatoric stresses produce distortion without change in volume.

The decomposition of the stress tensor, classified into two parts: a hydrostatic stress tensor and a deviatoric stress tensor, plays a significant role in material modeling, especially in studying the plastic behavior of structural materials [2.2, 2.3].

The invariants of the stress deviator tensor are obtained using a procedure similar to that used for the regular stress tensor. Thus, we have the following determinant similar to Equation 2.28 to determine the corresponding cubic equation for finding the principal deviatoric stresses:

$$\begin{vmatrix} s_x - s & \tau_{xy} & \tau_{xz} \\ \tau_{yx} & s_y - s & \tau_{yz} \\ \tau_{zx} & \tau_{zy} & s_z - s \end{vmatrix} = 0$$

Expanding the determinant, the following cubic equation is reached

$$S^3 - J_1S^2 - J_2S - J_3 = 0 \tag{2.41}$$

where J_1, J_2, and J_3 are the stress deviator invariants given by the following relations:

$$\begin{aligned} J_1 &= S_x + S_y + S_z = 0 \\ J_2 &= -(S_xS_y + S_yS_z + S_zS_x) + \tau_{xy}^2 + \tau_{yz}^2 + \tau_{zx}^2 \\ &= (1/6)[(\sigma_x - \sigma_y)^2 + (\sigma_y - \sigma_z)^2 + (\sigma_z - \sigma_x)^2] + \tau_{xy}^2 + \tau_{yz}^2 + \tau_{zx}^2 \\ &= (1/3)I_1^2 - I_2 = R \\ J_3 &= S_xS_yS_z - 2\tau_{xy}\tau_{yz}\tau_{zx} - S_x\tau_{yz}^2 - S_y\tau_{zx}^2 - S_z\tau_{xy} \\ &= (2/27)I_1^3 - (1/3)I_1I_2 + I_3 = -Q \end{aligned} \tag{2.42}$$

Here, I_j, $j = 1, 2, 3$ are the invariants of the regular stress Equations 2.30 and R and Q are the parameters introduced in Equation (b) of Section 2.6.1 and used in Equation 2.31.

An algebraic solution of the cubic equation in Equation 2.41 may be obtained by following the procedure used in solving Equation 2.29 for the ordinary stress state. Making use of the substitution, $S = 2(R/3)^{1/2}\,\text{Cos}\,\theta$, $0 < \theta < 60°$, it is readily confirmed that the angle θ is governed by

$$\cos 3\theta = -\frac{3\sqrt{3}}{2}\frac{Q}{R^{3/2}} \tag{2.43}$$

And the principal deviatoric stresses are

$$S_a = 2\sqrt{\frac{J_2}{3}}\cos\theta$$

$$S_b = 2\sqrt{\frac{J_2}{3}}\cos(\theta + 120^\circ) \qquad (2.44)$$

$$S_c = 2\sqrt{\frac{J_2}{3}}\cos(\theta + 120^\circ)$$

Since $J_2 = R$, the above solution is the first term in Equations 2.31 for the ordinary principal stresses. It follows that

$$\begin{Bmatrix} \sigma_1 \\ \sigma_2 \\ \sigma_3 \end{Bmatrix} = \begin{Bmatrix} S_1 \\ S_2 \\ S_3 \end{Bmatrix} + \begin{Bmatrix} \sigma_m \\ \sigma_m \\ \sigma_m \end{Bmatrix}$$

where $\sigma_m = (1/3)\, I_1$ is the hydrostatic stress.

In terms of the principal stresses, Equation 2.41 is equivalent to

$$(S - S_1)(S - S_2)(S - S_3) = 0$$

And the stress invariants reduce to

$$J_2 = -(S_1S_2 + S_2S_3 + S_3S_1) = (1/2)(S_1^2 + S_2^2 + S_3^2)$$

$$J_3 = S_1S_2S_3 = (1/3)(S_1^2 + S_2^2 + S_3^2)$$

It should be noted that because the hydrostatic stress is the same for all possible reference axes, the principal direction cosines of the stress deviator are identical to those of the ordinary stresses.

EXAMPLE 2.9

The state of stress at a point in a loaded member is given by

$$\sigma_{ij} = \begin{pmatrix} 40 & 20 & 30 \\ 20 & 60 & 15 \\ 30 & 15 & 50 \end{pmatrix} \text{MPa}$$

Find the stress invariants J_2 and J_3, and use them to determine the magnitudes of the principal stresses σ_j, $j = 1, 2, 3$.

Solution

Using Equations 2.30a, b and c, the stress invariants are

$$I_1 = 150\,\text{MPa}$$

$$I_2 = 5875\,(\text{MPa})^2$$

$$I_3 = 55{,}000\,(\text{MPa})^3$$

Next, Equations 2.42 give

$$J_2 = (1/3)I_1^2 - I_2 = 1625\,(\text{MPa})^2$$

$$J_3 = (2/27)I_1^3 - (1/3)I_1I_2 + I_3 = 11{,}250\,(\text{MPa})^3$$

The angle θ is obtained from Equation 2.43, $\theta = 21.17°$, and the principal deviatoric stresses are calculated using Equations 2.44. The results in the usual order of magnitudes are

$$S_1 = 43.407\,\text{MPa}$$
$$S_2 = -7.147\,\text{MPa}$$
$$S_3 = -36{,}259\,\text{MPa}$$

Finally, the principal stresses are obtained by adding the hydrostatic stress, $p = (1/3)(150) = 50$ MPa, to the previous values.

$$\sigma_1 = 93.341\,\text{MPa}$$
$$\sigma_2 = 42.853\,\text{MPa}$$
$$\sigma_3 = 13.741\,\text{MPa}$$

2.8 Octahedral Stresses

At any point in a stressed body, an inclined plane whose normal makes equal angles with the principal axes is known as the *octahedral* plane. The normal and resultant shearing stresses acting on this plane are the octahedral stresses. Such stresses are frequently used to predict the initiation of yield in structural components and other aspects of plasticity theory. If l, m, and n denote the direction cosines of the normal to an octahedral plane, then $l^2 = m^2 = n^2 = 1/3$ and $l = m = n = \pm 1/\sqrt{3}$. There are eight such planes in space. Figure 2.11 shows an octahedron element whose sides represent the octahedral planes of the state of stress at the origin of coordinates.

Equation 2.34 gives the value of the normal octahedral stress as

$$\sigma_{oct} = \frac{1}{3}(\sigma_1 + \sigma_2 + \sigma_3) = p \tag{2.45}$$

This stress expands or contracts the element without distorting its shape. For this reason, sometimes it is referred to as the volumetric or dilatational stress.

Similarly, Equation 2.35 yields the shear stress acting on the sides of the octahedron as

$$\tau_{oct} = \frac{1}{3}[(\sigma_1 - \sigma_2)^2 + (\sigma_2 - \sigma_3)^2 + (\sigma_3 - \sigma_1)^2]^{1/2} \tag{2.46a}$$

The octahedral shear stresses distort the shape of the octahedron without changing its volume. With reference to Cartesian axes, the expression of the octahedral shearing stress is derived as follows: first, express the right-hand side of Equations 2.46 in term of stress invariants, I_1, I_2, and I_3, and upon using the expressions of the invariants in Equations 2.30, it is found that

$$\tau_{oct} = \frac{1}{3}[(\sigma_x - \sigma_y)^2 + (\sigma_y - \sigma_z)^2 + (\sigma_z - \sigma_x)^2 + 6(\tau_{xy}^2 + \tau_{yz}^2 + \tau_{zx}^2)]^{1/2} = (2/3)^{1/2}J_2^{1/2} \tag{2.46b}$$

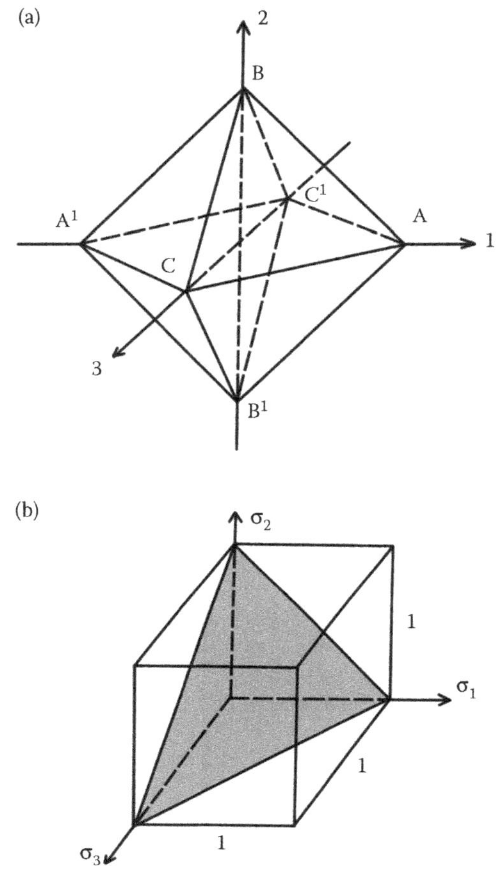

FIGURE 2.11
Octahedral planes in principle coordinate system, (a) eight octahedral planes and (b) octahedral plane in a cubical element.

Alternatively, the octahedral shear stress is also equivalent to

$$\tau_{oct} = \frac{2}{3}(\tau_1^2 + \tau_2^2 + \tau_3^2)^{1/2} \tag{2.46c}$$

where τ_1, τ_2, and τ_3 are the three principal shear stresses.

The octahedral stresses are invariants under a transformation of coordinate axes. They have particular significance in developing criterion for predicting the initiation of yield as will be discussed in Chapter 4.

EXAMPLE 2.10

A state of stress with reference to the x-, y-, and z-axes is given by the tensor

$$\sigma_{ij} = \begin{pmatrix} 60 & 30 & 40 \\ 30 & 20 & 50 \\ 40 & 50 & 10 \end{pmatrix} \text{ MPa}$$

Find:

1. The octahedral normal and shear stresses
2. The deviatoric stress tensor S_{ij}, i, j = x, y, z
3. The ratio $\frac{\tau_{oct}}{\tau_{max}}$

Solution

1. The normal octahedral stress is the average of the normal stresses, $\sigma_{oct} = p = 30$ MPa.

   The octahedral shear stress is computed from Equation 2.46b

$$\tau_{oct} = (1/3)[1600 + 100 + 2500 + 6(900 + 1600 + 2500)]^{1/2} = 61.644\,\text{MPa}$$

2. The deviatoric stress tensor is given by $S_{ij} = \sigma_{ij} - p\delta_{ij}$, that is,

$$S_{ij} = \begin{pmatrix} 60 & 30 & 40 \\ 30 & 20 & 50 \\ 40 & 50 & 10 \end{pmatrix} - 30 \begin{pmatrix} 1 & 0 & 0 \\ 0 & 1 & 0 \\ 0 & 0 & 1 \end{pmatrix} = \begin{pmatrix} 30 & 30 & 40 \\ 30 & -10 & 50 \\ 40 & 50 & -20 \end{pmatrix} \text{MPa}$$

3. In order to find the maximum shearing stress, we need to calculate the principal stresses.

We have $R = (1/3)\ I_1^2 - I_2 = 5700\ (\text{MPa})^2$, $Q = (1/3)\ I_1 I_2 - (2/27)\ I_1^2 - I_3 = -85{,}000\ (\text{MPa})^3$. $T = (R^3/27)^{1/2} = 82{,}819\ (\text{MPa})^3$. It follows that $\text{Cos}\ 3\theta = (-Q/T) = 0.5132$. $\theta = -19.71°$, and the principal stresses are $\sigma_1 = 112.072$ MPa, $\sigma_2 = 14.423$ MPa, and $\sigma_3 = -36.495$ MPa. The maximum shear stress is $\tau_{max} = (1/2)\ (\sigma_1 - \sigma_3) = 74.284$ MPa, and $\frac{\tau_{oct}}{\tau_{max}} = \frac{61.644}{74.258} = 0.83$.

2.9 Mohr's Stress Circles

Mohr's representation of the in-plane state of stress can be extended to 3D problems. It is convenient to use principal stresses σ_1, σ_2, and σ_3 as the starting state of stress at a point P in the body. If the principal axes are oriented along the x, y, and z coordinates with the origin at P, the normal and resultant shear stresses acting on a plane through P whose normal is $\mathbf{n} = l\mathbf{i} + m\mathbf{j} + n\mathbf{k}$ are given by the relations

$$\begin{aligned} \sigma_n &= l^2\sigma_1 + m^2\sigma_2 + n^2\sigma_3 \\ \tau_n^2 &= l^2\sigma_1^2 + m^2\sigma_2^2 + n^2\sigma_3^2 - \sigma_n^2 \end{aligned} \tag{2.47a}$$

The direction cosines l, m, and n satisfy the geometric identity

$$l^2 + m^2 + n^2 = 1 \tag{2.47b}$$

Solving Equations 2.47a, b for l^2, m^2, and n^2, furnish

$$\begin{aligned} l^2 &= \frac{\tau_n^2 + (\sigma_n - \sigma_2)(\sigma_n - \sigma_3)}{(\sigma_1 - \sigma_2)(\sigma_1 - \sigma_3)} \\ m^2 &= \frac{\tau_n^2 + (\sigma_n - \sigma_3)(\sigma_n - \sigma_1)}{(\sigma_2 - \sigma_3)(\sigma_2 - \sigma_1)} \\ n^2 &= \frac{\tau_n^2 + (\sigma_n - \sigma_1)(\sigma_n - \sigma_2)}{(\sigma_3 - \sigma_1)(\sigma_3 - \sigma_2)} \end{aligned} \tag{2.48}$$

Equations 2.48 give the admissible values of σ_n and τ_n on a specified inclined plane. Upon rearranging the terms, Equations 2.48 can be written as follows:

$$\tau_n^2 + \left(\sigma_n - \frac{\sigma_1 + \sigma_3}{2}\right)^2 = m^2(\sigma_2 - \sigma_3)(\sigma_2 - \sigma_1) + \left(\frac{\sigma_1 - \sigma_3}{2}\right)^2 \tag{2.49a}$$

$$\tau_n^2 + \left(\sigma_n - \frac{\sigma_1 + \sigma_2}{2}\right)^2 = n^2(\sigma_3 - \sigma_1)(\sigma_3 - \sigma_2) + \left(\frac{\sigma_1 - \sigma_2}{2}\right)^2 \tag{2.49b}$$

$$\tau_n^2 + \left(\sigma_n - \frac{\sigma_2 + \sigma_3}{2}\right)^2 = l^2(\sigma_1 - \sigma_2)(\sigma_1 - \sigma_3) + \left(\frac{\sigma_2 - \sigma_3}{2}\right)^2 \tag{2.49c}$$

It is clear that in the stress plane, τ_n versus σ_n, each of Equations 2.49a, b, c is of the form, $\tau_n^2 + (\sigma_n - C)^2 = R^2$, which represents a circle with center C and radius R. Since, $\sigma_1 > \sigma_2 > \sigma_3$, and $l^2 \geq 0$, $m^2 \geq 0$, $n^2 \geq 0$, it is readily concluded from the right-hand side of Equation 2.49a that the term $m^2(\sigma_2 - \sigma_3)(\sigma_2 - \sigma_1)$ is negative while in Equations 2.49b and 2.49c, the terms $n^2(\sigma_3 - \sigma_1)(\sigma_3 - \sigma_2)$ and $l^2(\sigma_1 - \sigma_2)(\sigma_1 - \sigma_3)$ are positive. Hence, Equations 2.49 imply the following inequalities:

$$\tau_n^2 + \left(\sigma_n - \frac{\sigma_1 + \sigma_3}{2}\right)^2 \leq \left(\frac{\sigma_1 - \sigma_3}{2}\right)^2 \tag{2.50a}$$

$$\tau_n^2 + \left(\sigma_n - \frac{\sigma_2 + \sigma_1}{2}\right)^2 \geq \left(\frac{\sigma_1 - \sigma_2}{2}\right)^2 \tag{2.50b}$$

$$\tau_n^2 + \left(\sigma_n - \frac{\sigma_3 + \sigma_2}{2}\right)^2 \geq \left(\frac{\sigma_2 - \sigma_3}{2}\right)^2 \tag{2.50c}$$

A plot of the three stress circles represented by the equality signs in the right-hand sides of Equations 2.50 are shown in Figure 2.12. The centers of the circles are denoted by C_1, C_2, and C_3. The centers are located at points with coordinates: $[(1/2)\ (\sigma_1 + \sigma_3), 0]$, $[(1/2)\ (\sigma_2 + \sigma_3), 0]$, $[(1/2)\ (\sigma_3 + \sigma_1), 0]$, respectively. The corresponding radii of the circles are the square roots of the quantities in the right-hand side of Equations 2.50, which correspond to the expressions in the right-hand side of Equations 2.49a, b, c for values of $l = m = n = 0$.

Note that for $m = 1$, the radius of circle C_1, obtained from Equation 2.49a, is equal to $(1/2)(\sigma_1 + \sigma_3) - \sigma_2$. Similarly, for $n = 1$, the radius of circle C_2 is $[(1/2)\ (\sigma_2 + \sigma_1) - \sigma_3]$, and that of circle C_3 is equal to $[(1/2)\ (\sigma_3 + \sigma_2) - \sigma_1]$.

The region of validity of the normal and resultant shear stresses across an inclined plane is confined to the area bounded by the circumferences of the circles. Such region is shown shaded in Figure 2.12. The coordinates of a point in this region, τ_n, σ_n, represent the stresses acting on the inclined plane.

In order to illustrate the graphical procedure required to determine the stresses σ_n, τ_n, suppose that we specify two of the direction cosines of the normal to the plane, say, l and n. From a vertical line passing through the point $(\sigma_1, 0)$ on the horizontal axis in Figure 2.12, draw a line at an angle $\alpha = \cos^{-1} l$ to intersect the stress circles. This line intersects circles C_2 and C_1 at points A and B, respectively. Next, with C_3 as center, draw an arc connecting points A and B as indicated in Figure 2.12. Note that the arc AB is part of the circle whose equation is represented by Equation 2.50c. Similarly, from a vertical line through point $(\sigma_3, 0)$, draw a line at an angle $\gamma = \cos^{-1} n$ with the vertical. This line intersects

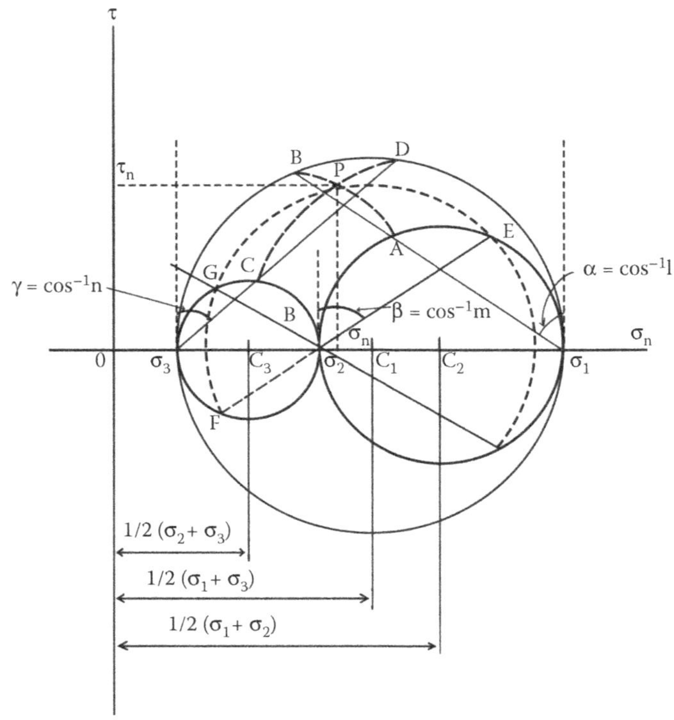

FIGURE 2.12
Mohr's stress circles.

the circles C_3 and C_1 at points C and D, respectively. With C_2 as center, draw an arc connecting the points C and D. The arc CD is part of the circle represented by Equation 2.50b. The two arcs AB and CD intersect at point P whose coordinates give the required values of σ_n, τ_n.

The value of the remaining direction cosine, m, can be used to check the location of point P as follows: draw an inclined line through point (σ_2, 0), which makes angle $\beta = \cos^{-1} m$ with the vertical. The line intersects circles C_2 and C_3 at points E and F, respectively. With C_1 as center draw an arc connecting points E and F, and point P lies on this arc. (Equation 2.50a represents the circle generating the arc EF.) It is interesting to note that an inclined line at angle $\beta = \cos^{-1} m$ on the other side of the vertical through (σ_2, 0) intersects circle C_3 at point G, which is also located on the arc EF.

In the degenerate case when $\sigma_3 = \sigma_2$, Mohr's circle reduces to a single circle as shown in Figure 2.13, and all admissible states of stress must lie on the circumference of the circle. When all three principal stresses are equal (case of hydrostatic pressure), Mohr's circle reduces to a point on the normal stress axis.

Figure 2.14 shows Mohr's circles of stress with the location of the deviatoric stress system discussed in Section 2.7. The origin of the deviatoric stress system is 0′ and the distance 00′ is given as

$$\sigma = \frac{\sigma_1 + \sigma_2 + \sigma_3}{3}$$

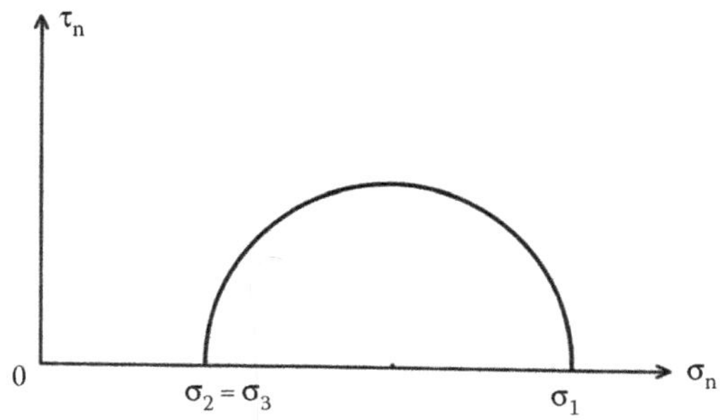

FIGURE 2.13
3D Mohr's circle ($\sigma_2 = \sigma_3$).

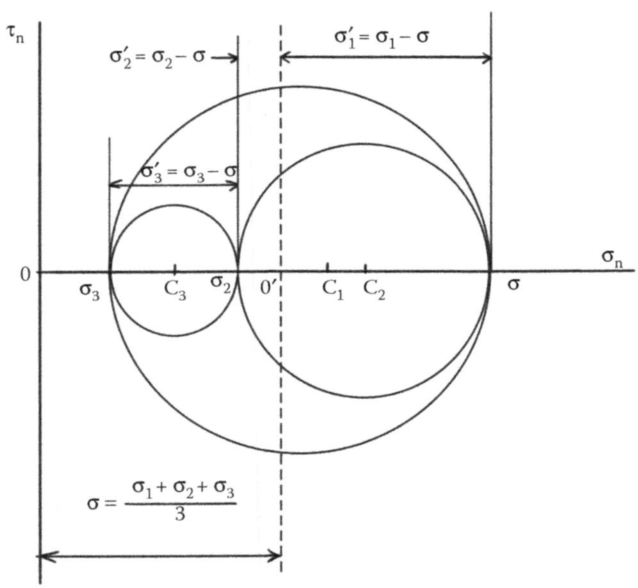

FIGURE 2.14
Origin (0′) of the deviatoric stress system.

The following examples illustrate the graphical procedure necessary to locate the normal and shearing stresses on a general inclined plane.

EXAMPLE 2.11

The state of stress at a point relative to the principal axes is given as

$$\sigma_1 = 10\,\text{MPa},\ \sigma_2 = 3\,\text{MPa},\ \sigma_3 = -1\,\text{MPa}$$

Use graphical construction to find σ_n and τ_n on a plane whose normal makes angles 60°and 45° with respect to the principal axes 1, 3, respectively.

Solution

Mohr's stress circles are constructed from centers C_1, C_2, and C_3 as seen in Figure 2.15.

The coordinates of circle C_1 are $((\sigma_1 + \sigma_3/2),\ 0) = (4.5,\ 0)$ MPa, and the circle has radius = 5.5 MPa.

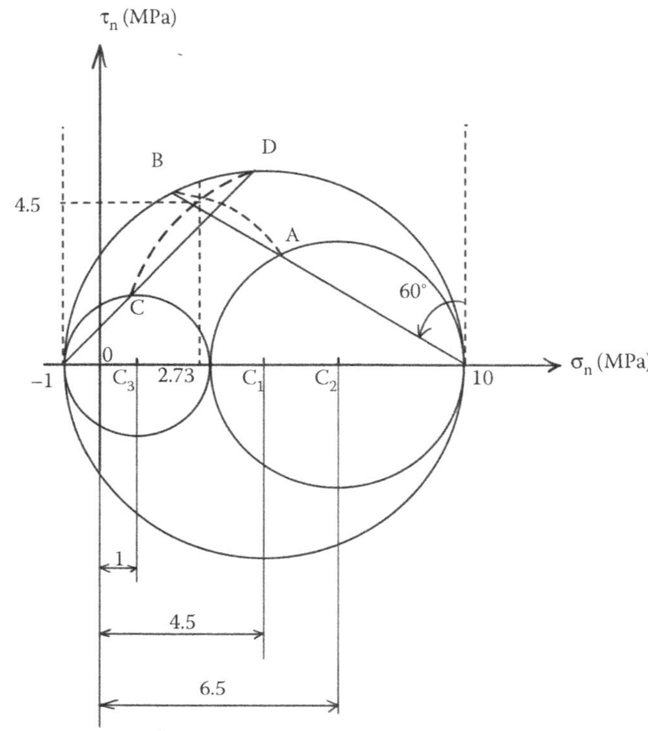

FIGURE 2.15
Example 2.11.

Also, the coordinates of circles C_2 and C_3 are $(\sigma_1 + \sigma_2/2, 0) = (6.5, 0)$ MPa, and $(\sigma_2 + \sigma_3/2, 0) = 1.5, 0$ MPa, respectively, and their corresponding radii are 3.5 and 2.0 MPa.

Draw a straight line at angle 60° to the left-hand side of the vertical line at point (10, 0). This line intersects stress circles C_2 and C_1 at points A and B, respectively. With C_3 as center, an arc is drawn joining points A and B. Another straight line is drawn at angle 45° with the vertical through point (−1, 0), which intersects circles C_3 and C_1 at points C and D. From center C_2, an arc joining points C and D is drawn. This arc intersects arc AB at point P whose coordinates give the required stresses on the inclined plane. Thus, it is found that $\sigma_n = 2.75$ MPa and $\tau_n = 4.75$ MPa. The third direction cosine, $m = \cos 45° = 1/\sqrt{2}$, is used to locate points E and F on C_2 and C_3, respectively. With C_1 as center, it is confirmed that arc EF intersects the previous arcs at point P.

EXAMPLE 2.12

The state of stress at a point is given as $\begin{pmatrix} \sigma_1 & 0 & 0 \\ 0 & \sigma_2 & 0 \\ 0 & 0 & \sigma_3 \end{pmatrix}$.

Use Mohr's stress circles and graphical construction to determine the octahedral stresses, σ_{oct} and τ_{oct}.

Solution

Mohr's stress circles with centers C_j, j = 1, 2, and 3 are shown in Figure 2.16. Since the octahedral normal stresses make equal angles with the principal axes, using the identity, $l^2 + m^2 + n^2 = 1$, it follows that $l = m = n = \frac{1}{\sqrt{3}}$. Hence, $\alpha = \beta = \gamma = 57.3°$.

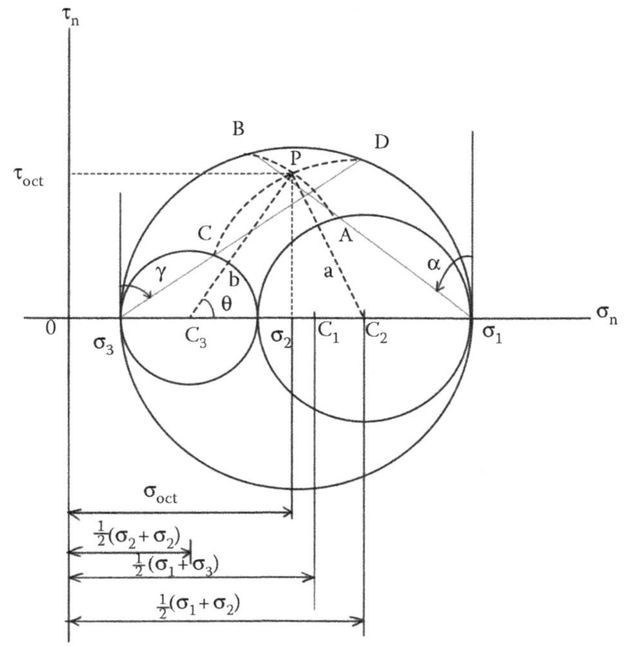

FIGURE 2.16
Example 2.12.

A straight line inclined at angle $\alpha = 57.3°$ to the left of the vertical at point (σ_1, 0) is drawn. This line intersects circles C_2 and C_1 at points A and B, respectively. From point C_3 as center, an arc joining points A and B is drawn. Similarly, an inclined line at angle $\gamma = 57.3°$ to the right of the vertical at point (σ_3, 0) locates points C and D on the stress circles and arc CD is drawn from point C_2 as center. The two arcs AB and CD intersect at point P as shown in Figure 2.16. The coordinates of point P are the required octahedral stresses.

To find the coordinates of point P, consider a triangle joining P to C_2 and C_3 as shown (dotted) in Figure 2.16. The lengths of the sides of this triangle are denoted by a, b, and c. Since $n^2 = l^2 = 1/3$, the right-hand sides of Equations 2.49b and c provide the following expressions for the magnitudes of sides a, b:

$$
\begin{aligned}
a^2 &= \frac{1}{3}(\sigma_3 - \sigma_1)(\sigma_3 - \sigma_2) + \left(\frac{\sigma_1 - \sigma_2}{2}\right)^2 \\
b^2 &= \frac{1}{3}(\sigma_1 - \sigma_2)(\sigma_1 - \sigma_3) + \left(\frac{\sigma_2 - \sigma_3}{2}\right)^2
\end{aligned} \tag{a}
$$

Also, length of side c of the triangle is $0C_2 - 0C_3 = \frac{1}{2}(\sigma_1 - \sigma_3)$.

Applying the cosine rule, $a^2 = b^2 + c^2 - 2bc\cos\theta$, and noting that $b\cos\theta = \sigma_n - 0C_3 = \sigma_n - \frac{1}{2}(\sigma_2 + \sigma_3)$, it is readily found that $\sigma_n = \frac{1}{3}(\sigma_1 + \sigma_2 + \sigma_3)$, which is the normal octahedral stress.

Also, $\tau_n^2 + [\sigma_n - \frac{1}{2}(\sigma_2 + \sigma_3)]^2 = b^2$.

Using the expression for b from Equation (a) and solving for τ_n, it can be confirmed that the expression of τ_{oct}, which is given in Equations 2.46, is obtained.

Problems

2.1 A small steel cube is placed at the bottom of a swimming pool. The unit weight of the water in the pool is $\gamma = 9817.5\ \text{N/m}^3$. If the depth of water is 2 m, find (a) the stresses acting on the faces of the cube; (b) the stresses acting on an arbitrary plane through the cube.

Ans: $19.74\ \text{kN/m}^2$.

2.2 The state of stress at a given point in a stressed body relative to rectangular axes is given by the array

$$\sigma_{ij} = \begin{pmatrix} 20 & 15 & -32 \\ 15 & -40 & 24 \\ -32 & 24 & 30 \end{pmatrix} (\text{ksi})$$

Compute the normal and maximum shear stresses on across a plane through the point whose normal makes angles 45° and 72° with the x- and y-axes, respectfully.

Ans: 5.5, 13.28 ksi.

2.3 Provide a geometrical approach to prove the following identity:

$$l^2 + m^2 + n^2 = 1$$

where l, m, and n are the direction cosines of a line segment in space. (Hint: assume the line segment is the diagonal of a parallelepiped with sides a, b, and c.)

2.4 The state of stress at a point in a structural member is given by

$$\sigma_x = Ax^2 + By^2,\ \sigma_y = Cy^2 + Dz^2,\ \sigma_z = Ez^2 + Fx^2$$
$$\tau_{xy} = Gy + Hz,\ \tau_{xz} = Iz + Jx,\ \tau_{yz} = Kx + Ly$$

where A, B,..., L are constants. Find expressions of the body forces to ensure the equilibrium. Units of the stresses are MPa.

Ans: $B_x = -2Ax - G - I$.

2.5 The stresses at a given point in a stressed body relative to rectangular axes are

$$\sigma_x = 20,\ \sigma_y = -40,\ \sigma_z = 30$$
$$\tau_{xy} = 15,\ \tau_{xz} = -32,\ \tau_{yz} = 24\ (\text{in units of ksi})$$

Compute the normal and maximum shear stresses on a plane through the point whose normal makes angles 45° with the x-axis and 72° with the y-axis.

Ans: 4.52 ksi, 10.12 ksi.

2.6 The stress state at point (P) in a structure relative to the Cartesian coordinates is given by the array

$$\begin{pmatrix} 10 & -25 & 30 \\ -25 & 20 & 0 \\ 30 & 0 & -15 \end{pmatrix} \text{MPa}$$

Determine the normal and resultant stresses acting on an oblique plane through (P) whose normal is in the direction $(1/10,\ 3/10,\ 3/\sqrt{10})$.

Ans: −7.4, 24.9 MPa.

2.7 Relative to the x-, y-, and z-axes, the state of stress at a point in a component is found to be proportional to the matrix

$$\begin{pmatrix} 4 & 0 & -3 \\ 0 & 5 & 2 \\ -3 & 2 & -3 \end{pmatrix} \text{MPa}$$

Determine the normal, shear, and resultant stresses that act on a plane whose normal is inclined at 45° to the z-axis and the projection of the normal on the xy-plane makes 50° with the yz-plane (Figure P2.7). Find also the direction of the resultant stress.

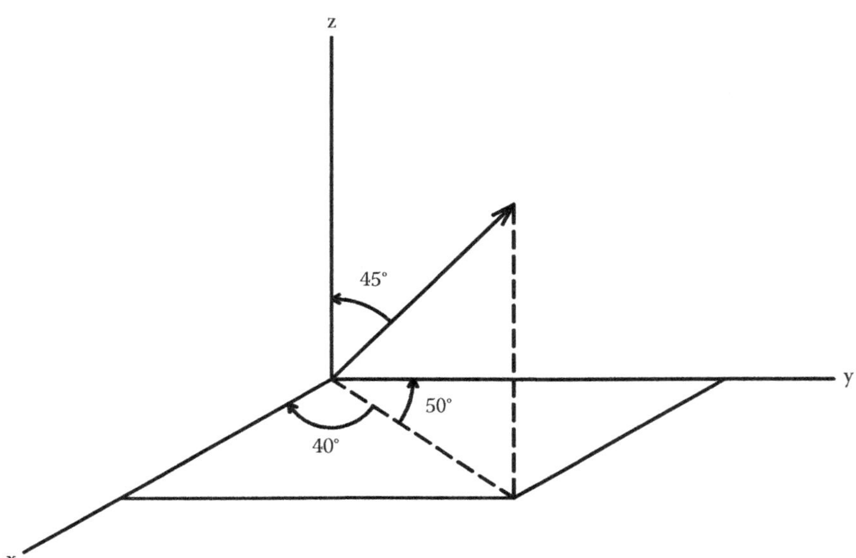

FIGURE P2.7

Ans: −0.42, 4.64 MPa.

2.8 The rectangular components of the stress at a point in a machine are given by

$$\begin{pmatrix} 2 & 3.6 & 2.4 \\ 3.6 & 4 & 1.8 \\ 2.4 & 1.8 & 4.5 \end{pmatrix} \text{MPa}$$

Find the normal, shear, and resultant stresses on a plane through the point parallel to the surface $4\,x - y + 2z = 6$.

Ans: 2.7, 4.4, 5.12.

2.9 In a loaded member, the x, y, and z stresses at a given point 0 in units of megapascals are

$$\sigma_x = 20,\ \sigma_y = -40,\ \sigma_z = 10$$
$$\tau_{xy} = 10,\ \tau_{xz} = -30,\ \tau_{yz} = 25$$

Determine the magnitude of the traction on a plane through 0 normal to the vector

$$2\mathbf{i} + \mathbf{j} + 3\mathbf{k}$$

Ans: 18.23 MPa.

2.10 The stress tensor at a given point 0 in a loaded component relative to the x-, y-, and z-axes is given by

$$\begin{pmatrix} 10 & 0 & 6 \\ 0 & -5 & 4 \\ 6 & 4 & 8 \end{pmatrix} \text{ksi}$$

Find the normal stress σ_n and shear stress τ_n at point 0 acting across plane 0BF in the parallelepiped shown in Figure P2.10.

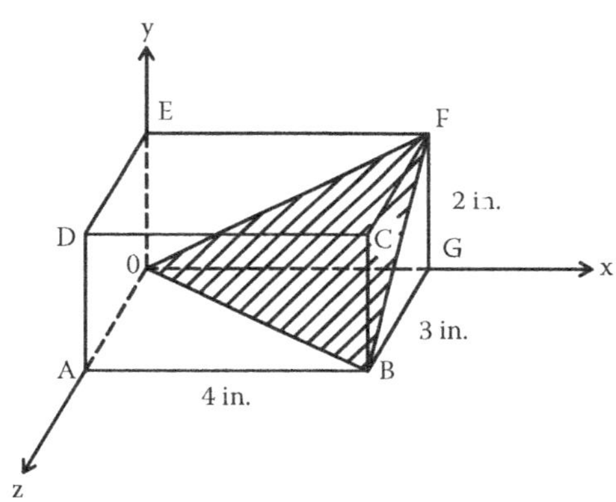

FIGURE P2.10

Ans: 4.1, 1.62.

2.11 For the state of stress given in Problem 2.10, find the normal and shear stresses at point 0 acting on a plane parallel to plane ABFE shown in Figure P2.11.

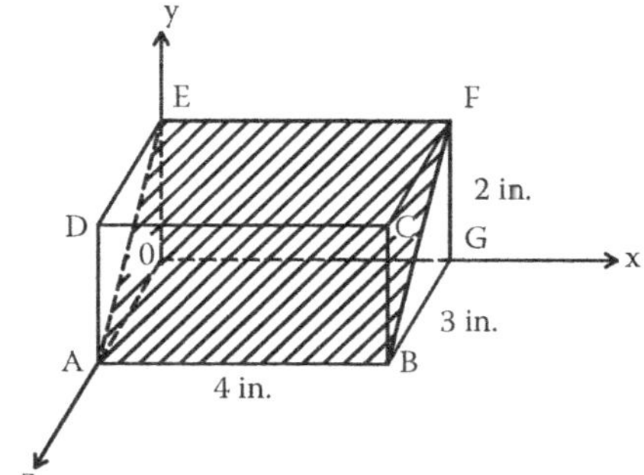

FIGURE P2.11

Ans: 2.69, 8.513 ksi.

2.12 The state of stress at point 0 in a structure is proportional to the matrix

$$\begin{pmatrix} -6 & -4 & 3 \\ -4 & 8 & 5 \\ 3 & 5 & 10 \end{pmatrix} \text{MPa}$$

Find the normal stress σ_n and shear stress τ_n acting across a plane through point 0 and parallel to the plane ABC shown in Figure P2.12.

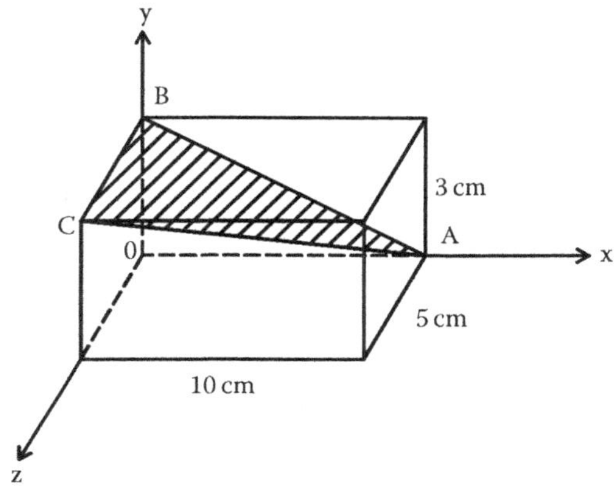

FIGURE P2.12

Ans: 3.3, 9.6 MPa.

2.13 In the previous problem what will be the directions of the resultant stress at point 0 acting on plane parallel to plane ABC as shown in Figure P2.12?

Ans: $\cos\theta_x = -0.584$, $\cos\theta_z = 0.565$.

2.14 The stresses at a point in a loaded component are given as follows:

$$\text{(a)} \begin{pmatrix} \sigma_x & 2 & 2 \\ 2 & 0 & 3 \\ 2 & 3 & 4 \end{pmatrix} \text{MPa} \quad \text{(b)} \begin{pmatrix} 2 & 0 & 3 \\ 0 & \sigma_y & 2 \\ 3 & 2 & 5 \end{pmatrix} \text{MPa}$$

In each case, find the unknown stress such that there will be a traction-free plane passing through the point, and determine the direction of the normal to the same plane.

Ans: 8/9 MPa, (1/11) (−9.−2, 6), 8 MPa, $1/(53)^{1/2}$ (6,−1,−4).

2.15 The resultant stress at a point 0 in a structural component has a value of 10 ksi and makes angles 45°, 80°, and 46.7° relative to the rectangular axes. Determine the normal (σ_n) and shear (τ_n) stresses acting on an oblique plane through the point 0 whose normal is inclined at 45° with the z-axis and makes equal acute angles with the x- and y-axes.

Ans: 9.26. 3.79 ksi.

2.16 At a given point 0 in a component, a resultant stress of 10 MPa is acting in the direction $(1/\sqrt{2}, 1/\sqrt{3}, 1/\sqrt{6})$ relative to the x-, y-, and z-axes. If the shear stresses acting on an element through point 0 with an oblique plane whose normal acts in the direction (1/3, 2/3, 2/3) are given as $\tau_{xy} = \tau_{xz} = 1.2$ MPa, and τ_{yz} determine the normal stresses σ_x, σ_y, and σ_z acting on the same element.

Ans: 16.4, 10.1, 7.5 MPa.

2.17 Show that the direction cosines of the resultant shear stress acting on an inclined plane derived in Equations 2.17 of the text satisfy the relation:

$$l_s^2 + m_s^2 + n_s^2 = 1$$

2.18 A state of stress at a point 0 of a body is given as follows:

$$\sigma_x = 3p, \quad \sigma_y = 2p, \quad \sigma_z = -4p$$
$$\tau_{xy} = 3p, \quad \tau_{xz} = 0, \quad \tau_{yz} = -p$$

where p is constant. (a) Find σ_n and τ_n on a plane whose normal is in the direction of the unit vector (2,2,1). (b) Show that τ_n acts in the direction of vector (4,1, –10).

Ans: 4p, 3.61p.

2.19 The stresses relative to an x, y, z coordinate system are given by

$$\begin{pmatrix} 10.5 & -4.5 & 3 \\ -4.5 & 6.4 & 2.5 \\ 3 & 2.5 & 4.1 \end{pmatrix} \text{ksi}$$

Determine the stresses relative to the x′, y′, and z′ coordinate system obtained by counterclockwise rotation of the xy-plane through an angle of 30° about the z-axis. Check your result by computing the stress invariants (Equations 2.30) for both coordinate systems.

2.20 The following Cartesian stresses exist at a point in an element of a machine component

$$\sigma_x = 7\,\text{MPa}, \quad \sigma_y = 9\,\text{MPa}, \quad \sigma_z = 12\,\text{MPa}$$
$$\tau_{xy} = 10\,\text{MPa}, \quad \tau_{xz} = \tau_{yz} = 0$$

Determine the stresses with respect to the x′, y′, z′ axes if

$$l_1 = \frac{12}{25}, \quad m_1 = -\frac{9}{25}, \quad n_1 = \frac{4}{5}$$
$$l_2 = \frac{3}{5}, \quad m_2 = \frac{4}{5}, \quad n_2 = 0$$
$$l_3 = -\frac{16}{25}, \quad m_3 = \frac{12}{25}, \quad n_3 = \frac{3}{5}$$

2.21 The stresses in the table represent the states of stress at specific points of a structural member in units of megapascals. For each case, determine the magnitude and direction of the principal stresses at the point.

Case	σ_x	σ_y	σ_z	τ_{xy}	τ_{xz}	τ_{yz}
(a)	10	−10	0	20	30	−5
(b)	4.2	1.5	−2.9	2.4	3.5	−2.3
(c)	6.5	2.5	5	3	4	1
(d)	4	4	3	0	2	2
(e)	10	0	6	3.5	3.5	0

Ans: (a) 6.419, −1.935, −36.75; (b) 6.419, 1.875, −5.194.

2.22 A state of stress at a point relative to rectangular coordinates is given by

$$\begin{pmatrix} 8 & 3 & 7 \\ 3 & -6 & 4 \\ 7 & 4 & 10 \end{pmatrix} \text{ksi}$$

Find the maximum shearing stress at the point.

Ans: 12.12 ksi.

2.23 The state of stress at point 0 in a bar subject to torsion is shown in Figure P2.23.

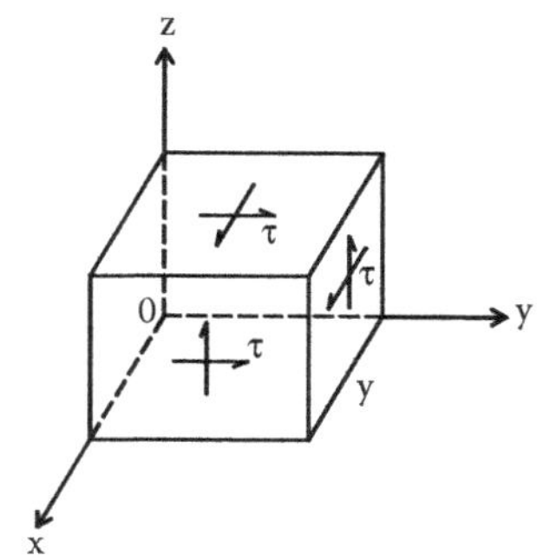

FIGURE P2.23

Show that the principal stresses are as follows: $\sigma_1 = 2\tau$, $\sigma_2 = \sigma_3 = -\tau$, and the principal directions of σ_1 are equal angles with the rectangular axes.

2.24 Let $\sigma_1 > \sigma_2 = \sigma_3$ be the principal stresses at a given point 0 in a stressed body. Show that

$$\sigma_1 = \sigma_n + \frac{\tau_n}{l}\sqrt{1 - l^2}$$

$$\sigma_2 = \sigma_3 = \sigma_n - \frac{\tau_n l}{\sqrt{1 - l^2}}$$

where σ_n, τ_n are the normal and shear stresses across a plane through 0, and l is the direction cosine of the normal relative to the x-axis.

2.25 At a point in a structural member, the principal stresses are 100, 150, and –200 MPa. If the principal stresses are oriented along the x-, y-, and z-axes, respectively, find the resultant stress across a plane whose normal makes 45° with the z-axis and the projection of the normal on the xy-plane makes 30° with the y-axis is shown in Figure P2.25.

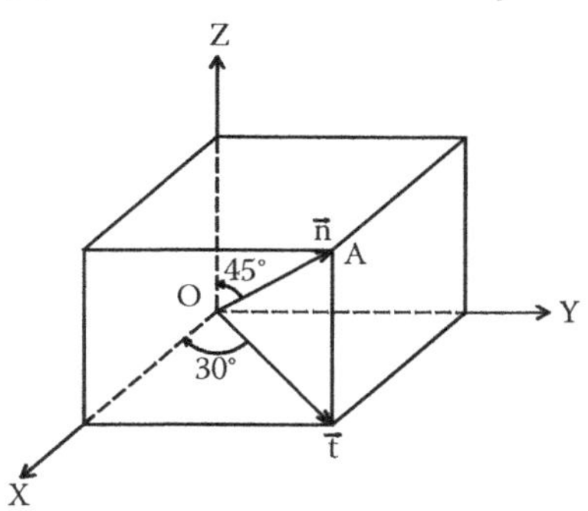

FIGURE P2.25

Ans: 173 MPa.

2.26 The stress tensor at a given point in a loaded component is given by the array

$$\begin{pmatrix} 1 & 2 & -1 \\ 2 & 0 & 3 \\ -1 & 3 & 1 \end{pmatrix} \text{ksi}$$

If one of the principal stresses has a value of 1.9 ksi, determine the corresponding three eigen vectors.

Ans: 0, 0.6, 0.8.

2.27 In a general state of stress, it is found that

$$\sigma_x = \sigma_y = p, \quad \sigma_z = 2p$$
$$\tau_{xy} = p, \quad \tau_{xz} = \tau_{yz} = -p$$

Show that the stress state is biaxial and find the principal stresses and their directions.

Ans: 3.414 p, 0.586 p, 0.

2.28 Relative to the principal axes, the stresses $\sigma_n = 10$ ksi and $\tau_n = 6$ ksi at a point of a body are found to have the directions, $1/\sqrt{3}, 1/\sqrt{2}, 1/\sqrt{6}$ and (0.807, −0.418, −0.418), respectively. Determine the magnitudes of the principal normal and shear stresses.

Ans: 18.39, 6.45, 3.86.

2.29 The state of stress at a point P in a body relative to rectangular coordinates is given by the matrix

$$\begin{bmatrix} 3 & -2 & 1 \\ -2 & 4 & -3 \\ 1 & -3 & 5 \end{bmatrix} \text{ksi}$$

Find

a. The octahedral stresses
b. The stress deviator tensor

Ans: 4, 3.16 ksi.

2.30 A state of stress relative to rectangular axes is given by

$$\sigma_{ij} = \begin{pmatrix} 20 & 40 & 30 \\ 40 & -10 & 20 \\ 30 & 20 & 50 \end{pmatrix} \text{MPa}$$

Find

a. The deviatoric stress tensor and the principal stresses
b. The ratio $\dfrac{\tau_{oct}}{\tau_{max}}$

Ans: 84.9, 12.8, −37.7 MPa; ratio = 0.87.

2.31 The principal stresses at a point in an elastic component are $\sigma_1 > \sigma_2 > \sigma_3$. For given values of σ_2 and σ_3, find the value of σ_1 such that the octahedral shearing stress has an extreme value.

Ans: $\sigma_2 + \sigma_3/2$.

2.32 Show that the deviatoric principal stresses, s_1, s_2, and s_3, at a point in a body are related to the regular principal stresses by the relation

$$s_1^2 + s_2^2 + s_3^2 = \frac{1}{3}[(\sigma_1 - \sigma_2)^2 + (\sigma_2 - \sigma_3)^2 + (\sigma_3 - \sigma_1)^2]$$

2.33 Prove the following results:

$$\frac{\sqrt{6}}{3} \leq \frac{\tau_{oct}}{\tau_{max}} \leq \frac{\sqrt{8}}{3}$$

2.34 The principal stresses at a point in a loaded member are

$$\sigma_1 = 60\,\text{MPa}, \quad \sigma_2 = 40\,\text{MPa}, \quad \sigma_3 = -10\,\text{MPa}$$

Compute the resultant stress on a plane whose direction cosines relative to the principal axes are 0.38, 0.56, and 0.74. Check your results by using Mohr's stress circles.

2.35 The principal stresses at a point in a stressed solid are $\sigma_1 = 9$ ksi, $\sigma_2 = 3$ ksi, and $\sigma_3 = -3$ ksi. Using Mohr's circle, find the normal and shear stresses on an oblique plane whose normal is defined by the following direction cosines:

$$l = \frac{1}{\sqrt{3}}, \quad m = \frac{1}{\sqrt{2}}, \quad n = \frac{1}{\sqrt{6}}$$

References

2.1. W. Johnson and P. B. Mellor, *Engineering Plasticity*, Ellis Horwood, John Wiley and Sons, New York, 1985.

2.2. J. Chakrabarty, *Theory of Plasticity*, McGraw-Hill Book Company, New York, 1987.

2.3. W. F. Chen and D. H. Han, *Plasticity for Engineers*, Springer-Verlag, New York, Berlin, 1988.

3

Analysis of Strain

Under the impact of physical actions such as external loads, changes in temperature, magnetic fields, etc., structural components exhibit deformations, that is, undergo changes in size and shape. In general, every particle of the component is displaced as shown in the dotted sketch in Figure 3.1. There are two kinds of displacements, a *rigid body* displacement in which the relative positions of the particles remain unchanged, and *deformations* or *straining* where changes occur in the relative positions of the particles. In this chapter, we are concerned with the deformation or straining of structural components modeled as continuous media. The rigid body displacements are of no interest and are typically studied in books dealing with the dynamics of rigid bodies.

The deformation of a component is measured by the *displacements* and the associated *strains* of its particles. The relations between strains and displacements describe the geometry of deformation of the component and constitute an important aspect of determining the response to external loads. The displacements describe the relative positions of each point of the body with reference to a particular coordinate system while the strains measure the relative changes in size and shape. Changes in size or lengths are measured by *normal* strains and changes in shapes or angles are measured by *shear* strains. An element of a component also experiences rotations, which can be related to the displacements. All these changes, which occur during deformation, vary from point to point throughout the body, and in order to account for the variation, the changes are defined at every point of the body.

3.1 Displacement at a Point

Consider an arbitrary point of the body, say, point P, in Figure 3.1. Before deformation, the position of point P relative to the x, y, and z coordinates is represented by the vector

$$\mathbf{r} = x\mathbf{i} + y\mathbf{j} + z\mathbf{k} \tag{3.1}$$

where **i**, **j**, and **k** are unit vectors along the reference axes. After deformation, the point P assumes the position P′ and its position vector is

$$\mathbf{r}' = x'\mathbf{i} + y'\mathbf{j} + z'\mathbf{k} \tag{3.2}$$

Here, the coordinates of point P′, that is, x′, y′, z′, are, in general, functions of the coordinates, x, y, z, before deformation. The displacement of point P due to the straining is then given by the vector $\mathbf{D} = \mathbf{r}' - \mathbf{r} = (x' - x)\mathbf{i} + (y' - y)\mathbf{j} + (z' - z)\mathbf{k}$. Let us suppose that the components of the displacement along the reference axes are denoted by arbitrary

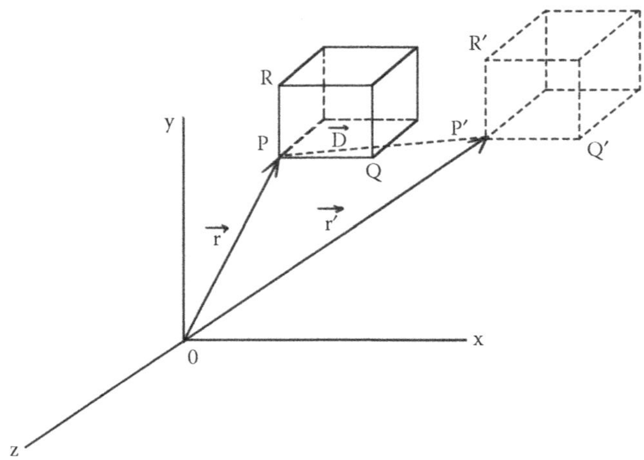

FIGURE 3.1
Displacement $\vec{D} = \overrightarrow{PP'}$.

functions of the coordinates, say, u(x, y, z), v(x, y, z), and w(x, y, z); then, the displacement of point P is expressed as

$$\mathbf{D} = u(x, y, z)\mathbf{i} + v(x, y, z)\mathbf{j} + w(x, y, z)\mathbf{k} \tag{3.3}$$

A small variation in the position vector of point P is $d\mathbf{r} = dx\,\mathbf{i} + dy\,\mathbf{j} + dz\,\mathbf{k}$, and a small variation in the displacement component u(x, y, z) is expressed by the first-order terms of Taylor's series, namely

$$du = \frac{\partial u}{\partial x}dx + \frac{\partial u}{\partial y}dy + \frac{\partial u}{\partial z}dz \tag{a}$$

Equation (a) is also expressed as the dot product, $du = d\mathbf{r}.\vec{\nabla}$, where $\vec{\nabla}$ stands for the gradient

$$\vec{\nabla} = \frac{\partial}{\partial x}\mathbf{i} + \frac{\partial}{\partial y}\mathbf{j} + \frac{\partial}{\partial z}\mathbf{k}$$

In a similar manner, the following expressions are written for small variations in the components v(x, y, z) and w(x, y, z):

$$dv = \frac{\partial v}{\partial x}dx + \frac{\partial v}{\partial y}dy + \frac{\partial v}{\partial z}dz = d\mathbf{r}.\vec{\nabla}\,v \tag{b}$$

$$dw = \frac{\partial w}{\partial x}dx + \frac{\partial w}{\partial y}dy + \frac{\partial w}{\partial z}dz = d\mathbf{r}.\vec{\nabla}\,w \tag{c}$$

In the next section, we define the normal and shear strains and derive the strain–displacement relations.

3.2 Normal and Shear Strains

We begin by considering the deformation of a line element PQ situated along the x-axis, as shown in Figure 3.2. The length of the element is dx, and after straining, the element retains its shape and experiences an axial extension (contraction) such that point P moves a distance u(x) to point P′, and point Q, which is located at a distance dx ahead of P, moves a distance u(x) + du(x) to point Q′. The normal strain, ε_x, at the point P, is defined as the limit of the change in length of the element divided by the original length, that is,

$$\varepsilon_x = \lim_{PQ \to 0} \frac{P'Q' - PQ}{PQ} \tag{3.4}$$

The final length is $P'Q' = (dx + u + du - u)$ and the original length is $PQ = dx$. Hence,

$$\varepsilon_x = \lim_{dx \to 0} \frac{(dx + u + du - u) - dx}{dx} = \frac{du}{dx} \tag{3.5a}$$

It follows that if a component of length, L_0, is stretched ***uniformly*** by an amount, δ, such that the final length is $L = L_0 + \delta$, the axial engineering strain is the extension of the element relative to the initial length,

$$\varepsilon = \frac{L - L_0}{L_0} = \frac{\delta}{L_0}$$

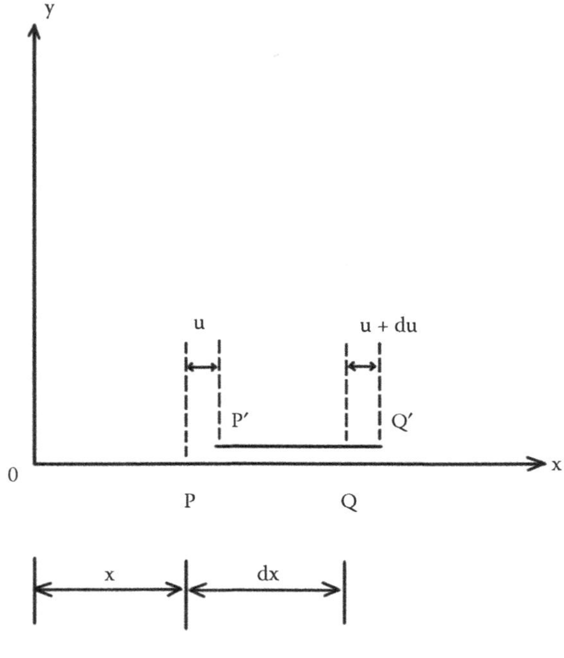

FIGURE 3.2
Elongation of a line element.

True strain: The true or natural strain of an element is defined as the increment in length divided by the current length, that is, $d\varepsilon = dL/L$, and the true strain is

$$\varepsilon_{true} = \int_{L_0}^{L} \frac{dl}{l} = Ln\frac{L}{L_0} = Ln(1+\varepsilon) \tag{3.5b}$$

The normal strain is a dimensionless parameter, commonly measured in terms of micrometer per meter (or inch per inch). For sign convention, extension is considered *positive* and contraction is *negative.*

The deformations of two-dimensional (2D) and three-dimensional (3D) elements consist of changes in length of the sides as well as changes in shape of the element measured by changes in the magnitude of right angles between the sides. This will be discussed in the following sections.

3.2.1 2D Strain

Consider an infinitesimal element PQRS in the xy-plane of an unstrained member, as shown in Figure 3.3. The deformation causes changes in the length and angles of the element, point P moves to point P′, Q to Q′, etc., the deformed element, P′Q′R′S′, is assumed to remain in the xy-plane.

To find the normal strains, let u(x, y) and v(x, y) represent the horizontal and vertical displacements of point P. The displacements of the other points Q, R, and S are as follows:

Point Q: $u + (\partial u/\partial x)dx$ in the x-direction and $v + (\partial v/\partial x)dx$ in the y-direction
Point R: $(u + (\partial u/\partial y)dy,\ v + (\partial v/\partial y)dy)$
Point S: $(u + (\partial u/\partial x)dx + (\partial u/\partial y)dy,\ v + (\partial v/\partial x)dx + (\partial v/\partial y)dy)$

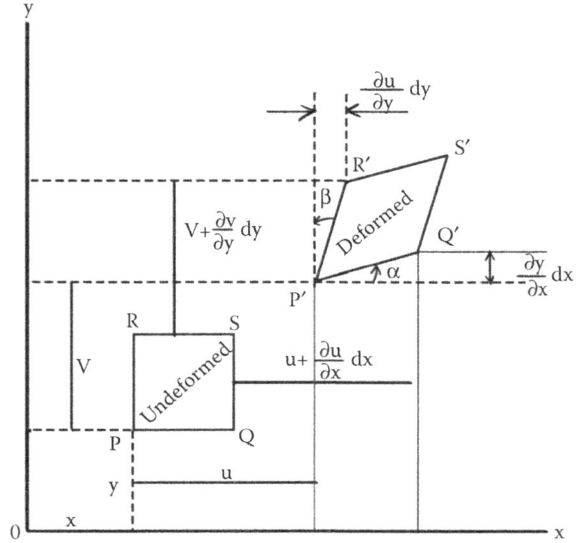

FIGURE 3.3
Deformation of a plane element normal and shear strains.

It is readily shown from Figure 3.3 that the lengths, P′Q′ and P′R′, are given by

$$P'Q' = \left\langle \left[u + \frac{\partial u}{\partial x}dx - (u - dx) \right]^2 + \left(\frac{\partial v}{\partial x}dx \right)^2 \right\rangle^{\frac{1}{2}} = dx \left[1 + 2\frac{\partial u}{\partial x} + \left(\frac{\partial u}{\partial x} \right)^2 + \left(\frac{\partial v}{\partial x} \right)^2 \right]^{\frac{1}{2}} \tag{3.6a}$$

$$P'R' = \left\langle \left[v + \frac{\partial v}{\partial y}dy - (v - dy) \right]^2 + \left(\frac{\partial u}{\partial y}dy \right)^2 \right\rangle^{\frac{1}{2}} = dy \left[1 + 2\frac{\partial v}{\partial y} + \left(\frac{\partial v}{\partial y} \right)^2 + \left(\frac{\partial u}{\partial y} \right)^2 \right]^{\frac{1}{2}} \tag{3.6b}$$

Expanding the right-hand sides of Equations 3.6 in Taylor's series and neglecting the second-order terms, it is readily shown that $P'Q' = dx(1 + (\partial u/\partial x))$ and $P'R' = dy(1 + (\partial v/\partial y))$. Hence, applying the definition of strain in Equation 3.4, it follows that the normal strains are

$$\varepsilon_x = \frac{\partial u}{\partial x} \tag{3.7a}$$

$$\varepsilon_y = \frac{\partial v}{\partial y} \tag{3.7b}$$

The engineering shearing strain at point P is a measure of the change in shape of the element and is defined as the limit of the relative change in the 90° angle at P before and after deformation, that is,

$$\gamma_{xy} = \lim_{dx \to 0, dy \to 0} = \frac{\pi}{2} - \theta^* = \alpha + \beta \tag{3.8}$$

where θ^* is the angle between P′Q′ and P′R′, α is the angle between P′Q′ and the horizontal, and angle β is between P′R′ and the vertical. Since angles α and β are assumed small, one can replace the angle by its tangent, that is,

$$\alpha = \tan\,\alpha = \frac{\frac{\partial v}{\partial x}dx}{dx\left(1 + \frac{\partial u}{\partial x}\right)} \tag{3.9a}$$

$$\beta = \tan\,\beta = \frac{\frac{\partial u}{\partial y}dy}{dy\left(1 + \frac{\partial v}{\partial y}\right)} \tag{3.9b}$$

Since the magnitudes of the strains, $\partial u/\partial x$ and $\partial v/\partial y$, are small compared to one, it follows that $1 + (\partial u/\partial x) \simeq 1$ and $1 + (\partial v/\partial y) \simeq 1$, and it follows from Equations 3.9 that

$$\alpha = \frac{\partial v}{\partial x}, \quad \beta = \frac{\partial u}{\partial y}$$

Hence, the engineering shearing strain between the x and y axes, a dimensionless quantity measured by radians, is given by

$$\gamma_{xy} = \frac{\partial u}{\partial y} + \frac{\partial v}{\partial x} \tag{3.10}$$

The sign convention for shearing strain is similar to that of the shear stress, positive shear stress produces positive shear strain. For the element in Figure 3.3, a positive shear strain γ_{xy}

indicates that the angle RPQ has decreased. The angle R′P′Q′ is equal to $(\pi/2) - \gamma_{xy}$ and angle $P'Q'S' = (\pi/2) + \gamma_{xy}$. For small strain, $\gamma_{xy} \ll 1$.

As a summary, the strains at each point of a plane component are related to the displacements by the relations

$$\varepsilon_x = \frac{\partial u}{\partial x}, \quad \varepsilon_y = \frac{\partial v}{\partial y}, \quad \gamma_{xy} = \frac{\partial u}{\partial y} + \frac{\partial v}{\partial x} \tag{3.11}$$

The displacements u and v are in general functions of x and y, and the relations in Equation 3.11 describe the geometry of deformation of the 2D element.

A 2D element can deform without any change in the length of its sides. Such deformation is basically a **pure shear**. Figure 3.4 shows the distortion of such an element in the xy-plane. The solid lines represent the original element and the deformed sides are shown dotted. The lengths of the sides remain the same while one diagonal has increased in length (OB become OB′) and the other diagonal has decreased by the same amount (AC became A′C′). A typical point P deforms into P′. The final coordinates of a typical point, x, y, of a typical point are related to the initial coordinates, x_0, y_0 by the equation

$$x = x_0 + y_0 \tan\alpha, \quad y = y_0$$

and the shearing strain is $\gamma = (\pi/2) - \angle C'OA = \alpha$.

It should be noted that there is a "mathematical" definition of shear strain, namely

$$\varepsilon_{xy} = \frac{1}{2}\left(\frac{\partial u}{\partial y} + \frac{\partial V}{\partial x}\right) = \frac{1}{2}\gamma_{xy} \tag{a}$$

The factor (1/2) is used in order to get uniformity when index notation for tensorial quantities is used.

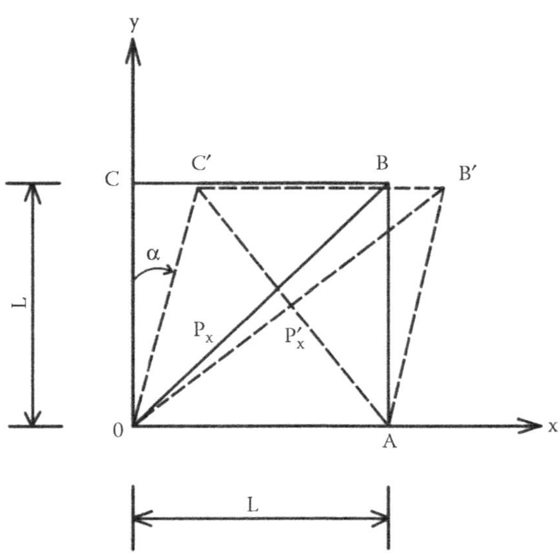

FIGURE 3.4
Distortion due to pure shear.

EXAMPLE 3.1

A plane sheet ABCD has sides AB = L = 2.0 m and AD = h = 1.0 m held fixed and corner C is stretched uniformly such that straight lines remain straight.

Find the strains and displacements developed in the element assuming that

a. Point C is stretched distance $e_1 = 15$ mm horizontally (Figure 3.5a).
b. In addition to above, point C experiences a vertical stretch of $e_2 = 12$ mm (Figure 3.5b).

Solution

a. Consider an arbitrary point P(x,y) and segments PQ = ds and PR = dy. Point P moves to P*, Q to Q*, and R to R*, as shown in Figure 3.5a. It is readily confirmed by considering linear proportional ratios that distance $PP^* = (e_1xy)/hL$, $QQ^* = [y\ e_1(x + dx)/hL]$, and $RR^* = [(y + dy)\ e_1x/hL]$. It follows that $P^*Q^* = PQ + QQ^* - PP^* = e_1y/hL + dx$.

   Hence, using the definitions of engineering strains

$$\varepsilon_x = \lim_{dx \to 0} \frac{P^*Q^* - PQ}{PQ} = \frac{e_1 y}{hL} \quad \text{(a)}$$

$$\varepsilon_y = 0$$

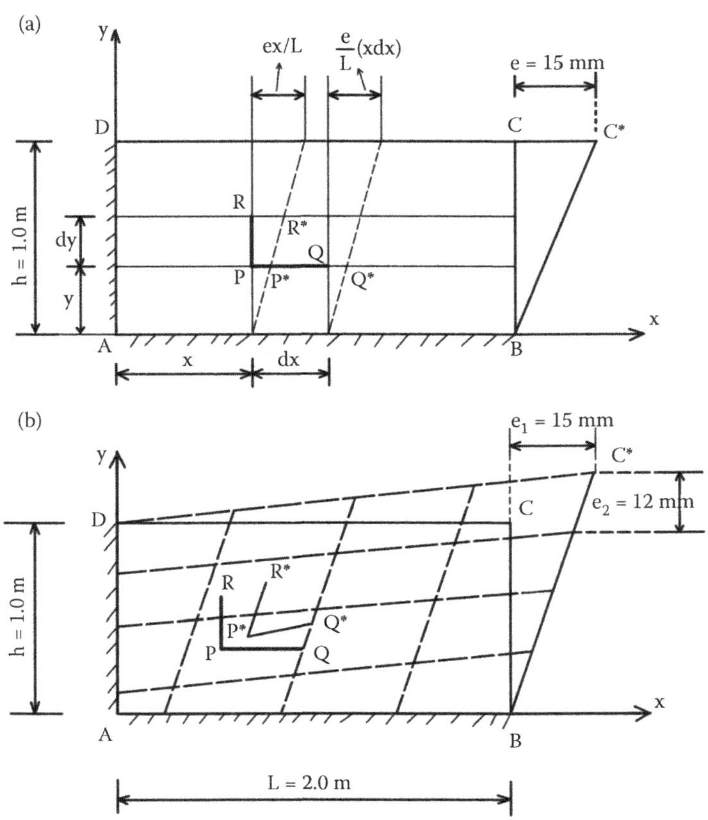

FIGURE 3.5
Example 3.1.

and the shearing strain at point P is given by

$$\gamma_{xy} = \frac{\pi}{2} - \angle R^*P^*Q^* = \frac{RR^* - PP^*}{RP} = \frac{e_1 x}{hL} \tag{b}$$

Inserting the numerical data, $e_1 = 15$ mm, $h = 1000$ mm, and $L = 2000$ mm, in expressions (a) and (b), the strains at any point in the element are

$$\varepsilon_x = 7.5 \times 10^{-6} y, \quad \varepsilon_y = 0 \quad \text{and} \quad \gamma_{xy} = 7.5 \times 10^{-6} x$$

The displacements are obtained as follows: $\varepsilon_x = \partial u/\partial x = 7.5(10^{-6})y$. Simple integration gives $u(x, y) = 7.5\ (10^{-6})\ xy + f(y)$, where $f(y)$ is the constant of integration.

Similarly, $\varepsilon_y = \partial v/\partial y = 0$, or $v(x, y) = g(x)$, and $\gamma_{xy} = \partial u/\partial y + \partial v/\partial x = 7.5\ (10^{-6})x$ furnishes $df/dy + dg/dx = 0$, that is, $df/dy = -(dg/dx) = A$ (constant). Thus, $f(y) = Ay + B$ and $g(x) = -Ax + C$ where A, B, and C are constants. Since sides $x = 0$ and $y = 0$ are fixed, $B = C = 0$, and the displacements are

$$u(x, y) = 7.5(10^{-6})xy + Ay$$

$$v(x, y) = 6.0(10^{-6})xy - Ax$$

The terms Ay and $-Ax$ represent rigid body motion, in this particular case represent rotation of the element, and do not influence the deformation.

b. Based on the results obtained in part (a), we can assume that the resulting displacements in the element are of the form

$$u(x, y) = C_1 xy \quad \text{and} \quad v(x, y) = C_2 xy$$

where C_1 and C_2 are constants to be determined from the prescribed conditions at point C of the element, namely, $u = 15$ mm and $v = 12$ mm at $x = 2000$ mm and $y = 1000$ mm. Thus, $C_1 = 7.5 \times 10^{-6}$ and $C_2 = 6 \times 10^{-6}$.

Hence, the resulting strains in this case are

$$\varepsilon_x = 7.5 \times 10^{-6} y$$

$$\varepsilon_y = 6 \times 10^{-6} x$$

$$\gamma_{xy} = 7.5 \times 10^{-6} x + 6.0 \times 10^{-6} y$$

and the displacements are

$$u = 7.5 \times 10^{-6} xy$$

$$v = 6.0 \times 10^{-6} xy$$

EXAMPLE 3.2

A thin circular wire of radius = a is heated nonuniformly so that the induced tangential strain at any point is given by the expression, $\varepsilon_\theta = C \cos^2\theta$, where C is a constant. Find the length of the deformed wire.

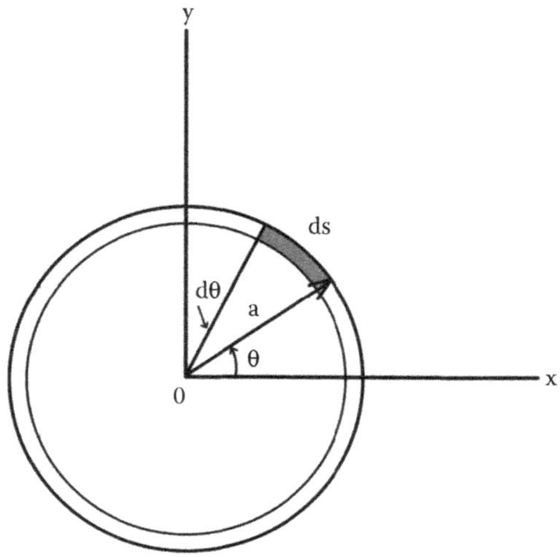

FIGURE 3.6
Example 3.2.

Solution

Consider a small element, $ds = a\,d\theta$, of the circumference as shown in Figure 3.6. The elongation of the element is equal to $\varepsilon_\theta\,ds$. Hence, the total elongation is given by

$$\Delta L = \int_0^{2\pi} \varepsilon_\theta\,ds = \int_0^{2\pi} aC\cos^2\theta\,d\theta$$

Evaluating the integral, it is found that

$$\Delta L = aC\pi$$

The circumference of the deformed wire is then equal to $\pi a(2 + C)$.

3.2.2 3D Strain

In the 3D case, we consider a rectangular prism element of sides dx, dy, dz, and follow a similar procedure to derive the relations between the strains and the displacement components u(x, y, z), v(x, y, z), and w(x, y, z) in each of the xy-, yz-, and xz-planes. The results are

$$\begin{aligned} \varepsilon_x &= \frac{\partial u}{\partial x}, & \varepsilon_y &= \frac{\partial v}{\partial y}, & \varepsilon_z &= \frac{\partial w}{\partial z} \\ \gamma_{xy} &= \frac{\partial u}{\partial y} + \frac{\partial v}{\partial x}, & \gamma_{yz} &= \frac{\partial v}{\partial z} + \frac{\partial w}{\partial y}, & \gamma_{zx} &= \frac{\partial w}{\partial x} + \frac{\partial u}{\partial z} \end{aligned} \tag{3.12}$$

The relations (3.12) describe the geometry of deformation of a continuum component irrespective of what causes the deformation and the nature of the material of the component. They are referred to as the *kinematic* relations. It is clear that the shearing strains are symmetrical, that is, $\gamma_{ij} = \gamma_{ji}$, i, j = x, y, and z. This is because the relative change in the right angle between sides i and j is identical to that between sides j and i, i, j = x, y, and z. The state of engineering strain at a point in a loaded member is represented by the six components of strain mentioned in Equations 3.12. The six components of strain are determined by three variables, namely, the displacements u, v, and w. It follows that the strains cannot be arbitrary functions of x, y, and z but must satisfy certain relations—known as compatibility conditions—which will be derived later in this chapter (see Equations 3.14).

In almost all cases of interest in engineering practice, the strains are small quantities. This means that the change in distance between two points of the component is small compared to the distance itself. In Equations 3.5, the distance du is small compared to dx, and the strain ε_x is a small quantity compared to unity. Similarly, the angles α and β are small compared to the original right angle, it follows that γ_{xy} is small quantity. If the component is subjected to small displacements, all components of strain are small. In few cases, however, small strains may accompany large displacements. An example is large deflection of a thin rod in which the ends move a large distance but the axial extension or contraction of the rod itself is small. The elastic strain for engineering materials in common use rarely exceeds 0.002. Similarly, the shearing strains are small such that the approximations $\sin\alpha \approx \alpha$ and $\cos\alpha \approx 1$ are valid.

Furthermore, it should be noted that the frame of reference (rectangular coordinates) used to derive formulas for strain is based on an undeformed state of the element while in Chapter 2, the frame of reference used to derive stress formulas is based on the deformed state. This creates an error when stress is related to strain. However, for small deformation, the assumption that both original frames of reference are based on undeformed element is adequate. In studying large deformation, a topic covered in books on finite elasticity, this approximate assumption is not valid and different formulation is required.

EXAMPLE 3.3

The corner C of a 0.4 m × 0.3 m × 0.2 m cubic element is displaced to point C*, whose coordinates are (0.398, 0.303, 0.199) m, as shown in Figure 3.7. Assuming all straight lines remain straight during deformation, find the state of strain at point C.

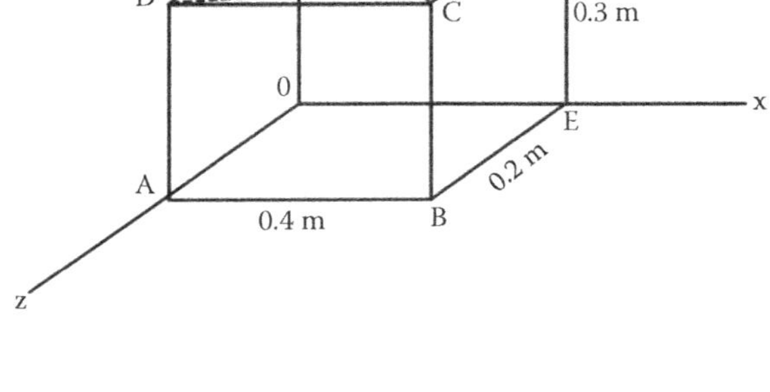

FIGURE 3.7
Example 3.3.

Solution

The displacements of point C along the x, y, and z axes are u = −2 mm, v = 3 mm, w = −1 mm. It follows that the normal strains at point C are

$$\varepsilon_x = -2/400 = -0.005, \quad \varepsilon_y = 3/300 = 0.01, \quad \varepsilon_z = -1/200 = -0.005$$

The shearing strains are computed as follows:

$$\gamma_{xy} = \frac{-2}{300} + \frac{3}{400} = 0.00083$$

$$\gamma_{xz} = \frac{-200}{200} + \frac{-1}{400} = -0.0125$$

$$\gamma_{yz} = \frac{3}{200} + \frac{-1}{300} = -0.017$$

An alternative method of obtaining the solution, similar to that used to solve part (b) of Example 3.1, is to assume the following expressions for the displacements in the element: u, v, w = A_j, j = 1, 2, 3 are constants to be determined from the prescribed displacements of point C. Thus,

$$A_1 = \frac{-2 \times 10^{-6}}{4 \times 3 \times 2} = -0.833 \times 10^{-6}$$

$$A_2 = \frac{3 \times 10^{-6}}{4 \times 3 \times 2} = 0.125 \times 10^{-6}$$

$$A_3 = \frac{-1 \times 10^{-6}}{4 \times 3 \times 2} = -0.0417 \times 10^{-6}$$

The strains at point C are computed using the strain displacement relations in Equation 3.12

$$\varepsilon_x = A_1yz = -0.093 \times 10^{-6}(300 \times 200) = -0.005$$

$$\varepsilon_y = A_2xz = 0.125 \times 10^{-6}(400 \times 200) = 0.01$$

$$\varepsilon_z = A_3xy = -0.0417 \times 10^{-6}(400 \times 300) = -0.005$$

$$\gamma_{xy} = A_1xz + A_2yz = -0.083 \times 10^{-6}(400 \times 200) + 0.125 \times 10^{-6}(300 \times 200) = -0.00083$$

and similar procedure for the remaining shearing strains.

3.3 Compatibility Conditions

The strain–displacement relations derived in Equations 3.11 and 3.12 describe the geometry of deformation of the body. As mentioned in the introduction, the geometry of deformation constitutes a basic component of determining the response of components to external loads. The other two components are equilibrium and mechanical properties (or stress–strain relations) of the material. It is necessary to satisfy the equations governing these aspects of the subject in order to arrive at the responses of structures and machine components to external loads.

Because the strain components are not independent functions of the coordinates, they are related by certain mathematical relations known as the *compatibility conditions*, first obtained by Saint Venant in 1860 [3.1]. For the 2D case involving ε_x, ε_y, and γ_{xy}, it is readily shown by differentiating the strain expressions in Equation 3.11, the following compatibility condition is reached:

$$\frac{\partial^2 \varepsilon_x}{\partial y^2} + \frac{\partial^2 \varepsilon_y}{\partial x^2} = \frac{\partial^2 \gamma_{xy}}{\partial x \partial y} \tag{3.13}$$

For the 3D elements, the strain–displacement relations given in Equation 3.12 can be differentiated to yield the following compatibility equations:

$$\begin{aligned}
&\frac{\partial^2 \varepsilon_x}{\partial y^2} + \frac{\partial^2 \varepsilon_y}{\partial x^2} = \frac{\partial^2 \gamma_{xy}}{\partial x \partial y} \\
&\frac{\partial^2 \varepsilon_y}{\partial z^2} + \frac{\partial^2 \varepsilon_z}{\partial y^2} = \frac{\partial^2 \gamma_{yz}}{\partial y \partial z} \\
&\frac{\partial^2 \varepsilon_z}{\partial x^2} + \frac{\partial^2 \varepsilon_x}{\partial z^2} = \frac{\partial^2 \gamma_{xz}}{\partial z \partial x} \\
&2\frac{\partial^2 \varepsilon_x}{\partial y \partial z} = \frac{\partial}{\partial x}\left(-\frac{\partial \gamma_{yz}}{\partial x} + \frac{\partial \gamma_{xz}}{\partial y} + \frac{\partial \gamma_{xy}}{\partial z}\right) \\
&2\frac{\partial^2 \varepsilon_y}{\partial z \partial x} = \frac{\partial}{\partial y}\left(\frac{\partial \gamma_{yz}}{\partial x} - \frac{\partial \gamma_{xz}}{\partial y} + \frac{\partial \gamma_{xy}}{\partial z}\right) \\
&2\frac{\partial^2 \varepsilon_z}{\partial x \partial y} = \frac{\partial}{\partial z}\left(\frac{\partial \gamma_{yz}}{\partial x} + \frac{\partial \gamma_{xz}}{\partial y} - \frac{\partial \gamma_{xy}}{\partial z}\right)
\end{aligned} \tag{3.14}$$

When a component is deformed, the induced strains must satisfy the compatibility conditions throughout the region occupied by the component. To illustrate this point, suppose that the strains in a given component are given by

$$\begin{aligned}
\varepsilon_x &= Axy^2 + Bx^2y \\
\varepsilon_y &= Cx^2y + Dxy^2 \\
\varepsilon_z &= 0 \qquad \text{(A, B, C, and D are constants)} \\
\gamma_{xy} &= x^3 + y^3 \\
\gamma_{yz} &= \gamma_{zx} = 0
\end{aligned} \tag{a}$$

Inserting the given strains into Equations 3.14, the first equation gives $2\,(Ax + Cy) = 0$ while the remaining equations vanish identically. Thus, Equations 3.14 are satisfied at $x = y = 0$ only but not throughout the region of the component. Hence, the strains in (a) are not compatible.

Suppose that the deformation of a component is described by

$$\begin{aligned}
\varepsilon_x &= \varepsilon_z = y^2 \\
\varepsilon_y &= x^2 \\
\gamma_{xy} &= 4xy \\
\gamma_{yz} &= 2yz \\
\gamma_{zx} &= 2(x + z)
\end{aligned} \tag{b}$$

Inserting (b) into Equations 3.14, the first gives: $4 = 4$, $2 = 2$, and the remaining equations vanish identically. Thus, the proposed strains in (b) are compatible throughout the region of the component.

Note that Equations 3.14 are equivalent to three independent fourth-order differential equations in term of strains. To illustrate this point, differentiate twice the following: first equation with respect to z; second equation with respect to x; third equation with respect to y; fourth equation with respect to y and z; fifth equation with respect to x and z; and the sixth equation with respect to x and y. It is readily confirmed that the result shows that the first group of three fourth-order differential equation is identical to the second group of three fourth-order differential equations.

The physical significance of the compatibility equations implies that when an element of a body deforms there are no voids in the element. Accordingly, Equations 3.14 are necessary and sufficient conditions to ensure that the strain components in the element produce single-valued continuous displacements for a simply connected body [3.1,3.2].

The meaning of a simply connected region is one in which any boundary curve can be continuously shrunk into a point without passing the outside boundary of the region.

3.4 Dilatation

The dilatation at a point of an element, usually denoted by e, is defined as the change in volume per unit volume of the element.

Consider a prismatic element of sides, dx, dy, and dz. The original volume is $V = dx\,dy\,dz$.

After deformation, the lengths of the sides of the element become $(1 + \varepsilon_x)dx$, $(1 + \varepsilon_y)dy$, and $(1 + \varepsilon_z)dz$, respectively, and the new volume is $V' = (1 + \varepsilon_x)dx(1 + \varepsilon_y)dy(1 + \varepsilon_z)dz$.

Hence, the change in volume is

$$\Delta V = V' - V = (\varepsilon_x + \varepsilon_y + \varepsilon_z + \cdots)dx\,dy\,dz$$

Where higher-order terms in strains, that is, $O(\varepsilon_{ij}^2)$, i, $j = x$, y, z, are neglected. It follows that

$$e = \frac{\Delta V}{V} = \varepsilon_x + \varepsilon_y + \varepsilon_z \tag{3.15}$$

The dilatation is related to the displacement by the relation

$$e = \operatorname{div}\vec{D} = \vec{\nabla} \cdot \vec{D} \tag{3.16}$$

3.5 Transformation of Strain

In Section 2.5, we discussed the rules governing the transformation of stress at a point in a structure. In this section, we examine the transformation of strain. Given the state of strain at a point relative to a set of orthogonal axes, x, y, and z, we determine the strain relative to another set of orthogonal axes, x', y', and z', through the point.

3.5.1 2D Strain

In this section, we consider the 2D element PQRS, which deforms into element P′Q′R′S′ as discussed in Section 3.2.1. Point P is displaced into point P′ and the displacement vector PP′ is shown in Figure 3.8. The x′ axis makes a counterclockwise angle θ with the x-axis. Referred to the x-, y-axes, the coordinates of P are (x, y) and its displacement has components u and v along the same axes while with reference to the rotated axes the coordinates of P are (x′, y′) and the displacements, say, u′ and v′, are shown in the same figure. It follows from the geometry of Figure 3.8 that

$$\begin{aligned} x &= OA = OB - AB = x' \cos\theta - y' \sin\theta \\ y &= AP = AD + DP = x' \sin\theta + y' \cos\theta \end{aligned} \tag{3.17}$$

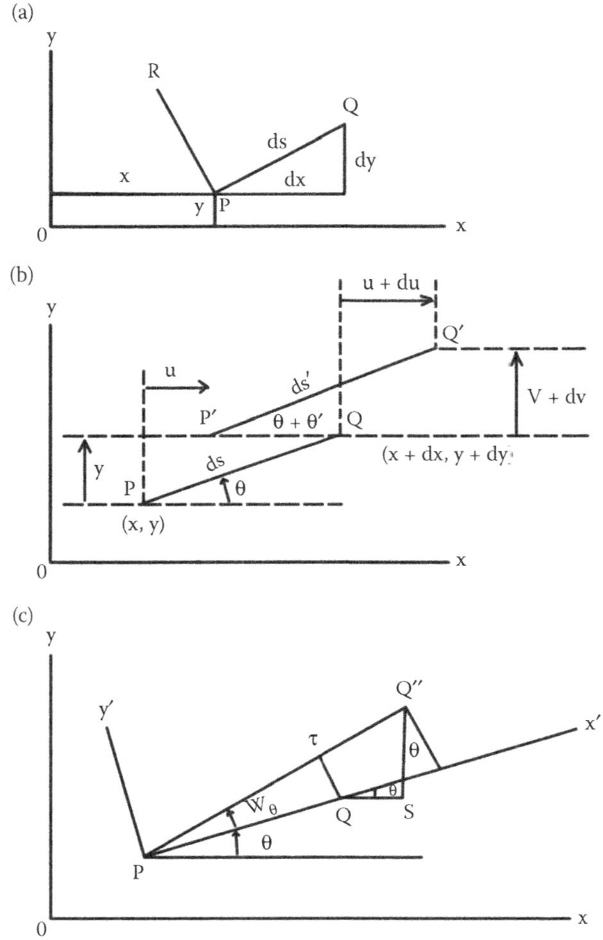

FIGURE 3.8
Transformation of two-dimensional strain.

Hence, $\partial x/\partial x' = \cos\theta$ and $\partial y/\partial y' = \sin\theta$. Similarly, the components of the displacements along both sets of axes are related by the relations:

$$\begin{aligned} u' &= u\cos\theta + v\sin\theta \\ v' &= v\cos\theta - u\sin\theta \end{aligned} \tag{3.18}$$

One method of determining the transformed strain is to use the chain rule of partial differentiation as follows:

$$\begin{aligned} \varepsilon_{x'} &= \frac{\partial u'}{\partial x'} = \left(\frac{\partial}{\partial x}\frac{\partial x}{\partial x'} + \frac{\partial}{\partial y}\frac{\partial y}{\partial x'}\right)u' = \left(\cos\theta\frac{\partial}{\partial x} + \sin\theta\frac{\partial}{\partial y}\right)(u\cos\theta + v\sin\theta) \\ &= \cos^2\theta\frac{\partial u}{\partial x} + \sin^2\theta\frac{\partial v}{\partial y} + \sin\theta\cos\theta\left(\frac{\partial v}{\partial x} + \frac{\partial u}{\partial y}\right) \end{aligned}$$

Making use of the strain–displacement relations in Equations 3.11, it follows that

$$\varepsilon_{x'} = \varepsilon_x\cos^2\theta + \varepsilon_y\sin^2\theta + \gamma_{xy}\sin\theta\cos\theta$$

Introducing the trigonometric identities: $\cos 2\theta = \cos^2\theta - \sin^2\theta$ and $\sin 2\theta = 2\sin\theta\cos\theta$ in the last expression, the transformed strain is found as

$$\varepsilon_{x'} = \frac{\varepsilon_x + \varepsilon_y}{2} + \frac{\varepsilon_x - \varepsilon_y}{2}\cos 2\theta + \frac{\gamma_{xy}}{2}\sin 2\theta \tag{3.19a}$$

Similarly, the following transformations are readily obtained:

$$\begin{aligned} \varepsilon_{y'} &= \frac{\partial v'}{\partial y'} = \varepsilon_y\cos^2\theta + \varepsilon_x\sin^2\theta - \gamma_{xy}\sin\theta\cos\theta \\ &= \frac{\varepsilon_x + \varepsilon_y}{2} - \frac{\varepsilon_x - \varepsilon_y}{2}\cos 2\theta - \frac{\gamma_{xy}}{2}\sin 2\theta \end{aligned} \tag{3.19b}$$

$$\begin{aligned} \gamma_{x'y'} &= \frac{\partial u'}{\partial y'} + \frac{\partial v'}{\partial x'} = (\varepsilon_y - \varepsilon_x)\sin\theta\cos\theta + \frac{\gamma_{xy}}{2}(\cos^2\theta - \sin^2\theta) \\ &= \frac{\varepsilon_y - \varepsilon_x}{2}\sin 2\theta + \frac{\gamma_{xy}}{2}\cos 2\theta \end{aligned} \tag{3.19c}$$

It is clear that the relations given in Equations 3.19 can be obtained from the corresponding relations for the transformation of stress, Equations 2.23, provided the normal stress and the shear strain in the latter equations are replaced by the normal strain and half the shear strain, respectively.

3.5.2 3D Strain

In this section, we consider the transformation of 3D strain. Given the strain relative to axes x, y, and z, we determine the strain relative to axes x′, y′, and z′. The direction cosines of the x′, y′, and z′ axes relative to the x, y, and z axes are given by the matrix

$$N = \begin{pmatrix} l_1 & m_1 & n_1 \\ l_2 & m_2 & n_2 \\ l_3 & m_3 & n_3 \end{pmatrix} \tag{a}$$

We consider a small element PQ of length dx' along the x' axis. The x' axis has direction cosines l_1, m_1, and n_1 relative to the original axes x, y, and z (Figure 3.9). P is taken as the origin of both sets of axes, and the coordinates of point Q are the projections of the element on the x, y, and z axes, that is, $dx = l_1\,dx'$, $dy = m_1\,dx'$, and $dz = n_1\,dx'$. After deformation, the displacement of point P has components u, v, and w along the x, y, and z axes, and the displacement of Q consists of the displacement of P plus infinitesimal increment, that is,

$$\begin{aligned} u_Q &= u + \frac{\partial u}{\partial x}dx + \frac{\partial u}{\partial y}dy + \frac{\partial u}{\partial z}dz \\ v_Q &= v + \frac{\partial v}{\partial x}dx + \frac{\partial v}{\partial y}dy + \frac{\partial v}{\partial z}dz \\ w_Q &= w + \frac{\partial w}{\partial x}dx + \frac{\partial w}{\partial y}dy + \frac{\partial w}{\partial z}dz \end{aligned} \tag{3.20}$$

Also, the coordinates of point Q are $(dx + u_Q - u,\ dy + v_Q - v,\ dz + w_Q - w)$. Hence, the deformation of the element is found by equating the square of its length to the sum of the squares of the coordinates of point Q, that is,

$$\begin{aligned} (dx' + \varepsilon_{x'}\,dx')^2 = {} & \left(dx + \frac{\partial u}{\partial x}dx + \frac{\partial u}{\partial y}dy + \frac{\partial u}{\partial z}dz\right)^2 + \left(dy + \frac{\partial v}{\partial x}dx + \frac{\partial v}{\partial y}dy + \frac{\partial v}{\partial z}dz\right)^2 \\ & + \left(dz + \frac{\partial w}{\partial x}dx + \frac{\partial w}{\partial y}dy + \frac{\partial w}{\partial z}dz\right)^2 \end{aligned} \tag{b}$$

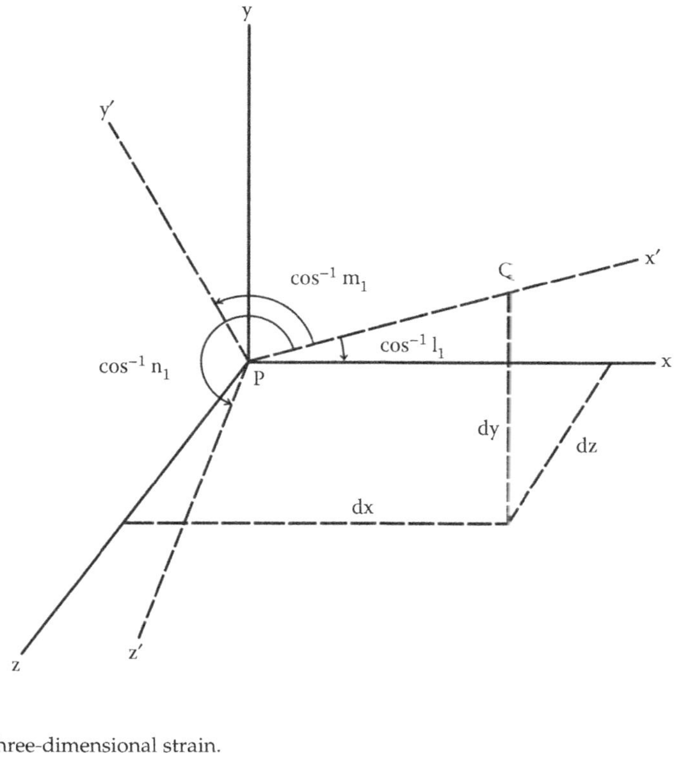

FIGURE 3.9
Transformation of three-dimensional strain.

Dividing Equation (b) by $(dx')^2$, and noting that $l_1 = dx/dx'$, $m_1 = dy/dx'$, and $n_1 = dz/dx'$ it follows that

$$(1+\varepsilon_{x'})^2 = \left(l_1 + l_1\frac{\partial u}{\partial x} + m_1\frac{\partial u}{\partial y} + n_1\frac{\partial u}{\partial z}\right)^2 + \left(l_1\frac{\partial v}{\partial x} + m_1 + m_1\frac{\partial v}{\partial y} + n_1\frac{\partial v}{\partial z}\right)^2 + \left(l_1\frac{\partial w}{\partial x} + m_1\frac{\partial w}{\partial y} + n_1 + n_1\frac{\partial w}{\partial z}\right)^2 \qquad \text{(c)}$$

Expanding Equation (c) in power series, and noting that $\varepsilon_{x'}$, $\partial u/\partial x$, $\partial u/\partial y$, ..., and $\partial w/\partial z$ are small quantities whose squares and products are neglected, $l_1^2 + m_1^2 + n_1^2 = 1$, it is readily shown that

$$\varepsilon_{x'} = l_1^2\frac{\partial u}{\partial x} + m_1^2\frac{\partial v}{\partial y} + n_1^2\frac{\partial w}{\partial z} + l_1 m_1\left(\frac{\partial u}{\partial y} + \frac{\partial v}{\partial x}\right) + m_1 n_1\left(\frac{\partial v}{\partial z} + \frac{\partial w}{\partial y}\right) + n_1 l_1\left(\frac{\partial w}{\partial x} + \frac{\partial w}{\partial z}\right) \qquad \text{(d)}$$

Making use of the notation in Equation 3.13, it is concluded that the relation (d) yields

$$\varepsilon_{x'} = l_1^2\varepsilon_x + m_1^2\varepsilon_y + n_1^2\varepsilon_z + l_1 m_1\gamma_{xy} + m_1 n_1\gamma_{yz} + n_1 l_1\gamma_{zx} \qquad (3.21a)$$

Following the same procedure used in deriving the previous equation, it can be shown that

$$\varepsilon_{y'} = l_2^2\varepsilon_x + m_2^2\varepsilon_y + n_2^2\varepsilon_z + l_2 m_2\gamma_{xy} + m_2 n_2\gamma_{yz} + n_2 l_2\gamma_{zx} \qquad (3.21b)$$

$$\varepsilon_{z'} = l_3^2\varepsilon_x + m_3^2\varepsilon_y + n_3^2\varepsilon_z + l_3 m_3\gamma_{xy} + m_3 n_3\gamma_{yz} + n_3 l_3\gamma_{zx} \qquad (3.21c)$$

To derive the expression for $\gamma_{x'y'}$, we need to investigate the change in the angle between the elements PQ and PR along the x′ and y′ axes, respectively, after deformation. After deformation, the unit vector in the direction of PQ, $\mathbf{e_Q} = (l_1', m_1', n_1')$, has direction cosines

$$l_1' = \frac{dx + u_Q - u}{dx'(1+\varepsilon_{x'})} = (1+\varepsilon_{x'})^{-1}\left(l_1 + l_1\frac{\partial u}{\partial x} + m_1\frac{\partial u}{\partial y} + n_1\frac{\partial u}{\partial z}\right)$$

$$m_1' = \frac{dy + v_Q - v}{dx'(1+\varepsilon_{x'})} = (1+\varepsilon_{x'})^{-1}\left(m_1 + l_1\frac{\partial v}{\partial x} + m_1\frac{\partial v}{\partial y} + n_1\frac{\partial v}{\partial z}\right) \qquad \text{(e)}$$

$$n_1' = \frac{dz + w_Q - w}{dx'(1+\varepsilon_{x'})} = (1+\varepsilon_{x'})^{-1}\left(n_1 + l_1\frac{\partial w}{\partial x} + m_1\frac{\partial w}{\partial y} + n_1\frac{\partial w}{\partial z}\right)$$

Expanding in power series and neglecting higher-order terms, it is found that the direction cosines of the unit vector $\mathbf{e_Q}$ reduce to

$$l_1' = l_1(1 + \varepsilon_x - \varepsilon_{x'}) + m_1\frac{\partial u}{\partial y} + n_1\frac{\partial u}{\partial z}$$

$$m_1' = l_1\frac{\partial v}{\partial x} + m_1(1 + \varepsilon_x - \varepsilon_{x'}) + n_1\frac{\partial v}{\partial z} \qquad \text{(f)}$$

$$n_1' = l_1\frac{\partial w}{\partial x} + m_1\frac{\partial w}{\partial y} + n_1(1 + \varepsilon_x - \varepsilon_{x'})$$

The direction cosines of the unit vector, $\mathbf{e_R} = (l_2, m_2, n_2)$, along the element PR are analogous to those given in Equations (f) provided the subscript 1 is replaced by 2 and the subscripts x and x′ by y and y′, respectively. The dot product $\mathbf{e_Q} \cdot \mathbf{e_R}$ yields

$$\mathbf{e_Q} \cdot \mathbf{e_R} = \cos\theta' = \cos\left(\frac{\pi}{2} - \gamma_{x'y'}\right) = \sin\gamma_{x'y'} = \gamma_{x'y'} \text{ (the shearing strain is small)}$$

Evaluating the dot product, making use of the orthogonality between the x′ and y′ axes, and utilizing the relations in Equations 3.12, it is found that

$$\begin{aligned}\gamma_{x'y'} &= 2(l_1l_2\varepsilon_x + m_1m_2\varepsilon_y + n_1n_2\varepsilon_z) + (l_1m_2 + l_2m_1)\gamma_{xy}\\ &\quad + (m_1n_2 + m_2n_1)\gamma_{yz} + (n_1l_2 + n_2l_1)\gamma_{zx}\end{aligned} \tag{3.21d}$$

Similarly, the expressions for $\gamma_{y'z'}$ and $\gamma_{x'z'}$ are readily obtained

$$\begin{aligned}\gamma_{y'z'} &= 2(l_2l_3\varepsilon_x + m_2m_3\varepsilon_y + n_2n_3\varepsilon_z) + (m_2l_3 + l_2m_3)\gamma_{xy}\\ &\quad + (n_2m_3 + m_2n_3)\gamma_{yz} + (l_2n_3 + n_2l_3)\gamma_{xz}\end{aligned} \tag{3.21e}$$

$$\begin{aligned}\gamma_{x'z'} &= 2(l_1l_3\varepsilon_x + m_1m_3\varepsilon_y + n_1n_3\varepsilon_z) + (l_1m_3 + m_1l_3)\gamma_{xy}\\ &\quad + (m_1n_3 + n_1m_3)\gamma_{yz} + (n_1l_3 + l_1n_3)\gamma_{xz}\end{aligned} \tag{3.21f}$$

It is clear that the transformation rules governing the strain are identical to those governing the transformation of stress provided that σ_i is replaced by ε_i and τ_{ij} by $(1/2)\gamma_{ij}, i, j = x, y, z$. In matrix notation, Equations 3.27a–f are written as $\varepsilon_{i'j'} = N\varepsilon_{ij}N^T$.

EXAMPLE 3.4

The deformation of a loaded component is described by the displacements

$$u = A(x^2 + y + z); \quad v = A(3y^2 + z + x); \quad w = A(2z^2 + x + y)$$

where $A = 10^{-3}$ mm.

Determine the elongation of a line element of length 20 cm that passes through the point (1,1,1) and makes equal angles with the x, y, and z axes.

Solution

We compute the strains at the point (1,1,1). Using Equations 3.12, we have

$$\varepsilon_x = \frac{\partial u}{\partial x} = 2Ax = 2A; \quad \varepsilon_y = \frac{\partial v}{\partial y} = 6Ay = 6A; \quad \varepsilon_z = \frac{\partial w}{\partial z} = 4Az = 4A$$

$$\gamma_{xy} = \frac{\partial u}{\partial y} + \frac{\partial v}{\partial x} = 2A; \quad \gamma_{yz} = \frac{\partial v}{\partial z} + \frac{\partial w}{\partial y} = 2A; \quad \gamma_{zx} = \frac{\partial w}{\partial x} + \frac{\partial u}{\partial z} = 2A$$

The line element that passes through the point (1,1,1) and makes equal angles with the rectangular axes has direction cosines $l = m = n = 1/\sqrt{3}$. The extensional strain along the element is determined from Equation 3.27a. Thus,

$$\varepsilon_{x'} = \frac{1}{3}(2A + 6A + 4A + 2A + 2A + 2A) = 6A = 6 \times 10^{-3}$$

The elongation of the element is therefore given by $\Delta L = \varepsilon_{x'} L = (0.006)(200) = 1.2\,\text{mm}$

EXAMPLE 3.5

The state of engineering strain at a point is given by

$$\varepsilon_x, \varepsilon_y, \varepsilon_z = (2.0, 1.2, -1.5) \times 10^{-3}$$

$$\gamma_{xy}, \gamma_{yz}, \gamma_{zx} = (1.0, -1.0, -0.5) \times 10^{-3}$$

For an element of the material located along the directions $\vec{n} = (0, \sqrt{3}/2, -1/2)$, find

a. The normal strain ε_n
b. The shear strain γ_{nt} where the unit vector t has components (−1, 0, 0)

Solution

a. Using Equation 3.21a with $l = 0$, $m = (3)^{1/2}/2$, $n = -1/2$, we get

$$\varepsilon_n = 0.96 \times 10^{-3}$$

Note that Equation 3.21a is identical to Equation 2.14 provided that the normal stress is replaced by the normal strain and the shearing strain by a half the shearing stress.

b. Using Equation 3.21d with l, m, n as given above and $l_2 = -1$, $m_2 = n_2 = 0$, the shear strain is given by

$$\gamma_n = -1.12 \times 10^{-3}$$

3.6 Strain Tensor

Just as in case of stress, strain is also a tensor of the second kind, which obeys certain mathematical equations. The engineering strain for fibers along the x and y axes is

$$\gamma_{xy} = \frac{\partial u}{\partial y} + \frac{\partial v}{\partial x} \tag{a}$$

The state of engineering strain at a point in the body is represented by

$$\varepsilon_{ij} = \begin{bmatrix} \varepsilon_x & \frac{1}{2}\gamma_{xy} & \frac{1}{2}\gamma_{xz} \\ \frac{1}{2}\gamma_{yx} & \varepsilon_y & \frac{1}{2}\gamma_{yz} \\ \frac{1}{2}\gamma_{zx} & \frac{1}{2}\gamma_{zy} & \varepsilon_z \end{bmatrix} \tag{3.22}$$

3.6.1 Mathematical Strain

There is also a mathematical definition of shear strain, $2\,\varepsilon_{xy} = \gamma_{xy}$, that is,

$$\varepsilon_{xy} = \frac{1}{2}\left(\frac{\partial u}{\partial y} + \frac{\partial v}{\partial x}\right) \tag{b}$$

The factor (1/2) is introduced to preserve a standard form of the mathematical equation when the coordinates are transformed from one system to the other. The corresponding array representing the state of mathematical strain is

$$\varepsilon_{ij} = \begin{bmatrix} \varepsilon_x & \varepsilon_{xy} & \varepsilon_{xz} \\ \varepsilon_{yx} & \varepsilon_y & \varepsilon_{yz} \\ \varepsilon_{zx} & \varepsilon_{zy} & \varepsilon_z \end{bmatrix} \tag{3.23}$$

The symbols ε_{ij}, ij = x, y, and z are known as tensorial shear strains. Because the strain and stress are tensors of the second kind, all formulas concerning the stress derived in Chapter 2 including stresses on an inclined plane, transformation of stress, principal values of stress, etc. can be used for strains provided that σ is replaced by ε and τ by $(1/2)\gamma$ or by its equivalent in the mathematical definition of strain. The mathematical definition of strain is rarely used in engineering work.

Strain deviator tensor: The strain deviator tensor is obtained by subtracting the spherical part of the strain (average of the three normal strains) from the normal strain [2.2]. The result is

$$(\varepsilon_{ij}) = \begin{pmatrix} \frac{1}{3}(2\varepsilon_x - \varepsilon_y - \varepsilon_z) & \frac{1}{2}\gamma_{xy} & \frac{1}{2}\gamma_{xz} \\ \frac{1}{2}\gamma_{yx} & \frac{1}{3}(2\varepsilon_y - \varepsilon_z - \varepsilon_x) & \frac{1}{2}\gamma_{yz} \\ \frac{1}{2}\gamma_{zx} & \frac{1}{2}\gamma_{zy} & \frac{1}{3}(2\varepsilon_z - \varepsilon_x - \varepsilon_y) \end{pmatrix} \tag{3.24}$$

Since the sum of the normal strains in Equation 3.24 is zero, the strain deviator is associated with distortion of the element, the same as in the case of pure shear stress.

The strain deviator tensor and its properties are used in the next section to determine the principal strains and the associated directions (see Section 2.7).

EXAMPLE 3.6

For the plane element PQR in Section 3.5.1, derive Equations 3.19a–c based on the following definitions:

$$\varepsilon_{x'} = \frac{\partial u'}{\partial x'}, \quad \varepsilon_{y'} = \frac{\partial v'}{\partial y'}, \quad \varepsilon_{x'y'} = \frac{1}{2}\left(\frac{\partial v'}{\partial x'} + \frac{\partial u'}{\partial y'}\right)$$

where axes x′ and y′ make counterclockwise angle θ with the x and y axes.

Solution

If point P has coordinates (x, y) and (x′, y′) relative to the original and rotated axes, Figure 3.10a, then

$$x = x'\cos\theta - y'\sin\theta, \quad y = y'\cos\theta + x'\sin\theta \tag{a}$$

After deformation, point P moves to P′. If the displacement PP′ has components u and v along the x and y axes, respectively, then relative to axes x′ and y′, the displacement components are as follows (Figure 3.10b):

$$u' = u\cos\theta + v\sin\theta, \quad v' = v\cos\theta - u\sin\theta \tag{b}$$

Also, we have

$$\varepsilon_{x'} = \varepsilon_x \cos^2\theta + \varepsilon_y \sin^2\theta + 2\varepsilon_{xy}\sin\theta\cos\theta$$

or, equivalently

$$\varepsilon_{x'} = \frac{\varepsilon_x + \varepsilon_y}{2} + \frac{\varepsilon_x - \varepsilon_y}{2}\cos 2\theta + \varepsilon_{xy}\sin 2\theta \tag{c}$$

Replacing θ by θ + π/2 in the last equation, we obtain

$$\varepsilon_{y'} = \frac{\varepsilon_x + \varepsilon_y}{2} - \frac{\varepsilon_x - \varepsilon_y}{2}\cos 2\theta - \varepsilon_{xy}\sin 2\theta \tag{d}$$

And, it is readily confirmed that

$$\varepsilon_{x'y'} = \frac{1}{2}\left(\frac{\partial u'}{\partial y'} + \frac{\partial v'}{\partial x'}\right) = \varepsilon_{xy}\cos 2\theta - \frac{\varepsilon_x - \varepsilon_y}{2}\sin 2\theta \tag{e}$$

as in Equations 3.19a through 3.19c.

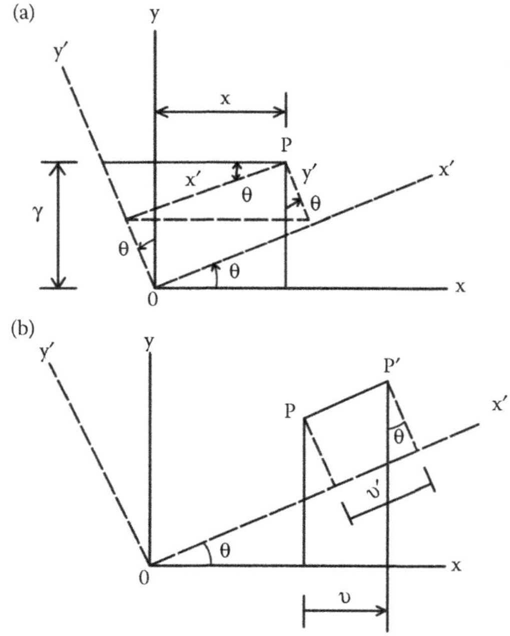

FIGURE 3.10
Example 3.6.

3.7 Principal Strains

In the previous article, we have seen that the transformation properties of stress and strain are identical. This is not surprising since both parameters are tensors of the second kind and obey the same rules of transformation, rules of finding principal strains, and so on.

At every point in a deformed body, there exist three principal planes on which the shear strains vanish and the normal strains have extreme values, that is, *principal strains*. Similar to Equations 2.28 and 2.29 for the stresses, the principal engineering strains are the roots of a cubic equation obtained by expanding the following determinant:

$$\begin{vmatrix} \varepsilon_x - \varepsilon & \frac{1}{2}\gamma_{xy} & \frac{1}{2}\gamma_{xz} \\ \frac{1}{2}\gamma_{yx} & \varepsilon_y - \varepsilon & \frac{1}{2}\gamma_{yz} \\ \frac{1}{2}\gamma_{zx} & \frac{1}{2}\gamma_{zy} & \varepsilon_z - \varepsilon \end{vmatrix} = 0 \tag{a}$$

Expanding Equation (a) and writing the principal strain as ε_p, the principal strains in three dimensions are the roots of the cubic equation

$$\varepsilon_p^3 - E_1\varepsilon_p^2 + E_2\varepsilon_p - E_3 = 0 \tag{3.25}$$

where E_j, $j = 1, 2, 3$ are the strain invariants and are given by

$$\begin{aligned} E_1 &= \varepsilon_x + \varepsilon_y + \varepsilon_z \\ E_2 &= \varepsilon_x\varepsilon_y + \varepsilon_y\varepsilon_z + \varepsilon_z\varepsilon_x - \frac{1}{4}\left(\gamma_{xy}^2 + \gamma_{yz}^2 + \gamma_{zx}^2\right) \\ E_3 &= \begin{vmatrix} \varepsilon_x & \frac{\gamma_{xy}}{2} & \frac{\gamma_{xz}}{2} \\ \frac{\gamma_{yx}}{2} & \varepsilon_y & \frac{\gamma_{yz}}{2} \\ \frac{\gamma_{zx}}{2} & \frac{\gamma_{zy}}{2} & \varepsilon_z \end{vmatrix} \\ &= \varepsilon_x\varepsilon_y\varepsilon_z + \frac{1}{4}\left(\gamma_{xy}\gamma_{yz}\gamma_{zx} - \varepsilon_x\gamma_{yz}^2 - \varepsilon_y\gamma_{zx}^2 - \varepsilon_z\gamma_{xy}^2\right) \end{aligned} \tag{3.26}$$

The principal strains, ε_1, ε_2, ε_3, and the associated principal directions as well as the principal shear strains, γ_1, γ_2, γ_3, and their directions are determined by using the procedure illustrated in Section 2.7 for the principal stresses. Thus, the three principal strains are obtained from the following equations:

$$\varepsilon_a = 2\left(\frac{R}{3}\right)^{\frac{1}{2}}\cos\theta + \frac{1}{3}E_1$$

$$\varepsilon_b = 2\left(\frac{R}{3}\right)^{\frac{1}{2}}\cos(\theta + 120^\circ) + \frac{1}{3}E_2 \qquad (3.27a)$$

$$\varepsilon_c = 2\left(\frac{R}{3}\right)^{\frac{1}{2}}\cos(\theta - 120^\circ) + \frac{1}{3}E_3$$

where the following abbreviations are used:

$$R = (1/3)E_1^2 - E_2$$

$$\cos 3\theta = -Q/(2T) \qquad (3.27b)$$

$$Q = (1/3)E_1E_2 - (2/27)E_1^3 - E_3$$

$$T = (R^3/27)^{1/2}$$

Once the numerical values of the principal strains are obtained, they can be arranged in the usual order $\varepsilon_1 > \varepsilon_2 > \varepsilon_3$. The direction cosines of the principal strains are determined from equations similar to Equations 2.32a or 2.32b provided that the normal stresses are replaced by the normal strains and the shear stresses by half the corresponding shearing strains.

Note that in terms of principal strains, the strain invariants are

$$E_1 = \varepsilon_1 + \varepsilon_2 + \varepsilon_3$$

$$E_2 = \varepsilon_1\varepsilon_2 + \varepsilon_2\varepsilon_3 + \varepsilon_3\varepsilon_1 \qquad (b)$$

$$E_3 = \varepsilon_1\varepsilon_2\varepsilon_3$$

Also, the normal and shear strains across an arbitrary plane with normal $\mathbf{n} = l\mathbf{I} + m\mathbf{j} + n\mathbf{k}$ are given by

$$\varepsilon_n = l^2\varepsilon_1 + m^2\varepsilon_2 + n^2\varepsilon_3$$

$$\gamma_n = 2\left(l^2\varepsilon_1^2 + m^2\varepsilon_2^2 + n^2\varepsilon_3^2 - \varepsilon_n^2\right)^{1/2} \qquad (c)$$

In terms of the principal deviatoric strains, $\varepsilon_1' = \varepsilon_1 - \varepsilon_m$, $\varepsilon_2' = \varepsilon_2 - \varepsilon_m$, and $\varepsilon_3' = \varepsilon_3 - \varepsilon_m$, ε_m is the mean normal strain, Equation 3.25 is

$$\varepsilon_p'^3 - E_1'\varepsilon_p'^2 + E_2'\varepsilon_p' - E_3' = 0 \qquad (3.28)$$

where the deviatoric strain invariants are given by

$$E_1' = 0 \tag{3.29a}$$

$$E_2' = \frac{1}{6}[(\varepsilon_x - \varepsilon_y)^2 + (\varepsilon_y - \varepsilon_z)^2 + (\varepsilon_z - \varepsilon_x)^2] + \frac{1}{4}\left(\gamma_{xy}^2 + \gamma_{yz}^2 + \gamma_{zx}^2\right) \tag{3.29b}$$

$$E_3' = \frac{1}{6}\begin{vmatrix} 2(2\varepsilon_x - \varepsilon_y - \varepsilon_z) & 3\gamma_{xy} & 3\gamma_{xz} \\ 3\gamma_{yx} & 2(2\varepsilon_y - \varepsilon_z - \varepsilon_x) & 3\gamma_{yz} \\ 3\gamma_{zx} & 3\gamma_{zy} & 2(2\varepsilon_z - \varepsilon_x - \varepsilon_y) \end{vmatrix} \tag{3.29c}$$

Using the procedure applied in Section 2.7 (Chapter 2) to derive the solution of Equation 2.29, the solution of Equation 3.27 for the principal deviatoric strains is obtained as follows:

$$\varepsilon_1' = 2\left(\frac{E_2'}{3}\right)^{\frac{1}{2}}\cos\beta, \quad \varepsilon_2' = 2\left(\frac{E_2'}{3}\right)^{\frac{1}{2}}\cos(\beta + 120^\circ), \quad \varepsilon_3' = 2\left(\frac{E_2'}{3}\right)^{\frac{1}{2}}\cos(\beta + 240^\circ)$$

where $\cos 3\beta = (E_3'/2)(3/E_2')^{\frac{3}{2}}$. And the principal strains are

$$\varepsilon_j = \varepsilon_j' + \frac{1}{3}E_1', \quad j = 1,2,3 \tag{3.30}$$

The corresponding principal directions, $l_j'(j = 1,2,3)$, are found as was done for the principal directions of the stresses (Equation 2.36).

The principal shear strains are obtained from the relations (see the derivation of Equation 2.43) for the principal shearing stresses.

$$\begin{aligned} \gamma_1 &= |\varepsilon_2 - \varepsilon_3| = |\varepsilon_2' - \varepsilon_3'| \\ \gamma_2 &= |\varepsilon_3 - \varepsilon_1| = |\varepsilon_3' - \varepsilon_1'| \\ \gamma_3 &= |\varepsilon_1 - \varepsilon_2| = |\varepsilon_1' - \varepsilon_2'| \end{aligned} \tag{3.31}$$

EXAMPLE 3.7

The strains at a point across a given plane are $\varepsilon_n = 0.0042$ and $\gamma_n = 0.0026$. If the direction cosines of ε_n and γ_n relative to the principal directions are

$$\left(\frac{1}{\sqrt{14}}, \frac{2}{\sqrt{14}}, \frac{3}{\sqrt{14}}\right), \quad \left(\frac{2}{\sqrt{5}}, -\frac{1}{\sqrt{5}}, 0\right)$$

respectively, find the principal strains.

Solution

The direction cosines of ε_n are $l = 0.2673$, $m = 0.5345$, and $n = 0.8018$, and those of γ_n are $l_s = 0.8944$, $m_s = -0.4472$, and $n_s = 0$.

The next step is to observe that relative to the principal axes, the normal and shear strains and their corresponding direction cosines may be expressed by equations analogous to those derived in equation (a) of solved Example 2.8 in Chapter 2.

The corresponding strain equations are obtained by replacing in the just mentioned equations the normal stress by the normal strain and the shearing stress by half the shearing strain. The result is

$$l_s = 2l\left(\frac{\varepsilon_1 - \varepsilon_n}{\gamma_n}\right)$$
$$m_s = 2m\left(\frac{\varepsilon_2 - \varepsilon_n}{\gamma_n}\right) \qquad \text{(a)}$$
$$n_s = 2n\left(\frac{\varepsilon_3 - \varepsilon_n}{\gamma_n}\right)$$

It follows that

$$\varepsilon_1 = \varepsilon_n + \frac{l_s\gamma_n}{2l} = 0.00855$$
$$\varepsilon_2 = \varepsilon_n + \frac{m_s\gamma_n}{2m} = 0.00311$$
$$\varepsilon_3 = \varepsilon_n + \frac{\gamma_n n_s}{2n} = 0.0042$$

EXAMPLE 3.8

Given the following state of strain:

$$\varepsilon_{ij} = \begin{Bmatrix} 3 & 0.6 & 0.4 \\ 0.6 & 2 & 0 \\ 0.4 & 0 & 1 \end{Bmatrix} \times 10^{-3}$$

Determine

a. The deviatoric strain tensor, ε'_{ij}, i, j – x, y, z
b. The associated deviatoric strain invariants, E'_j, j = 1, 2, 3
c. The magnitudes of the principal deviatoric strains, ε'_j, j = 1, 2, 3

Solution

a. The average normal strain is $\varepsilon_{av} = 2 \times 10^{-3}$. Subtracting the average normal strain from the normal strains gives the deviatoric strain tensor

$$\varepsilon'_{ij} = \begin{pmatrix} 1 & 0.6 & 0.4 \\ 0.6 & 0 & 0 \\ 0.4 & 0 & -1 \end{pmatrix} \times 10^{-3}$$

b. Equations 3.26 give the regular strain invariants as

$$E_1 = 6 \times 10^{-3}, \quad E_2 = 10.48 \times 10^{-6}, \quad E_3 = 5.32 \times 10^{-9}$$

The associated deviatoric strain components are

$$E'_1 = 0, \quad E'_2 = (1/3)E_1^2 - E_2 = 1.52 \times 10^{-6},$$
$$E'_3 = (2/27)E_1^3 - (1/3)E_1E_2 + E_3 = 0.36 \times 10^{-9}$$

c. Equations 3.29 give the principal deviatoric strains

$$\varepsilon_1' = 1.34 \times 10^{-3}$$
$$\varepsilon_2' = -0.25 \times 10^{-3}$$
$$\varepsilon_3' = -1.09 \times 10^{-3}$$

3.8 Octahedral Strains

As discussed in Section 2.8, the normal to the octahedral plane makes equal angles with the principal axes 1, 2, and 3. Across the octahedral plane, the normal strain is

$$\varepsilon_{oct} = \frac{1}{3}(\varepsilon_1 + \varepsilon_2 + \varepsilon_3) \tag{3.32}$$

where ε_1, ε_2, and ε_3 are the principal strains. Similarly, the shear strain, γ_{oct}, is given by

$$\gamma_{oct} = \frac{2}{3}\left[(\varepsilon_1 - \varepsilon_2)^2 + (\varepsilon_2 - \varepsilon_3)^2 + (\varepsilon_3 - \varepsilon_2)^2\right]^{\frac{1}{2}} \tag{3.33}$$

In terms of rectangular coordinates, the octahedral strains are

$$\varepsilon_{oct} = \frac{1}{3}(\varepsilon_x + \varepsilon_y + \varepsilon_z) \tag{3.34a}$$

$$\gamma_{oct} = \frac{2}{3}\left[(\varepsilon_x - \varepsilon_y)^2 + (\varepsilon_y - \varepsilon_z)^2 + (\varepsilon_z - \varepsilon_x)^2 + \frac{3}{2}\left(\gamma_{xy}^2 + \gamma_{yz}^2 + \gamma_{zx}^2\right)\right]^{\frac{1}{2}} \tag{3.34b}$$

While in term of strain invariants, we have

$$\begin{aligned} \varepsilon_{oct} &= \frac{1}{3}E_1 \\ \gamma_{oct} &= \frac{2\sqrt{2}}{3}\left(E_1^2 - 3E_2\right)^{\frac{1}{2}} \end{aligned} \tag{3.35}$$

3.9 Mohr's Circle of Strain

Since the transformation properties of strain are identical to those of stress, it follows that with minor change in notation, Mohr's circles of strain can be constructed in a similar manner (see Section 2.9). Denoting the principal strains by ε_1, ε_2, and ε_3, and using a plot of $\gamma_n/2$ (γ_n is the shear strain) versus ε_n (normal strain), Mohr's circles of strain consist of three circles having centers located at points C_1 $((\varepsilon_1 + \varepsilon_3)/2, 0)$, C_2 $((\varepsilon_1 + \varepsilon_2)/2, 0)$ and C_3 $((\varepsilon_2 + \varepsilon_3)/2, 0)$ as shown in Figure 3.11. The radii of the circles are $(\varepsilon_1 - \varepsilon_3)/2$, $(\varepsilon_1 - \varepsilon_2)/2$, and $(\varepsilon_2 - \varepsilon_3)/2$, respectively. The coordinates of points in the common area bounded by the circumferences of the circles, which is shown shaded in Figure 3.11, represent the normal and shear strains at points associated with inclined planes. To determine graphically the strains ε_n and γ_n at a

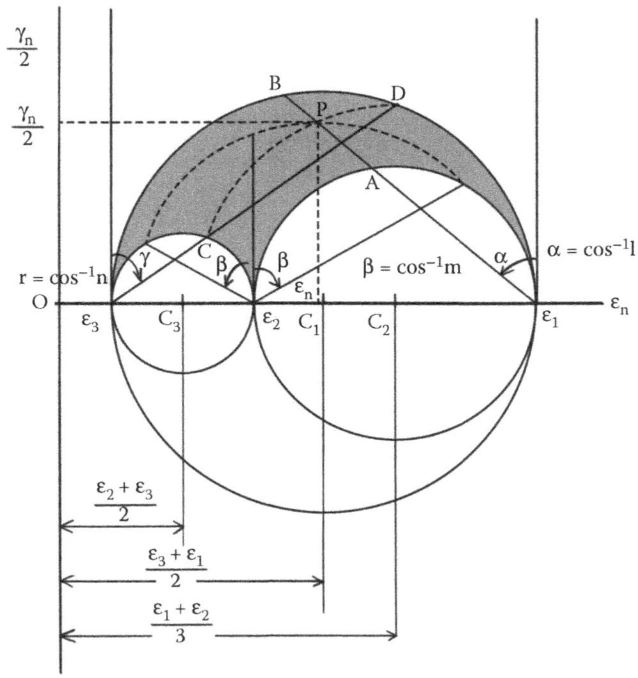

FIGURE 3.11
Mohr's circles of strain shaded region represent coordinates $\varepsilon_n, 1/2\gamma_n$.

point on an inclined plane with direction cosines (l, m, n), draw a line at angle $\alpha = \cos^{-1}l$ with the vertical at point (ε_1, 0), the line intersects circles C_2 and C_1 at points A and B. With C_3 as center, an arc is drawn joining points A and B. Another inclined line CD is set at angle $\gamma = \cos^{-1}n$ with the vertical at point (ε_3, 0), and with C_2 as center, another arc is drawn joining points C and D. The intersection of the two arcs locates point P whose coordinates give the required strains. The third angle, $\beta = \cos^{-1}m$, set on both sides of the vertical through point (ε_2, 0) may be used to check the location of point P.

Superimposed on Figure 3.11, Figure 3.12 displays the location of the origin O′ of the deviatoric strain system and the corresponding principal deviatoric strains. The distance between the two origins is

$$OO' = \frac{1}{3}(\varepsilon_1 + \varepsilon_2 + \varepsilon_3)$$

EXAMPLE 3.9

The principal strains at a point in a body are $\varepsilon_1 = 11\ \mu$, $\varepsilon_2 = 5\ \mu$, and $\varepsilon_3 = 2\ \mu$. Using Mohr's circles of strain, find the normal and shearing octahedral strains.

Solution

Figure 3.13 illustrates the three circles of strains. The direction cosines of the normal to an octahedral plane are

$$\frac{1}{\sqrt{3}} = 0.5774$$

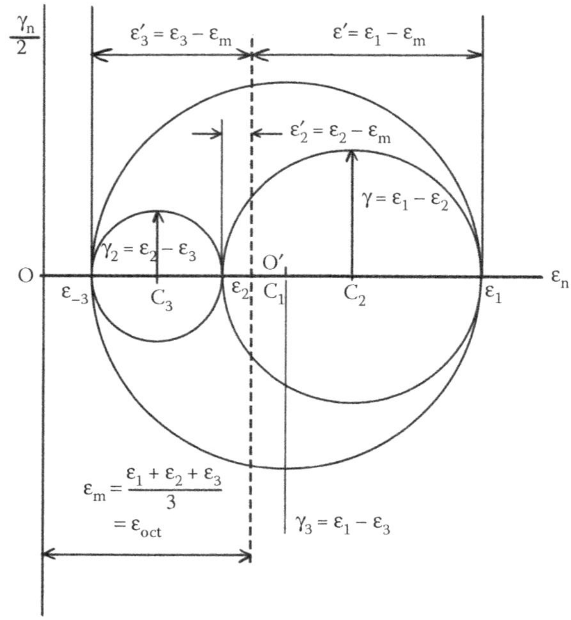

FIGURE 3.12
Mohr's circles of regular (origin O) and deviatoric (origin O′) strain.

$l = m = n = 4$. It follows that $\alpha = \beta = \gamma = \cos^{-1} 0.5774 = 54.7°$.

The center of the outer circle is point C_1, where $OC_1 = (1/2)\ (\varepsilon_1 + \varepsilon_3) = 6.5 \times 10^{-6}$. Its radius $= 4.5 \times 10^{-6}$.

The center of the larger inner circle is point C_2, where $OC_2 = (1/2)\ (\varepsilon_1 + \varepsilon_2) = 8 \times 10^{-6}$. Its radius $= 3 \times 10^{-6}$.

The center of the third circle is point C_3, where $OC_3 = (1/2)\ (\varepsilon_2 + \varepsilon_3) = 3.5 \times 10^{-6}$. Its radius $= 1.5 \times 10^{-6}$.

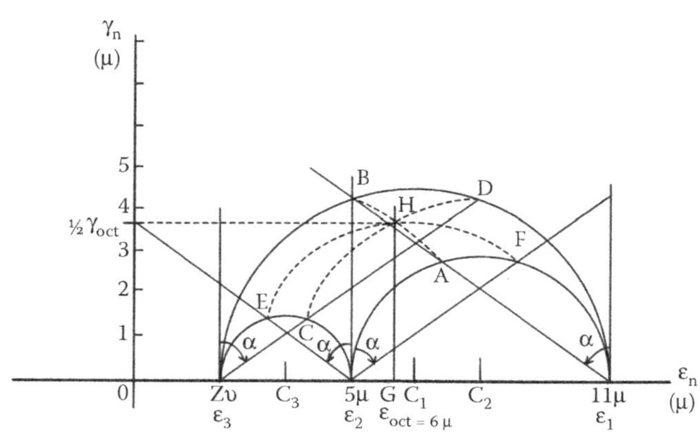

FIGURE 3.13
Example 3.9.

At the locations of ε_1 and ε_3 on the horizontal axis lines, making angles $\alpha = 54.7°$ with the vertical are drawn to intersect the circles. The first line intersects circles C_2, C_1 at points A and B, respectively. With C_3 as center, arc AB is drawn. The second line intersects circles C_3 at point C and circle C_1 at point D, and with C_2 as center arc CD is drawn to intersect arc AB at point H. The coordinates of point H give the required values of the normal and shearing octahedral strains.

As a check on the accuracy of the above graphical construction for location of point H, two inclined lines making angle α with the vertical are drawn at point with coordinates (ε_2, 0) as shown in Figure 3.13. These points locate points E on circle C_3 and point F on circle C_2, then arc EF is drawn from center C_1. It is clear that arc EF passes through point H confirming the accuracy of the graphical construction.

The coordinates of point H provide the normal and shearing octahedral strains

$$\varepsilon_{oct} = 6 \times 10^{-6}$$

$$\gamma_{oct} = 7.5 \times 10^{-6}$$

3.10 Strain Gages

Strain gages are used to measure the average strains at a point on the surface of a test specimen and in components under service conditions. The type of strain gage commonly used in practice is the *bonded electrical resistance gage*. It consists of a grid of thin wires, 0.025 mm diameter, or rectangular filament foil, 0.0025 mm thick, inserted between nonconductive sheets. Figure 3.14 illustrates a sketch of such gage. The back sheet is securely cemented to the component under investigation. As the component is deformed, its surface strain is stamped on the wire/filament foil in the gage and the electrical gage will stretch or shorten accordingly [3.3].

The electrical resistance of a wire/filament of length L is given by the formula

$$R = \frac{\rho L}{A} \tag{3.36}$$

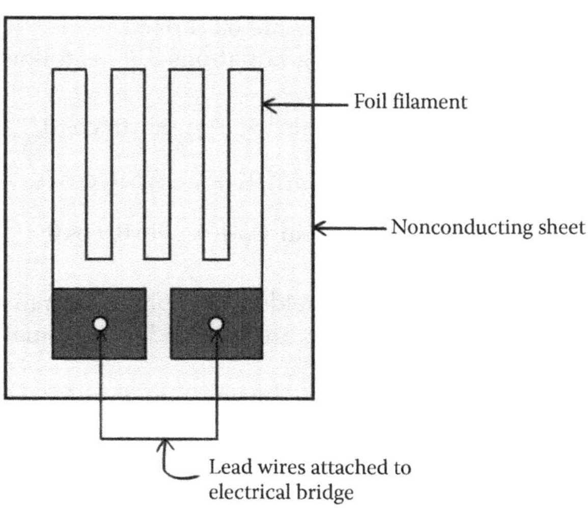

FIGURE 3.14
Strain gage.

where ρ is the specific resistance (resistivity) and A is the cross-sectional area. Any change in length is accompanied by change in area and consequently results in change in electrical resistance. Since the cross-sectional area of the filament foil is rectangular, the ratio of the surface area to the cross-sectional area is greater than that of the round bar, most electrical strain gages use filament foils. An electrical bridge circuit, attached to the strain gage through the lead wire, is used to translate the changes in resistance to strains.

For many metal alloys, the change in resistance ΔR varies linearly with the changes in length ΔL, that is,

$$\frac{\Delta R}{R} = K\left(\frac{\Delta L}{L}\right) = K\varepsilon \tag{3.37}$$

Here, K is the gage factor and ε is the strain of the gage. It follows that

$$K = \left(\frac{\Delta R}{R}\right)\frac{1}{\varepsilon} \tag{3.38}$$

Typical values of K for common strain gages vary between 2 and 2.2. A copper–nickel alloy, known as constantan (also known as Eurecka) consisting of 55% copper and 45% nickel, is commonly used for the filament foil because its resistivity does not change much when heated and it has better corrosion resistance. It has a gage factor K = 2. The changes in electrical resistance by normal strain levels experienced in engineering structures are very small, and accordingly, sensitive equipment is required to measure the strain. One common device used in practice is the Wheatstone bridge device circuit, which is capable of strains as small as $1 \times 10^{-6} = \mu$.

Strain rosette: The plane strains, ε_x, ε_y, and γ_{xy}, at a point on the surface of a component are measured by mounting assemblies of strain gages, known as *rosettes*, to measure strains along three directions or more. Typical rosettes are shown in Figure 3.15. Figure 3.15a shows a general rosette with three strain gages at angles θ_a, θ_b, and θ_c with the x-axis. Figure 3.15b shows a rectangular strain gage rosette in which the axes are 45° apart, and Figure 3.15c shows a rosette where the three strain gages are 60° apart.

The measured strains are computed using Equations 3.19 as follows:

$$\begin{aligned}
\varepsilon_a &= \varepsilon_x \cos^2\theta_a + \varepsilon_y \sin^2\theta_a + \gamma_{xy}\sin\theta_a\cos\theta_a \\
\varepsilon_b &= \varepsilon_x \cos^2\theta_b + \varepsilon_y \sin^2\theta_b + \gamma_{xy}\sin\theta_b\cos\theta_b \\
\varepsilon_c &= \varepsilon_x \cos^2\theta_c + \varepsilon_y \sin^2\theta_c + \gamma_{xy}\sin\theta_c\cos\theta_c
\end{aligned} \tag{3.39}$$

Solving Equations 3.39 simultaneously yields the in-plane normal and shear strains.

The principal strains and their orientations are obtained using Equations 3.25 through 3.27 or the 2D Mohr's circle of strain.

The out-of-plane principal stress is obtained using Equation 6.2, which is

$$\varepsilon_z = \frac{\nu}{1-\nu}(\varepsilon_x + \varepsilon_y) \tag{3.40}$$

where ν is Poisson's ratio.

The following numerical example illustrates the procedure involved.

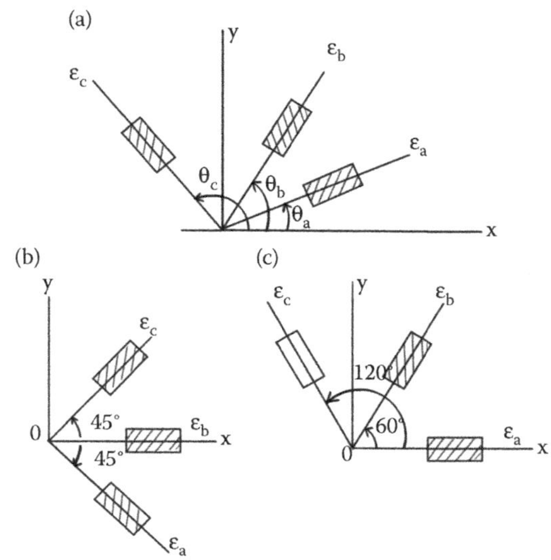

FIGURE 3.15
Strain gage rosettes.

EXAMPLE 3.10

A hollow cylindrical pressure vessel is subjected to combined internal pressure and torque of unknown magnitudes. In order to assess the strength of the vessel under service conditions, a rectangular strain gage rosette is mounted on the outside of the vessel such that the center gage of the rosette is along the axial axis of the vessel. The strain gage readings in units of $\mu = 10^{-6}$ are $\varepsilon_a = 500$, $\varepsilon_b = 260$, and $\varepsilon_c = 200$.

Find the principal strains and the associated directions. Assume Poisson's ratio = 0.3.

Solution

Applying Equations 3.39, we have

$$\theta_a = -45^\circ \text{ gives } 500 = (1/2)\varepsilon_x + (1/2)\varepsilon_y - (1/2)\gamma_{xy} \tag{1}$$

$$\theta_b = 0 \text{ gives } 260 = \varepsilon_x \tag{2}$$

$$\theta_c = 450 \text{ gives } 200 = (1/2)\varepsilon_x + (1/2)\varepsilon_y + (1/2)\gamma_{xy} \tag{3}$$

Solving the above equations simultaneously, we obtain $\varepsilon_y = 440$ and $\gamma_{xy} = -300$.

The in-plane principal strains are computed using Equations 3.27. The results are

$$\varepsilon_1 = 402 \times 10^{-6}, \quad \varepsilon_2 = 298 \times 10^{-6}$$

And the associated directions are obtained using the analogy with stress discussed in Section 3.7, $\theta_p = 29.5^\circ$ and 119.5°

The out-of-plane principal normal strain is computed using Equation 3.40 as

$$\varepsilon_3 = -300 \times 10^{-6}$$

The maximum shearing strains are

$$\gamma_{xy} = \pm(\varepsilon_1 - \varepsilon_2) = \pm 105 \times 10^{-6}$$

at 45° to the in-plane principal strains.

Problems

3.1 A member of a 3D truss is located between the origin of rectangular coordinates and point P (1,1,0.5) m, as shown in Figure P3.1. If the member remains straight and end P experiences small displacements, $u = 3$ mm, $v = 2$ mm, and $w = -1.5$ mm, find the average extensional strain of the member.

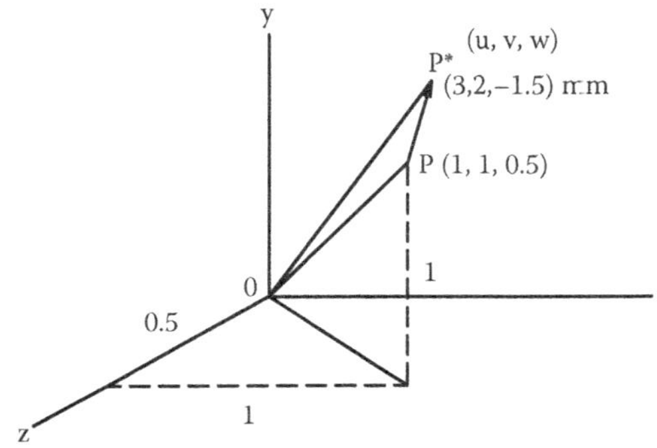

FIGURE P3.1

Ans: 0.156.

3.2 A rigid bar AB is supported by two elastic members BC and BD (Figure P3.2). If the bar AB rotates clockwise through an angle of 1° in the vertical plane, find the average extensional strain of BD.

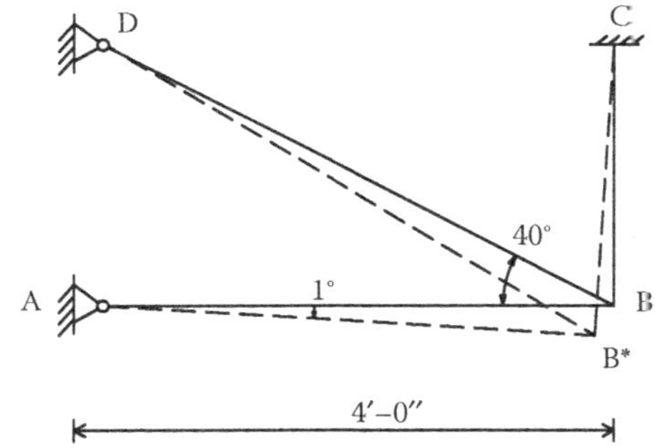

FIGURE P3.2

Ans: 8.6×10^{-3}.

3.3 A steel wire of length $L = 20$ cm is subjected to nonuniform heating such that the extensional strain is $\varepsilon_x = k(x/L)^3$, where $k = 2 \times 10^{-2}$ and x is the distance of a point from one end of the wire. Find the length of the deformed wire.

Ans: 20.1 cm.

3.4 A 0.4 m by 0.3 m thin plate ABCD is fixed along sides AB and AD. The plate is subject to deformation such that point C is displaced to C′ and all straight lines in the plate remain straight (Figure P3.4). Find the strain components ε_x, ε_y, and γ_{xy} at point C.

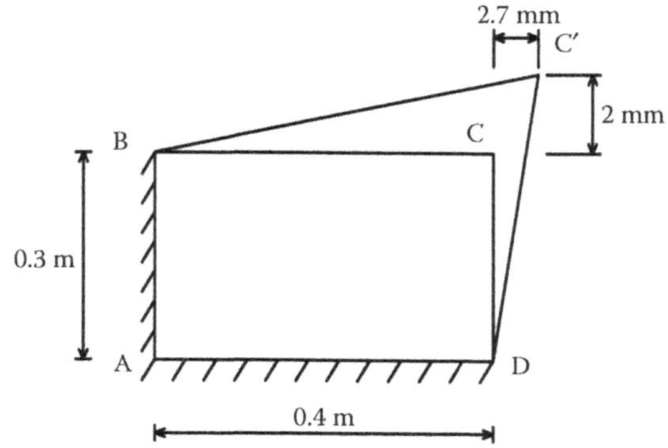

FIGURE P3.4

Ans: 6.8,6,7,1.4− $\times 10^{-3}$.

3.5 A thin triangular plate APB is fixed on side AB. The plate is strained such that apex P is displaced a distance of 5 cm to point P′ and all straight lines in the plate remain straight after deformation, as shown in Figure P3.5. Find the average extensional strains of sides AP and BP and the shear strain at point P.

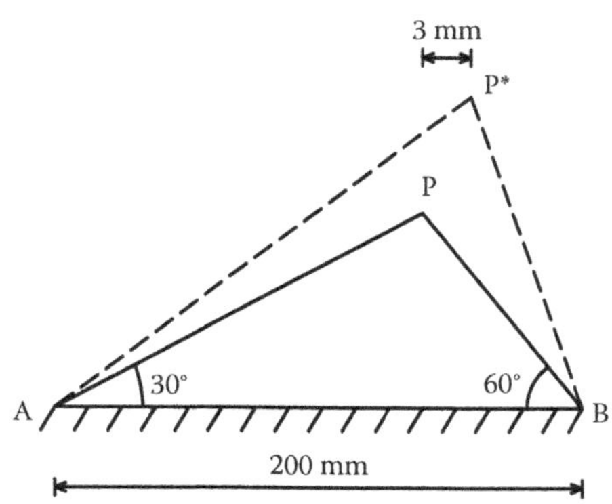

FIGURE P3.5

Ans: Shear = 1.3.

3.6 Consider the deformation of element OABC shown in Figure 3.4. Find the logarithm of the length ratio of diagonals OB′ and AC′ (lines of maximum extension and contraction).

Ans: $2\tanh^{-1}\alpha/2$.

3.7 The fibers of a plane element shown in Figure P3.7 are subjected to a deformation given by $x = a\,x_0$, $y = b\,y_0$, where (x_0, y_0) and (x, y) are the initial and final coordinates of an arbitrary point, respectively, a and b are constants such that $a > 1 > b$. Show that (a) the straight lines that remain straight during deformation make angles $+\eta$ and $+\eta_0$, with the x-axis in the final and initial states, respectively, given by the relation $\tanh(\eta) = \frac{b}{a}\tanh(\eta_0)$; (b) the deformation results in a change of volume of the element unless $ab = 1$ (when $ab = 1$, the deformation corresponds to pure shear).

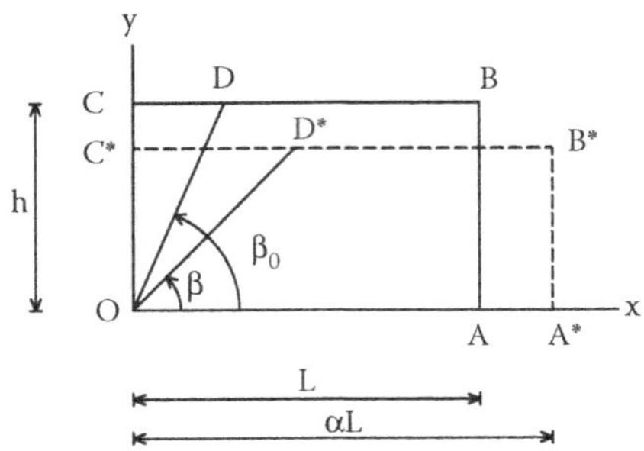

FIGURE P3.7

3.8 A thin membrane is stretched between two concentric circles of radii a and b, $b > a$ as shown in Figure P3.8. The inner boundary is held fixed and the outer boundary is rotated counterclockwise through an angle β such that all radial lines remain straight. Show that for small angle β, the shear strain is $\gamma_{rt} = b\beta/(b - a)$.

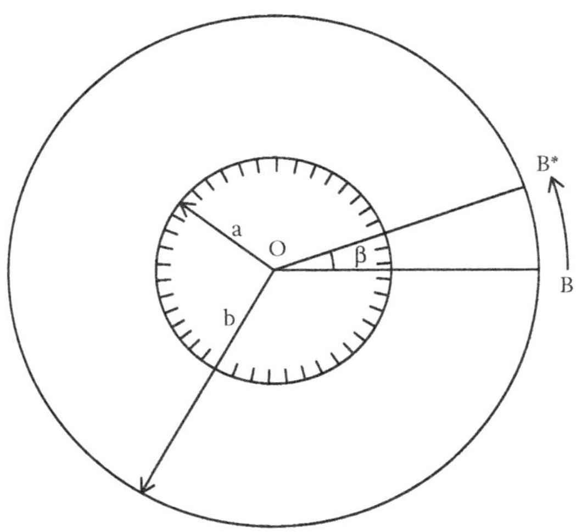

FIGURE P3.8

3.9 A circular plate of radius = a is fixed at points A and B and deforms under load into an ellipse

$$\frac{x^2}{(1+\beta)^2} + y^2 = a^2$$

where β is a small number compared to 1. The deformation is such that all points of the circle are displaced in the radial direction only (Figure P3.9). Use polar coordinates, r,

θ, shown in the figure to find the shear strain between the radial and tangential directions at an arbitrary point of the plate.

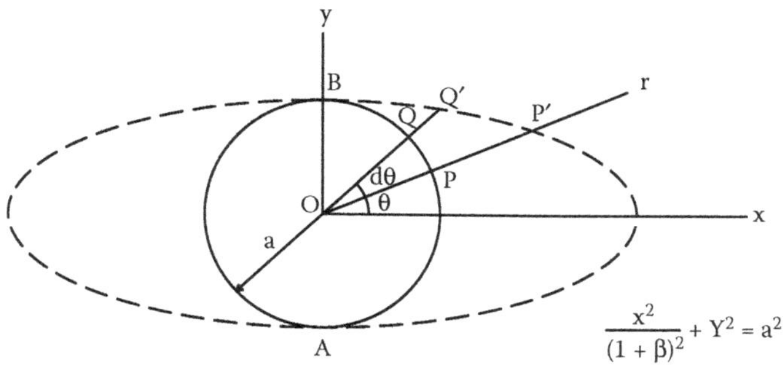

FIGURE P3.9

Ans: $\gamma_{r\theta} = -2\beta\,(xy/a^2)$.

3.10 A thin ring of radius R is subject to internal pressure such that the radius is increased by an amount ΔR. Find the circumferential strain ε_θ.

Ans: $\Delta R/Ri$.

3.11 Derive an expression for the change in volume of an element of unit volume subjected to the action of small pure shear strain γ_{xy}.

Ans: zero.

3.12 The deformation of a component relative to rectangular axes is described by the displacements

$$u = A\left[\left(\frac{x}{L}\right)^2 + \frac{y}{L} + \frac{z}{L}\right]$$

$$v = A\left[\frac{x}{L} + 3\left(\frac{y}{L}\right)^2 + \frac{z}{L}\right]$$

$$w = A\left[\frac{x}{L} + 2\left(\frac{y}{L}\right) + 2\left(\frac{z}{L}\right)^2\right]$$

where A and L are constants. Find the elongation of a line element of length L = 100 mm that passes through the point (1,1,1) mm and makes equal angles with the axes. Assume $A = 10^{-1}$ mm.

3.13 Determine whether the following strain field can exist in a loaded body:

$$\begin{aligned}
\varepsilon_x &= 3x^2 + 2y^2 \\
\varepsilon_y &= 3y^2 + 2z^2 \\
\varepsilon_z &= y^2 + x^2 \\
\gamma_{xy} &= 6xy \\
\gamma_{yz} &= 4yz \\
\gamma_{zx} &= xz
\end{aligned}$$

Ans: No.

3.14 Repeat the previous problem for the following system of strains:

$$\begin{aligned}\varepsilon_x &= 2x^2 + 4y^2\\ \varepsilon_y &= y^2 + 2z^2\\ \varepsilon_z &= x^2 + 3z^2\\ \gamma_{xy} &= 8xy\\ \gamma_{yz} &= 4yz\\ \gamma_{zx} &= 2xz\end{aligned}$$

Ans: Yes.

3.15 The strains in a 2D element is given as follows:

$$\begin{aligned}\varepsilon_x &= Ax^2(x^2 + y^2)\\ \varepsilon_y &= By^2(x^2 + y^2)\\ \gamma_{xy} &= Cxy(x^2 + y^2)\end{aligned}$$

where A, B, and C are constants. Find (a) a necessary relationship between the constants; (b) expressions for the displacements u(x, y) and v(x, y).

Ans: $A = B = (3/2)\,C$.

3.16 The state of strain in a loaded element is given by

$$\varepsilon = \begin{bmatrix} 0.002 & 0.001 & 0 \\ 0.001 & 0.003 & 0 \\ 0 & 0 & -0.003 \end{bmatrix}$$

Find the principal strains.

Ans: $3.21,\ 1.8,\ -3 \times 10^{-3}$.

3.17 The state of strain at a point is given as

$$\varepsilon = \begin{bmatrix} 1000 & 0 & -1000 \\ 0 & 2000 & -500 \\ -1000 & -500 & -1000 \end{bmatrix}\mu, \quad \mu = 10^{-6}$$

Find the principal strains.

Ans: $2.41,\ 0.044,\ -0.045 \times 10^{-3}$.

3.18 At a point O in a structure, the principal strains are 0.005, 0.002, and –0.001. If the x, y, and z axes are the principal directions, find the normal and shear strains at point O across a plane whose normal is inclined at 30° to the z-axis and the projection of the normal on the xy-plane is inclined at 40° to the xz-plane.

Ans: $0.06 \times 10^{-3},\ 3.96 \times 10^{-3}$.

3.19 The strains at a point across an oblique plane are $\varepsilon_n = 0.005$ and $\gamma_n = 0.003$. Given that the directions of ε_n and γ_n relative to the principal directions at the point are

$(1/10, 3/10, 3/\sqrt{10})$ and $(-0.678, -0.678, 0.286)$, respectively, determine the principal strains.

Ans: 0.00545, −0.00517, −0.00161.

3.20 The deviatoric principal strains at a point in a loaded member are $\varepsilon_1', \varepsilon_2', \varepsilon_3'$. Show that

$$\varepsilon_1' + \varepsilon_2' + \varepsilon_3' = 0$$

$$(\varepsilon_1')^2 + (\varepsilon_2')^2 + (\varepsilon_3')^2 = \frac{1}{3}\left[(\varepsilon_1 - \varepsilon_2)^2 + (\varepsilon_2 - \varepsilon_3)^2 + (\varepsilon_3 - \varepsilon_1)^2\right]$$

3.21 For the state of strain given in Problem 3.17, determine the octahedral strains and confirm your results by using the strain invariants.

Ans: 0.67×-10^{-3}, 2.656×10^{-3}.

3.22 A strain gage rosette bonded to the surface of a loaded component gave the following reading: $\varepsilon_0 = 490 \times 10^{-6}$, $\varepsilon_{45} = 300 \times 10^{-6}$, $\varepsilon_{90} = -400 \times 10^{-6}$.

Find the principal strains and their directions.
(Ans.: 562×10^{-6}, -492×10^{-6}, 15°, 105°)

References

3.1. I. S. Sokolnikoff, *Mathematical Theory of Elasticity*, McGraw-Hill Book Co., New York, 1956.
3.2. P. C. Chou and N. J. Pagano, *Elasticity*, D. van Nostrand Co., Princeton, New Jersey, 1967.
3.3. M. Heteny, Editor, *Handbook of Experimental Stress Analysis*, 2nd edition, John Wiley and Son, New York, 1987.

4

Stress–Strain Relations

This chapter deals with determining the stress–strain relations of materials commonly used in engineering. Elastic and plastic stress–strain relations are discussed. The basic relations are based on experimental data obtained in the laboratory by performing tests on specimen of the material. The most common test is the static tensile test of a round bar monotonically stretched to failure at room temperature. Other tests include subjecting a specimen to twisting and normal moments to reveal relations between shear stresses and strains as well as the moment–curvature relations. This is the third aspect of determining the response of structural components to external loads, and complements the equilibrium and deformation aspects covered in Chapters 2 and 3, respectively.

Most of the common engineering materials such as steel and aluminum are ductile. They exhibit elastic and plastic deformations when subjected to external forces. Elastic deformation is characterized as being reversible; the material recovers its original shape upon removal of the force. Elastic unit elongation is typically small compared to unity, for example, in structural steel, it is usually less than 0.001. Plastic deformations occur once the applied force exceeds the elastic limit, the induced strains are irrecoverable and much larger than elastic strains.

A material is said to be *homogeneous* if its mechanical properties are the same at every point of the component. Thus, a beam of steel is considered homogeneous. However, if the beam is attached to a piece of wood along its length, the resulting component is considered nonhomogeneous. A material is said to be *isotropic* (derived from the Greek *isos* meaning equal and *tropos* meaning direction) if it has the same mechanical properties in all directions. Thus, wood is anisotropic material because it has different mechanical properties along the direction of the grains and normal to them. Steel and most metals are considered isotropic.

First, we discuss the algebraic relations for elastic behavior and then we consider plastic and nonlinear elastic behaviors. Finally, the associated yield and plastic failure criteria required to solve plasticity problems are also discussed.

4.1 Elastic Behavior

A standard specimen of ductile material such as mild steel is placed between the jaws of a testing machine and tensile load is applied under static loading condition till failure. For each increment of load, the corresponding strain is recorded. The conventional or engineering definitions of stress and strain are used, that is, stress is defined as current load divided by original area (Equation 1.2) and strain is the elongation divided by original length (Equation 3.4). The true stress (load divided by current area) and true strain (increment in length divided by current length) are discussed in Chapters 2 and 3, respectively, and in more detail in Section 4.3.4. A typical stress–strain curve is shown in Figure 4.1. The basic features of the curve are discussed in the next section.

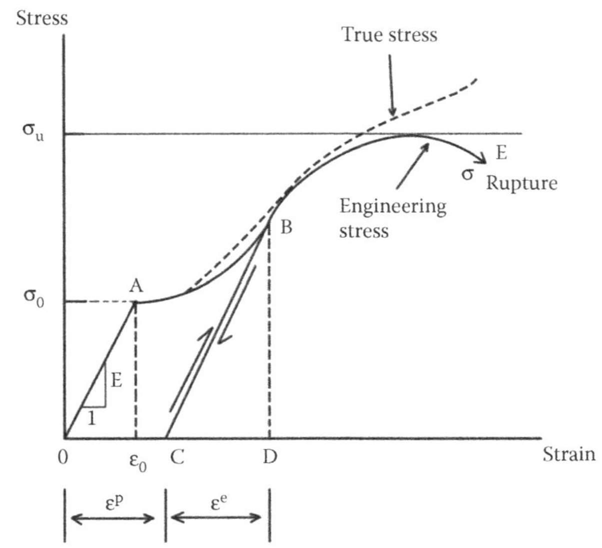

FIGURE 4.1
Stress–strain diagonal of ductile material.

4.1.1 Hooke's Law

Initially, the experimental data in Figure 4.1 generate a straight line 0A. Along this line, the normal stress is related to the normal strain by what is commonly referred to as Hooke's law. If the x-coordinate is taken along the axis of the specimen, then Hooke's law is

$$\sigma_x = E\varepsilon_x \tag{4.1}$$

where E denotes the modulus of elasticity in tension, also referred to as Young's modulus. The axial loading induces lateral contraction or reduction in the cross-sectional area of the specimen, and the lateral strains in the y and z coordinates are linearly proportional to the axial strain. Hence,

$$\varepsilon_y = \varepsilon_z = -\nu\varepsilon_x = -\nu\left(\frac{\sigma_x}{E}\right) \tag{4.2}$$

Here, the dimensionless constant ν is Poisson's ratio with typical values between 0.1 and 0.5. For structural steel, Poisson's ratio is 0.3. The linear portion of the stress–strain curve terminates at point A, which is known as the *proportional limit*. The point of initiation of plastic or irreversible deformation is called the *yield point* or the *elastic limit*, which occurs slightly above point A. However, for all intents and purposes, point A can be taken as the yield point of the material as indicated in Figure 4.1. The yield stress is denoted by σ_0 and the corresponding yield strain by ε_0. In this range of loading, the deformation of the material is elastic because the specimen returns to its original shape upon removal of the load, that is, loading, unloading, and reloading generate line 0A. Beyond A, there is an extension of constant load indicating the beginning of plastic flow and plastic deformation.

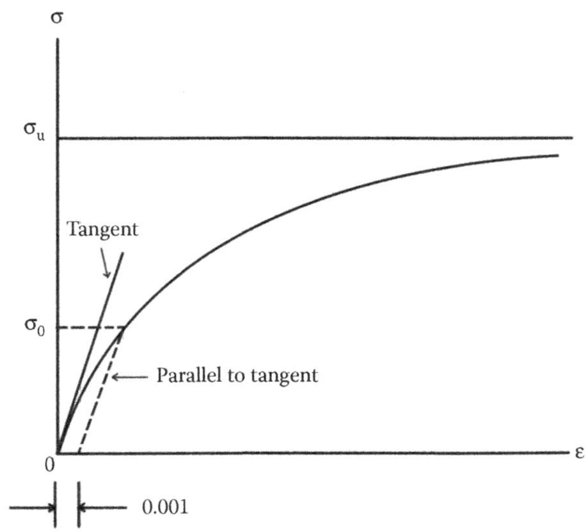

FIGURE 4.2
Estimate of yield stress by offset method.

For some metals, the experimental data generate a continuous stress–strain curve without recognizable initial straight line, as shown in Figure 4.2. In such cases, an offset approximate method is used to determine the yield stress. At the origin, a line is drawn tangent to the curve, and from a point with strain equal to 0.001 (in some cases, 0.002 is used), a line is drawn parallel to the tangent, and at the point of intersection with the curve, a horizontal line is drawn to determine the yield stress.

Another simple test consists of applying equal and opposite twisting moments (torques) at the ends of a specimen to determine the shear stress–shear strain relation by finding the change in a right angle of a typical element on the surface of the specimen undergoing pure shear. In the elastic range, we have the following Hooke's law in shear:

$$\tau_{xy} = G\gamma_{xy} \tag{4.3}$$

where G is the shear modulus of elasticity or modulus of rigidity. The applied torques do not create any normal stress.

Note that Equations 4.1 through 4.3 are also applicable to specimen undergoing simple compression, the same material moduli and Poisson's ratio can be used in compression as in tension. It is interesting to note that the magnitudes of E and G are very large in comparison to the allowable stresses.

For a 2D elastic, homogeneous, and isotropic element in the xy-plane subjected to biaxial stresses uniformly distributed on the x and y sides, numerous test measurements have confirmed that the resulting strains can be obtained by superposing the relevant strains from Equations 4.1 and 4.2. Thus, we have Hooke's law in two dimensions

$$\varepsilon_x = \frac{\sigma_x}{E} - \upsilon\frac{\sigma_y}{E}, \quad \varepsilon_y = \frac{\sigma_y}{E} - \upsilon\frac{\sigma_x}{E} \tag{4.4}$$

Similarly, for an element subjected to triaxial loading, the 3D Hooke's law is obtained as follows:

$$\begin{aligned}
\varepsilon_x &= \frac{1}{E}[\sigma_x - \upsilon(\sigma_y + \sigma_z)] \\
\varepsilon_y &= \frac{1}{E}[\sigma_y - \upsilon(\sigma_z + \sigma_x)] \\
\varepsilon_z &= \frac{1}{E}[\sigma_z - \upsilon(\sigma_x + \sigma_y)] \\
\gamma_{xy} &= \frac{\tau_{xy}}{G} \\
\gamma_{yz} &= \frac{\tau_{yz}}{G} \\
\gamma_{zx} &= \frac{\tau_{zx}}{G}
\end{aligned} \tag{4.5}$$

It should be noted that to arrive at the results in Equations 4.4 and 4.5, the principle of superposition has been used to calculate the strains and stresses produced by several forces. This basic principle of elastic behavior is valid as long as the strains are small and the associated displacements do not substantially affect the action of external loads. Moreover, the small changes in the dimensions of the deformed element and any displacements of the points of application of external loads are neglected. In few cases, however, such small deformations cannot be neglected and need to be included in the analysis. The simultaneous action of axial forces and lateral loads deforming a thin member is such an example. The axial forces produce tension or compression only; however, in the presence of lateral forces, they have to be included in calculating the lateral deflections of the member [4.1]. It turns out that the deflection is not linearly proportional to the forces and the principle of superposition in such cases is not valid.

The elastic constants E, G, and ν for an isotropic homogeneous material are not independent. They satisfy the relation

$$E = 2G(1 + \nu) \tag{4.6}$$

The validity of Equation 4.6 can be confirmed by considering the deformation of a square element in the xy-plane, 0ABC, subjected to shear stresses τ_{xy}, as in Figure 4.3. In this element, the normal stresses and strains vanish, $\sigma_x = \sigma_y = \varepsilon_x = \varepsilon_y = 0$. Consider axes $0x'$ and $0y'$ that make angles 45° with respect to the x- and y-axes. The normal strain in the x' direction is obtained from Equation 3.25a with $\theta = 45°$ as

$$\varepsilon_{x'} = \frac{\varepsilon_x + \varepsilon_y}{2} + \frac{\varepsilon_x - \varepsilon_y}{2}\cos 90° + \frac{\gamma_{xy}}{2}\sin 90° = 0 + 0 + \frac{\tau_{xy}}{2G} \tag{a}$$

Also, making use of stress transformation, Equations 2.23a and b, gives $\sigma_{x'} = -\sigma_{y'} = \tau_{xy}$, and Hooke's law in two dimensions yields

$$\varepsilon_{x'} = \frac{1}{E}(\sigma_{x'} - \upsilon\sigma_{y'}) = \frac{(1 + \upsilon)\tau_{xy}}{E} \tag{b}$$

The relation (4.6) is obtained upon equating Equations (a) and (b).

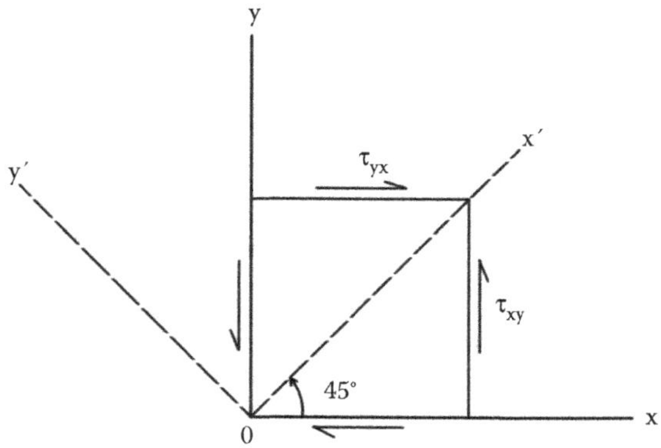

FIGURE 4.3
Plane element in pure shear.

In order to express the normal stresses in Equation 4.5 in terms of normal strains and elastic constants, it is convenient to make use of the dilatation (Equation 3.15), $e = \varepsilon_x + \varepsilon_y + \varepsilon_z$. Adding the normal stresses gives

$$\sigma_x + \sigma_y + \sigma_z = \frac{E}{1 - 2\upsilon} e \tag{4.7}$$

and solving for the normal stresses from Equations 4.5

$$\begin{aligned} \sigma_x &= \frac{\upsilon E}{(1+\upsilon)(1-2\upsilon)} e + \frac{E}{1+\upsilon} \varepsilon_x \\ \sigma_y &= \frac{\upsilon E}{(1+\upsilon)(1-2\upsilon)} e + \frac{E}{1+\upsilon} \varepsilon_y \\ \sigma_z &= \frac{\upsilon E}{(1+\upsilon)(1-2\upsilon)} e + \frac{E}{1+\upsilon} \varepsilon_z \end{aligned} \tag{4.8a}$$

And the shear stresses are

$$\tau_{xy} = G\gamma_{xy}, \quad \tau_{yz} = G\gamma_{yz}, \quad \tau_{zx} = G\gamma_{zx} \tag{4.8b}$$

A more concise way of writing Equations 4.8 is to use Lame's elasticity constants for isotropic materials, λ and μ, where

$$\lambda = \frac{\upsilon E}{(1+\upsilon)(1-2\upsilon)} = \frac{2\mu\upsilon}{1-2\upsilon}, \quad \mu = G \tag{4.9}$$

In terms of Lame's constants, Equation 4.5 becomes

$$\begin{aligned} \sigma_x &= \lambda e + 2\mu\varepsilon_x \\ \sigma_y &= \lambda e + 2\mu\varepsilon_y \\ \sigma_z &= \lambda e + 2\mu\varepsilon_z \end{aligned} \tag{4.10}$$

$$\tau_{xy} = \mu\gamma_{xy}, \quad \tau_{yz} = \mu\gamma_{yz}, \quad \tau_{zx} = \mu\gamma_{zx}$$

Note that Young's modulus of elasticity and Poisson's ratio in terms of Lame's constants are

$$E = \frac{\mu(3\lambda + 2\mu)}{\lambda\mu}$$

$$\upsilon = \frac{\lambda}{2(\lambda + \mu)}$$

Generalize Hooke's law: Equations 4.5 show that for a homogeneous elastic isotropic material each of the six strain components is a linear function of the six components of stress. Similar observation can be made about Equations 4.8 for the stresses. For general homogeneous elastic material, the generalized Hooke's law relates each of the stress (strain) components in terms of six strain (stress) components. In matrix notation, we have $\{\sigma_{ij}\} = \{C_{ij}\}\{\varepsilon_{ij}\}$ or $\{\varepsilon_{ij}\} = \{D_{ij}\}\{\sigma_{ij}\}$, i, j = 1,2, …, 6, indicating that each matrix is of order 6 × 6 containing 36 material-dependent constants.

Thus, for any homogeneous elastic material, the generalized Hooke's law is

$$\begin{bmatrix} \sigma_x \\ \sigma_y \\ \sigma_z \\ \tau_{xy} \\ \tau_{yz} \\ \tau_{zx} \end{bmatrix} = \begin{bmatrix} C_{11} & C_{12} & C_{13} & C_{14} & C_{15} & C_{16} \\ C_{21} & C_{22} & C_{23} & C_{24} & C_{25} & C_{26} \\ C_{31} & C_{32} & C_{33} & C_{34} & C_{35} & C_{36} \\ C_{41} & C_{42} & C_{43} & C_{44} & C_{45} & C_{46} \\ C_{51} & C_{52} & C_{53} & C_{54} & C_{55} & C_{56} \\ C_{61} & C_{62} & C_{63} & C_{64} & C_{65} & C_{66} \end{bmatrix} \begin{bmatrix} \varepsilon_x \\ \varepsilon_y \\ \varepsilon_z \\ \gamma_{xy} \\ \gamma_{yz} \\ \gamma_{zx} \end{bmatrix} \tag{4.11}$$

where C_{ij}, i, j = 1, 2, …, 6 are 36 material-dependent constants. Each of the 36 constants has the same value at every point of a homogeneous element. For elastic isotropic material, the number of independent constants is 2 as indicated in Equation 4.8 or Equation 4.10. In Section 4.2, it is shown that the C's matrix for anisotropic crystalline materials can have at most 21 independent material constants. The inverse of Equation 4.11 is the matrix $\{\varepsilon_{ij}\} = \{D_{ij}\}\{\sigma_{ij}\}$.

4.1.2 Incompressible Material

If an element is subjected to a uniform hydrostatic pressure of magnitude p, then $\sigma_x = \sigma_y = \sigma_z = -p$ and Equation 4.7 gives the following relation between the dilatation or unit volume expansion e and p:

$$e = -\frac{3(1 - 2\upsilon)p}{E} \tag{4.12}$$

The ratio (−p/e) is called the modulus of volume expansion or the bulk modulus of elasticity, commonly denoted by K. Thus,

$$K = \frac{E}{3(1 - 2\upsilon)} = \lambda + \frac{2}{3}\mu \tag{4.13}$$

For $\nu = 1/2$, $K = \infty$ and the material is called incompressible, while for $\nu = 0$, $K = E/3$.

4.1.3 Equations of Equilibrium in Terms of Displacements

For purposes of solving 3D problems in elasticity, it is necessary to express the basic equations of equilibrium, Equation 2.5 or Equation 2.7, in terms of the displacements u, v, and w.

The first step is to express Hooke's law, Equation 4.10, in terms of the displacements. Consider the following equations in the xy-plane:

$$\sigma_x = \lambda e + 2G\frac{\partial u}{\partial x}$$
$$\tau_{xy} = G\left(\frac{\partial u}{\partial y} + \frac{\partial v}{\partial x}\right) \tag{a}$$
$$\tau_{xz} = G\left(\frac{\partial w}{\partial x} + \frac{\partial u}{\partial z}\right)$$

Inserting Equations (a) into the first equation of equilibrium, Equation 2.7, provides

$$(\lambda + G)\frac{\partial e}{\partial x} + G\left(\frac{\partial^2 u}{\partial x^2} + \frac{\partial^2 u}{\partial y^2} + \frac{\partial^2 u}{\partial z^2}\right) + B_x = 0 \tag{b}$$

It is easy to see that the remaining two equations of equilibrium assume a form identical to that in Equation (b) provided that B_x is replaced B_y and B_z, and the derivatives of e are with respect to y and z, respectively. Making use of Laplace's operator in three dimensions, $\nabla^2 = (\partial^2/\partial x^2) + (\partial^2/\partial y^2) + (\partial^2/\partial z^2)$, the equations of equilibrium in terms of displacements become

$$(\lambda + G)\frac{\partial e}{\partial x} + G\nabla^2 u + B_x = 0$$
$$(\lambda + G)\frac{\partial e}{\partial y} + G\nabla^2 v + B_y = 0 \tag{4.14}$$
$$(\lambda + G)\frac{\partial e}{\partial z} + G\nabla^2 w + B_z = 0$$

For constant body forces, differentiating the first of Equations 4.14 with respect to x, the second with respect to y, the third with respect to z, and adding the result yield $(\lambda + 2G)\nabla^2 e = 0$, that is, the volume expansion of the element satisfies the equation $\nabla^2 e = 0$.

Also, expressing the elastic constants in Equations 4.14 in terms of Poisson's ratio, the equilibrium equations for zero-body forces become

$$\nabla^2 u + \frac{1}{1 - 2\upsilon}\frac{\partial e}{\partial x} = 0$$
$$\nabla^2 v + \frac{1}{1 - 2\upsilon}\frac{\partial e}{\partial y} = 0 \tag{4.15}$$
$$\nabla^2 w + \frac{1}{1 - 2\upsilon}\frac{\partial e}{\partial z} = 0$$

EXAMPLE 4.1

Find the change in volume of a parallelepiped element, 4 ft × 2 ft × 1.5 ft, when subjected to a hydrostatic pressure $p = 5000$ psi. Assume $E = 30 \times 10^6$ psi and $\nu = 0.3$.

Solution

Initial volume of the element is $V_0 = 48 \times 24 \times 18 = 20{,}736\ \text{in.}^3$

The bulk modulus is given by Equation 4.13: $K = (30 \times 10^6)/(3 \times 0.4) = 25 \times 10^6$ psi.

The unit volume change or dilatation is $e = -p/K = (-5 \times 10^3)/25 \times 10^6 = -0.2 \times 10^{-3}$.

Hence, the change in volume of the element is $\Delta V = e\,V_0 = -0.2 \times 10^{-3} \times 20{,}736 = -4.15\ \text{in.}^3$ (the minus indicates that the element experienced contraction).

EXAMPLE 4.2

Express the strain compatibility Equations 3.14 in terms of stresses.

Solution

The first step is to use the following form of Hooke's law for the normal strains:

$$\begin{aligned}\varepsilon_x &= \frac{1}{E}[(1+\upsilon)\sigma_x - \upsilon I_1]\\ \varepsilon_y &= \frac{1}{E}[(1+\upsilon)\sigma_y - \upsilon I_1]\\ \varepsilon_z &= \frac{1}{E}[(1+\upsilon)\sigma_z - \upsilon I_1]\end{aligned} \tag{a}$$

where I_1 stands for the stress invariant $I_1 = \sigma_x + \sigma_y + \sigma_z$. Inserting the first two of the above equations in the first compatibility condition, Equations 3.14, results in

$$(1+\upsilon)\left(\frac{\partial^2\sigma_x}{\partial y^2} + \frac{\partial^2\sigma_y}{\partial x^2}\right) - \upsilon\left(\frac{\partial^2 I_1}{\partial x^2} + \frac{\partial^2 I_1}{\partial y^2}\right) = 2(1+\upsilon)\frac{\partial^2\tau_{xy}}{\partial x\partial y} \tag{b}$$

Next, by using equilibrium of the element, the term in the right-hand side of Equation (b) can be expressed by terms involving derivatives of normal stresses only. Thus, differentiating the first and second equations of Equation 2.5 with respect to x and y, respectively, and simplifying the algebra yields

$$2\frac{\partial^2\tau_{xy}}{\partial x\partial y} = -\frac{\partial^2\sigma_x}{\partial x^2} - \frac{\partial^2\sigma_y}{\partial y^2} - \frac{\partial}{\partial z}\left(\frac{\partial\tau_{xz}}{\partial x} + \frac{\partial\tau_{yz}}{\partial y}\right) \tag{c}$$

Moreover, differentiating the third of Equations 2.5 with respect to z and inserting the result in relation (c) results in

$$2\frac{\partial^2\tau_{xy}}{\partial x\partial t} = \frac{\partial^2\sigma_z}{\partial z^2} - \frac{\partial^2\sigma_x}{\partial x^2} - \frac{\partial^2\sigma_y}{\partial y^2} \tag{d}$$

Inserting Equation (d) into (b), one obtains

$$\nabla^2 I_1 - (1+\upsilon)\nabla^2\sigma_z - \frac{\partial^2 I_1}{\partial z^2} = 0 \tag{e}$$

Following the identical procedure, the second and third equations of compatibility conditions provide the following:

$$\begin{aligned}\nabla^2 I_1 - (1+\upsilon)\nabla^2\sigma_x - \frac{\partial^2 I_1}{\partial x^2} &= 0\\ \nabla^2 I_1 - (1+\upsilon)\nabla^2\sigma_y - \frac{\partial^2 I_1}{\partial y^2} &= 0\end{aligned}$$

It is readily confirmed that by adding the last three equation, we have $\nabla^2 I_1 = 0$.

Hence, the compatibility conditions in terms of normal stresses are

$$\nabla^2\sigma_x + \frac{1}{1+\upsilon}\frac{\partial^2 I_1}{\partial x^2} = 0$$
$$\nabla^2\sigma_y + \frac{1}{1+\upsilon}\frac{\partial^2 I_1}{\partial y^2} = 0 \qquad \text{(f)}$$
$$\nabla^2\sigma_z + \frac{1}{1+\upsilon}\frac{\partial^2 I_1}{\partial z^2} = 0$$

A similar procedure can be used to obtain the remaining three compatibility conditions in terms of stresses. Inserting Hooke's law in the fourth equation of Equation 3.14 results in the following relation:

$$\frac{\partial^2}{\partial y \partial z}\left(I_1 - \sigma_y - \sigma_z - \frac{\upsilon I_1}{1+\upsilon}\right) + \frac{\partial^2 \tau_{yz}}{\partial x^2} = \frac{\partial^2 \tau_{zx}}{\partial x \partial y} + \frac{\partial^2 \tau_{xy}}{\partial x \partial z} \qquad \text{(g)}$$

Using equilibrium of the element, Equation 2.5, to replace the shear stress terms in the right-hand side of Equation (g) by terms containing stresses similar to those in the left-hand side of Equation (g), and simplifying the algebra results in the required compatibility condition in terms of stresses only. Identical procedure can be used in the remaining equations of Equation 3.14. The summary of the results is

$$\nabla^2\tau_{yz} = \frac{1}{1+\upsilon}\frac{\partial^2 I_1}{\partial y \partial z}$$
$$\nabla^2\tau_{zx} = \frac{1}{1+\upsilon}\frac{\partial^2 I_1}{\partial z \partial x} \qquad \text{(h)}$$
$$\nabla^2\tau_{xy} = \frac{1}{1+\upsilon}\frac{\partial^2 I_1}{\partial x \partial y}$$

Equations (f) and (h) are the required compatibility conditions in terms of stresses. They are commonly referred in the literature as the Michell–Beltrami compatibility equations.

4.2 Strain Energy

This section deals with the strain energy stored in deformable bodies. When external loads are applied to structures, the work done by internal forces is stored in the form of strain energy. The total strain energy stored in the element is denoted by U(x, y, z). The strain energy density (strain energy per unit volume) is denoted by $u_0(x, y, z)$. When a material is stressed by external loads to the elastic limit, the mechanical work is stored as strain energy, upon unloading the work is consumed to recover the original shape.

In order to determine the total strain energy stored in a body, the strain energy density is integrated over the initial volume of the body

$$U = \int_V u_0(x, y, z)\,dxdydz \qquad (4.16)$$

To illustrate the use of Equation 4.16, consider an element in the shape of a rectangular prism of sides dx, dy, and dz subjected to normal stresses, σ on side AC and $\sigma + (\partial\sigma/\partial x)$ dx across side BD, as in Figure 4.4. The corresponding displacements of the sides are u

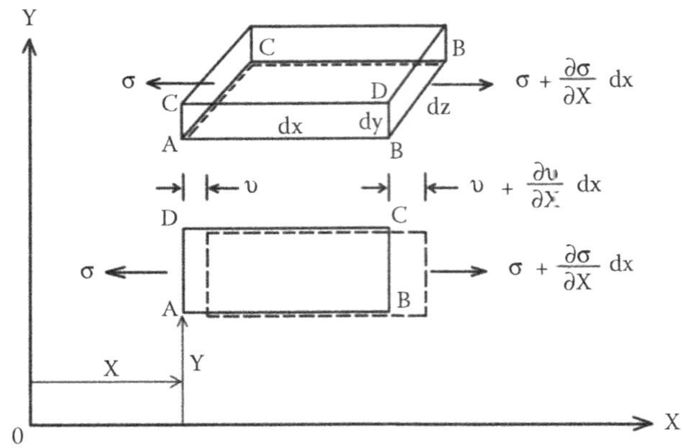

FIGURE 4.4
Element subject to normal stress.

and $u + (\partial u/\partial x)dx = u + \varepsilon dx$. Hence, the net work done by opposite forces through displacement u is

$$dW = dU = \int_0^{\varepsilon} \sigma d\left(\frac{\partial u}{\partial x}\right) dydz = \int_0^{\varepsilon} (\sigma d\varepsilon)\, dxdydz$$

It follows that the strain energy density is

$$u_0 = \int_0^{\varepsilon} \sigma\, d\varepsilon = E\int_0^{\varepsilon} \varepsilon\, d\varepsilon = \frac{1}{2}E\varepsilon^2 = \frac{1}{2}\sigma\varepsilon = \frac{\sigma^2}{2E} \tag{4.17}$$

The strain energy density is the area under the stress–strain curve. Linear elastic behavior is shown in Figure 4.5a and nonlinear elastic behavior is shown in Figure 4.5b. The area above the stress–strain curve is the *complementary energy density* and is denoted by u_0^*. Thus,

$$u_0^* = \int_0^{\sigma} \varepsilon\, d\sigma = \frac{\sigma^2}{2E} = \frac{E\varepsilon^2}{2} \tag{4.18}$$

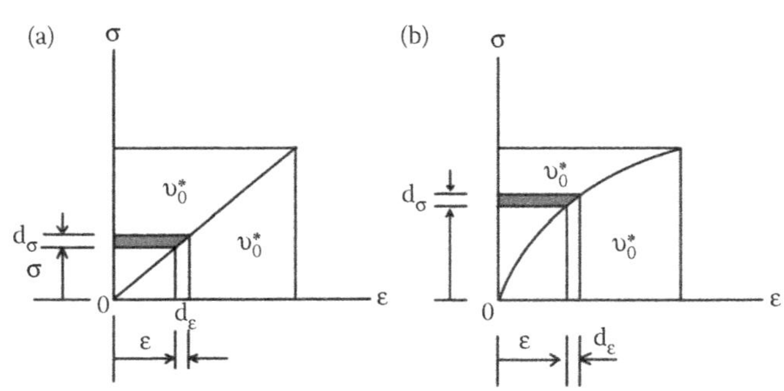

FIGURE 4.5
Strain energy due to universal stress, (a) linear elastic material and (b) nonlinear elastic material.

For linear elastic material $u_0 = u_0^*$, for nonlinear elastic material they are not equal, however, the sum in both cases is $u + u_0 = \sigma\varepsilon$. In terms of units, the strain energy density is expressed in unit of work per unit volume, in the customary US units, the unit is inch–pound per cubic inch (in.–lb/in.^{3}) or psi while in the SI system it is Joules per cubic meter (J/m^3) or Pascals.

The ability of the material to store energy without permanent deformation is a measure of its resilience. Thus, the *modulus of resilience* is defined as the stored elastic strain energy density at yield. Similarly, the area under the stress–strain curve up to the point of rupture is referred to as the *modulus of toughness* of the material.

Similarly, if the element is subjected to shear stresses only, then the strain energy density in the xy-plane is

$$u_0 = \int_0^{\gamma_{xy}} \tau_{xy}\, d\gamma_{xy} = \frac{G}{2}\gamma_{xy}^2 = \frac{\tau_{xy}^2}{2G} \tag{4.19}$$

When all six components of stress act simultaneously on the element, the strain energy density and the complementary energy density are obtained by adding the contributions of all stresses, that is,

$$\begin{aligned} u_0(\varepsilon_{ij}) &= \int_0^{\varepsilon_{ij}} \sigma_{ij}\, d\varepsilon_{ij} \\ u_0*(\sigma_{ij}) &= \int_0^{\sigma_{ij}} \varepsilon_{ij}\, d\sigma_{ij} \end{aligned} \qquad (i, j = x, y, z) \tag{4.20}$$

It is evident that the sum of the two strain energies in Equation 4.20 is equal to $\sigma_{ij}\,\varepsilon_{ij}$. Also, the following relations hold:

$$\frac{\partial u_0(\varepsilon_{ij})}{\partial \varepsilon_{ij}} = \sigma_{ij} \quad \text{and} \quad \frac{\partial u_0*(\sigma_{ij})}{\partial \sigma_{ij}} = \varepsilon_{ij} \tag{4.21}$$

Relation (4.21) implies that the elastic stresses are gradients of the strain energy density while the strains are gradients of the complementary energy density. In other words, the strain energy and complementary strain energy densities are potential function.

For linear elastic isotropic element subjected to three dimensions state of stress, the strain energy density is

$$u_0 = \frac{1}{2}(\sigma_x\varepsilon_x + \sigma_y\varepsilon_y + \sigma_z\varepsilon_z + \tau_{xy}\gamma_{xy} + \tau_{yz}\gamma_{yz} + \tau_{zx}\gamma_{zx}) \tag{4.22a}$$

Using Hooke's law (Equation 4.5), the strain energy density in terms of stresses only is

$$u_0 = \frac{1}{2E}\left(\sigma_x^2 + \sigma_y^2 + \sigma_z^2\right) - \frac{\upsilon}{E}(\sigma_x\sigma_y + \sigma_y\sigma_z + \sigma_z\sigma_x) + \frac{1}{2G}\left(\tau_{xy}^2 + \tau_{yz}^2 + \tau_{zx}^2\right) \tag{4.22b}$$

And using Equations 4.10, the strain energy density in terms of strains only is

$$u_0 = \frac{1}{2}\left[\lambda e^2 + 2\mu\left(\varepsilon_x^2 + \varepsilon_y^2 + \varepsilon_z^2\right) + \mu\left(\gamma_{xy}^2 + \gamma_{yz}^2 + \gamma_{zx}^2\right)\right] \tag{4.22c}$$

Note that Equations 4.22b and 4.22c give the following relations:

$$\frac{\partial u_0}{\partial \sigma_x} = \frac{1}{2E}(2\sigma_x) - \frac{\upsilon}{E}(\sigma_y + \sigma_z) = \varepsilon_x$$

$$\frac{\partial u_0}{\partial \varepsilon_x} = \frac{1}{2}(2\lambda e + 4\mu\varepsilon_x) = \sigma_x$$

Similar relations also exist for other components of u_0 as indicated by Equation 4.21. Such relations enable us to determine the number of independent constants in the generalized Hooke's law for anisotropic homogeneous elastic material (Equation 4.11). The matrix in Equation 4.11 contains 36 terms, which represent the material constants between the stresses and strains, however, it turns out that utmost only 21 constants are independent. To confirm this fact, we note that the first row of the matrix is

$$\sigma_x = \frac{\partial u_0}{\partial \varepsilon_x} = C_{11}\varepsilon_x + C_{12}\varepsilon_y + C_{13}\varepsilon_z + C_{14}\gamma_{xy} + C_{15}\gamma_{yz} + C_{16}\gamma_{zx} \tag{4.23a}$$

which gives $(\partial^2 u_0/\partial\varepsilon_x\partial\varepsilon_y) = C_{12}$.

The second row of the matrix, that is,

$$\sigma_y = \frac{\partial u_0}{\partial \varepsilon_y} = C_{21}\varepsilon_x + C_{22}\varepsilon_y + C_{23}\varepsilon_z + C_{24}\gamma_{xy} + C_{25}\gamma_{yz} + C_{26}\gamma_{zx} \tag{4.23b}$$

and it yields $(\partial^2 u_0/\partial\varepsilon_x\partial\varepsilon_y) = C_{21}$.

Hence, $C_{12} = C_{21}$. Similar procedure can be used to show that $C_{ij} = C_{ji}$, $i, j = 1, 2, \ldots, 6$, that is the matrix is symmetric and the number of independent constants are 6 in the first row, 5 in the second row,..., and 1 in the sixth row. Hence, the number of independent constants for anisotropic elastic material in Equation 4.11 is $6 + 5 + 4 + 3 + 2 + 1 = 21$. Of course, for isotopic elastic material, the number of independent constants is 2.

EXAMPLE 4.3

The stress–strain curves of certain ductile materials are as follows:

a. $\sigma = A\,\varepsilon^n$
b. $\sigma = \sigma_0\,(1 + A\,\varepsilon^n)$

where A and n are constants and σ_0 is the yield stress. Find the ratio u_0^*/u_0.

Solution

a. Using Equations 4.17 and 4.18, we have

$$u_0 = \int_0^\varepsilon \sigma\, d\varepsilon = A\int_0^\varepsilon \varepsilon^n\, d\varepsilon = \frac{\sigma\varepsilon}{n+1}$$

$$u_0^* = \int_0^\sigma \varepsilon\, d\sigma = An\int_0^\sigma \varepsilon^n\, d\varepsilon = \frac{n\sigma\varepsilon}{n+1}$$

Hence, the ratio $u_0^*/u_0 = n$, which is constant (n is known as the strain hardening index of the material, see Section 4.3).

Note that an alternative method of finding the complementary strain energy density is to write $u_0^* = \sigma\varepsilon - u_0$.

b. In this case, we have

$$u_0 = \int_0^{\varepsilon} \sigma_0(1 + A\varepsilon^n)d\varepsilon = \sigma_0\varepsilon\left(1 + \frac{A\varepsilon^n}{n+1}\right)$$

And for the complementary energy, it is easier to use the total area under the stress–strain curve minus u_0, or

$$u_0^* = \sigma_0(1 + A\varepsilon^n)\varepsilon - \sigma_0\varepsilon\left(1 + \frac{A\varepsilon^n}{n+1}\right) = \frac{nA}{n+1}\sigma_0\varepsilon^{n+1}$$

Hence, the ratio is

$$\frac{u_0^*}{u_0} = \frac{nA\varepsilon^n}{n + 1 + A\varepsilon^n}$$

which varies with the strain and material constants A and n.

4.2.1 Dilatational and Distortional Strain Energy Densities

In Section 2.7, it is stated that any state of stress σ_{ij} can be decomposed into a hydrostatic stress tensor consisting of diagonal terms $\sigma_m = (1/3)\ (\sigma_x + \sigma_y + \sigma_z)$ and a deviatoric stress tensor, $S_{ij} = \sigma_{ij} - \sigma_m$ (Equation 2.40). The hydrostatic stress (also known as the volumetric or dilatational stress) produces changes in the element dimensions without distorting its shape while the deviatoric or distortional stresses produce distortion in the shape of the element without change in its volume. Similarly, the strain energy density consists of two parts, one part measures the change in volume of the element without changing its shape, and the other measures the distortion in shape of the element without changing its shape. Thus, $u_0 = u_v + u_d$.

The dilatational strain energy density responsible for changes in the volume of the element is given by

$$u_v = \frac{3}{2}\sigma_m\varepsilon_m = \frac{\sigma_m^2}{2K} = \frac{I_1^2}{18K} \tag{4.24}$$

where the mean strain is $\varepsilon_m = e/3 = 1/3(\varepsilon_x + \varepsilon_y + \varepsilon_z)$, e is the dilatation, K is the bulk modulus (Equations 4.13) and I_1 is the first stress invariant (Equation 2.30a).

The distortional strain energy density, u_d, is obtained by subtracting the expression in Equation 4.24 from that in Equation 4.22b. After simplifying the algebra, the result is

$$u_d = u_0 - u_v = \frac{1}{12G}\left[(\sigma_x - \sigma_y)^2 + (\sigma_y - \sigma_z)^2 + (\sigma_z - \sigma_x)^2 + 6\left(\tau_{xy}^2 + \tau_{yz}^2 + \tau_{zx}^2\right)\right] \tag{4.25a}$$

Making use of Equations 2.42 and 2.46b, the distortional strain energy density is also equivalent to

$$u_d = \frac{J_2}{2G} = \frac{3\tau_{oct}^2}{4G} \tag{4.25b}$$

Here, J_2 is the second stress deviator invariant (Equation 2.42) and τ_{oct} is the octahedral shearing stress (Equation 2.46b).

The total strain energy density in the element is

$$u_0 = \frac{I_1^2}{18K} + \frac{J_2}{2G} = \frac{(1 - 2\upsilon)I_1^2}{6E} + \frac{(1 + \upsilon)J_2}{2E} \tag{4.26}$$

For isotropic linear elastic behavior, $u_0^* = u_0$. Also, since I_1^2 and J_2 are positive quantities, the strain energy and complementary strain energy densities are positive definite quadratic functions.

EXAMPLE 4.4

A solid aluminum sphere (radius = 10 in.) is subjected to a hydrostatic pressure p so that its volume is reduced by 0.04%. Compute the strain energy stored in the sphere. Assume elastic behavior, $E = 10 \times 10^6$ psi and $\nu = \frac{1}{3}$.

Solution

In this example,

$$\sigma_x = \sigma_y = \sigma_z = -p; \quad \tau_{xy} = \tau_{yz} = \tau_{xz} = 0$$

From Equation 4.7, the dilatation $e = [(1-2\upsilon)(-3p)]/E = -(p/E)$ and the change in volume is $\Delta V/V = -0.0004$. Equating the two quantities gives $p = 10 \times 10^6 \times 0.0004 = 4000$ psi.

The distortional strain energy, $u_d = 0$, and the dilatational energy is computed from Equations 4.25 $u_v = (-3p)^2/18\,K$.

Substituting for $K = E/[3(1-2\upsilon)]$ and p, it follows that $u_v = (-12{,}000)^2/18(10^7) = 0.8$ psi.

Hence, the total strain energy stored in the sphere is $U = u_v$ x volume of sphere $= (0.8\ \text{psi}) \times \left[\frac{4}{3}\pi(10)^3 \text{in.}^3\right] = 3351$ lb-in.

EXAMPLE 4.5

The complementary strain energy density of a nonlinear elastic material is given by the function

$$u_0^* = aI_1^2 + bJ_2^2$$

where a, b are constants, I_1 and J_2 are the usual stress invariants. The strain–stress relation of the material in the uniaxial tension test is represented by the Ramberg–Osgood equation

$$10^2\varepsilon = \sigma/10 + (2/3)(\sigma/10)^3, \quad (\sigma \text{ in units of ksi})$$

Find numerical values of the constants a and b.

Solution

The complementary strain energy density in terms of principal stresses is

$$u_0^* = a(\sigma_1 + \sigma_2 + \sigma_3)^2 + (b/36)[(\sigma_1 - \sigma_2)^2 + (\sigma_2 - \sigma_3)^2 + (\sigma_3 - \sigma_1)^2]^2 \tag{a}$$

Differentiating Equation (a) with respect to σ_1 and noting that for uniaxial loading, $\sigma_2 = \sigma_3 = 0$, Equation 4.22b yields

$$\begin{aligned} \varepsilon_1 &= 2a\sigma_1 + (b/36)(2)\left(\sigma_1^2 + 0 + \sigma_1^2\right)(2\sigma_1 + 0 + 2\sigma_1) \\ &= 2a\sigma_1 + (4/9)b\sigma_1^3 \end{aligned} \tag{b}$$

Comparing Equation (b) with the given tension stress test data, we have

$a = 0.5 \times 10^{-3}$

$b = 0.15 \times 10^{-4}$

4.2.2 Impact Loading

The concept of strain energy density in conjunction with conservation of energy can be used to determine the stresses induced by suddenly applied loads or impact loads. Impact load is an example of dynamic loading, which varies with time, and in general, involves elaborate analysis to determine its response. An obvious example of impact loading is an object striking a structure. For low-velocity impact loading, which causes mainly elastic behavior, simple energy balance calculations give reasonable result for the stresses induced by the loads.

Consider the problem illustrated in Figure 4.6. A weight, W, is dropped from height, h, onto a column made of ductile material, length, L, and cross-sectional area, A. It is required to determine the normal stress produced by the load assuming elastic behavior. As the weight falls, its potential energy of magnitude, W h, is converted into kinetic energy of magnitude (1/2) (W/g) V^2, where the velocity $V = (2gh)^{1/2}$, and g is the acceleration due to gravity ($g = 9.81\ m/s^2$ in SI units or $32.2\ ft/s^2$ in US customary units). At the instant of impact, the potential energy is zero and the kinetic energy is transformed into other energies including elastic energy absorbed by the column, energy causing localized plastic strains, and other energies dissipated into sound and heat. Neglecting all energies except the strain energy absorbed by the member, conservation of energy at impact implies that

$$W(h + \Delta L) = u_0(AL) = \frac{\sigma^2}{2E}(AL) \tag{4.27a}$$

where u_0 is the strain energy density for uniaxial load, σ is the axial stress induced in the column, and the contraction of the column is $\Delta L = \varepsilon L = (\sigma/E)L$, E is the modulus of elasticity.

Simplifying Equation 4.27a results in

$$\sigma^2 - 2\frac{W}{A}\sigma - \frac{2WhE}{AL} = 0 \tag{4.27b}$$

The roots of Equation 4.27b are

$$\sigma = \frac{W}{A} \pm \sqrt{\left(\frac{W}{A}\right)^2 + \frac{2EhW}{AL}} \tag{4.28}$$

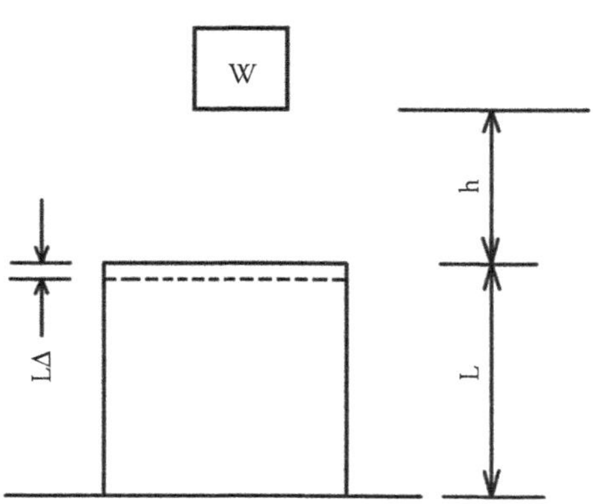

FIGURE 4.6
Impact loading.

The positive sign in Equation 4.28 provides the compressive stress at maximum contraction of the column while the negative sign gives the stress at the end of rebound of the top surface.

When $h = 0$, that is, load W is suddenly applied to the surface, $\sigma = 2(W/A)$, which is twice the static stress induced by the load gradually placed on the column. This simple example illustrates the difference of the response of structural members to static and dynamic loads.

4.3 Plastic Behavior

The stress–strain relations derived previously are applicable to strains before the material is stressed beyond its elastic limit. They are based on stress–strain data obtained from a simple tension test using *engineering* stress as ordinate versus *engineering* strain as abscissa, as indicated in Figure 4.1. The same data are used to illustrate the plastic behavior of the material. However, before discussing a workable theory to model the plastic behavior of structural elements, the basic features of materials beyond the elastic or reversible range are summarized below:

1. Plastic deformation is irreversible, that is, permanent.
2. Magnitude of plastic strain is many times more than that of elastic strain.
3. Experimental evidence shows that the yield point and plastic flow of ductile material is unaffected by all-round pressure or hydrostatic stress (average of normal stresses). Stress deviator or reduced stress causes initiation of yield and subsequent plastic flow. Volume of the material remains constant during plastic flow (dilatation = 0). Some materials are found to suffer permanent volume change when subjected to pressure of extremely high magnitude [4.2,4.3]. In normal applications, the change in volume is neglected.
4. In order to simplify the modeling of plastic behavior of structural elements, the *"Deformation theory of plasticity"* is used. In this theory, which is essentially a nonlinear theory of elasticity with specific stress–strain relations, the loading and unloading paths of the stress–strain curve are the same.
5. In general, the stresses and strains in the plastic range depend not only on the current state of stress but also on the entire history of loading and deformation, that is, it is *path-dependent*, and there is no one-to-one correspondence between stress and strain in plastically deformed element except when the deformation theory of plasticity is used.

Next, the standard tensile test of a specimen of the material introduced in Section 4.1 is discussed in some detail, especially beyond the proportional limit.

4.3.1 Static Tension Test

The stress versus strain data obtained from a static tensile test of a specimen of ductile material is shown in Figure 4.1. Initially, the behavior is elastic represented by straight line 0A, the slope of this line is the modulus of elasticity E. Point A defines the elastic limit (yield point) beyond which irreversible deformation occurs. The yield stress is denoted by σ_0. Immediately after point A there is a small distance of constant stress accompanied by permanent plastic deformation, followed by strain-hardening region where each increment of load is accompanied by increase in strain. The slope of the stress–strain curve decreases progressively until the load reaches the ultimate value, the highest point on the stress–strain curve where the slope = 0 and the stress = σ_u. Afterward, the load quickly decreases due to necking or thinning of the cross-sectional area and the specimen breaks apart at the rupture or failure stress below the ultimate load. If one plots true stress (load divided by instantaneous cross-sectional area) instead of engineering stress (load divided by original area) versus engineering strain, the dotted line in Figure 4.1 is generated. Unloading from any point in the plastic region, say point B, the path is a straight line BC parallel to line 0A. At zero stress level, there is a residual permanent plastic strain ε^P represented by segment 0C. Note that the total strain at zero stress level corresponding to point B consists of elastic component CD (recoverable), plus plastic component 0C (permanent). Upon reloading the specimen from point C, line CB is approximately retraced, the stress at point B is the *subsequent yield stress*, which is higher than the initial yield stress σ_0. Beyond point B, further plastic deformation is induced, the material is referred to as strain-hardened or work-hardened, and the curve follows approximately similar path until failure at the ultimate stress.

The longitudinal extension in the uniaxial tension test is accompanied by a shortening in the lateral direction, the ratio of lateral strain to longitudinal strain is known as the *contraction ratio,* usually denoted by η. For an isotropic elastic material, $\eta = \nu$, Poisson' ratio. Beyond the elastic limit (yield point), the plastic part of the lateral strain is equal to ½ of the longitudinal plastic strain, that is, $-1/2\ d\varepsilon$, and the contraction ratio is $\eta = d\varepsilon^{\ni}/d\varepsilon$. The total lateral strain increment consists of elastic plus plastic parts (see Problem P4.12).

Bauschinger effect: If one performs a static compression test on a new specimen of the same ductile material, an almost identical stress–strain curve to that in tension is obtained. The initial yield stress in compression has the same magnitude as that in tension, that is, $-\sigma_0$, and further compressive loading will cause the material to strain harden as in tensile loading.

Suppose that the previous specimen, which has been loaded in tension beyond the yield point and unloaded elastically from point B in the strain-hardening range to point C at zero stress level and residual plastic strain 0C, is loaded in compression starting from point C, the resulting stress–strain curve is quite different (see Figure 4.7). Initially, the path is a straight line of slope E until yield is observed at point A′. The magnitude of the yield stress at point A′ is σ_0', which is smaller than the initial yield stress in tension and much smaller than the subsequent yield stress at point B. Such phenomenon is known as Bauschinger effect [4.4]. It occurs in ductile metals whenever there is reversal of stress. Bauschinger effect can be reduced and removed completely by mild annealing of the metal. In normal practice of determining the plastic deformation of structural elements, it is usual to neglect Bauschinger effect and assume the yield stresses in compression and tension are equal.

Based on the above discussion, it is evident that the stress–strain relations of ductile material are nonlinear and determining the deformation of structural elements in the plastic range, in general, requires an incremental approach. The material behavior is also path-dependent and one needs to include the loading history in the analysis.

4.3.2 Basic Stress–Strain Relations

In this section, we consider an incremental approach to describe the plastic behavior of ductile materials. Figure 4.8a and b shows typical stress–strain curves of ductile metals.

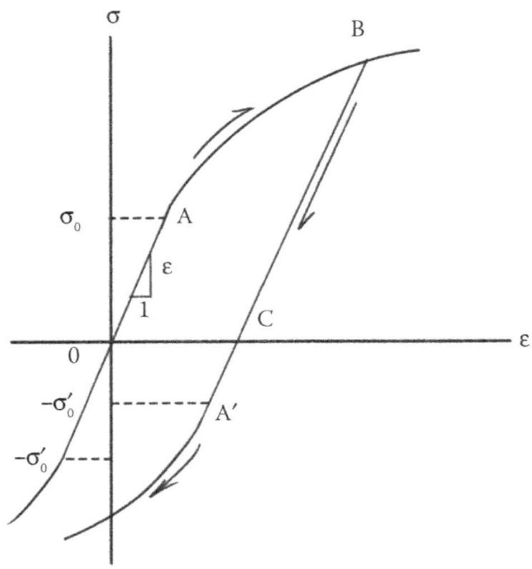

FIGURE 4.7
Reversed loading.

We consider a typical stage in the stress–strain curve and a small strain increment $d\varepsilon$ represented by curve CD in Figure 4.8a. The increment consists of an elastic strain increment $d\varepsilon^e$ and a plastic strain increment $d\varepsilon^p$, that is,

$$d\varepsilon = d\varepsilon^e + d\varepsilon^p \tag{a}$$

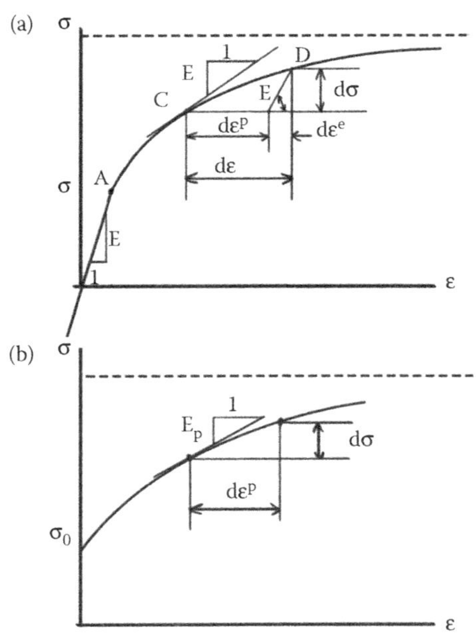

FIGURE 4.8
Stress and strain increments, (a) tangent modulus and (b) plastic modulus.

The corresponding stress increment in the interval is given by the relation

$$d\sigma = E_t d\varepsilon \tag{4.29}$$

where E_t is the tangent modulus, which varies during plastic deformation. In Figure 4.8a, the tangent modulus is the slope of the tangent to the stress–strain curve at point C, the initial point of the strain increment. Separating the plastic component of strain from the total increment, Figure 4.8b, the stress component is related to the plastic strain increment by the relation

$$d\sigma = E_p d\varepsilon^p \tag{4.30}$$

where E_p is the plastic modulus, which is the slope of the tangent to the stress–strain curve excluding the elastic part, as shown in Figure 4.8b.

The elastic component of the increment is related to the stress increment by the usual Hooke's law

$$d\sigma = E d\varepsilon^e \tag{4.31}$$

Substituting Equations 4.29 through 4.31 in relation (a), the following relation is obtained:

$$\frac{1}{E_t} = \frac{1}{E} + \frac{1}{E_p} \tag{4.32}$$

Solving for E_t and E_p, we have

$$E_t = \frac{EE_p}{E + E_p} \tag{4.33}$$

$$E_p = \frac{EE_t}{E - E_t}$$

The tangent and plastic moduli are functions of the deformation history, their magnitudes can be determined from the static tensile test data based on a given strain-hardening rule.

4.3.3 Models of Uniaxial Stress–Strain Relations

The uniaxial stress–strain relations of ductile materials are determined in the laboratory using the static tensile test. Best-fit equations are derived to represent the whole strain range covering elastic and strain-hardening behaviors. For some materials, the elastic part is neglected and empirical equations can be used to describe the strain-hardening part. In this section, few commonly used models are described.

Elastic–perfectly plastic material: In this model, the strain-hardening effect is neglected, plastic flow is assumed to begin at the yield stress σ_0, which remains constant as shown by the horizontal line in Figure 4.9a. The stress is as follows:

$$\begin{aligned} \sigma &= E\varepsilon, \quad \sigma < \sigma_0 \\ \sigma &= \sigma_0 \end{aligned} \tag{4.34}$$

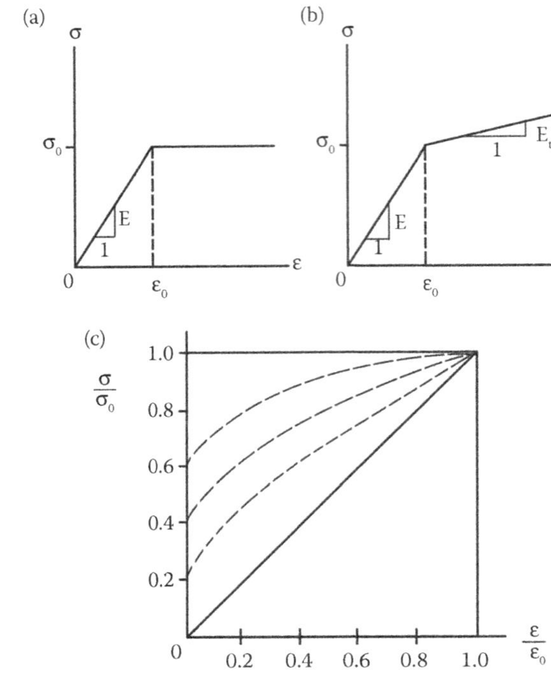

FIGURE 4.9
Idealized stress–strain curves, (a) elastic–perfect plastic, (b) elastic–linearly hardening, and (c) power law.

Here, E is the modulus of elasticity. In an element, the strain is elastic until the material yields, the strain is $\sigma_0/E = \varepsilon_0$, the stress remains constant at the yield level and the corresponding strain continues to grow until it is confined by the surrounding elastic material. The magnitude of the plastic strain can be determined from extent of the elastic material restricting the plastic flow.

Elastic–linear-hardening material: After the material yields, the strain hardening part is assumed to be a straight line with slope, E_t, the tangent modulus of the material. The stress–strain diagram consists of two straight lines, as shown in Figure 4.9b. The magnitude of the tangent modulus is usually a fraction of Young's modulus, $E_t = kE$, k is constant. The stress–strain relation is given by

$$\begin{aligned} \varepsilon &= \frac{\sigma}{E}, & \text{for } \sigma \leq \sigma_0 \\ \varepsilon &= \frac{\sigma}{E} + \frac{\sigma - \sigma_0}{E_t}, & \text{for } \sigma \geq \sigma_0 \end{aligned} \tag{4.35}$$

Power law: When the elastic behavior of a material is negligible, the material is basically rigid/plastic, and it is convenient to represent its stress–strain relation by Ludwik's power law [4.5]

$$\sigma = K\varepsilon^n \tag{4.36}$$

where K and n are constants, K is the coefficient of stress or strength, and n is the strain-hardening exponent. Typical values of n are between 0 and 0.5. This relation gives zero initial stress and infinite slope, as shown in Figure 4.9c. When $n = 0$, the relation represents a rigid-perfectly plastic material with yield stress $= K$, the initial slope increases with values of n and in the limit as n approaches unity, the relation reduces to Hooke's law with $K = E$.

A generalized power law can be obtained from Equation 4.36 by shifting the stress axis by a constant amount m as follows [4.6]:

$$\sigma = K(m + \varepsilon)^n \tag{4.37}$$

Here, K, m, and n are constants.

Ramberg–Osgood equation: This equation is a nonlinear stress–strain expression used to represent the entire test data of an elastic–plastic material [4.7]. It has the basic form

$$\varepsilon = \frac{\sigma}{E} + a\left(\frac{\sigma}{b}\right)^n \tag{4.38a}$$

where a, b, and n are material constants (n designates the strain-hardening component). If σ_0 and ε_0 are reference or nominal stress and strain, respectively, then using a as a nondimensional constant, Ramberg–Osgood equation in nondimensional form is

$$\frac{\varepsilon}{\varepsilon_0} = \frac{\sigma}{\sigma_0} + a\left(\frac{\sigma}{\sigma_0}\right)^n \tag{4.38b}$$

Instead of using the yield stress as reference, the flow stress, defined as half the yield stress plus the ultimate stress, that is, $\frac{1}{2}(\sigma + \sigma_u)$, is sometimes used. Note that at the nominal stress, σ_0, the plastic strain is a times the elastic strain.

The tangent modulus, Equation 4.29, is obtained by differentiating Equation 4.38b

$$\frac{1}{\varepsilon_0} = \frac{1}{\sigma_0}\frac{d\sigma}{d\varepsilon} + \frac{an}{\sigma_0}\left(\frac{\sigma}{\sigma_0}\right)^{n-1}\frac{d\sigma}{d\varepsilon}$$

Solving for $(d\sigma/d\varepsilon)$ and noting that $E = \sigma_0/\varepsilon_0$, the tangent modulus is

$$E_t = \frac{E}{1 + an\left(\frac{\sigma}{\sigma_0}\right)^{n-1}}$$

Similarly, the plastic modulus is determined from Equation 4.33

$$E_p = \frac{E}{an\left(\frac{\sigma}{\sigma_0}\right)^{n-1}}$$

The variation of σ/σ_0 versus $\varepsilon/\varepsilon_0$ for $a = 3/7$ and σ_0 is a nominal yield stress at which $E_s =$ 0.7 E, which reasonably represents common ductile materials [4.8], as shown in Figure 4.10. For nonhardening material, $(n = 00)$, the stress–strain curve degenerates into that of elastic–perfectly plastic material.

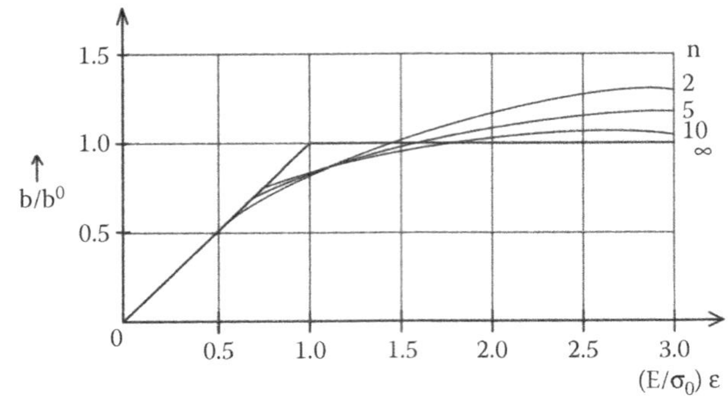

FIGURE 4.10
Ramberg–Osgood's stress–strain curve.

EXAMPLE 4.6

A member AB of length $L = a + b$ is fixed at both ends and subjected to an axial force P at section C distance, a, from end A ($a < b$), as seen in Figure 4.11. The member is made of an elastic–perfectly plastic material with yield stress σ_0 and yield strain ε_0. The axial force is increased from $P = 0$ until plastic flow occurs throughout the member at $P = P_p$, then unloaded to $P = 0$, and followed by reloading in the reversed direction until plastic flow occurs throughout the member at $P = P'_p$. Find

a. The maximum elastic force P_e and the fully plastic force P_p during loading.
b. The residual stress and strain when P_p is unloaded completely.

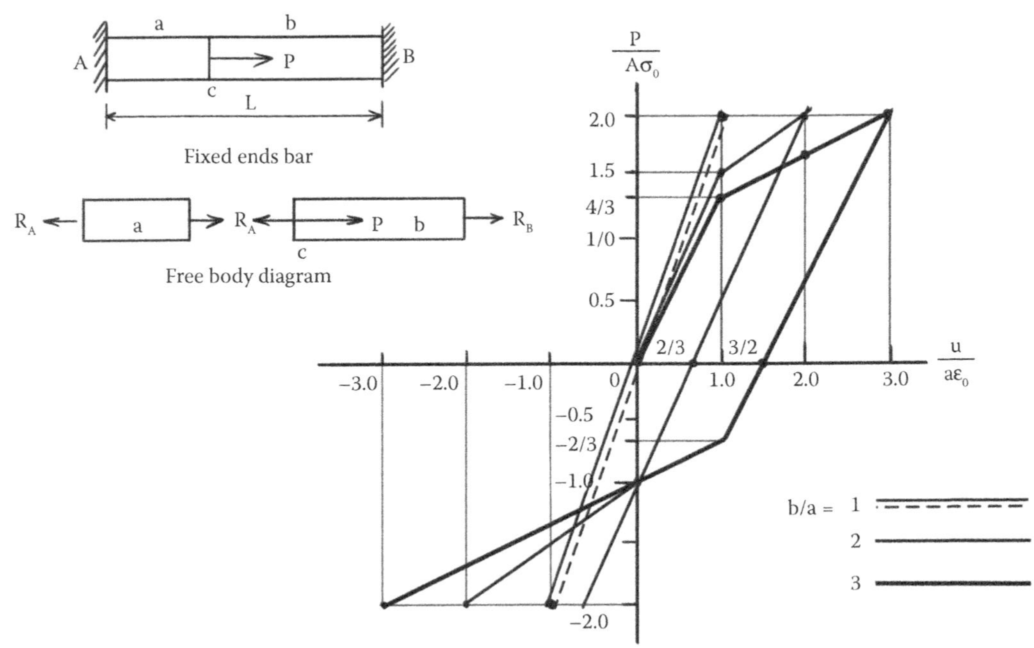

FIGURE 4.11
Load–displacement (u/ε_0) curve (Example 4.6).

c. The plastic limit load P'_p during reversed loading.
d. For $b/a = 1, 2$, and 3, sketch the curves $P/(\sigma_0 A)$ versus $u/(a\varepsilon_0)$ and $P/(\sigma_0 A)$ versus u/L, where A is the cross-sectional area and u is the axial displacement of section C of the member.
e. Find numerical values of the forces, stresses, and strains in parts (a) through (c) assuming $a = 10$ in., $b = 3a$, $E = 30 \times 10^3$ ksi, and $\sigma_0 = 30$ ksi.

Solution

The first step is to determine the equations of force equilibrium and compatibility of strain (geometry of deformation) in the member. A free body diagram of the forces is shown in the top left corner of Figure 4.11. We have the forces acting on the member, $R_A - R_B = P$. In terms of stresses

$$\sigma_a - \sigma_b = \frac{P}{A} \tag{a}$$

Because both ends of the member are fixed, the sum of horizontal displacements of both sections of the member is zero, $\delta_a + \delta_b = 0$. In terms of strain, we have

$$a\varepsilon_a + b\varepsilon_b = 0 \tag{b}$$

Equations (a) and (b) are valid through all stages of the deformation.

a. As P is increased from zero, the initial deformation is elastic, the strains are $\varepsilon_a = (\sigma_a/E)$ and $\varepsilon_b = (\sigma_b/E)$. Equations (a) and (b) provide

$$\sigma_a = \frac{bP}{LA}, \quad \sigma_b = -\frac{aP}{LA}$$
$$\varepsilon_a = \frac{bP}{ELA}, \quad \varepsilon_b = -\frac{aP}{ELA}$$

Since $b > a$, the magnitude of the stress in part (a) of the member is greater than that in part (b). It follows that part (a) yields first. The maximum elastic load that initiates yield is

$$\frac{P_e}{A} = \frac{L}{b}\sigma_0$$

The corresponding stress in part (b) is elastic, of magnitude $\sigma_b = -(a/b)\sigma_0$. The displacement of section C is $u = a\varepsilon_a = a\varepsilon_0$, and $(u/L) = (a/L)\varepsilon_0$. Note that part (a) of the member stretches by this amount while part (b) shrinks by the same. Also, when $a = b$, both parts of the member yields at the same time.

As P is increased beyond P_e, the stress in part (a) of the member remains at the yield level and the stress in part (b) increases until it yield in compression, $\sigma_b = -\sigma_0$. Using equilibrium equation, the plastic limiting load is

$$\frac{P_P}{A} = 2\sigma_0$$

The corresponding strains in the member are $\varepsilon_b = -\sigma_0/E = -\varepsilon_0$ and using Equation (b), $\varepsilon_a = (b/a)\varepsilon_0$.

It follows that the displacement of section C is $u = b\varepsilon_0$ and $(u/L) = (b/L)\varepsilon_0$.

b. Unloading from $P = P_p$ to $P = 0$

The unloading path is elastic and the residual stresses are obtained by adding to the existing stresses the increment in elastic stress caused by $-P_p = -2\sigma_0 A$, that is,

$$\sigma_a^R = \sigma_0 + \Delta\sigma_a = \sigma_0 + \frac{b}{L}(-2\sigma_0) = \left(1 - \frac{2b}{L}\right)\sigma_0$$

$$\sigma_b^R = -\sigma_0 + \Delta\sigma_b = -\sigma_0 - \frac{a}{L}(-2\sigma_0) = -\left(1 - \frac{2a}{L}\right)\sigma_0$$

The above expressions are identical. Hence, the residual stress in the member is

$$\sigma_a^R = \sigma_b^R = \left(1 - \frac{2b}{L}\right)\sigma_0$$

which is less than the yield stress. Because only elastic strains have been unloaded, the plastic strain in the member at $P = 0$ is identical to that at P_p. At P_p, the plastic strain in section (b) of the member is zero. Hence, $\varepsilon_b = -\sigma_0/E = -\varepsilon_0$, using the compatibility equation, we have

$$\varepsilon_a = -\frac{b}{a}\varepsilon_b = \frac{b}{a}\varepsilon_0$$

In order to determine the plastic strain corresponding to force P_p (and $P = 0$), the strain in section (a) above is separated into elastic and plastic components as follows:

$$\varepsilon_a = \varepsilon_0 + \left(\frac{b}{a} - 1\right)\varepsilon_0$$

Hence, the plastic strain at this stage of the deformation is

$$\varepsilon_a^p = \left(\frac{b}{a} - 1\right)\varepsilon_0$$

and the unloading path intersects the zero stress axis at a distance,

$$\frac{b}{a} - \frac{2b}{L} = \frac{b}{a}\left(1 - \frac{2a}{L}\right)$$

c. Reversed loading: denote the reversed axial force by P', initially the increments in the reversed loading are elastic. It follows that the stresses in the member are given by

$$\sigma_a = \sigma_a^R + \Delta\sigma_a = \left(1 - \frac{2b}{L}\right)\sigma_0 - \frac{bP'}{LA}$$

$$\sigma_b = \sigma_b^R + \Delta\sigma_b = \left(1 - \frac{2b}{L}\right)\sigma_0 + \frac{aP'}{LA}$$

Since the magnitude of the stress in part (a) of the member is still larger than that in part (b), part (a) yields first in compression. Hence, the load-initiating yield in reversed loading is

$$\frac{P'_e}{A} = -2\frac{a}{b}\sigma_0$$

The corresponding strain is obtained by adding the strain increment to the initial strain of the current deformation stage, that is,

$$\varepsilon_a = \varepsilon_a^p - \varepsilon_0 = \left(\frac{b}{a} - 2\right)\varepsilon_0$$

The corresponding displacement of section C is $u/(a\varepsilon_0) = (b/a) - 2$, or $u/L = (b/a - 2)(a/L)\varepsilon_0$.

As the force increases further, the stress in part (a) remains at the yield level and the elastic stress in part (b) carries the load till it yields in tension. The fully plastic force in reversed direction is obtained from equilibrium as

$$\frac{P_p'}{A} = -2\sigma_0$$

Since the reversed loading path from P_e' to P_p' is parallel to the loading path from P_e to P_p, it follows that the expressions of u at section C are $(u/a\varepsilon_0) = -(b/a)$ and $(u/L) = -(b/L)\varepsilon_0$.

d. The nondimensional load versus displacement curves are sketched (Figure 4.12) using $b/a = 1$, 2, and 3, and the following system of coordinates:

$$\left[\frac{P}{A\sigma_0}, \frac{u}{a\varepsilon_0}\right] = [0, 0] \rightarrow \left[\frac{L}{b}, 1\right] \rightarrow \left[2, \frac{b}{a}\right] \rightarrow \left[0, \frac{b}{a}\left(1 - \frac{2a}{L}\right)\right]$$
$$\rightarrow \left[-2\frac{a}{b}, \left(\frac{b}{a} - 2\right)\right] \rightarrow \left[-2, -\frac{b}{a}\right]$$

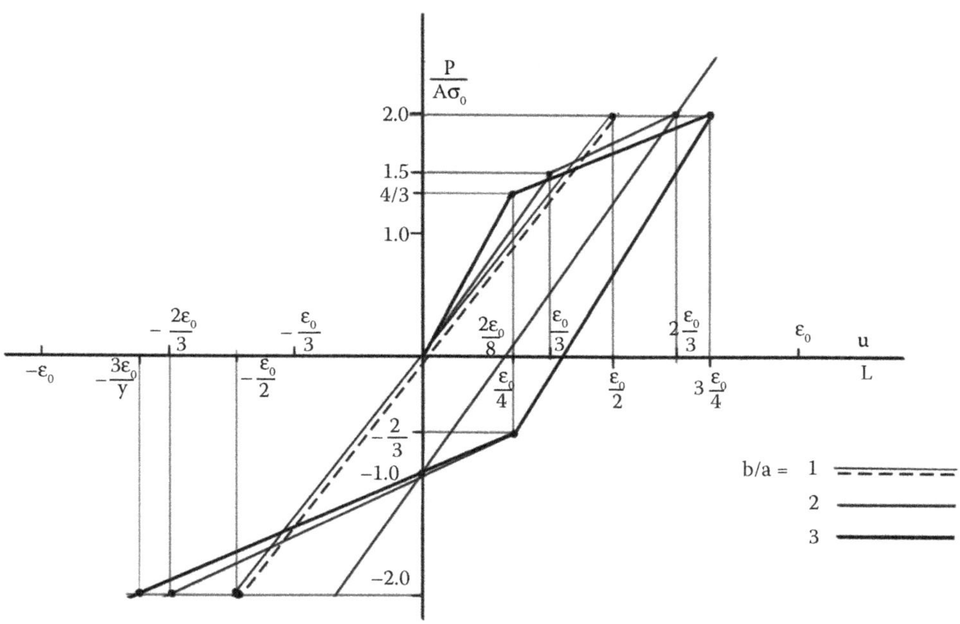

FIGURE 4.12
Load–displacement curve (u/L) (Example 4.6).

$$\left[\frac{P}{A\sigma_0}, \frac{u}{L}\right] = [0,0] \rightarrow \left[\frac{L}{b}, \frac{a}{L}\varepsilon_0\right] \rightarrow \left[2, \frac{b}{L}\varepsilon_0\right] \rightarrow \left[0, \left(1 - \frac{2a}{L}\right)\left(\frac{b}{L}\varepsilon_0\right)\right]$$

$$\rightarrow \left[-\frac{2a}{b}, \left(\frac{b}{a} - 2\right)\frac{a}{L}\varepsilon_0\right] \rightarrow \left[-2, \frac{b}{L}\varepsilon_0\right]$$

e. For $b/a = 3$, $a = 10$ in., $b = 30$ in., $L = 40$ in., yield stress = 30 ksi, and $E = 30 \times 10^3$ ksi

Force initiating yield: $(P_e/A) = (4/3) \times 30 = 40$ ksi

The corresponding axial displacement of section C = 10 in. × [(30 ksi)/(30 × 10^3 ksi)] = 0.01 in.

Fully plastic force: $(P_p/A) = 2 \times 30$ ksi = 60 ksi

The corresponding displacement of section C = 30 in. × [(30 ksi)/(30 × 10^3 ksi)] = 0.03 in.

Residual stress:

$$\sigma_a^R = \sigma_b^R = \left(1 - \frac{2 \times 30}{40}\right) \times 30\text{ ksi} = -15\text{ ksi}$$

Plastic strain = (3 − 1) × 0.001 = 0.002

Force initiating yield in reversed loading: $(P'_e/A) = -2(1/3) \times 30$ ksi = −20 ksi

Fully plastic force in reversed loading: $(P'_p/A) = -60$ ksi

EXAMPLE 4.7

Consider the fixed ends member of the previous example. Assume that the material is elastic–linear strain hardening (Figure 4.9b) with the following stress–strain relations:

$$\sigma = E\varepsilon, \qquad \sigma < \sigma_0$$
$$\sigma = \sigma_0 + E_t(\varepsilon - \varepsilon_0), \quad \sigma > \sigma_0$$

where σ_0 is the yield stress, ε_0 is the yield strain, and E_t is the tangent modulus, $E_t = 0.2$ E.

The force P is increased from $P = 0$ to $P = P_1$, which initiate yield in part (b) of the member, increased further to $P = P_2 = (\beta\sigma_0 A)$ where β is an arbitrary constant, $\beta > 1$, A is the cross-sectional area of the member, and then the force is unloaded to $P = 0$. Assume that the magnitude of the yield stress in compression is the same as that in tension.

Find the residual stress and strains and the plastic strains in the member after unloading. Also, for $\beta = 4$ and $b/a = 1, 2, 3$, determine the numerical values of these parameters in terms of yield stress and strain of the material.

Solution

The basic equations of equilibrium and compatibility of strain, which are valid at all stages of the deformation, are those derived in Example 4.6, namely

$$\sigma_a - \sigma_b = \frac{P}{A}$$
$$a\varepsilon_a + b\varepsilon_b = 0$$

As P is increased from zero, the deformation in the member is elastic until yield is initiated in part (a) of the member with force $P_e/A = (L/b)\,\sigma_0$, and the corresponding stress in part (b) of the member is $(-a/b)\,\sigma_0$. Further increase in the force is absorbed by part (b) of the member until it yields in compression at $P = P_1$. The force P_1 is determined by finding the increments in stress in both sections of the member.

For an increment in force ΔP, the increments in the stresses are $\Delta\sigma_a = E_t\Delta\varepsilon_a = 0.2E\varepsilon_a$, and $\Delta\sigma_b = E\Delta\varepsilon_b$. Inserting these values in the basic equations gives

$$\Delta\sigma_a = \frac{b}{5a+b}\frac{\Delta P}{A}$$

$$\Delta\sigma_b = -\frac{5a}{5a+b}\frac{\Delta P}{A}$$

Since the increment in P is equal to $P_1 - P$, the stresses in the member are

$$\sigma_a = \sigma_0 + \frac{b}{5a+b}\frac{P_1 - P_e}{A}$$
$$\sigma_b = -\frac{a}{b}\sigma_0 - \frac{5a}{5a+b}\frac{P_1 - P_e}{A} \tag{a}$$

Expressions (a) are valid until P_1 reaches a level that causes the stress in part (b) to reach $\sigma_b = -\sigma_0$. Inserting this stress and the value of P_e in the second equation of Equations (a), the expression for P_1 is

$$\frac{P_1}{A} = \frac{9a+b}{5a}\sigma_0 \tag{b}$$

and the stress in part (a) of the member becomes $\sigma_a = [(4a+b)/5a]\sigma_0$.

Next, an increase in the force from $P = P_1$ to $P = P_2 = (\beta\ \sigma_0\ A)$ causes plastic deformation in both parts of the member, implying that the increments in stresses at this stage of the deformation are functions of the tangent modulus, $\Delta\sigma_a = E_t\Delta\varepsilon_a = 0.2E\Delta\varepsilon_a$ and $\Delta\sigma_b = E_t\Delta\varepsilon_b = 0.2E\Delta\varepsilon_b$. Inserting the stress increments in the basic equations of equilibrium and compatibility of strains, solving the resulting equations simultaneously, the result is

$$\Delta\sigma_a = \frac{b}{L}\frac{\Delta P}{A}$$

$$\Delta\sigma_b = -\frac{a}{L}\frac{\Delta P}{A}$$

where the increment in force is $P_2 - P_1$. It follows that the stresses in the member when $P = P_2$ are

$$\sigma_a = \frac{4a+b}{5a}\sigma_0 + \frac{b}{L}\left(\beta - \frac{9a+b}{5a}\right)\sigma_0$$

$$\sigma_b = -\sigma_0 - \frac{a}{L}\left(\beta - \frac{9a+b}{5a}\right)\sigma_0$$

Upon simplification, the stresses become

$$\sigma_0 = \frac{1}{5L}(4a - 4b + 5b\beta)\sigma_0$$
$$\sigma_b = \frac{1}{5L}(4a - 4b - 5a\beta)\sigma_0 \tag{c}$$

The plastic strains at stage P_2 can be obtained by subtracting yield stress in each section of the bar from expressions (c) and dividing the result by the plastic modulus, E_p. Using

Equation 4.33, $E_p = 0.25$ E, and we have

$$\varepsilon_a^p = \frac{\frac{1}{5L}(4a - 4b + 5b\beta)\sigma_0 - \sigma_0}{0.25E} = \frac{4}{5L}(-a - 9b + 5b\beta)\varepsilon_0$$
$$\varepsilon_b^p = \frac{\frac{1}{5L}(4a - 4b - 5a\beta)\sigma_0 + \sigma_0}{0.25E} = \frac{4}{5L}(9a + b - 5a\beta)\varepsilon_0 \quad \text{(d)}$$

The final stage of deformation is to unload the force from $P = P_2$ to $P = 0$. During the unloading process, the deformation is elastic so that corresponding to an increment of $\Delta P = -(\beta\sigma_0 A)$, the increments in stresses in both parts of the member are

$$\Delta\sigma_a = -\frac{b\beta}{L}\sigma_0$$
$$\Delta\sigma_b = -\frac{a\beta}{L}\sigma_0$$

Adding these increments to the stresses in Equation (c), the residual stresses in parts (a) and (b) of the member are found to be equal. Hence, the residual stress in the member is

$$\sigma_a^{res} = \sigma_b^{res} = \frac{4(a - b)}{5(a + b)}\sigma_0 \quad \text{(e)}$$

which is independent of the parameter β.

The residual strains consist of two components, elastic and plastic, as follows:

$$\varepsilon_a^{res} = \frac{\sigma_a^{res}}{E} + \varepsilon_a^p = \frac{4(a - b)}{5L}\varepsilon_0 + \frac{4}{5L}(-a - 9b + 5b\beta)\varepsilon_0 = \frac{4b}{L}(\beta - 2)\varepsilon_0$$
$$\varepsilon_b^{res} = \frac{\sigma_b^{res}}{E} + \varepsilon_b^p = \frac{4(a - b)}{5L}\varepsilon_0 + \frac{4}{5L}(9a + b - 5a\beta)\varepsilon_0 = \frac{4a}{L}(2 - \beta)\varepsilon_0 \quad \text{(f)}$$

For $\beta = 4$ and $b/a = 1, 2,$ and 3, numerical values of the parameters are listed in the table below:

b/a	$P_1/A\sigma$	σ_{res}/σ_0	$\varepsilon_a^{res}/\varepsilon_0$	$\varepsilon_b^{res}/\varepsilon_0$	$\varepsilon_a^p/\varepsilon_0 Z$	$\varepsilon_b^p/\varepsilon_0$
1	2	0	4	–4	4	–4
2	2.2	–4/15	16/3	–8/3	5.6	–2.4
3	2.4	–6/15	18/3	–2	6.4	–1.6

EXAMPLE 4.8

A three-bar truss supports a vertical load P at joint D (Figure 4.13). The length of the middle bar is L, the inclined bars make angle α with the vertical, and the cross-sectional area of each bar is A. The bars are made of an elastic–perfectly plastic material with yield stress σ_0 and elastic modulus E. The load is increased to the plastic limit P_p, unloaded to $P = 0$, reloaded in reversed direction (compression) to the plastic limit P_p^*, and finally unloaded to zero.

Find

a. The elastic and plastic limits P_e and P_p in the initial loading
b. The elastic and plastic limits P_e^* and P_p^* in the reversed loading
c. The residual stresses and strains of the bars
d. The residual vertical and horizontal displacements of joint D at the end of the deformation

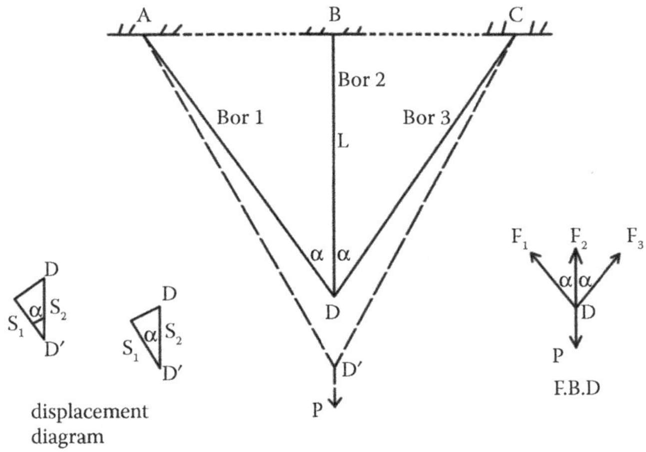

FIGURE 4.13
Example 4.8.

Solution

A free body diagram of the forces in the truss and a sketch of the displacement of joint D are shown in Figure 4.13. Equilibrium of the forces requires $F_1 = F_3$ and $2F_1 \cos\alpha + F_2 = 0$. Denoting the vertical displacement of joint D by δ_2, it follows that the displacements of bars AD and CD are $\delta_1 = \delta_3 = \delta_2 \cos\alpha$. Hence, in terms of stresses and strains, the basic equations of equilibrium and compatibility of strains, which are valid throughout the deformations, are

$$2\sigma_1 \cos\alpha + \sigma_2 = \frac{P}{A}$$

$$\sigma_3 = \sigma_1$$

$$\varepsilon_1 = \varepsilon_3 = \varepsilon_2 \cos^2\alpha$$

The initial deformation is governed by Hooke's law giving $\sigma_1 = \sigma_3 = \sigma_2 \cos^2\alpha$. Using the equilibrium equation, we have

$$\sigma_1 = \sigma_3 = \frac{\cos^2\alpha}{1 + 2\cos^3\alpha}\frac{P}{A}$$

$$\sigma_2 = \frac{1}{1 + 2\cos^3\alpha}\frac{P}{A}$$

a. *Initial loading:* Since the magnitude of the stress in the middle bar is greater than that in the inclined bars, the middle bar yields first, the stresses are $\sigma_2 = \sigma_0$, $\sigma_1 = \sigma_3 = \sigma_0 \cos^2\alpha$, and the elastic limit is

$$\frac{P_e}{A} = (1 + 2\cos^3\alpha)\sigma_0 \quad \text{(a)}$$

The corresponding strains are $\varepsilon_2 = \varepsilon_0$, $\varepsilon_1 = \varepsilon_3 = (\cos^2\alpha)\,\varepsilon_0$, and the vertical and horizontal displacements of joint D are $v = L\,\varepsilon_0$, $u = 0$.

As P is increased beyond the elastic limit, the stress in the middle bar remains at the yield level, the additional load is carried by the inclined bars until they yield at the plastic limit, which is obtained from Equation (a) as

$$\frac{P_p}{A} = (1 + 2\cos\alpha)\sigma_0 \tag{b}$$

The corresponding strains are $\varepsilon_1 = \varepsilon_3 = \varepsilon_0$, and the strain in the middle bar, using compatibility of strains equation, is $\varepsilon_2 = \varepsilon_0/\cos^2\alpha$.

At this stage of the deformation, the displacements of joint D are $u = 0$ and $v = (L/\cos^2\alpha)\,\varepsilon_0$.

The unloading from P_p to $P = 0$ is elastic where the increment in stresses is given by

$$\Delta\sigma_1 = \Delta\sigma_3 = -\frac{\cos^2\alpha}{1 + 2\cos^3\alpha}\frac{\Delta P}{A}$$

$$\Delta\sigma_2 = -\frac{1}{1 + 2\cos^3\alpha}\frac{\Delta P}{A}$$

Here, $(\Delta P/A) = (1 + 2\cos\alpha)\sigma_0$. Hence, the initial residual stresses in the bars are

$$\sigma_1 = \sigma_3 = \sigma_0 - \frac{\cos^2\alpha(1 + 2\cos\alpha)}{1 + 2\cos^3\alpha}\sigma_0 = \frac{\sin^2\alpha}{1 + 2\cos^3\alpha}\sigma_0$$

$$\sigma_2 = \sigma_0 - \frac{(1 + 2\cos\alpha)}{1 + 2\cos^3\alpha}\sigma_0 = -\frac{2\sin^2\alpha\cos\alpha}{1 + 2\cos^3\alpha}\sigma_0$$

The corresponding residual strains are

$$\varepsilon_1 = \varepsilon_3 = \frac{\sin^2\alpha}{1 + 2\cos^3\alpha}\varepsilon_0$$

$$\varepsilon_2 = \frac{\varepsilon_1}{\cos^2\alpha} = \frac{\tan^2\alpha}{1 + 2\cos^3\alpha}\varepsilon_0$$

And the displacements of joint D are $u = 0$, $v = \varepsilon_2 L = [(\tan^2\alpha)/(1 + 2\cos^3\alpha)](L\varepsilon_0)$.

b. *Reversed loading:* In reversed loading the middle bar yields first in compression, the elastic limit is obtained as follows:

$$-\frac{2\sin^2\alpha\cos\alpha}{1 + 2\cos^3\alpha}\sigma_0 + \frac{1}{1 + 2\cos^3\alpha}\frac{P_e^*}{A} = -\sigma_0$$

or

$$\frac{P_e^*}{A} = -(1 + 2\cos\alpha\cos 2\alpha)\sigma_0 \tag{c}$$

Further loading causes the inclined bars to yield and the plastic limit in reversed loading is

$$\frac{P_p^*}{A} = -(1 + 2\cos\alpha)\sigma_0 \tag{d}$$

and $\sigma_1 = \sigma_2 = \sigma_3 = -\sigma_0$, $\varepsilon_1 = \varepsilon_3 = -\varepsilon_0$, $\varepsilon_2 = -\varepsilon_0/\cos^2\alpha$
$u = 0$, $V = -(L/\cos^2\alpha)\,\varepsilon_0$

c. Finally, the unloading to $P = 0$ produces the following residual stresses and strains in the truss

$$\sigma_1^{res} = \sigma_3^{res} = -\sigma_0 - \frac{\cos^2\alpha}{1 + 2\cos^3\alpha}\left(\frac{P_p^*}{A}\right) = -\frac{\sin^2\alpha}{1 + 2\cos^3\alpha}\sigma_0$$

$$\sigma_2^{res} = -\sigma_0 - \frac{1}{1 + 2\cos^3\alpha}\left(\frac{P_p^*}{A}\right) = \frac{2\cos\alpha\sin^2\alpha}{1 + 2\cos^3\alpha}\sigma_0 \quad \text{(e)}$$

and

$$\varepsilon_1^{res} = \varepsilon_3^{res} = -\frac{\sin^2\alpha}{1 + 2\cos^3\alpha}\varepsilon_0$$

$$\varepsilon_2^{res} = -\frac{\tan^2\alpha}{1 + 2\cos^3\alpha}$$

d. The residual displacements of joint D are

$$u = 0$$

$$V = L\varepsilon_2^{res} = -\frac{\tan^2\alpha}{1 + 2\cos^3\alpha}L\varepsilon_0$$

Numerical values of the results in Equations (a) through (e) for $\alpha = 30°$, 45°, and 60° are listed in the table below:

α	$P_e/A\sigma_0$	$P_p/A\sigma_0$	$P_e^*/A\sigma_0$	$P_p^*/A\sigma_0$	σ_1^{res}/σ_0	σ_2^{res}/σ_0	$\varepsilon_1^{res}/\varepsilon_0$	$\varepsilon_2^{res}/\varepsilon_0$
30°	2.30	2.73	–1.87	–2.73	–0.11	0.19	–0.11	–0.15
45°	1.71	2.4	–1.0	–2.4	–0.29	0.41	–0.29	–0.59
60°	1.25	2.0	–0.5	–2.0	–0.6	0.6	–0.6	–2.4

4.3.4 True Stress and Strain

The true stress is defined as the load divided by the current cross-sectional area, $\sigma = P/A$. The conventional or engineering stress is the load divided by the original area, $\sigma^* = P/A_0$. (In order to distinguish between the two definitions, the engineering stress and strain in this section are denoted by σ^* and ε^*.) The corresponding true or natural strain, originally introduced by Ludwick [4.5], is defined as the increment in length divided by the current length, $\varepsilon = dL/L$, while the engineering strain is the increment in length divided by the original length, $\varepsilon^* = (L - L_0)/L_0$. A_0 and L_0 refer to the original cross-sectional area and length.

For successive deformations, the natural strain is the sum of all increments in length. Hence, the trues strain in tension is

$$\varepsilon = \int_{L_0}^{L} \frac{dL}{L} = \text{Ln}\frac{L}{L_0} = \text{Ln}(1 + \varepsilon^*) \quad (4.39a)$$

The true strain in compression is

$$\varepsilon = -\int_{L_0}^{L} \frac{dL}{L} = -\text{Ln}\frac{L}{L_0} = -\text{Ln}\frac{L_0 - (L_0 - L)}{L_0} = -\text{Ln}(1 - \varepsilon^*) \quad (4.39b)$$

The difference between true and engineering strains can be attributed to the large strains encountered beyond the yield point especially in the necking down region of the tensile test specimen just before rupture (Figure 4.1). If ε^* is small compared to unity, the logarithmic term in Equation 4.39 can be expanded in series to show that the true and engineering strains are equivalent. As the deformation progresses in the plastic range, the true strain becomes progressively lower in tension and higher in compression.

Note that unlike the conventional engineering strain, the logarithmic strain is additive, for example, consider a bar of length l_1 stretched to length l_2, the strain is $\varepsilon_1 = \text{Ln}\ (\varepsilon_2/\varepsilon_1)$. Suppose that the bar is stretched further to length l_3, then $\varepsilon_2 = \text{Ln}\ (\varepsilon_3/\varepsilon_2)$, the total strain is $\varepsilon = \text{Ln}\ (\varepsilon_2/\varepsilon_1) + \text{Ln}\ (\varepsilon_3/\varepsilon_2) = \text{Ln}\ (\varepsilon_3/\varepsilon_1)$.

For studying plastic deformation of materials, the true stress–strain curve provides more information. Since the volume of material is constant during plastic deformation, that is for 1D element, $A_0L_0 = AL$, it follows that if P is the current loading the true stress is

$$\sigma = \frac{P}{A} = \frac{P(L_0 + dL)}{L_0} = \sigma^*(1 + \varepsilon^*) \tag{4.40}$$

For compressive loading, $\sigma = \sigma^*(1 - \varepsilon^*)$.

The true and engineering stresses become identical if the strain is small compared to unity.

Instability condition: When a tensile test specimen experiences large strain, there is a decrease in the cross-sectional area (necking) because the increase in load due to strain hardening of the material is not sufficient enough to balance the ensuing reduction in the area. The load attains its maximum value at the onset of necking and instability occurs. Since $P = \sigma A$, the instability condition is

$$\mathbf{dP} = \mathbf{0} = \sigma\, dA + d\sigma A \tag{4.41a}$$

Using the constant volume condition, $AL = A_0L_0$, or $AdL + LdA = 0$, we have at the instability condition $d\sigma/\sigma = -\, dA/A = dL/L = d\varepsilon$. Hence,

$$\begin{aligned} \frac{d\sigma}{d\varepsilon} &= \sigma \\ \frac{d\sigma}{d\varepsilon^*} &= \frac{\sigma}{1 + \varepsilon^*} = \sigma^* \end{aligned} \tag{4.41b}$$

As an example, consider a tensile test specimen made of a material whose true stress–strain curve is $\sigma = K\,\varepsilon^n$, where K is the strength coefficient and n is the strain-hardening exponent. It follows that

$$\frac{d\sigma}{d\varepsilon} = nK\varepsilon^{n-1} = \frac{nK\varepsilon^n}{\varepsilon} = \frac{n\sigma}{\varepsilon}$$

Using Equation 4.41b, the true strain at the instant of instability due to plastic flow is

$$\varepsilon = n \tag{4.42a}$$

which indicates that the true strain has the same numerical value as the strain-hardening index. The corresponding stresses and strains in the specimen are

$$\sigma_1 = Kn^n,\ \sigma_2 = \sigma_3 = 0, \quad \text{and} \quad \varepsilon_1 = n,\ \varepsilon_2 = \varepsilon_3 = -n/2$$

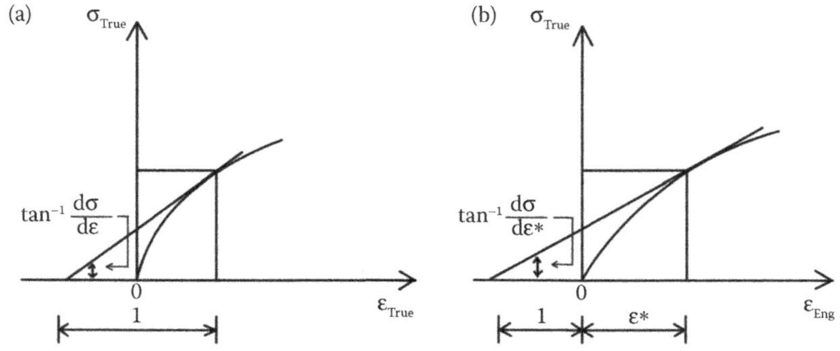

FIGURE 4.14
Graphical method for determining the point of instability.

If the power law is expressed in terms of true stress versus engineering strain, $\sigma = K(\varepsilon^*)^n$, the engineering strain at instability is obtained from Equation 4.41b as

$$\varepsilon^* = \frac{n}{1-n} \tag{4.42b}$$

Note that Equation 4.41b provides graphical methods to find the true stress and strain at instability. If the true stress versus true strain curve is available, by drawing a tangent to the curve such that the length of the subtangent along the strain axis is unity, then the coordinates of the point of intersection provide the required values of the true stress and strain at instability (Figure 4.14a). Similar construction can be used if the true stress versus engineering strain curve is used, as shown in Figure 4.14b.

EXAMPLE 4.9

A three-bar truss, identical to the one in Example 4.8, supports a vertical force at joint D (Figure 4.15). It is required to determine the initial angle of inclination (α) of the inclined members such that plastic instability occurs simultaneously in the three bars assuming the following material properties:

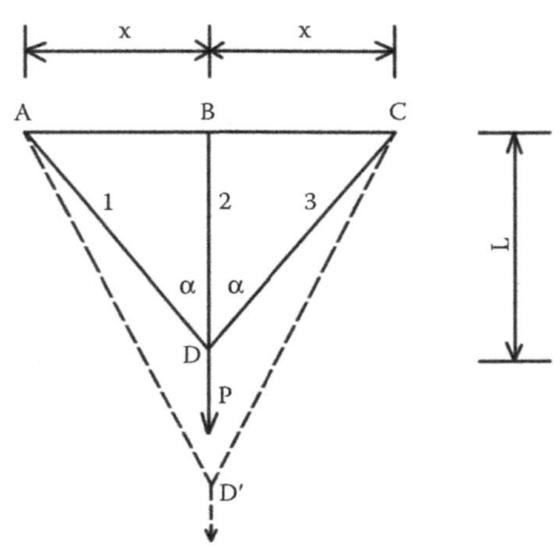

FIGURE 4.15
Example 4.9.

a. The true stress–true strain curves of the inclined bars AC and AD are $\sigma_1 = \sigma_3 = K_1\ (\varepsilon_1)_1^n$ and that of the middle bar BD are $\sigma_2 = K_2\ (\varepsilon_2)_2^n$. K_1 and K_2 denote the strength coefficients in unit of stress and n_1 and n_2 are the strain-hardening indices.

b. Repeat above, assuming that an engineering strain is used instead of true strain and the remaining parameters remain the same.

Solution

a. We assume that at plastic instability joint D moves to point D′ and the inclined members AD and CD are the dotted lines in Figure 4.15. According to the result in Equation 4.42a, we have $\varepsilon_1 = n_1 = \text{Ln}\ (AD'/AD)$, $\varepsilon_3 = \varepsilon_1$, and $\varepsilon_2 = n_2 = \text{Ln}\ (BD'/BD)$. Hence,

$$AD' = AD(e_1^n) \quad \text{and} \quad BD' = BD(e_2^n) = L(e_2^n) \tag{a}$$

Denoting distance AB in Figure 4.15 by x, it follows that $(AD')^2 = x^2 + (BD')^2 = x^2 + L^2\ e_2^{2n}$

Substituting for AD′ from relation (a), we have

$$(AD)^2 e_1^{2n} = x^2 + L^2 e_2^{2n} \tag{b}$$

Using the identity $x^2 = (AD)^2 - L^2$, relation (b) gives for the ratio L/AD

$$\cos\alpha = \frac{L}{AD} = \sqrt{\frac{e_1^{2n} - 1}{e_2^{2n} - 1}}$$

b. In this case instead of Equation 4.42a, we use Equation 4.42b to give

$$\varepsilon_1^* = \frac{n_1}{1 - n_1}, \quad \varepsilon_2^* = \frac{n_2}{1 - n_2}$$

Following the identical steps as above, the initial angle of the inclination is

$$\cos\alpha = \frac{L}{AD} = \frac{1 - n_2}{1 - n_1}\sqrt{\frac{(2 - n_1)n_1}{(2 - n_2)n_2}}$$

4.4 Yield Criterion

The yield criterion is a hypothesis defining the initiation of yield in a structural member subject to general state of stress. In the simple tension test of a specimen, the limit of proportionality or elastic limit is usually taken as the yield stress in tension. Thus, the yield criterion for such members is $\sigma = \sigma_0$. In general, for members made of isotropic materials subject to arbitrary loadings, it is sufficient to describe the state of stress by the principal stresses, $\sigma_1, \sigma_2, \sigma_3$, or their invariants, I_1, I_2, I_3 and the yield criterion is assumed to have the functional form $f(\sigma_1, \sigma_2, \sigma_3, k_1, k_2, k_3, \ldots) = 0$, where $k_1, k_2, \ldots$ are material constants to be determined experimentally. Since hydrostatic pressure is known to produce negligible plastic deformation, it is assumed that initiation of yield depends only on stress deviators and material constants [4.2,4.3], and others. Thus, the functional form needed to describe yield criteria is

$$f(J_2, J_3, k_1, k_2, \ldots) = 0 \tag{4.43}$$

Several criteria have been proposed to predict the yield under general states of stress. Two widely used criteria, *maximum shearing stress* proposed by Tresca (1864) and *strain energy of distortion (octahedral shearing stress criterion)* introduced by von Mises (1913), are discussed. It is assumed that the material is homogeneous and isotropic, the magnitudes of yield stress in tension and compression are equal. Moreover, thermal effects and strain rates are neglected. It turns out that for such materials hydrostatic pressure loading produces insignificant plastic deformations and can be neglected. The validity of any yield criterion hypothesis must be verified by experiments [4.9].

4.4.1 Maximum Shearing Stress Criterion

The maximum shearing stress criterion, commonly referred to as Tresca's criterion, is based on experimental observations that ductile metals subjected to general state of stress yield when the maximum shearing stress at a point reaches a critical value k, $\tau_{max} = k$. The material constant is the shear stress at yield, τ_0, which is approximately half the yield stress in tension, $\tau_0 = (1/2)\,\sigma_0$.

In terms of principal stresses arranged in the usual order of magnitudes, $\sigma_1 > \sigma_2 > \sigma_3$, the maximum shearing stress is $\tau_{max} = \frac{1}{2}|\sigma_1 - \sigma_3|$, and the criterion is $\tau_{max} = (1/2)\,\sigma_0$, that is, $|\sigma_1 - \sigma_3| = \sigma_0$, $\sigma_1 - \sigma_3 = \pm\sigma_0$, or

$$\frac{\sigma_1}{\sigma_0} - \frac{\sigma_3}{\sigma_0} = \pm 1 \tag{4.44}$$

For 2D states of stress, one of the principal stresses vanishes, $\sigma_3 = 0$, and $\sigma_1 > \sigma_2$. There are two cases to be considered:

a. When σ_1 and σ_2 are of opposite signs, one stress is tensile and the other is compressive, the maximum shearing stress is $(1/2)\,(\sigma_1 - \sigma_2)$, and the criterion is

$$\frac{\sigma_1}{\sigma_0} - \frac{\sigma_2}{\sigma_0} = \pm 1 \tag{4.45a}$$

b. When σ_1 and σ_2 are of opposite signs, and $\sigma_1 > \sigma_2$, the maximum shearing stress is equal to $(1/2)\,\sigma_0$, and the yield criterion is

$$\sigma_1 = \pm\sigma_0 \tag{4.45b}$$

Similarly, if $\sigma_2 > \sigma_1$, the yield criterion implies

$$\sigma_2 = \pm\sigma_0 \tag{4.45c}$$

Figure 4.16 presents plots of σ_1 versus σ_2 of the criteria in Equations 4.45a–c, which form a hexagon shape. Points inside the hexagon indicate elastic states of stress, whereas the outside points designate plastic stresses. The stresses coinciding with the boundary represent the onset of yield in the member according to Tresca's criterion. In the first and third quadrants, the vertical and horizontal lines represent plots of Equations 4.45b and 4.45c, respectively. The slanted lines in the second and fourth quadrants of the hexagon are plots of Equation 4.45a.

Note that the maximum shearing stress criterion involves the maximum and minimum principal stresses only and does not reflect the influence of the intermediate principal stress

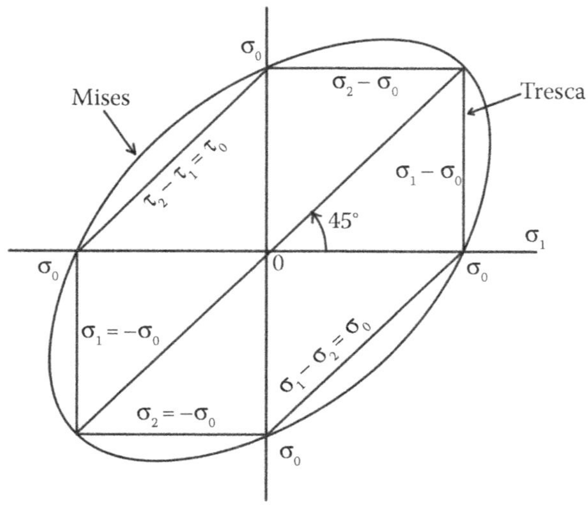

FIGURE 4.16
Tresca and Mises' biaxial yield criteria ($\sigma_3 = 0$).

on initiation of yield. The maximum distortion energy density criterion or the octahedral shear stress criterion, discussed in the next section, involves the three principal stresses.

4.4.2 Maximum Distortion Energy Criterion

This criterion, commonly referred to as von Mises' criterion, stipulates that yield is initiated in a member subjected to general state of stress when the *maximum strain energy density of distortion*, u_d (Equation 4.25a), or *the octahedral shearing stress*, τ_{oct} (Equation 2.46b), reaches a critical value k^2, that is for u_d we have

$$\frac{1}{12G}\left[(\sigma_x - \sigma_y)^2 + (\sigma_y - \sigma_z)^2 + (\sigma_z - \sigma_x)^2 + 6\left(\tau_{xy}^2 + \tau_{yz}^2 + \tau_{zx}^2\right)\right] = k^2 \qquad \text{(a)}$$

In terms of principal stresses only, we have

$$\frac{1}{12G}\left[(\sigma_1 - \sigma_2)^2 + (\sigma_2 - \sigma_3)^2 + (\sigma_3 - \sigma_1)^2\right] = k^2 \qquad \text{(b)}$$

The critical value k^2 is obtained by equating u_d in Equation (b) to its associated value when a specimen of the material begins to yield in the tension test, that is, u_d for the specimen = $(1/6G)\,\sigma_0^2$. It follows that Mises' criterion in terms of principal stresses is

$$\left[(\sigma_1 - \sigma_2)^2 + (\sigma_2 - \sigma_3)^2 + (\sigma_3 - \sigma_1)^2\right] = 2\sigma_0^2 \qquad (4.46a)$$

The strain energy of distortion is also related to the octahedral shearing stress and the second stress deviator invariant, Equation 4.25b, $u_d = \left(3\tau_{oct}^2\right)/4G = J_2/2G$. Thus, yielding begins according to von Mises' criterion when the octahedral shearing stress or J_2 reaches a critical value k^2. According to Equation 4.43, the functional form of this criterion is

$$f(J_2) = J_2 - k^2 = 0 \qquad (4.46b)$$

Using Equation 2.42 for J_2, the criterion in terms of principal stresses becomes

$$(\sigma_1 - \sigma_2)^2 + (\sigma_2 - \sigma_3)^2 + (\sigma_3 - \sigma_1)^2 = 6k^2 \quad (4.46c)$$

Here, k is the yield stress in shear.

Initiation of yield in the simple tension test occurs when $\sigma_1 = \sigma_0$ and $\sigma_2 = \sigma_3 = 0$. Inserting these values in Equation 4.46c, the constant k is given by

$$k = \frac{\sigma_0}{\sqrt{3}} \quad (4.47)$$

The value of k in Equation 4.47 is greater than the corresponding value in Tresca's criterion, $k = (1/2)\,\sigma_0$, the ratio of the two values is $(2/\sqrt{3}) = 1.155$. This implies that the magnitude of the stresses used to predict initiation of yield in both criteria does not vary by more than (12%–15%) [2.1].

As mentioned earlier, unlike Mises' criterion, Tresca's criterion does not account for the intermediate principal stress. The influence of the intermediate principal stress on initiation of yield in ductile metals is considered in Reference 2.1.

For the 2D plane stress case, $\sigma_3 = 0$, Mises' criterion, Equation 4.46a is reduced to

$$\sigma_1^2 - \sigma_1\sigma_2 + \sigma_2^2 = \sigma_0^2 \quad (4.48)$$

Equation 4.48 when plotted on rectangular axes represents an ellipse whose major and minor axes make 45° with the σ_1 and σ_2 axes as shown in Figure 4.16. Points within the ellipse represent elastic stresses, whereas points on the boundary of the ellipse indicate initiation of yield.

EXAMPLE 4.10

A thin-walled cylindrical tank with closed ends is made of structural steel having yield stress $\sigma_0 = 250$ MPa. The mean radius of the tank is r = 25 cm, and the thickness is t = 8 mm. The tank is subjected to an axial force F and a gradually increasing pressure p (Figure 4.17).

Find the values of p necessary to initiate yield in the tank using Tresca's and von Mises' criteria for the following axial force conditions:

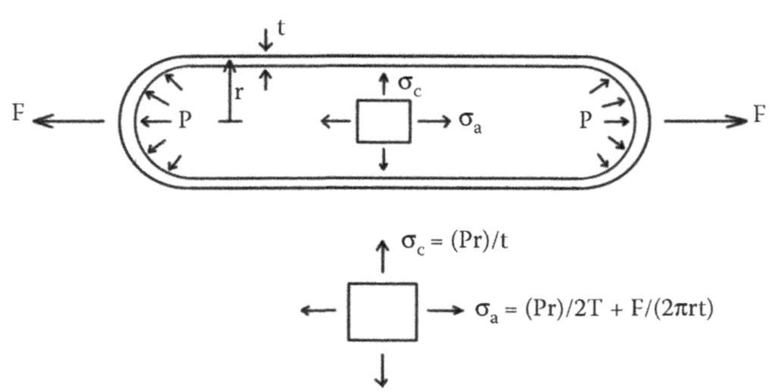

FIGURE 4.17
Example 4.10.

a. $F = 0$
b. $F = 0.2\ F_e$, F_e is the axial force necessary to initiate yield when acting alone, $F_e = 2\pi r t \sigma_0$
c. $F = 0.8\ F_e$

Solution

For loads F and p acting on the tank, an element of the wall is subjected to the following axial, circumferential, and radial stresses:

$$\begin{aligned} \sigma_a &= \frac{F}{2\pi rt} + \frac{pr}{2t} = \frac{F}{F_e}\sigma_0 + \frac{pr}{2t} \\ \sigma_c &= \frac{pr}{t} \\ \sigma_r &= 0 \end{aligned} \tag{a}$$

Note that the internal pressure causes a radial compressive stress $\sigma_r = p$. Since $r \gg t$, the radial stress is neglected compared to the axial and circumferential stresses. Thus, the element is in a state of biaxial stresses.

Application of Tresca's criterion, Equation 4.44, hinges on the largest principal stress, which in this case is either the axial or circumferential stress. If we assume $\sigma_1 = \sigma_c$, then $(p_e\, r/t) = \sigma_0$, provided that $(pr/t) > (pr/2t) + (F/F_e)\sigma_0$, or at yield $p_e r/t\sigma_0 \geq 2(F/F_e)$.

Hence, if the principal stress is the axial stress, we have $p_e r/t\sigma_0 = 1$, provided that $F/F_e \leq 1/2$.

If $\sigma_1 = \sigma_a$, then

$$\frac{F}{F_e}\sigma_0 + \frac{p_e r}{2t} = \sigma_0, \quad \text{and} \quad \frac{p_e r}{t\sigma_0} = 2\left(1 - \frac{F}{F_e}\right), \quad \frac{1}{2} \leq \frac{F}{F_e} \leq 1$$

In sum, Tresca's criterion gives

$$\begin{aligned} \frac{p_e r}{t\sigma_0} &= 1, \quad \text{if } 0 \leq \frac{F}{F_e} \leq \frac{1}{2} \\ \frac{p_e r}{t\sigma_0} &= 2\left(1 - \frac{F}{F_e}\right), \quad \text{if } \frac{1}{2} \leq \frac{F}{F_e} \leq 1 \end{aligned} \tag{b}$$

and Mises' criterion gives

$$\frac{p_e r}{t\sigma_0} = \frac{2}{\sqrt{3}}\left[1 - \left(\frac{F}{F_e}\right)^2\right]^{1/2} \tag{c}$$

a. When $F = 0$, Tresca's criterion gives from Equation (b)

$$p_e = \frac{0.008\ \text{m}}{0.25\ \text{m}} = 250\ \text{MPa} = 8\ \text{MPa}$$

and Mises' criterion using Equation (c)

$$p_e = \frac{2}{\sqrt{3}}\left(\frac{0.008\ \text{m}}{0.25\ \text{m}}\right)(250\ \text{MPa}) = 9.24\ \text{MPa}$$

b. When $F = 0.2\ F_e$, using Equations (b) and (c), Tresca's criterion provides $p_e = 8$ MPa and according to Mises' criterion,

$$p_e = \frac{0.008\ \text{m}}{0.25\ \text{m}} \times 250\ \text{MPa}\left(\frac{2}{\sqrt{3}}\right)\sqrt{0.96} = 9.05\ \text{MPa}$$

c. $F = 0.8\ F_e$: According to Tresca's criterion

$$p_e = \frac{0.008\,\text{m}}{0.25\,\text{m}}(250\,\text{MPa})(0.4) = 3.2\,\text{MPa}$$

whereas Mises' criterion gives

$$p_e = \frac{0.008\,\text{m}}{0.25\,\text{m}}(250\,\text{MPa})\left(\frac{2}{\sqrt{3}}\right)(0.6) = 5.54\,\text{MPa}$$

EXAMPLE 4.11

An aluminum solid shaft of radius r and yield strength σ_0 is subjected to a combined loading of axial tensile force F and a torque T, as shown in Figure 4.18.

a. Plot the yield curves σ/σ_0 versus τ/σ_0 using Tresca and Mises' criteria.
b. Determine the magnitudes of the torque required to initiate yield assuming that $r = 1.0$ in., $F = 100$ kip, and $\sigma_0 = 38$ ksi.

Solution

a. The state of stress in the bar is biaxial consisting of the following normal and shear stresses:

$$\sigma_x = \sigma = \frac{F}{\pi r^2}$$

$$\tau_{xy} = \tau = \frac{2T}{\pi r^2}$$

The principal stresses are

$$\sigma_1 = \frac{\sigma}{2} + \sqrt{\frac{\sigma^2}{4} + \tau^2}$$
$$\sigma_2 = \frac{\sigma}{2} - \sqrt{\frac{\sigma^2}{4} + \tau^2} \qquad \text{(a)}$$

Inserting Equations (a) into 4.45a, Tresca's yield condition is

$$\sigma^2 + 4\tau^2 = \sigma_0^2 \qquad \text{(b)}$$

Similarly Equation 4.48 furnishes Mises' yield condition

$$\sigma^2 + 3\tau^2 = \sigma_0^2 \qquad \text{(c)}$$

In nondimensional form, Equations (b) and (c) become

$$\left(\frac{\sigma}{\sigma_0}\right)^2 + 4\left(\frac{\tau}{\sigma_0}\right)^2 = 1$$
$$\left(\frac{\sigma}{\sigma_0}\right)^2 + 3\left(\frac{\tau}{\sigma_0}\right)^2 = 1 \qquad \text{(d)}$$

Equations (d) is two ellipses representing the maximum shearing stress and maximum distortional strain energy yield criteria, and only the first quadrant of the ellipses is shown in Figure 4.18. Points within each ellipse represent elastic

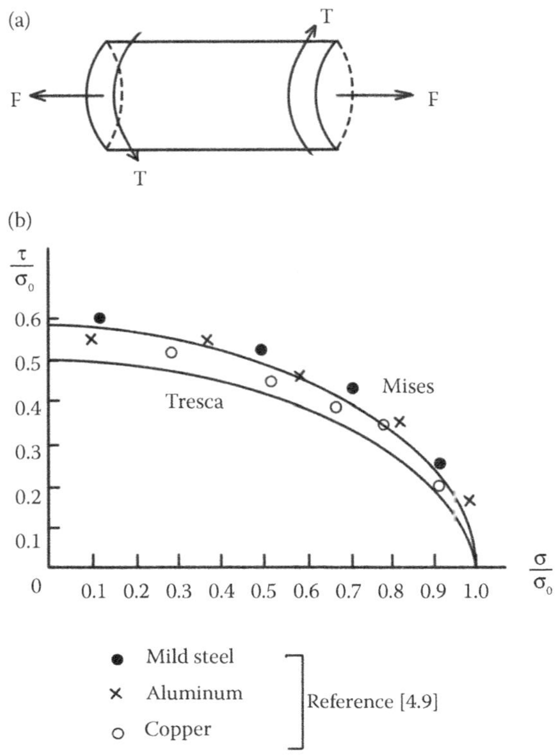

FIGURE 4.18
Example 4.11.

behavior whereas points on the boundaries indicate initiation of yield. For ease of reference, numerical values required to plot the curves are listed below:

Numerical Values of Equations (d)

$\frac{\sigma}{\sigma_0}$	$\frac{\tau}{\sigma_0}$ – **Tresca**	$\frac{\tau}{\sigma_0}$ – **Mises**
0.0	0.5	0.577
0.1	0.498	0.575
0.2	0.490	0.566
0.3	0.477	0.551
0.4	0.450	0.529
0.5	0.433	0.5
0.6	0.4	0.462
0.7	0.375	0.412
0.8	0.3	0.364
0.9	0.218	0.252
1.0	0.0	0.0

Few data points indicating initiation of yield for mild steel, aluminum, and copper, based on the classical results of Taylor and Quinny [4.9], are shown in Figure 4.18. Most of the data points seem to be situated near Mises' criterion.

b. If $F = 100$ kip and $r = 1$ in., then the stresses acting on the member are $\sigma = 31.831$ ksi and $\tau = 0.637$ T ksi.

   Using Equations (b) and (c) with 38 ksi yield stress, we find the torques required to initiate the yield are

   Tresca's criterion: $T = 16.3$ kip-in.

   Mises' criterion: $T = 18.8$ kip-in.

4.4.3 Yield Surfaces in Haigh–Wesregaard Stress Space

The Haigh–Westergaard stress space consists of three mutually perpendicular axes with three principal stresses as positional coordinates, which are used to represent any admissible state of stress in a loaded component (Section 2.10). In this section, we utilize the same space to determine the geometrical representation of yield surfaces when plastic deformation first appears in the component.

Figure 4.19 shows such a space with origin 0, the coordinates of any arbitrary point P in the space are the principal stresses, σ_1, σ_2, σ_3. Line 0H makes equal angles with the coordinate axes. It is termed the hydrostatic axis. Any plane normal to 0H is called a deviatoric plane.

The stress vector 0P is resolved into two components, 0Q along 0H and 0R normal to 0H with component 0Q representing a hydrostatic or spherical stress I_1. At any point on this line, the principal stresses are equal and the deviatoric stresses vanish, $S_1 = S_2 = S_3 = 0$. The magnitude of 0Q is $\sqrt{3}\sigma_0$.

Component 0R represents the deviatoric stress, components S_1, S_2, S_3, and magnitude $\left[S_1^2 + S_2^2 + S_3^2\right]^{1/2} = (2J_2)^{1/2}$, where J_2 is the second invariant of the stress deviator (Section 2.7).

The deviatoric plane containing 0R has the form

$$\sigma_1 + \sigma_2 + \sigma_3 = \left(\sqrt{3}\right)\rho \tag{4.49a}$$

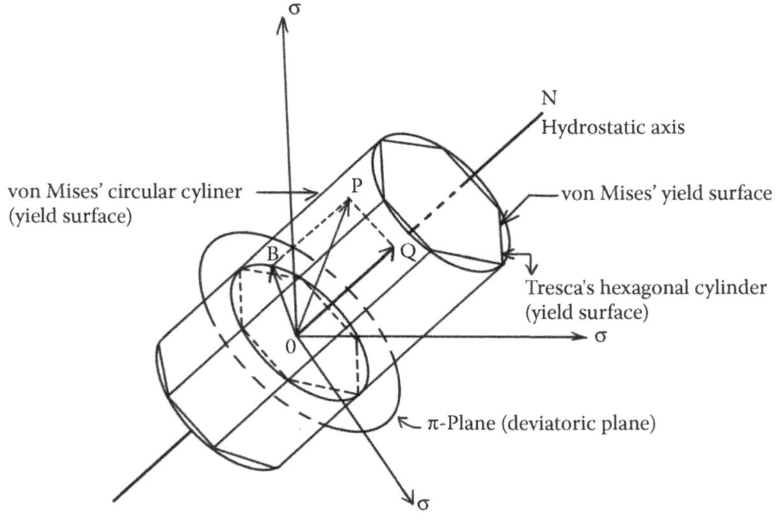

FIGURE 4.19
Yield surface representing yield criteria in principal stress space.

where ρ is the distance from the origin to the plane containing 0R along 0H. The particular deviatoric plane passing through the origin

$$\sigma_1 + \sigma_2 + \sigma_3 = 0 \tag{4.49b}$$

is called the π-plane.

Because a hydrostatic or spherical stress has no effect on yielding [4.2,4.3], yielding occurs in the material due to the magnitude and direction of stress deviator. Based on the above-mentioned result of the magnitude, the yield surface is a right cylinder whose axis is line 0H and generators are the normals to any deviatoric plane (Figure 4.19). Any point inside the cylinder represents an elastic state of stress, whereas points on the surface designate a state of yield. Next, we determine the specific surfaces in the stress space representing Tresca and Mises' criteria.

Tresca's criterion: Assuming $\sigma_1 > \sigma_2 > \sigma_3$ and using the expressions of the principal stresses derived in Chapter 2, Equation 2.59, we arrive at the result

$$\frac{\sigma_1 - \sigma_3}{2} = \sqrt{\frac{J_2}{3}}[\cos\theta - \cos(\theta + 120^\circ)], \quad 0 \leq \theta \leq 60^\circ \tag{a}$$

Inserting expression (a) into Equation 4.44, simplifying the algebra gives

$$2\sqrt{J_2}\sin(\theta + 60^\circ) = \sigma_0, \quad 0 \leq \theta \leq 60^\circ \tag{4.50}$$

The result in Equation 4.50, which is independent of the hydrostatic stress, represents a yield surface consisting of a cylinder with axis 0H and generators parallel to 0H. On a deviatoric plane normal to line 0H, Equation 4.50 is a straight line passing through two points defined by $\theta = 0$ and $\theta = 60^\circ$. This line forms one sector of the yield locus. Similar lines are obtained by using five more possible ordering of the magnitudes of the principal stresses. Thus, the yield surface is a regular hexagonal prism in the Haigh–Westergaard stress space shown in Figure 4.19. The yield locus of the biaxial state of stress, which is shown in Figure 4.16, is the intersection of the hexagonal yield surface with the coordinate plane $\sigma_3 = 0$.

Mises' criterion: In this criterion, yielding begins when u_d reaches a critical value k. The functional form of the criterion, Equation 4.46b, is

$$f(J_2) = J_2 - k^2 = 0 \tag{4.51}$$

The value of k, which is the yield stress in pure shear, is given in Equation 4.47. It follows that the yield surface represented by Equation 4.51 is a circular cylinder whose generators are parallel to line 0H (Figure 4.19). In this case, the yield locus of the biaxial stress state is the ellipse shown in Figure 4.16.

4.4.4 Strain Hardening Rules

When the load acting on a structural member is increased beyond the initial yield stress, the material experiences hardening of the plastic strain, a phenomenon known as strain hardening or work hardening. To describe the history of such phenomenon, a hardening parameter denoted by κ is used. The plastic modulus is assumed to be a function of this parameter $E_p(\kappa)$, which is used to describe subsequent stages of plastic deformation whether in tension or in compression. The functional form of the plastic modulus is determined from the uniaxial tension test.

There are several ways of defining the hardening parameter to describe the stages of plastic deformation [2.1,2.2]. They include the following.

The hardening parameter may be taken as the plastic strain

$$\varepsilon^{p} = \int d\varepsilon^{p} \tag{a}$$

However, since the plastic strain is reversible and cannot be accumulated, using the accumulated plastic strain increments is a more realistic choice for the hardening parameter. Thus,

$$\kappa = \varepsilon_{p} = \int \sqrt{(d\varepsilon^{p}\, d\varepsilon^{p})^{2}} \tag{4.52a}$$

Equation 4.52c is the sum of the effective or equivalent plastic strain defined as

$$|d\varepsilon_{p}| = \sqrt{d\varepsilon^{p}\, d\varepsilon^{p}} \tag{4.52b}$$

Note that the effective or equivalent plastic strain in terms of principal plastic strains is

$$\varepsilon^{p}_{Eq} = \sqrt{\frac{2}{3}}\left[\left(\varepsilon^{p}_{1}\right)^{2} + \left(\varepsilon^{p}_{2}\right)^{2} + \left(\varepsilon^{p}_{3}\right)^{2}\right]^{\frac{1}{2}} \tag{4.52c}$$

The numerical factor in expression (4.52c) is chosen so that the magnitude of the equivalent plastic strain is reduced to ε^{p}_{1}, which is the value of the specimen in the uniaxial tension test.

Another definition of a possible strain-hardening parameter, which is also widely used, is taken as the plastic work done, that is,

$$\kappa = W_{p} = \int \sigma\, d\varepsilon^{p} \tag{4.53}$$

Next, three types of idealized strain-hardening rules are described.

4.4.4.1 Isotropic Hardening Rule

According to this rule, the progressively increasing yield stresses in tension and compression are always equal, in other words, the reversed compressive yield stress is equal to the tensile yield stress before load reversal. This implies that Bauschinger effect, discussed in Section 4.3.2, is neglected.

Mathematically, the rule is expressed as follows:

$$|\sigma| = |\sigma(\kappa)| \tag{4.54}$$

4.4.4.2 Kinematic Hardening Rule

In this rule, it is assumed that the difference between yield stresses in tension and yield stresses in compression is a constant. Hence,

$$[\sigma_{0}(\kappa)]_{ten.} - [\sigma_{0}(\kappa)]_{com} = 2\sigma_{0} \tag{4.55}$$

where σ_0 is the yield stress in tension and $-\sigma_0$ is the corresponding yield stress in compression.

4.4.4.3 Independent Hardening Rule

The independent hardening rule assumes that plastic deformation occurring during tensile loading affects tensile yield stresses and plastic deformation occurring under compressive loading affects compressive yield stresses. It follows that two independent hardening parameters, κ^1 and κ^2, describing tensile and compressive plastic deformation, respectively, are needed to express this rule. Mathematically, the expressions of this rule are

$$\begin{aligned} \sigma &= \sigma_0^{\text{tens.}}(\kappa^1), \quad \sigma > 0 \\ \sigma &= \sigma_0^{\text{comp.}}(\kappa^2), \quad \sigma < 0 \end{aligned} \tag{4.56}$$

EXAMPLE 4.12

Consider the fixed ends member used in Examples 4.6 and 4.7. After unloading the force P from magnitude $P_2 = \beta\sigma_0 A$ to $P = 0$, the member is reloaded in the reversed direction.

Find P at which the member will subsequently yield in the reversed direction assuming isotropic, kinematic, and independent hardening rules.

Solution

In the initial loading stage discussed in Example 4.7, the following stresses are induced in both sections of the member by force $P_2 = \beta\sigma_0 A$

$$\begin{aligned} \sigma_a &= \frac{1}{5L}(4a - 4b + 5b\beta)\sigma_0 \\ \sigma_b &= \frac{1}{5L}(4a - 4b - 5a\beta)\sigma_0 \end{aligned} \tag{a}$$

After unloading to $P = 0$, the residual stress is

$$\sigma_a^{\text{res}} = \sigma_b^{\text{res}} = \frac{4(a-b)}{5L}\sigma_0$$

During reloading in the reversed direction, the initial stress in the member consists of the residual stress plus an elastic increment due to P. Hence,

$$\begin{aligned} \sigma_a &= \frac{4(a-b)}{5L}\sigma_0 + \frac{bP}{LA} \\ \sigma_b &= \frac{4(a-b)}{5L}\sigma_0 - \frac{aP}{LA} \end{aligned} \tag{b}$$

Subsequent yielding in the member depends on the strain-hardening rule used. The three rules discussed in Section 4.4.4 are considered.

a. *Isotopic hardening rule:* Making use of Equation 4.54, the equations that determine the subsequent yield conditions in the member are obtained from Equations (a) and (b) as follows:

$$\begin{aligned} \frac{4(a-b)}{5L}\sigma_0 + \frac{bP}{LA} &= -\frac{1}{5L}(4a - 4b + 5b\beta)\sigma_0 \\ \frac{4(a-b)}{5L}\sigma_0 - \frac{aP}{LA} &= -\frac{1}{5L}(4a - 4b - 5a\beta)\sigma_0 \end{aligned} \tag{c}$$

It turns out that section "a" of the member yields first at a value of P furnished by the first equation in Equation (c)

$$\frac{P}{A\sigma_0} = \frac{1}{5b}(-8a + 8b - 5b\beta) \tag{1}$$

b. *Kinematic hardening rule:* According to Equation 4.55, the equations determining the yield of the member are

$$\begin{aligned} \frac{4(a-b)}{5L}\sigma_0 + \frac{bP}{LA} &= \frac{1}{5L}(4a - 4b + 5b\beta)\sigma_0 - 2\sigma_0 \\ \frac{4(a-b)}{5L}\sigma_0 - \frac{aP}{LA} &= \frac{1}{5L}(4a - 4b - 5a\beta)\sigma_0 - 2\sigma_0 \end{aligned} \tag{d}$$

Solving for P, it is readily confirmed that part "a" of the member yields first at a value of P, obtained from the first equation in Equation (d), as

$$\frac{P}{a\sigma_0} = \frac{1}{5b}(-10a - 10b + 5b\beta) \tag{2}$$

c. *Independent hardening rule:* According to Equation 4.56, part "a" of the member yields first in compression. This implies that

$$\frac{4(a-b)}{5L}\sigma_0 + \frac{bP}{LA} = -\sigma_0$$

which gives

$$\frac{P}{A\sigma_0} = -\frac{1}{5b}(9a + b) \tag{3}$$

Note that part "b" will yield in tension at a higher value of P.

Numerical values of P necessary to cause yield in reversed loading can be obtained using the table below, which gives numerical values of the right-hand sides of Equations (1) through (3) for $\beta = 4$ and $b/a = 1, 2, 3,$ and 4.

b/a	Isotropic Hardening	Kinematic Hardening	Independent Hardening
1	−4	0	−2
2	−3.2	1	−1.1
3	−2.9	1.3	−0.8
4	−2.8	1.5	−0.65

EXAMPLE 4.13

The tensile test of a specimen made of an elastic–linear-hardening plastic material furnishes the following stress–strain relations:

$$\begin{aligned} \sigma &= E\varepsilon, & \sigma &< \sigma_0 \\ \sigma &= \sigma_0 + E_p\varepsilon^P, & \sigma &\geq \sigma_0 \end{aligned}$$

where $E = 30 \times 10^3$ ksi, $E_p = 0.25\,E$, the yield stress of the material is equal to 36 ksi, and ε^P is the plastic strain (Equation 4.30).

An element of the material is stretched to a stress of magnitude $\sigma = 1.5\ \sigma_0$, then unloaded to zero stress, reversely loaded in compression to a stress of magnitude $\sigma = -1.5\sigma_0$, and finally unloaded to its initial state of zero stress. Assuming that the material hardens according to each of the three hardening rules discussed in Section 4.4.4:

a. Sketch the stress–strain curves of the loading history.
b. Compute the accumulated plastic strains and plastic work in the element at the end of each loading path.

Solution

a. Initially the deformation is elastic, the element first yields in tension, yield stress $\sigma_0 = 36$ ksi and yield strain $\varepsilon_0 = 36/(30 \times 10^3) = 0.0012$, shown as point A in Figure 4.20.

Further loading causes plastic deformation. The tangent modulus is computed using Equation 4.33 as $E_t = 0.2\ E = 6 \times 10^3$ ksi. At point B, the stress is specified as $1.5\ \sigma_0 = 54$ ksi and the corresponding strain consists of the strain at point A plus the increment in strain, that is,

$$\varepsilon_B = \varepsilon_0 + \frac{\sigma_B - \sigma_0}{E_t} = 0.0012 + \frac{18}{6 \times 10^3} = 0.0042$$

Next, the element is unloaded elastically to point C, where the stress is zero and the corresponding strain is

$$\varepsilon_C = \varepsilon_B - \frac{\sigma_B}{E} = 0.0042 - \frac{54}{30 \times 10^3} = 0.0024$$

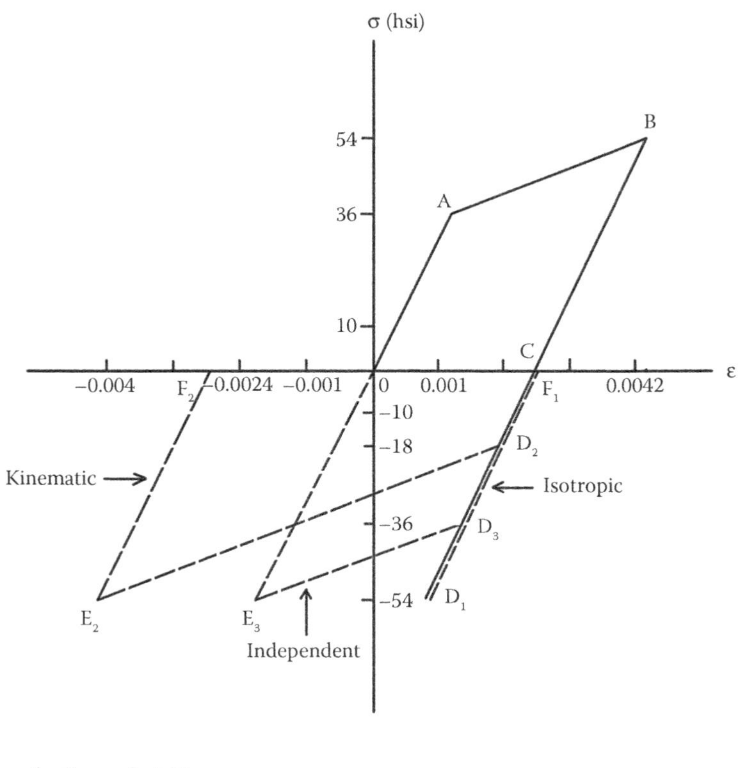

FIGURE 4.20
Stress–strain curves for Example 4.13.

The loading paths of the element, from origin to points A, B, and C, are identical, the remaining stages of the loading history are discussed for each of the specified strain-hardening rules.

i. *Isotropic hardening:* Beyond point C, the element is loaded in compression. According to the isotopic hardening, Equation 4.54, it subsequently yields at point D_1 without plastic deformation. The stress and strain at point D_1 are

$$\sigma_{D_1} = -\sigma_B = -54\,\text{ksi}$$

$$\varepsilon_{D_1} = \varepsilon_C - \frac{|\sigma_{D_1}|}{E} = 0.0024 - \frac{54}{30 \times 10^3} = 0.0006$$

Finally, the element unloads elastically from point D_1 to zero stress at point F_1 where the strain at F_1 is

$$\varepsilon_{F_1} = \varepsilon_{D_1} + \frac{|\sigma_{D_1}|}{E} = 0.0006 + \frac{54}{30 \times 10^3} = 0.0024$$

Point F_1 coincides with C. The stress–strain loading history using the isotropic hardening rule is as follows:

$$[0.0] \rightarrow [0.0012, 36\,\text{ksi}]_A \rightarrow [0.0042, 54\,\text{ksi}]_B \rightarrow [0.0034, 0]_C$$
$$\rightarrow [0.0006, -54\,\text{ksi}]_{D_1} \rightarrow [0.0024, 0]_{F_{1=C}}$$

ii. *Kinematic hardening:* Based on this rule, Equation 4.55, after unloading to point C, the element yields again under reversed compressive loading at point D_2 where the stress and strain at point D_2 are

$$\sigma_{D_2} = -(2\sigma - \sigma_B) = -(72 - 54) = -18\,\text{ksi}$$

$$\varepsilon_{D_2} = \varepsilon_C + \frac{\sigma_{D_2}}{E} = 0.0024 + \frac{-18}{30 \times 10^3} = 0.0018$$

Beyond point D_2 the element experiences plastic deformation due to compressive loading until point E_2 is reached. The stress and strain E_2 are

$$\sigma_{E_2} = -(2\sigma_0 + \sigma_{D_2}) = -(72 - 18) = -54\,\text{ksi}$$

$$\varepsilon_{E_2} = \varepsilon_{D_2} + \frac{\sigma_{E_2} - \sigma_{D_2}}{E_t} = 0.0018 + \frac{-54 + 18}{0.3 \times 30 \times 10^3} = -0.0042$$

Finally, the element unloads elastically from point E_2 to zero stress at point F_2 where the strain is

$$\varepsilon_{F_2} = \varepsilon_{E_2} - \frac{\sigma_{E_2}}{E} = -0.0024 + \frac{54}{30 \times 10^3} = -0.0024$$

The stress–strain loading history according to the kinematic hardening rule is

$$[0,0] \rightarrow [0.0012, 36\,\text{ksi}]_A \rightarrow [0.0042, 54\,\text{ksi}]_B \rightarrow [0.0024, 0]_C \rightarrow [0.0018, -18\,\text{ksi}]_{D_2}$$
$$\rightarrow [-0.0042, -54\,\text{ksi}]_{E_2} \rightarrow [-0.0024, 0]_{F_2}$$

iii. *Independent hardening:* According to the independent hardening rule, Equation 4.56, the element yields under compressive loading at point D_3 with stress = −36 ksi, and the corresponding strain is

$$\varepsilon_{D_3} = \varepsilon_C + \frac{\sigma_{D_3}}{E} = 0.0024 + \frac{-36}{30 \times 10^3} = 0.0012$$

The next stage from point D_3 to point E_3; plastic deformation occurs due to reversed compressive loading. The stress and strain at E_3 are

$$\sigma_{E_3} = -\sigma_B = -54\,\text{ksi}$$

$$\varepsilon_{E_3} = \varepsilon_{D_3} + \frac{\sigma_{E_3} - \sigma_{D_3}}{E_t} = 0.0012 + \frac{-54 + 36}{6 \times 10^3} = -0.0018$$

The final stage of the loading, the element unloads elastically to zero stress at point F_3. The corresponding strain at point F_3 is

$$\varepsilon_{F_3} = \varepsilon_{E_3} + \frac{\sigma_{E_3}}{E} = -0.0018 + \frac{54}{30 \times 10^3} = 0$$

Point F_3 coincides with the origin of coordinates, and the stress–strain loading history according to the independent hardening rule is

$$[0,0] \rightarrow [0.0012, 36\,\text{ksi}]_A \rightarrow [0.0042, 54\,\text{ksi}]_B \rightarrow [0.0024, 0]_C \rightarrow [0.0016, -18\,\text{ksi}]_{D_3}$$
$$\rightarrow [-0.0042, -54\,\text{ksi}]_{E_3} \rightarrow [-0.0024, 0]_{F_3}$$

The strain–stress curves of the three hardening rules are sketched in Figure 4.20.

b. For a given plastic stress increment, the increment in plastic strain is (Equation 4.30)

$$d\varepsilon^p = \frac{d\sigma}{E_p} \tag{a}$$

and the increment of plastic work corresponding to a finite stress increment is

$$\Delta W_p = \frac{1}{E_p}\int_{\sigma}^{\sigma+\Delta\sigma} \sigma\, d\sigma = \frac{1}{E_p}\left[\frac{\sigma^2}{2}\right]_{\sigma}^{\sigma+\Delta\sigma} = \frac{1}{2E_p}[\Delta\sigma(2\sigma + \Delta\sigma)] \tag{b}$$

where E_p is the plastic modulus. Next, we consider the stages of the loading paths in Figure 4.20:

From point 0 to point A, the behavior is elastic. There are no plastic strain and plastic work.

From point A to point B, plastic deformation occurs, the plastic strain and plastic work at B are

$$\varepsilon_B^p = \frac{\sigma_B - \sigma_A}{E_p} = \frac{54 - 36}{0.25 \times 30 \times 10^3} = 0.0024$$
$$W_{pB} = W_{pA} + \Delta W_p = 0 + \frac{\Delta\sigma}{2E_p}(2\sigma + \Delta\sigma) = 0.108\,\text{kip-in./in.}^3 \tag{c}$$

Note that in arriving at the result in Equation (c), the stress is 36 ksi, the increment in stress is 18 ksi, and $E_p = 7.5 \times 10^3$ in.^{2}

The next stage in the path is to unload from point B to point C, the element unloads elastically. Hence, there is no plastic nor plastic work. The remaining stages of the loading history need to be considered separately for each hardening rule.

1. *Isotropic hardening rule:* The material yields due to the reversed compressive loading at point D_1 with a stress of magnitude equal to that at point B, there are no plastic deformations at this stage of the loading.
   The next stage is to unload elastically from point D_1 to point F_1 without any plastic deformation.
2. *Kinematic hardening rule:* In this case, the material yields in reversed compression at point D_2. There is no plastic deformation from point C to point D_2.
   The next stage involves plastic deformation due to the compressive loading from point D_2 to point E_2. The resulting plastic strain is given by

$$\varepsilon^p_{E_3} = \varepsilon^p_B + \frac{\sigma_{E_3} - \sigma_{D_2}}{E_p} = 0.0024 + \frac{-54 + 18}{7.5 \times 10^3} = -0.0024 \quad \text{(d)}$$

   And the plastic work consists of the component at B, $W_{pB} = 0.108$ kip-in./in.^{3}—Equation (c) above, plus the increment in work, shown in Equation (b) with $\sigma = 18$ ksi, $\Delta\sigma = \sigma_{D_2} - \sigma_{E_2} = 36$ ksi.
   Hence,

$$W_{pE_2} = W_{pB} + \Delta W_p = 0.108 + 0.173 = 0.281 \text{ kip-in./in.}^3 \quad \text{(e)}$$

3. *Independent hardening rule:* According to this hypothesis, the material yields in compression at point D_3, yield stress is −36 ksi, corresponding strain is 0.0012. There is no plastic deformation in this stage.
   The next stage consists of a compressive loading from point D_3 to point E_3. The accumulate plastic is as follows:

$$\varepsilon^p_{E_3} = \varepsilon^p_B + \frac{\sigma_{E_3} - \sigma_{D_3}}{E_p} = 0.0024 + \frac{-54 - (-36)}{7.5 \times 10^3} = 0 \quad \text{(f)}$$

   and the plastic work is given by

$$W_{pE_3} = W_{pB} + \Delta W_p = W_{pB} + \frac{\Delta\sigma(2\sigma + \Delta\sigma)}{2E_p} \quad \text{(g)}$$

   where W_{pB} is given in Equation (c) above, $\sigma = -36$ ksi and $\Delta\sigma = \sigma_{E_3} - \sigma_{D_3} = -18$ ksi.
   It follows that the plastic work is $W_{pE_3} = 0.216$ kip-in./in.^{3}
   The final loading stage, from point E_3 to F_3, the material unloads elastically without plastic deformation, point F_3 coincides with the origin.

4.5 Plastic Stress–Strain Relations

Engineering materials strain hardens beyond the yield point, that is, additional load is required to induce increment in strain. The strain consists of elastic and plastic components. The stress–strain relations of a member subject to a uniaxial loading are straightforward and easy to use (Figure 4.1). However, the plastic behavior of members subject to a complex state of loading, which involves six stress components and six strain components, is not straightforward. The main question to answer is how to use the experimental evidence observed in the static tension test to predict the behavior under any general state of stress.

This section contains a brief review of the plastic stress–strain relations used to analyze the practical problems. A more detailed discussion is available in books dealing specifically with the theory of plasticity, for example, References 4.11 through 4.14 and many others. Plastic stress–strain relations suitable to components made of *elastic–perfectly plastic* materials and relations used in the *Deformation theory of plasticity (J_2 material)* are discussed.

4.5.1 Stress–Strain Relations for Perfectly Plastic Material

In many situations, the ductile structural material is modeled to have elastic–perfectly plastic behavior, as indicated in Figure 4.9a. The 1D stress–strain relation consists of two straight lines in which plastic deformation beyond the yield stress σ_0 occurs under constant flow stress, and effect of strain hardening is neglected. In order to discuss components subject to multiaxial state of stress and determine the proper elastic–perfectly stress–strain relations to be used, the concepts of plastic potential function, flow equation and rules, and loading and unloading conditions are needed.

Plastic potential function and flow equation: It is recalled that in the elastic case the strain ε_{ij}, $i, j = x, y, z$ can be obtained by differentiating an elastic potential function or the complementary strain energy density, u_0^*, with respect to the stress σ_{ij} as indicated in Equation 4.21. For the plastic case, von Mises [4.15] proposed a similar concept, a plastic potential function in the form of a scalar function of the stresses, $g(\sigma_{ij})$, which defines the components of the plastic strain increments.

Thus, the plastic flow equation is

$$d\varepsilon_{ij}^p = d\lambda \frac{\partial g}{\partial \sigma_{ij}} \tag{4.57}$$

where $d\lambda$ is a positive scalar factor of proportionality with a nonzero value only when plastic deformation occurs.

Associated flow rule: When the potential function is equal to the yield function, $g(\sigma_{ij}) = f(\sigma_{ij})$, then Equation 4.57 becomes the associated flow rule as follows:

$$d\varepsilon_{ij}^p = d\lambda \frac{\partial f}{\partial \sigma_{ij}} \tag{4.58}$$

and plastic flow develops along the normal to the yield surface.

Loading and unloading criteria: When the stress level reaches the yield surface, plastic deformation occurs, and this condition is known as "load condition." However, if after a stress increment the resulting stress moves to the inside of the yield surface, then the condition is termed "unloading condition."

For a perfectly plastic material with yield surface $f(\sigma_{ij}) = 0$ and additional loading $d\sigma_{ij}$, the loading and unloading criteria are

$$\begin{aligned} &\text{Loading: } f(\sigma_{ij}) = 0, \quad df = \frac{\partial f}{\partial \sigma_{ij}} d\sigma_{ij} = 0 \\ &\text{Unloading: } f(\sigma_{ij}) = 0, \quad df = \frac{\partial f}{\partial \sigma_{ij}} d\sigma_{ij} < 0 \end{aligned} \tag{4.59}$$

The Prandtl–Reuss plastic stress–strain equation: Consider Mises' yield criterion, Equation 4.46b, as the plastic potential function, that is,

$$f(\sigma_{ij}) = J_2 - k^2 = 0, \quad k = \frac{\sigma_0}{\sqrt{3}} \tag{4.60}$$

and noting that $J_2 = (1/2)\, S_{ij}\, S_{ij}$, the associated plastic flow rule gives for the plastic strain increment

$$d\varepsilon_{ij}^{P} = d\lambda \frac{\partial f}{\partial \sigma_{ij}} = d\lambda S_{ij} \tag{4.61}$$

where S_{ij} is the stress deviator and $d\lambda$ is the factor of proportionality given as

$$d\lambda = 0,\ J_2 < k^2 \quad \text{or} \quad J_2 = k^2,\ dJ_2 < 0$$

$$d\lambda > 0,\ J_2 = k^2 \quad \text{and} \quad J_2 = 0$$

Equation 4.61 furnishes the Prandtl–Reuss stress–strain relations for perfectly plastic material [4.16].

$$\frac{d\varepsilon_x^P}{s_x} = \frac{d\varepsilon_y^P}{s_y} = \frac{d\varepsilon_z^P}{s_z} = \frac{d\gamma_{xy}^P}{2\tau_{xy}} = \frac{d\gamma_{yz}^P}{2\tau_{yz}} = \frac{d\gamma_{zx}^P}{2\tau_{zx}} = d\lambda \tag{4.62}$$

which indicate that the plastic strain increments are proportional to the stress deviators and shear stresses.

Note that Equation 4.61 indicates that

$$d\varepsilon_{ii}^{P} = d\lambda s_{ii} = 0$$

In other words, there is no volumetric deformation. The response of components made of perfectly plastic material to external load is discussed in Chapter 7.

4.5.2 Deformation Theory of Plasticity (J_2—Material)

The deformation theory of plasticity (J_2—material) treats the strain hardened ductile material as if it is a nonlinear elastic material. There is a one-to-one correspondence between stress and strain, the loading is monotonic, unloading and reloading retrace the initial loading path. The resulting stress–strain relations are commonly referred to as Hencky's plastic stress–strain relations [4.15].

The total strain consists of elastic and plastic parts, that is in index notations

$$\varepsilon_{ij} = \varepsilon_{ij}^{e} + \varepsilon_{ij}^{P}, \quad i, j = x, y, z \tag{4.63}$$

The elastic strains are given by Hooke's law. Because the plastic strains are associated with changes in shape of the body, they are taken to be proportional to deviatoric stresses as follows:

$$\varepsilon_{ij}^{P} = \lambda S_{ij}, \quad i, j = x, y, z \tag{4.64}$$

where λ is a material constant, positive during loading, zero during unloading, and S_{ij} are the deviatoric stresses discussed in Section 2.7, Equations 2.40. The constant λ for the material is determined from a best-fit procedure with the stress–strain curve of the tensile test data. To illustrate the procedure, consider the effective or equivalent stress given in terms of the principal stresses by

$$\sigma_e = \frac{1}{\sqrt{2}}[(\sigma_1 - \sigma_2)^2 + (\sigma_2 - \sigma_3)^2 + (\sigma_3 - \sigma_1)^2]^{\frac{1}{2}} \tag{4.65a}$$

The effective or equivalent stress is also known as "Mises" stress. It can be applied not only to initiation of yield but also to any stage of the plastic deformation in the body. For uniaxial loading, we have $\sigma_e = \sigma_1$.

The effective stress is also related to the second deviatoric stress invariant J_2 by the following relation (Equation 2.42):

$$\sigma_e = \sqrt{3J_2} \tag{4.65b}$$

It can be shown using Equation 2.42 that the effective stress is also a function of the deviatoric stresses S_{ij}. In terms of indicial notation, we have

$$\sigma_e = \left(\frac{3}{2} S_{ij} S_{ij}\right)^{\frac{1}{2}} \tag{4.65c}$$

The corresponding effective or equivalent strain in terms of principal strains is given by

$$\varepsilon_e = \frac{\sqrt{2}}{3}\left[(\varepsilon_1 - \varepsilon_2)^2 + (\varepsilon_2 - \varepsilon_3)^2(\varepsilon_3 - \varepsilon_1)^2\right]^{\frac{1}{2}} \tag{4.66a}$$

which is identical to (see Problem 4.36)

$$\varepsilon_e = \left[\frac{2}{3}\left(\varepsilon_1^2 + \varepsilon_2^2 + \varepsilon_3^2\right)\right]^{\frac{1}{2}} \tag{4.66b}$$

Note that the numerical factor in Equations 4.66 is chosen so that for uniaxial loading we have $\varepsilon_e = \varepsilon_1$. Note that for uniaxial loading, $\varepsilon_2 = \varepsilon_3 = -(1/2)\varepsilon_1$. Also in indicial notation, relation (4.66b) is

$$\varepsilon_e = \left(\frac{2}{3} \varepsilon_{ij} \varepsilon_{ij}\right)^{\frac{1}{2}} \tag{4.66c}$$

Multiplying relation (4.66c) by itself and using Equation 4.58, we have

$$(\varepsilon_{ij}\varepsilon_{ij})^{\frac{1}{2}} = \lambda(S_{ij}S_{ij})^{\frac{1}{2}} \tag{a}$$

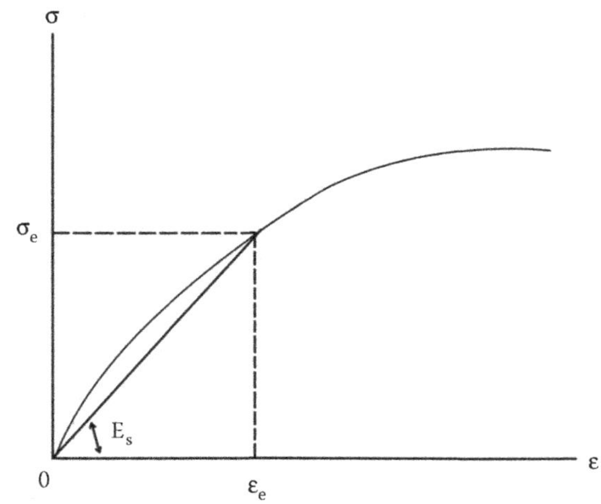

FIGURE 4.21
Secant modulus $E_s = \sigma_e/\varepsilon_e$.

And using the relations in Equations 4.66b and 4.66c, the material constant in Equation (a) is

$$\lambda = \frac{3\varepsilon_e}{2\sigma_e} = \frac{3}{2E_s}, \quad E_s = \frac{\sigma_e}{\varepsilon_e} \tag{4.67}$$

where E_s is the plastic modulus or secant modulus of the material, as shown in Figure 4.21.

It follows that Hencky's plastic stress–strain relations of the deformation theory of plasticity are

$$\begin{aligned}
\varepsilon_x &= \frac{1}{E_s}\left[\sigma_x - \frac{1}{2}(\sigma_y + \sigma_z)\right] \\
\varepsilon_y &= \frac{1}{E_s}\left[\sigma_y - \frac{1}{2}(\sigma_z + \sigma_x)\right] \\
\varepsilon_z &= \frac{1}{E_s}\left[\sigma_z - \frac{1}{2}(\sigma_x + \sigma_y)\right] \\
\gamma_{xy} &= \frac{3}{E_s}\tau_{xy} \\
\gamma_{yz} &= \frac{3}{E_s}\tau_{yz} \\
\gamma_{zx} &= \frac{3}{E_s}\tau_{zx}
\end{aligned} \tag{4.68}$$

Equations 4.68 are valid for proportional loading because of the condition $dJ_2 > 0$.

The validity of the deformation theory of plasticity for loading paths other than the proportional loading paths discussed above has been studied by Budiansky [4.16].

EXAMPLE 4.14

A circular shaft, 20 cm diameter, is subjected to a combined state of loading consisting of torque T = 150 kNm and bending moment M = 225 kNm. The stress–strain relation in simple tension of the J_2 material of the shaft is given by

$$\varepsilon = \frac{\sigma}{E}, \qquad \sigma \le \sigma_0$$

$$\varepsilon = \frac{\sigma}{E} + \frac{\sigma - \sigma_0}{m}, \qquad \sigma > \sigma_0$$

where Young's modulus $E = 200$ GPa, yield stress $\sigma_0 = 250$ MPa, constant $m = 25$ GPa, and Poisson's ratio $\upsilon = 0.3$. Find the components of normal and shear strains.

Solution

The loading causes the following normal and shear stresses:

$$\tau_{xy} = \tau = \frac{Tr}{J} = \frac{16T}{\pi d^3} = \frac{16(150)}{\pi(0.2)^3} = 95.5\,\text{MPa}$$

$$\sigma_x = \sigma = \frac{My}{I} = \frac{32M}{\pi d^3} = \frac{32(225)}{\pi(0.2)^3} = 286.5\,\text{MPa}$$

It follows that the principal stresses in the shaft are

$$\sigma_{1_2} = \frac{\sigma}{2} \pm \sqrt{\frac{\sigma^2}{4} + \tau^2}$$

and von Mises' yield criterion for J_2 material, Equation 4.48, gives

$$\sigma^2 + 3\tau^2 = \sigma_0^2$$

That is $(286.5)^2 + 3\,(95.5)^2 = (330.22)^2 > (250)^2$, indicating that an element of the shaft has yielded under the applied state of loading. The strains in the element include elastic and plastic components.

The elastic components are obtained using the usual Hooke's law

$$\varepsilon_x^e = \frac{\sigma}{E} = \frac{286.5}{200 \times 10^3} = 1.433 \times 10^{-3}$$

$$\varepsilon_y^e = \varepsilon_z^e = -\frac{\upsilon\sigma}{E} = -0.43 \times 10^{-3}$$

and

$$\gamma_{xy}^e = \frac{2(1+\upsilon)}{E}\tau = \frac{2(1.3)(95.5)}{200 \times 10^3} = 1.242 \times 10^{-3}$$

$$\gamma_{yz}^e = \gamma_{zx}^e = 0$$

The plastic strains are found from Equation 4.68 where $E_s = \sigma_e/\varepsilon_e$.

The effective stress is given by Equation 4.65b. Using Equation 2.42 in Chapter 2, we have

$$\sigma_e = \sqrt{\sigma^2 + 3\tau^2} = \sqrt{(286.5)^2 + 3(95.5)^2} = 330.822\,\text{MPa}$$

The effective strain is obtained from the plastic part of the given stress–strain relation of the material in the simple tension test as follows:

$$\varepsilon_e = \frac{\sigma_e - \sigma_0}{m} = \frac{330.822 - 250}{25 \times 10^3} = 3.233 \times 10^{-3}$$

Hence,

$$\frac{1}{E_s} = \frac{\varepsilon_e}{\sigma_e} = \frac{3.233 \times 10^{-3}}{330.822\,\text{MPa}} = 9.773 \times 10^{-6}\frac{1}{\text{MPa}}$$

The plastic strains are calculated using Equation 4.68

$$\varepsilon_x^P = 9.773 \times 10^{-6}\frac{1}{\text{MPa}}(286.5\,\text{MPa}) = 27 \times 10^{-4}$$

$$\varepsilon_y^P = \varepsilon_z^P = \frac{1}{E_s}\left(-\frac{1}{2}\sigma_x\right) = -13.5 \times 10^{-4}$$

$$\gamma_{xy}^P = \frac{3\tau_{xy}}{E_s} = 3\left(0.773 \times 10^{-6}\frac{1}{\text{MPa}}\right)(95.5\,\text{MPa}) = 2.8 \times 10^{-3}$$

$$\gamma_{yz}^P = \gamma_{zx}^P = 0$$

The total strains are obtained by adding the elastic and plastic strains. Hence,

$$\varepsilon_x = 1.433 \times 10^{-3} + 27 \times 10^{-4} = 4.133 \times 10^{-3}$$

$$\varepsilon_y = \varepsilon_z = -0.43 \times 10^{-3} - 13.5 \times 10^{-4} = -1.78 \times 10^{-3}$$

$$\gamma_{xy} = 1.242 \times 10^{-3} + 2.8 \times 10^{-3} = 4.042 \times 10^{-3}$$

$$\gamma_{yz} = \gamma_{zx} = 0$$

Problems

4.1 A square element ABCD, unit thickness, length of side = 10 in. is subject to the stresses shown in Figure P4.1. Assume elastic behavior with $E = 30 \times 10^3$ ksi and $G = 12 \times 10^3$ ksi. Find, (a) the lengths of the deformed diagonals AC and BD, (b) the deformed angle DAB.

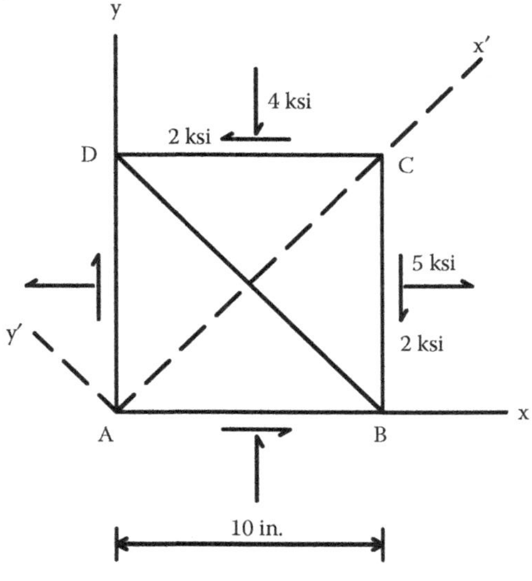

FIGURE P4.1

Ans: 14.142 in., 14.141 in.; 90.01°.

4.2 At a point in a loaded member, the state of stress relative to rectangular coordinates is given by the array

$$\begin{bmatrix} 100 & 50 & 30 \\ 50 & -200 & 20 \\ 30 & 20 & 300 \end{bmatrix} \text{MPa}$$

Find the state of strain in the body. Assume elastic behavior, E = 210 GPa and Poisson's ratio = 0.3.

Ans: Normal strains – $(333, -1524, 1571) \times 10^{-6}$.

4.3 In a loaded elastic member, a parallelepiped element 0AB CDE FG has the dimensions given in Figure P4.3. The element is subject to the following stresses in MPa units:

$$\begin{vmatrix} 10 & 4 & -3 \\ 4 & -6 & 2 \\ -3 & 2 & -2 \end{vmatrix} \text{MPa}$$

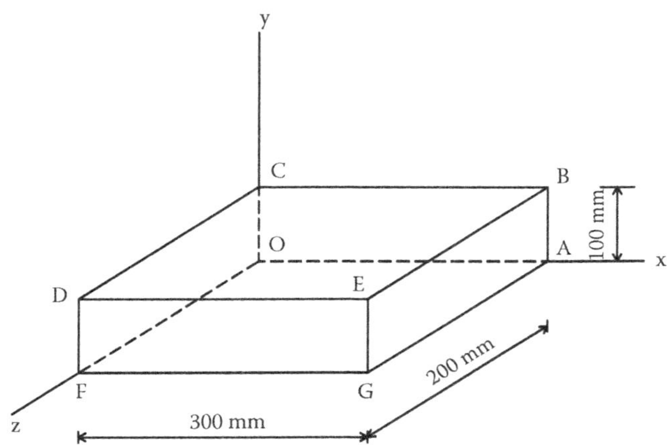

FIGURE P4.3

Find the change in volume of the element. Assume E = 200 MPa and Poisson's ratio = 0.3.

Ans: 0.

4.4 Find expressions for Young's modulus of elasticity and Poisson's ratio in terms of the shear modulus (G) and the bulk modulus (K).

Ans:

$$E = \frac{9KG}{3K + G}, \quad \upsilon = \frac{3K - 2G}{2(3K + G)}$$

4.5 A stainless-steel circular bar of length L = 2 ft. and diameter d = 1 in. is subjected to an axial load P = 40 kip, as shown in Figure P4.5. Assume elastic behavior,

$E = 30 \times 10^3$ ksi, Poisson's ratio = 0.3. (a) Find the changes in length, diameter, and volume of the bar, (b) what are the changes if in addition to the axial load the bar is subjected to two equal and opposite torques of magnitude T = 30 kip-in. at the ends.

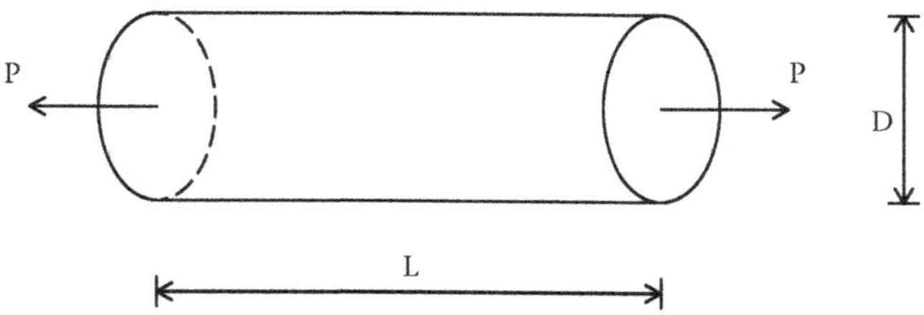

FIGURE P4.5

Ans: (a) Change in diameter = −0.0123 in., (b) additional changes = 0.

4.6 Cork is one of the materials whose Poisson's ratio is essentially zero. Find the largest possible value of Poisson's ratio for an elastic material.

Ans: $(0 \leq \upsilon \leq 1/2)$.

4.7 At a point in an elastic body, the principal strains ε_3, ε_2, ε_1 are in the ratio 3:4:5. The largest principal stress is $\sigma_1 = 140$ MPa. Find the ratio $\sigma_3{:}\sigma_2{:}\sigma_1$ and values of σ_2 and σ_3. Assume Poisson's ratio = 0.3 and E = 200 GPa.

Ans: 12:13:14; 130 MPa, 120 MPa.

4.8 The displacements of a 2D element in a structural component along the x and y axes are as follows:

$$u(x, y) = C_1x^3 + C_2xy^2$$
$$v(x, y) = C_3x^2 y$$

where C_j, j = 1, 2, 3 are constants. For a valid elasticity solution, find the relations among the constants in terms of Poisson's ratio υ. For Poisson's ratio = 0, find the stresses.

Ans: $6\,C_1 + (1 - \nu)\,(C_2 + C_3) = 0$, $\sigma_x = Ek\,y^2$, k = constant.

4.9 The state of stress at a point in a loaded member is given by the matrix

$$\begin{bmatrix} 15 & 5 & 5 \\ 5 & -10 & 0 \\ 5 & 0 & 12 \end{bmatrix} \text{MPa}$$

Determine what fraction of the strain energy density changes the shape of an element in the member without distortion. Assume Poisson's ratio = 0.3.

Ans: 11%.

4.10 A structural member is subjected to an axial load. If, during a certain application, Poisson's ratio of the material increases from 0.2 to 0.3, what can you conclude about the deformation of the member?

Ans: u_v/u decreases while u_d/u increases.

4.11 A uniformly tapered member of circular cross section is fixed at one end and subjected to an axial load P at the free end as shown in Figure P4.11. The length of the member is L and the diameters at the ends are d_1 and d_2 ($d_2 > d_1$). Assume elastic behavior with E as the modulus of elasticity, determine the strain energy stored in the member and its elongation.

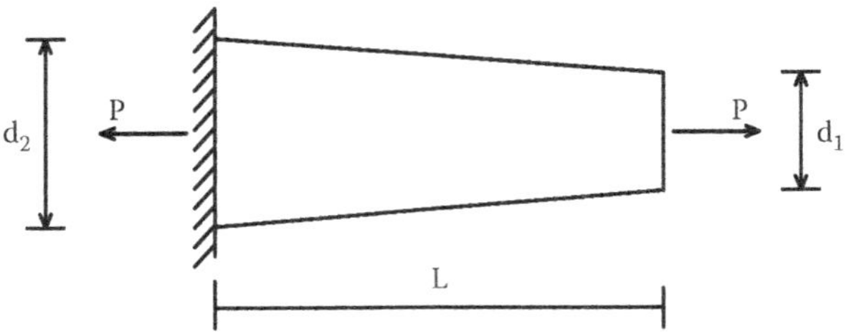

FIGURE P4.11

Ans: $U = (2P^2L/\pi d_1 d_2 E)$, $\delta = 4PL/\pi E d_1 d_2$.

4.12 Show that the lateral contraction of an isotropic material in the strain-hardening region of the tension test is given by

$$\eta = \frac{1}{2} - \frac{1}{E}\left(\frac{1}{2} - \upsilon\right)E_t$$

4.13 The stress–strain curve in the tensile test of a ductile material is described by the following Ramberg–Osgood's relation:

$$\varepsilon = \frac{\sigma}{E} + 0.3\varepsilon_0\left(\frac{\sigma}{\sigma_0}\right)^n$$

where σ_0 is the reference stress (yield stress of the material) and n is the strain-hardening exponent. Find expressions of the tangent modulus, E_t, plastic modulus, E_p, and work done, W_p, in terms of the stress, n, and the plastic strain.

Ans: $E_t = \dfrac{E\sigma}{\sigma + nE\sigma_0\varepsilon^p}$, $W_p = \dfrac{n\sigma\varepsilon^p}{n+1}$.

4.14 For the Ramberg–Osgood stress–strain curve of the previous example, find an expression of the strain that causes instability in the tensile test and compare it to the instability strain of a material obeying the power law $\sigma = C\varepsilon^{1/n}$, where C is constant. If $n = 6$ and $\varepsilon_0 = 0.002$, find the percentage error in the strains.

Ans: 0.51%.

4.15 A compound member of length L is made up of a solid rod of material 1 surrounded by a thick cylinder of material 2 (Figure P4.15). The cross-sectional areas are A_1 and A_2, respectively. Both rod and cylinder are bonded together to form one member.

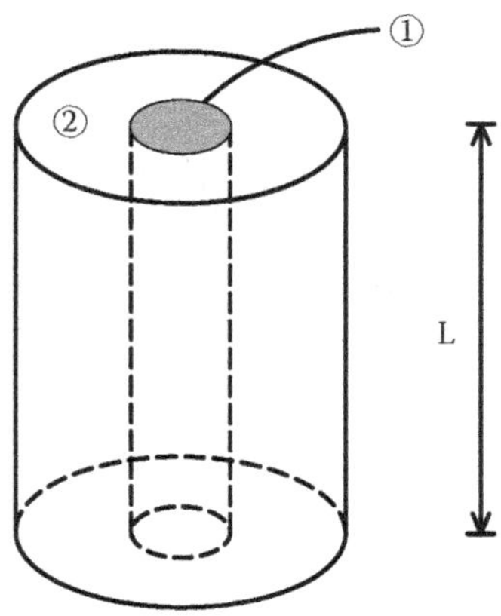

FIGURE P4.15

The true stress–true strain curves in the tensile test of materials are

$$\sigma_1 = K_1\varepsilon^{n_1} \quad \text{and} \quad \sigma_2 = K_2\varepsilon^{n_2}$$

The compound member is pulled in tension by a force P until instability occurs. Find the strain at instability in terms of K_1, K_2, n_1, and n_2. If $K_1 = 20{,}000$ MPa, $n_1 = 0.20$, $K_2 = 80{,}000$ MPa, $n_2 = 0.40$, calculate the ratio A_1/A_2 if the true strain at instability is $\varepsilon = 0.35$.

Ans: $A_1/A_2 = 1.65$.

4.16 Repeat Problem 4.15 if the tensile test data of the materials are described by the following true stress–engineering strain relations:

$$\sigma_1 = K_1 e^{n_1} \quad \text{and} \quad \sigma_2 = K_2 e^{n_2}$$

Assuming the same numerical data for K_1, K_2, n_1, and n_2 as in P4.15, find the ratio A_1/A_2 if the engineering strain at instability is $e = 0.5$.

Ans: $A_1/A_2 = 1.74$.

4.17 Show that a rigid/plastic material may be represented by the following true stress σ–true strain ε equation:

$$\sigma = A(1 - me^{-n\varepsilon})$$

where A, m, n are constants and e is the exponential constant (e = 2.71862). Find the true stress and true strain at the onset of instability in the tensile tension test.

Ans: $\sigma = \dfrac{nA}{n+1}$, $\varepsilon = \dfrac{1}{n}[\mathrm{Ln}(1+n)m]$.

4.18 The cross section of a prismatic beam subjected to pure bending moment M is an isosceles triangle of base = b and height = h. The material is elastic–perfectly plastic with yield stress = σ_0. Find the ratio of the fully plastic moment divided by the moment that initiates yield, M_p/M_y.

Ans: $M_p/M_y = 2.31$.

4.19 The state of stress at a point in an element of a loaded member is given by the matrix

$$\varepsilon = \frac{1}{n}[\text{Ln}(1+n)m]$$

If the member yields according to Tresca's criterion, find the yield strength of the material.

Ans: 11.43 ksi.

4.20 The state of stress at a point in a structural component is given by

$$\begin{bmatrix} 400 & 0 & 100 \\ 0 & 300 & 100 \\ 100 & 100 & 200 \end{bmatrix} \text{MPa}$$

Given that the yield stress of the material = 180 MPa, find the factor of safety (f) against failure by yield assuming that yield occurs in accordance with (a) maximum shearing stress theory, (b) maximum distortional energy theory.

Ans: (a) f = 1.7, (b) = 2.25.

4.21 A thin-walled tube of mean radius r and thickness t is closed at the ends and subjected to an axial tensile load F, which is less than the necessary value to cause yield in the tube. The material is elastic–perfectly plastic with yield stress σ_0. If a generally increasing internal pressure p is applied, find the value of p_e to initiate yield in the tube according to (i) maximum shear stress theory, and (ii) maximum distortional energy theory.

Find the numerical values of p_e when t = 0.25 in., r = 10 in., F = 200 kip, and the yield stress = 36 ksi.

Ans: (i) 1163 psi, (ii) 1034 psi.

4.22 A thin-walled cylindrical pressure vessel with 30 cm diameter, wall thickness 3 mm, and closed ends is fabricated from a material with yield strength 350 MPa. The vessel is subject to an internal pressure of 5 MPa. Find the additional tensile axial load F, which causes yield in the vessel assuming (i) Tresca's criterion and, (ii) von Mises' criterion.

Ans: (i) 636 KN, (ii) 777 kN.

4.23 The state of stress at a point in a loaded component is

$$\begin{bmatrix} 20 & 5 & 0 \\ 5 & 25 & 0 \\ 0 & 0 & -15 \end{bmatrix} \text{ksi}$$

The material is high strength steel with yield stress = 50 ksi. Find the factor of safety against failure by yield based on (a) the Tresca's criterion, (b) the von Mises' criterion.

Ans: (a) 1.16, (b) 1.69.

4.24 An element of a ductile material is subjected to the following 2D state of stress:

$$\sigma_1 = \sigma$$
$$\sigma_2 = A\sigma$$
$$\sigma_3 = 0$$

where A is an arbitrary constant and σ is positive. If the material yields when the maximum shearing stress reaches the critical value k, find the values of stress σ in terms of k and A where $A < 0$, $0 < A < 1$ and $A > 1$.

Ans: $2k/(1 - A)$, $2k$, $2k/A$.

4.25 A fixed-ends prismatic beam, 80 cm length and 2 cm × 6 cm rectangular cross section, is subjected to an axial force F at section C distance 20 cm from the left end as shown in Figure P4.25. The beam is made of an elastic–strain-hardening material with the following stress–strain relation:

$$\sigma = E\varepsilon, \qquad \sigma < \sigma_0$$
$$\sigma = \sigma_0 + E_p\varepsilon^p, \quad \sigma > \sigma_0$$

where $E = 200$ GPa, yield stress $\sigma_0 = 250$ MPa, plastic modulus $E_p = (1/3)$ E, and ε^p is the plastic strain. The axial force is increased from $F = 0$ to $F_2 = 3\ \sigma_0 A$ and then unloaded to $F = 0$ (A is the cross-sectional area). (a) Find Fe (force initiating yield in the beam), F1 (fully plastic force), plastic strains accumulated in the beam, and the residual stresses and strains when the beam is completely unloaded, (b) plot the curve F versus horizontal displacement u of section C.

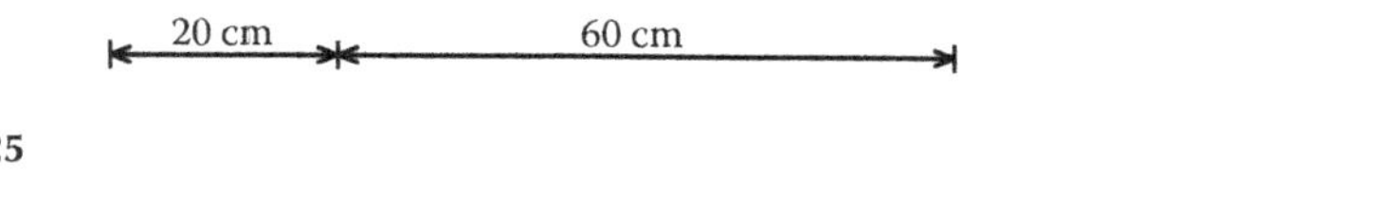

FIGURE P4.25

Ans: $F_e = 400$ kN, $F_1 = 750$ kN, $\varepsilon_a^p = 3.28 \times 10^{-3}$, $\varepsilon_b^p = -0.47 \times 10^{-3}$ $\sigma_a^{res} = \sigma_b^{res} = -93.75$ MPa $\varepsilon_a^{res} = 2.81 \times 10^{-3}$, $\varepsilon_b^{res} = -0.94 \times 10^{-3}$.

4.26 For the fixed ends beam of the previous problem, the force F after complete unloading is applied in the reverse direction. Find the values of F that will subsequently initiate yield in the beam assuming (i) isotropic, (ii) kinematic, and (iii) independent hardening rules.

Ans: (i) –600 kN, (ii) 100 kN, (iii) –250 kN.

4.27 The three-bar truss shown in Figure P4.27 is subjected to a horizontal load Q at joint D. The length of the middle bar is L and the inclined bars make angle α with the vertical. The bars are made of an elastic–perfectly plastic material with yield stress, σ_0, modulus, E, and cross-sectional area, A.

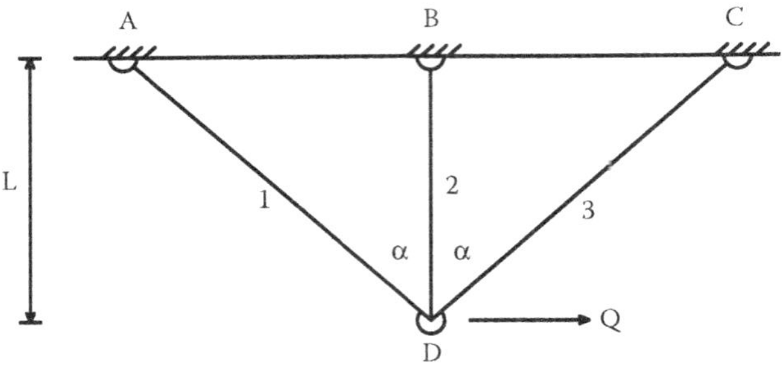

FIGURE P4.27

The load is increased from $Q=0$ to the plastic limit load Q_p, unloaded to $Q=0$, reloaded in the reversed direction to the plastic limit Q_p^*, and finally unloaded to $Q=0$. Find (a) the elastic limit load Q_e and the plastic limit load Q_p in the initial loading, (b) the elastic and plastic limit loads, Q_e^* and Q_p^*, in the reversed loading state, (c) the residual stresses and strains in the bar and the residual horizontal and vertical displacements, u and v, of joint D at the end of the deformation.

Ans: $Q_e = Q_p = (2A\sin\alpha)\sigma_0, \quad Q_e^* = Q_p^* = -(2A\sin\alpha)\sigma_0 \frac{u}{L} = \frac{\varepsilon_0}{\sin\alpha\cos\alpha}, \quad V = 0.$

4.28 The same three-bar truss of the previous problem is subjected to the combined effects of a vertical load P and horizontal load Q at joint D as shown in Figure P4.28.

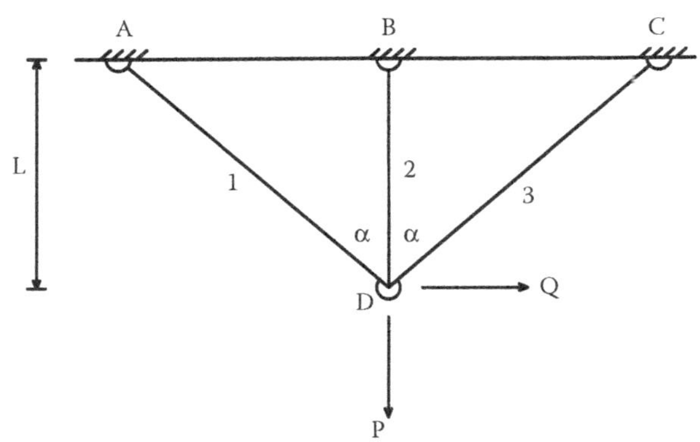

FIGURE P4.28

Determine the residual horizontal and vertical displacements of joint D, u, and v respectively, at the end of the deformation for each of the following loading paths: (a) the load P is increased from zero to the plastic limit load P_p, maintained at this level while load Q is increased from zero to the plastic collapse load of the truss, (b) the load Q is first increased to Q_p, maintained at this level while load P in increased to the collapse load.

$$\textbf{Ans: } a-a-\frac{u}{L}=\left(\frac{4\cos^2\alpha+1}{4\sin\alpha\cos^3\alpha}\right)\varepsilon_0,\ \frac{V}{L}=\left(\frac{\varepsilon_0}{4\cos^4\alpha}\right)$$

$$\frac{u}{L}=\left(\frac{5\varepsilon_0}{4\sin\alpha\cos\alpha}\right),\ \frac{V}{L}=\left(\frac{\varepsilon_0}{4\cos^2\alpha}\right).$$

References

4.1. S. Timoshenko, *Strength of Materials*, 3rd edition, Vol. 2, Von Nostrand Co., New York, 1956.
4.2. P. W. Bridgman, *J. Appl. Phys.*, vol. 18, p. 246, 1947.
4.3. B. Crossland, *Proc. Inst. Eng.*, vol. 169, p. 935, 1954.
4.4. J. Bauschinger, *Zivilingenieur*, vol. 27, p. 289, 1981.
4.5. P. Ludwik, *Elemente der Technologisschen Mechanik*, Springer Verlag, Berlin, 1909.
4.6. H. W. Swift, *J. Mech. Phys. Solids*, vol. 1, p. 1, 1952.
4.7. W. Ramberg and W. R. Osgood, NACA Tech. Note, 902, 1943.
4.8. B. Budiansky and O. L. Mangasarmian, Plastic stress concentration at a circular hole in an infinite sheet subjected to equal biaxial tension, *J. Appl. Mech. ASME*, vol. 26, p. 59, 1960.
4.9. G. I. Taylor and H. Quincey, *Philos. Trans. R. Soc. A*, vol. 300, p. 323, 1931.
4.10. W. F. Chen and D. J. Han, *Plasticity for Structural Engineers*, Springer-Verlag, New York, 1988.
4.11. W. Lode, *Z. Angew Math. Mech.*, vol., 5, p. 142, 1926.
4.12. A. Nadai, *Theory of Flow and Fracture of Solids*, McGraw-Hill Book Company, New York, 1950.
4.13. R. von Mises, *Z. Angew Math. Mech.*, vol. 8, p. 61, 1928.
4.14. L. Prandtl, *Proc. 1st International Congress of Applied Mechanics*, Delft, The Netherlands, p. 43, 1924.
4.15. H. Hencky, *Z. Angew Math. Mech.*, vol. 4, p. 223, 1924.
4.16. B. Budiansky, A reassessment of deformation theory of plasticity, *J. Appl. Mech. ASME*, vol. 26, pp. 259–264, 1959.

5

Torsion of Prismatic Bars

5.1 Introduction

In this chapter, we discuss methods to determine the stress and deformation in prismatic bars of arbitrary cross sections subject to equal and opposite end torques. A typical member is shown in Figure 5.1. The origin of rectangular coordinates is located at the center of one end of the member and the z-axis is aligned along an axis parallel to a generator of the cross section. Under the action of end torques, T, every cross section rotates about the z-axis, and except for bars with circular profiles, experience out-of-plane deformation or warping (Figure 5.2b). The nonvanishing stresses are τ_{xz} and τ_{yz}, which vary with the x and y coordinates, and the remaining stresses, σ_x, σ_y, σ_z, and τ_{xy}, vanish throughout the member. The angle of twist ϕ is a measure of the rotation of one end of the member relative to the other. The unit angle of twist is $\theta = \phi/L$, where L is the length of the member. The goal is to determine the shear stresses, τ_{xz} and τ_{yz}, and the displacements u, v, and w, which determine the rotation and warping of the cross sections. The first correct solution of the torsion problem was given by Saint-Venant [5.1].

On any cross section of the member, the distribution of the shear stresses must satisfy the requirements of equilibrium. This implies that the summation of forces along the x and y directions must vanish

$$\iint_A \tau_{zx}\, dxdy = \iint_A \tau_{zy}\, dxdy = 0 \tag{a}$$

Also, the sum of the moments of the shear stresses about the z-axis must balance the external torque T according to the relation

$$\iint_A (x\tau_{zy} - y\tau_{zx})\, dxdy = T \tag{b}$$

The equilibrium equation governing the shear stresses is (Equation 2.5)

$$\frac{\partial \tau_{xz}}{\partial x} + \frac{\partial \tau_{yz}}{\partial y} = 0 \tag{5.1}$$

The corresponding kinematics relations for the shearing strains are given by Equation 3.12.

$$\begin{aligned} \gamma_{xz} &= \frac{\partial w}{\partial x} + \frac{\partial u}{\partial z} \\ \gamma_{yz} &= \frac{\partial w}{\partial y} + \frac{\partial v}{\partial z} \end{aligned} \tag{5.2}$$

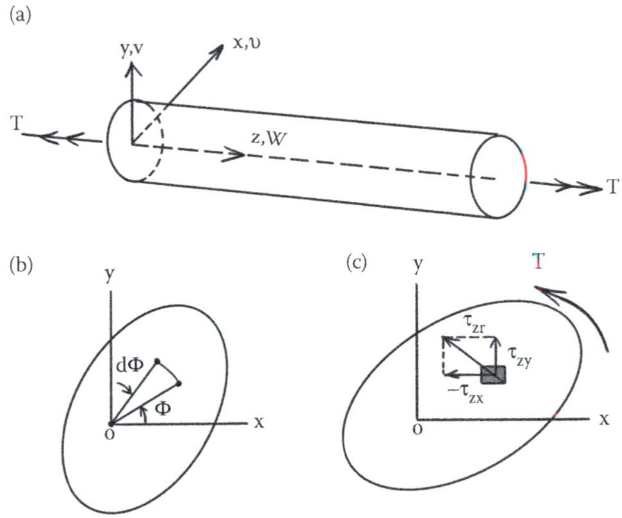

FIGURE 5.1
(a) Torsion of a prismatic bar of arbitrary cross section; (b) displacements; (c) stresses.

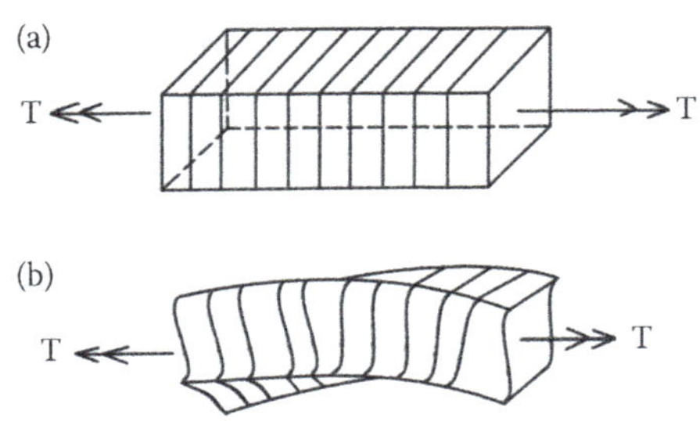

FIGURE 5.2
Torsion of a rectangle bar. (a) Rectangular bar subjected to torsion without warping; (b) bar undergoing torsion with warping.

In addition to Equations 5.1 and 5.2, the shear stress–shear strain behavior or constitutive equation of the material needs to be specified to complete the analysis.

5.2 Torsion of a Cylindrical Bar

Consider a solid cylindrical member of radius, a, subjected to end moments, T. For elastic behavior, it is proved in elementary courses on mechanics of materials that the shear stress $\tau_{z\theta}$ in polar coordinates, or simply τ, and the shear strain γ acting over any element of the cross section are proportional to the radial distance r from the central axis; Equations 1.24 through 1.27.

$$\tau = Gr\theta = \frac{2Tr}{\pi a^4} \tag{5.3}$$

$$\gamma = r\theta$$

The torsional rigidity of the member is

$$\frac{T}{\theta} = \frac{\pi a^4 G}{2} = GJ, \quad J = \frac{\pi a^4}{2}$$

where G is the shear modulus of the material and J is the polar moment of inertia. The displacement components are computed by considering the deformation of point P, as in Figure 5.1b. After subjecting the bar to end torques, point P moves to point P′, and for small values of the angle of twist $d\phi$, $\sin d\phi \approx d\phi$ and $\cos d\phi \approx 1$, the increment in the tangential displacement du_θ is given as $du_\theta = r\, d\phi$, and the increments along the Cartesian axes are computed as

$$du = r\cos(\phi + d\phi) - r\cos\phi = -y\, d\phi$$

$$dv = r\sin(\phi + d\phi) - r\sin\phi = x\, d\phi$$

$$dw = 0$$

Since, $d\phi = \theta\, dz$, it follows that

$$u = -\theta yz, \quad v = \theta xz, \quad w = 0 \tag{5.4}$$

Denoting the yield stress of the material by τ_0, $\tau_0 = G\gamma_0$, the member begins to yield at the location of greatest shear stress, that is, at radius $r = a$ when the applied torque reaches the value of the elastic limit

$$T_e = \frac{1}{2}\pi\tau_0 a^3 \tag{a}$$

The corresponding value of the unit angle of twist is $\theta_e = \gamma_o/a = \tau_0/Ga$. Note that τ_0 is approximately equal to $\sigma_y/2$, where σ_y is the yield stress in tension.

5.2.1 Elastic–Perfectly Plastic Material

In the elastic–perfectly plastic model of the material, an increase in the torque beyond the elastic limit would create a plastic zone in the form of an annulus spreading from the boundary and surrounding an elastic core of radius = c, as shown in Figure 5.3. In this case, the boundary of the plastic zone is known *a priori* because of the symmetry of the cross section. The distribution of the shear stress becomes

$$\begin{aligned} \tau &= \tau_0\frac{r}{c}, \quad 0 \le r \le c \\ \tau &= \tau_0, \quad c \le r \le a \end{aligned} \tag{5.5}$$

and the twisting moment is computed as $T = \iint_A \tau r dA$ with $dA = r dr\, d\theta$. Hence,

$$T = 2\pi\int_0^a \tau r^2\, dr = 2\pi\frac{\tau_0}{c}\int_0^c r^3\, dr + 2\pi\tau_0\int_c^a r^2\, dr$$

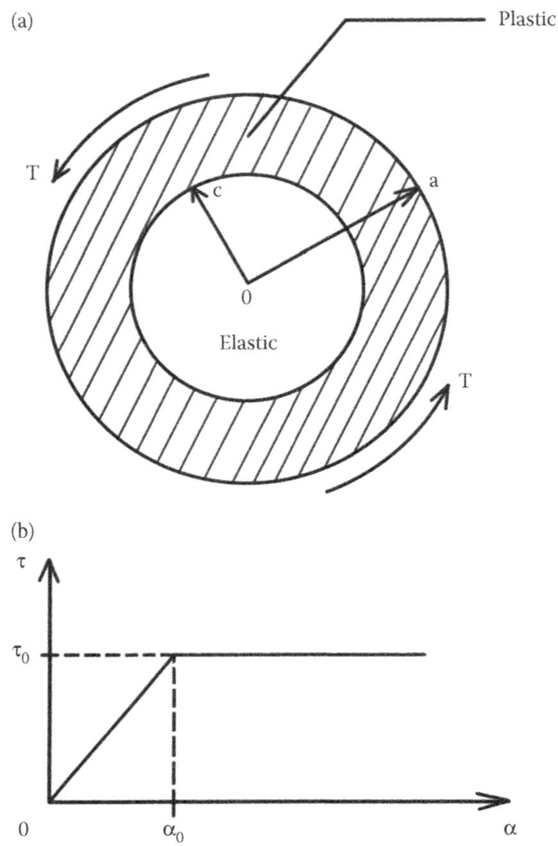

FIGURE 5.3
Elastic–plastic torsion of a cylindrical bar. (a) Plastic zone; (b) stress diagram.

Evaluating the integrals yield, the expression

$$T = \frac{T_e}{3}\left(4 - \frac{c^3}{a^3}\right) \tag{5.6}$$

Since the geometrical relation $\gamma = r\,\theta$ holds throughout the cross section, it follows that at $r = c$, $\gamma_o = c\,\theta$, and because $\gamma_o = a\,\theta_e$, it follows that $c/a = \theta_e/\theta$, and expression (5.6) is equivalent to

$$T = \frac{T_e}{3}\left[4 - \left(\frac{\theta_e}{\theta}\right)^3\right] \tag{5.7}$$

The value of the ultimate or fully plastic torque is obtained by setting $c \to 0$ in Equation 5.6, which yields

$$T_p = \frac{4}{3}T_e = \frac{2\pi}{3}\tau_0 a^3 \tag{b}$$

For, $T_e \leq T \leq T_p$, the unit angle of twist is obtained from Equation 5.7 as

$$\theta = \frac{\theta_e}{(4 - 3(T/T_e))^{1/3}} \quad \text{(c)}$$

Note that when c tends to zero, θ tends to infinity ($\theta = \gamma_0/c$), and the magnitude of the fully plastic torque is attained in the limit as the angle of twist increases, $T \to T_p$. Therefore, an elastic core of material must exist for all finite values of the angle of twist. In the limit, the elastic core reduces to a single point.

Residual stresses: Suppose that the twisting moment is unloaded from the level given by Equation 5.6. Since the unloading path is elastic, the residual stresses are obtained by subtracting the elastic value of the shear stress from the expressions in Equation 5.5. Hence, in the elastic region $0 \leq r \leq c$

$$\tau^R = \tau_0 \frac{r}{c} - \frac{T_e}{3}\left(4 - \frac{c^3}{a^3}\right)\frac{2r}{\pi a^4}$$

which can be simplified to yield

$$\frac{\tau^R}{\tau_0} = \frac{r}{c}\left(1 - \frac{4}{3}\frac{c}{a} + \frac{1}{3}\frac{c^4}{a^4}\right), \quad 0 \leq r \leq c \quad \text{(5.8a)}$$

In an identical manner, the residual shear stress in the plastic region is computed as

$$\tau^R = \tau_0 - \frac{T_e}{3}\left(4 - \frac{c^3}{a^3}\right)\frac{2r}{\pi a^4}$$

and

$$\frac{\tau^R}{\tau_0} = 1 - \frac{r}{3a}\left(4 - \frac{c^3}{a^3}\right), \quad c \leq r \leq a \quad \text{(5.8b)}$$

The residual unit angle of twist is obtained as

$$\theta^R = \theta - \frac{T}{GJ} = \frac{\gamma_o}{c} - \frac{T_e}{3G}\left(4 - \frac{c^3}{a^3}\right)\frac{2}{\pi a^4}$$

which can be simplified to yield

$$\theta^R = \frac{\gamma_o}{c}\left[1 - \frac{c}{3a}\left(4 - \frac{c^3}{a^3}\right)\right] \quad \text{(5.9)}$$

EXAMPLE 5.1

A 2 in. diameter steel bar is fixed at one end and the free end is twisted through an angle of 10° and then released (Figure 5.4). The length of the bar is 50 in., and the material is considered elastic–perfectly plastic with yield stress of 15 ksi.

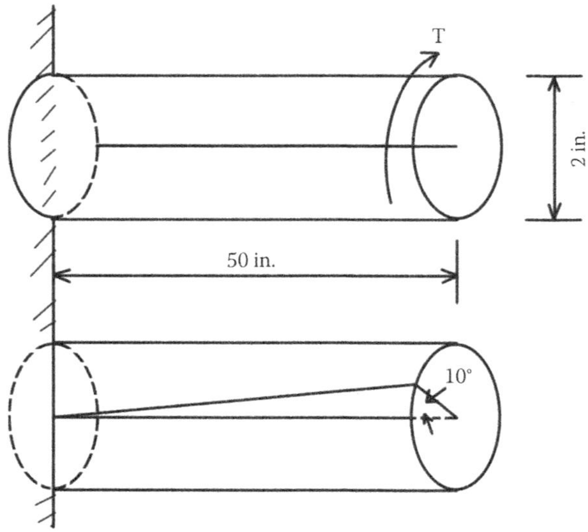

FIGURE 5.4
Example 5.1.

Assuming $G = 10 \times 10^3$ ksi, find

a. The torque that is applied at the free end
b. The residual shear stress and angle of twist upon unloading to $T = 0$

Solution

The torque that initiates yield is $T_e = 1/2\pi\tau_0 a^3 = \pi/2(15)(1)^3 = 23.56$ ksi, and the corresponding angle of twist is $\theta_e = \gamma_o/a = \tau_0/Ga = 0.0015$ rad/in.

The angle of twist experienced by the bar is $\theta = \phi/L = 10(\pi/180)/50 = 0.00349$ rad/in.

Hence, the applied torque is greater than T_e, and it induces a plastic zone in the bar.

a. The applied torque at the free end of the bar is computed from Equation 5.7 as

$$T = \frac{T_e}{3}\left[4 - \left(\frac{\theta_e}{\theta}\right)^3\right] = \frac{23.56}{3}\left[4 - \left(\frac{0.0015}{0.00349}\right)^3\right] = 30.8 \text{ kip-in.}$$

b. The plastic zone is an annulus extending from the outer boundary to radius c where

$$c = a\left(\frac{\theta_e}{\theta}\right) = 0.43 \text{ in.}$$

The residual shearing stress is computed from Equations 5.8a and 5.8b as

$$\tau^R = 15.28r, \qquad 0 \le r \le 0.43$$

$$\tau^R = 15 - 19.6r, \quad 0.43 \le r \le 1$$

Finally, the residual angle of twist is calculated using Equation 5.9 as

$$\phi^R = \theta^R L = 4.38°$$

EXAMPLE 5.2

A hollow circular shaft of inner radius b and outer radius a ($b < a$) is made of an elastic-perfectly plastic material with yield stress in shear, τ_0. Show that the ratio of the plastic and elastic torque is given by

$$\frac{T_P}{T_e} = \frac{4}{3}\left[1 - \frac{b^3}{a^3}\right] \Big/ \left[1 - \frac{b^4}{a^4}\right]$$

Solution

The cross section of the member is shown in Figure 5.5. The shear stress is given by $\tau = Tr/J$ where $J = \pi/2(a^4 - b^4)$. The maximum stress occurs at the external boundary $r = a$, and the torque that initiates yield at the boundary is given by

$$T_e = \frac{\tau_0 J}{a} = \frac{\pi\tau_0}{2} a^3 \left(1 - \frac{b^4}{a^4}\right) \quad \text{(a)}$$

When the applied torque is increased further, an annular plastic zone is developed and spreads inward, as shown in Figure 5.5. Denoting the internal radius of the plastic zone by c, the shear stress for the elastic–perfectly plastic material is given by

$$\begin{aligned} \tau &= \tau_0 && \text{for } c \le r \le a \\ \tau &= \tau_0 \frac{r}{c} && \text{for } b \le r \le c \end{aligned} \quad \text{(b)}$$

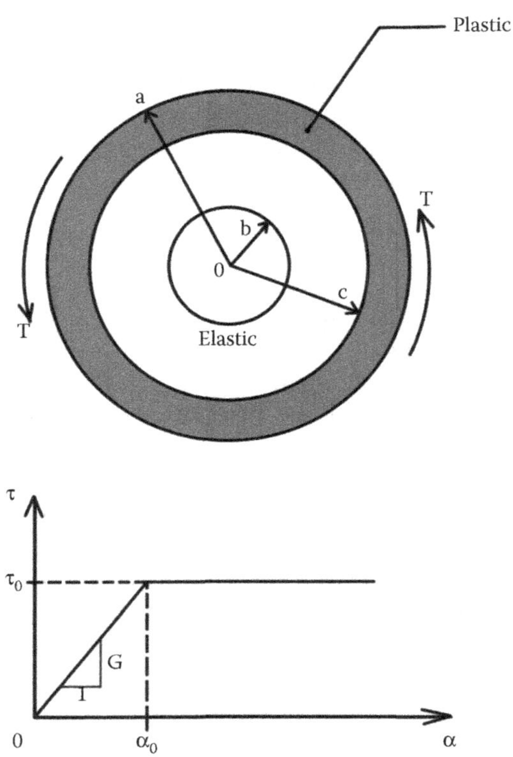

FIGURE 5.5
Example 5.2.

And the elastic/plastic torque supported by the member is computed as follows:

$$T = \iint \tau r dA = 2\pi \int_b^a \tau r^2 dr$$

Using the stress–strain relations (b), we find

$$T = 2\pi\tau_0 \left[\int_b^c \frac{r^3}{c} dr + \int_c^a r^2 dr \right]$$

$$T = \frac{2\pi}{3}\tau_0 a^3 \left[1 - \frac{c^3}{4a^3}\left(1 + 3\frac{b^4}{c^4}\right) \right] \quad \text{(c)}$$

When $c = a$, the torque in Equation (c) coincides with the yield torque in Equation (a). The ultimate torque is reached when $c = b$, $T_p = (2\pi/3)\tau_0 a^3(1 - b^3/a^3)$, and $T_p/T_e = 4/3[1 - b^3/a^3]/[1 - b^4/a^4]$.

In the particular case of $a = 2b$, $T_p/T_e = 1.24$, which indicates that the member can support 24% more torque than what the elastic theory predicts.

5.2.2 Work-Hardening Material

In this section, instead of assuming elastic–perfectly plastic material to determine the response of cylindrical bars to torsion, we consider work-hardening or strain-hardening material. The cylindrical bar in Example 5.2 is used to illustrate the procedure.

A hollow shaft of outer radius a, inner radius b, $b < a$, made of strain-hardening material, is subjected to end torques T. The plastic strain in the static tensile test is a straight line of slope G_p, usually a fraction of the modulus G, as shown in Figure 5.6. The basic equations needed to determine the response include an equation of geometry, $\gamma = r\theta$, equilibrium equation to determine the applied torque:

$$T = \iint_A \tau(r) r dA$$

Here, A is the cross-sectional area of the bar, and an equation that relates the shear stress to the shear strain of the material at all stages of the deformation.

In the elastic stage of deformation when the stress in the outer fibers reaches the yield stress τ_0, the torque that initiates yield is the maximum elastic torque given by (Example 5.2)

$$T_e = \frac{\pi\tau_0 a^3}{2}\left(1 - \frac{b^4}{a^4}\right)$$

and the corresponding unit angle of twist is $\theta_e = (\tau_0/Ga)$. As $T > T_e$, the shear stress increases with the radius r in both elastic and plastic regions.

The total equivalent or effective strain in any plastic element, which can be obtained from Equation 4.66c, gives the necessary stress–strain relation of the assumed strain hardening of the material shown in Figure 5.6, which is as follows:

$$\tau = \tau_0 + \frac{G_p}{3}\left(r\theta - \frac{\tau}{G}\right) \quad (5.10)$$

where $\theta = \tau_0/Gc$, and c is the radius to the elastic–plastic (strain hardening) boundary shown in red in Figure 5.6. Substituting for θ in Equation 5.10 and rearranging the

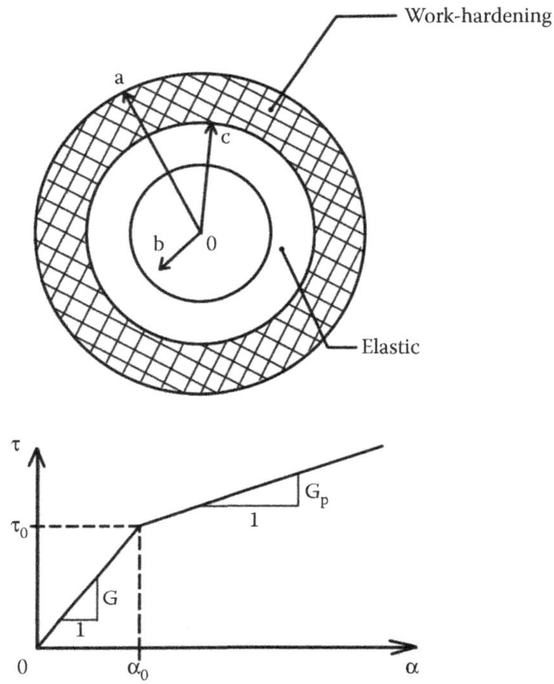

FIGURE 5.6
Torsion of a cylindrical bar made of work-hardening material.

terms lead to the stresses

$$\tau = \tau_0\left[1 + (rG_p/3cG)/1 + (G_p/3G)\right], \quad c \le r \le a \tag{5.11a}$$

and the shear stress in the elastic region is (see Equation (b) in Example 5.2)

$$\tau = \tau_0(r/c), \quad b \le r \ge c \tag{5.11b}$$

It follows that the applied torque is

$$T = \iint\limits_A \tau(r)rdA = \int \tau(r)r(2\pi r)dr = 2\pi\tau_0\left\{\int_b^c \left(\frac{r}{c}\right)r^2dr + \int_c^a \left[\frac{1 + rG_p/3Gc}{1 + G_p/3G}\right]r^2\,dr\right\}$$

Performing the integration and simplifying the algebra, the applied torque is

$$\left(1 + \frac{G_p}{3G}\right)T = \frac{2\pi\tau_0 a^3}{3}\left\{1 + \frac{1}{4}\left[-\frac{c^3}{a^3}\left(1 + \frac{3b^4}{c^4}\right) + \frac{aG_p}{cG}\left(1 - \frac{b^4}{a^4}\right)\right]\right\} \tag{5.12a}$$

which gives the relation between the applied torque and extent of the strain-hardening plastic zone during the twisting action.

By setting $b = 0$ in Equation 5.12a, the corresponding formula for the applied torque in the torsion of a solid cylindrical shaft of strain-hardening material and radius a is obtained

$$\left(1 + \frac{G_p}{3G}\right)T = \frac{2\pi\tau_0 a^3}{3}\left[1 + \frac{1}{4}\left(-\frac{c^3}{a^3} + \frac{aG_p}{cG}\right)\right] \tag{5.12b}$$

The fully plastic torque T_p for the strain-hardening hollow bar is obtained by setting $c = b$ in Equation 5.12a. The result is

$$\left(1 + \frac{G_p}{3G}\right)T_p = \frac{2\pi a^3}{3}\tau_0\left[1 - \frac{b^3}{a^3} + \frac{G_p}{4G}\left(\frac{a}{b} - \frac{b^3}{a^3}\right)\right] \tag{5.13}$$

and the fully plastic angle of twist is $\theta_p = \tau_0/Gc$.

Note that if $G_p = 0$, Equation 5.12a is reduced to Equation (c) in Example 5.2 while Equation 5.12b is reduced to Equation 5.7.

Residual stresses: If the bar is unloaded from an elastic–plastic stage of deformation, the residual stresses in the elastic and plastic zones are obtained by subtracting the elastic stress distribution, namely, $2Tr/\pi(a^4 - b^4)$, where T is given by Equation 5.12a, from the stresses given in Equations 5.11.

It turns out that the residual stress is negative in the outer zone of the plastic annulus and positive in the remainder of the cross section, Reference 2 2.

The residual unit angle of twist is $\theta = \tau_0/Gc$—the elastic unit angle of twist, which is given by

$$\frac{2T}{\pi(a^4 - b^4)G},$$

where T is given by Equation 5.12a.

Variations of T/T_e *with* θ/θ_e: Consider the hollow cylindrical bar discussed above. When the material is elastic–perfectly plastic, $G_p = 0$, making use of Equations (a) and (c) in Example 5.2, and noting that $a/c = \theta/\theta_e$, it is readily confirmed that the torque–unit angle of twist relation is

$$\frac{T}{T_e} = \frac{1}{1 - (b^4/a^4)}\left[\frac{4}{3} - \frac{1}{3}\left(\frac{\theta}{\theta_e}\right)^{-3} - \left(\frac{b^4}{a^4}\right)\left(\frac{\theta}{\theta_e}\right)\right] \tag{5.14a}$$

For solid cylindrical bar, $b = 0$, and the relation is

$$\frac{T}{T_e} = \frac{1}{3}\left[4 - \left(\frac{\theta}{\theta_e}\right)^{-3}\right] \tag{5.14b}$$

Similarly, for work-hardening material, making use of Equation 5.12a, the torque–unit angle of twist relation for hollow cylindrical bar is

$$\left(1 + \frac{G_p}{3G}\right)\frac{T}{T_e} = \frac{G_p}{3G}\left(\frac{\theta}{\theta_e}\right) + \frac{1}{1 - (b/a)^4}\left[\frac{4}{3} - \frac{1}{3}\left(\frac{\theta}{\theta_e}\right)^{-3} - \left(\frac{b}{a}\right)^4\left(\frac{\theta}{\theta_e}\right)\right] \tag{5.15a}$$

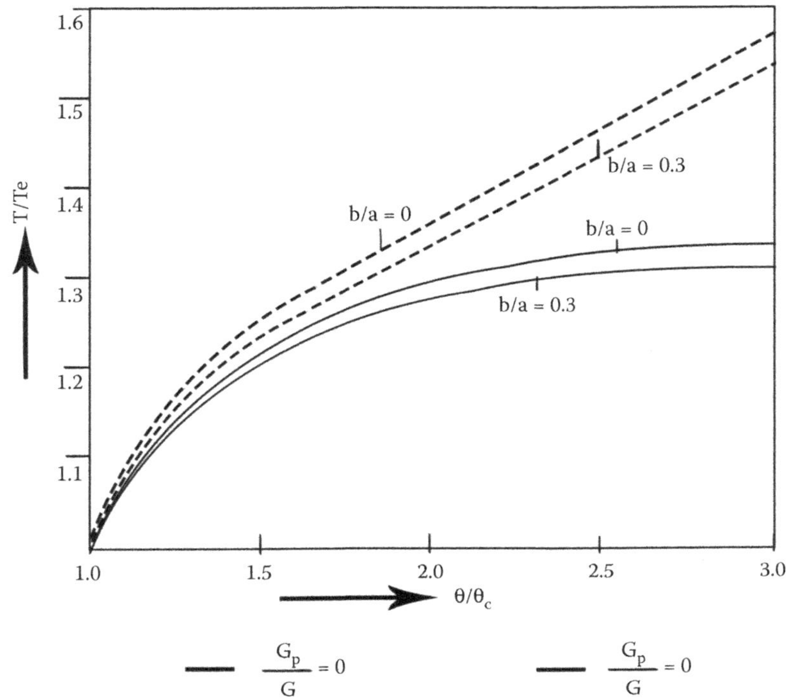

FIGURE 5.7
Torque versus unit angle of twist in the elastic–plastic range.

And for solid cylindrical bar, $b = 0$, the corresponding relation is

$$\left(1 + \frac{G_p}{3G}\right)\frac{T}{T_e} = \frac{1}{3}\left[4 - \left(\frac{\theta}{\theta_e}\right)^{-3} + \frac{G_p}{G}\left(\frac{\theta}{\theta_e}\right)\right] \tag{5.15b}$$

Figure 5.7 shows the torque versus unit angle of twist for torsion of cylindrical bars with $b/a = 0.3$ and $G_p/G = 0.5$ in the elastic/plastic range.

5.3 Torsion of Noncircular Bars

Consider a prismatic bar of arbitrary cross section subject to equal and opposite torques applied at the ends, as shown in Figure 5.8. Every cross section rotates about the z-axis, which is taken along a longitudinal axis of the member. The point of the cross section intersecting the z-axis is referred to as *the center of twist*. The location of the center of twist is a function of the shape of the cross section. At this point of the cross section, the displacements u and v are zero. All cross sections of the member will warp due to the twisting action.

We begin by using *Saint-Venant's semi-inverse approach* to derive the displacement expressions in the member. The components of the displacement in a circular bar subject to end torques were derived in Equation 5.4 as $u = -\theta yz$, $v = \theta xz$, $w = 0$, where the constant θ is

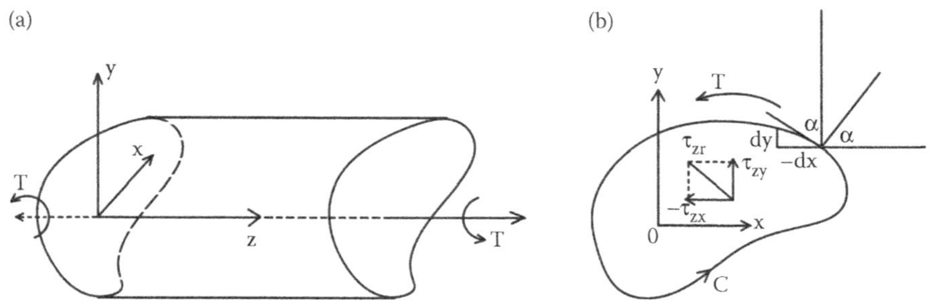

FIGURE 5.8
Torsion of noncircular bar. (a) Noncircular bar; (b) stresses.

the unit angle of twist. Based on the result, it is *assumed* that the rotation and warping of noncircular cross sections are represented by the following displacement functions:

$$u = -\theta yz$$
$$v = \theta xz \tag{5.16}$$
$$w = \theta\omega(x, y)$$

Here, $\omega(x, y)$ is the warping function and the constant θ is introduced for mathematical convenience. It follows that $\varepsilon_x = \varepsilon_y = \varepsilon_z = \gamma_{xy} = 0$ and the nonvanishing shear strains are

$$\begin{aligned} \gamma_{xz} &= \frac{\partial u}{\partial z} + \frac{\partial w}{\partial x} = -\theta y + \theta\frac{\partial \omega}{\partial x} \\ \gamma_{yz} &= \frac{\partial v}{\partial z} + \frac{\partial \omega}{\partial y} = \theta x + \theta\frac{\partial \omega}{\partial y} \end{aligned} \tag{5.17}$$

The normal stresses σ_x, σ_y, σ_z and the shearing stress τ_{xy} are equal to zero. The nonvanishing stresses τ_{xz} and τ_{yz}, which are functions of x and y only, are determined by obtaining a solution of the equation of equilibrium, $\partial\tau_{xz}/\partial x + \partial\tau_{yz}/\partial y = 0$, subject to satisfying the boundary conditions, and the stress–strain behavior of the material. Next, elastic and the plastic solutions are discussed separately.

5.4 Elastic Solutions

For an elastic material, the nonvanishing stresses are related to the strains by Hooke's law

$$\begin{aligned} \tau_{xz} &= G\gamma_{xz} = G\theta\left(-y + \frac{\partial \omega}{\partial x}\right) \\ \tau_{yz} &= G\gamma_{yz} = G\theta\left(x + \frac{\partial \omega}{\partial y}\right) \end{aligned} \tag{5.18}$$

Eliminating the warping function $\omega(x, y)$ by differentiating the first of Equations 5.18 with respect to y and the second with respect to x, and subtracting the first from the second yields the strain-compatibility equation in terms of the stresses

$$\frac{\partial \tau_{yz}}{\partial x} - \frac{\partial \tau_{xz}}{\partial y} = 2G\theta \tag{5.19}$$

The equation governing the warping function itself is found by inserting the stresses in Equation 5.18 into the equilibrium Equation 5.1. The result is

$$\frac{\partial^2 \omega}{\partial x^2} + \frac{\partial^2 \omega}{\partial y^2} = 0 \tag{5.20}$$

Equation 5.20 indicates that the warping function is governed by Laplace's differential equation.

Stress function: In order to solve for the required shear stresses, it is convenient to introduce a stress function, $\phi(x, y)$, such that the equilibrium equation is identically satisfied. This can be achieved by writing

$$\tau_{zx} = \frac{\partial \phi}{\partial y}, \quad \tau_{zy} = -\frac{\partial \phi}{\partial x} \tag{5.21}$$

It follows from Equation 5.19 that the differential equation governing the stress function is

$$\frac{\partial^2 \phi}{\partial x^2} + \frac{\partial^2 \phi}{\partial y^2} = -2G\theta \tag{5.22}$$

Equation 5.22 is known as Poisson's equation and its solution furnishes the distribution of stresses in the bar.

Boundary conditions: Because the external torques are applied at the ends of the bar, the lateral surface of the member is free from stress. If the normal to the boundary of the cross section makes angle α with the x-axis (Figure 5.8b), then, $l = \cos\alpha$, $m = \sin\alpha$. Using the third relation in Equation 2.13, which is $p_z = l\tau_{zx} + m\tau_{xy} + n\sigma_z$, the following is the boundary condition on C:

$$l\tau_{zx} + m\tau_{zy} = 0 \tag{5.23}$$

Now, $l = \cos\alpha = dy/ds$, $m = \sin\alpha = -dx/ds$, and substituting in Equation 5.23 results in

$$\frac{dy}{ds}\left(\frac{\partial \phi}{\partial y}\right) + \frac{dx}{ds}\left(\frac{\partial \phi}{\partial x}\right) = 0 \quad \text{or} \quad \frac{d\phi}{ds} = 0 \tag{a}$$

It follows that $\phi(x, y) = k$ (constant), on contour C. Since the shear stresses are obtained from Equation 5.21, it is clear that any constant added to the stress function does not affect the shear stresses. It turns out that for solid cross sections, the constant k can be chosen to be zero. For multiconnected cross sections, that is, members containing more than one hole in the cross section, or tubular members, the constant k is assigned to one hole only, and one needs to use different mathematical tools to determine the shear stresses. One useful approach is to use complex variables to solve such problems [5.2].

At any point of the cross section of the bar, the maximum shear stress τ_{max} is

$$\tau_{max} = [\tau_{xz}^2 + \tau_{yz}^2]^{1/2} = \left[\left(\frac{\partial\phi}{\partial x}\right)^2 + \left(\frac{\partial\phi}{\partial y}\right)^2\right]^{1/2} \tag{5.24}$$

Since the x and y components of the unit normal to the curve $\phi(x, y) =$ constant are proportional to the quantities $\partial\phi/\partial x$ and $\partial\phi/\partial y$, respectively, the maximum shear stress is tangential to the curve $\phi =$ constant. For a solid bar or any member with singly connected cross section, $k = 0$, and the boundary condition is

$$\phi = 0, \quad \text{on contour C} \tag{5.25}$$

It is readily verified that the summation of forces on any cross section vanishes by observing that

$$\begin{aligned} \iint_A \tau_{zx} dxdy &= \iint_A \frac{\partial\phi}{\partial y} dxdy = \int_C \phi dx = k\int_C dx = 0 \\ \iint_A \tau_{zy} dxdy &= \iint_A -\frac{\partial\phi}{\partial x} dxdy = -\int_C \phi dy = -k\int_C dy = 0 \end{aligned} \tag{b}$$

Here, A represents the area of the cross section and the line integrals are evaluated along closed contours.

The twisting moment is obtained as

$$T = \iint_A (x\tau_{zy} - y\tau_{zx}) dxdy = -\iint_A \left(x\frac{\partial\phi}{\partial x} + y\frac{\partial\phi}{\partial y}\right) dxdy$$

Performing an integration by parts on the last double integral yields

$$T = -\int_C \phi(xdy - ydx) + 2\iint_A \phi dxdy = -\oint_C \phi(xl + ym)ds + 2\iint_A \phi dxdy \tag{c}$$

For simply connected domains, the first integral vanishes since $\phi = 0$ on the boundary C, and the twisting moment reduces to the evaluation of the integral

$$T = 2\iint_A \phi dxdy \tag{5.26}$$

Thus, the torque is numerically equal to twice the volume under the surface $z = \phi(x, y)$. For members with multiconnected cross section, the stress function can take different constant values on the outer and inner boundaries, and both integrals in Equation (c) contribute to the applied torque.

Note that if one end of the member is fixed, the angle of twist and the warping will vary with z, and the longitudinal fibers experience tensile and compressive stresses. However, for members with solid cross sections such as circular, rectangular, etc., the solutions generated from Equations 5.22 and 5.25, which is based on the assumption that the resisting torque is due to pure shearing stresses developed across all cross sections including the ends, provide satisfactory results in regions away from the restrained section.

In sum, the elastic solution of the torsion problem for singly connected domain is reduced to finding the stress function ϕ, which satisfies Poisson's Equation 5.22 and boundary condition (Equation 5.25). Equations 5.22 and 5.25 are subsequently used to find the distribution of shear stresses. The applied torque and the warping function are then determined from Equations 5.18, 5.20, and 5.26. This completes the formulation and shows that the displacement expressions in Equation 5.16 yield a unique solution, which satisfies the governing equations of the problem.

Several methods are available to obtain the solution of the elastic torsion problem. Solid bars with common geometrical shapes, for example, elliptical, triangular, rectangular, etc., are discussed next using simple algebraic polynomials and infinite trigonometric series.

5.4.1 Polynomial Solution

Saint-Venant [5.1] discussed several solutions to the elastic torsion problem by using algebraic polynomials.

Elliptic cross section: Consider a solid bar with an elliptical cross section subject to end torques T. The origin of coordinates is taken at the center O of the ellipse and the z-axis along the longitudinal axis of the bar, as shown in Figure 5.9. The boundary of the cross section is given by

$$\frac{x^2}{a^2} + \frac{y^2}{b^2} = 1$$

A suitable stress function is the polynomial

$$\phi = B\left(1 - \frac{x^2}{a^2} - \frac{y^2}{b^2}\right), \quad \text{B is a constant} \tag{5.27}$$

It is readily confirmed that the stress function ϕ vanishes on the boundary of the cross section, C, and provides a solution to Equation 5.22 provided that the constant B is given by

$$B = \frac{G\theta a^2 b^2}{a^2 + b^2} \tag{a}$$

The applied torque T is evaluated from Equation 5.26 as

$$T = \frac{2G\theta a^2 b^2}{a^2 + b^2} \iint_A \left(1 - \frac{x^2}{a^2} - \frac{y^2}{b^2}\right) dxdy = \frac{2G\theta a^2 b^2}{a^2 + b^2}\left[A - \frac{I_y}{a^2} - \frac{I_x}{b^2}\right]$$

Here, A is the cross-sectional area, and I_x and I_y are the moments of inertia of the cross section about the x and y axes, respectively. Since $A = \pi ab$, $I_x = \pi ab^3/4$, and $I_y = \pi a^3 b/4$, it follows that

$$T = \frac{\pi G\theta a^3 b^3}{a^2 + b^2} \tag{5.28}$$

The torsional rigidity of the bar is $T/\theta = (\pi G a^3 b^3/a^2 + b^2) = (GA^4/4\pi^2 I_p)$ with $I_p = I_x + I_y$ designates the polar moment of inertia of the cross section.

Using Equations (a) and 5.28, the constant $B = T/\pi ab$, and the stress function (5.27) is

$$\phi = \frac{T}{\pi ab}\left(1 - \frac{x^2}{a^2} - \frac{y^2}{b^2}\right) \tag{b}$$

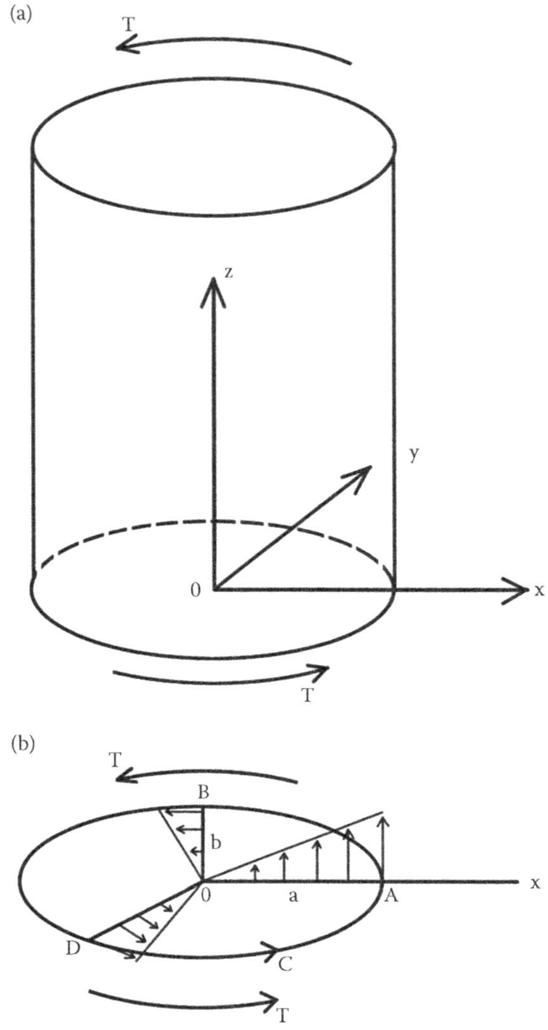

FIGURE 5.9
Elastic torsion of an elliptical member. (a) Elliptical bar; (b) stresses.

Differentiating the stress function with respect to y and x gives the following shear stresses:

$$\begin{aligned}\tau_{zx} &= -\frac{2Ty}{\pi ab^3} = -\frac{Ty}{2I_x}\\ \tau_{zy} &= \frac{2Tx}{\pi a^3 b} = \frac{Tx}{2I_y}\\ |\tau_{zr}| &= \frac{2T}{\pi ab}\left(\frac{x^2}{a^4}+\frac{y^2}{b^4}\right)^{1/2} = \frac{2T}{A}\left(\frac{x^2}{a^4}+\frac{y^2}{b^4}\right)^{1/2}\end{aligned} \tag{5.29}$$

At $x = 0$, τ_{zy} is zero and τ_{zx} attains its maximum value at $y = b$. Similarly, at $y = 0$, τ_{zx} is zero and τ_{zy} has its maximum value at $x = a$. The resultant shear stress, τ_{zr}, is readily shown to have its maximum value at the ends of the minor axis ($x = 0$, $y = \pm b$) where

$$\tau_{max} = \frac{2T}{\pi ab^2} \tag{c}$$

At the cylindrical boundary of the bar, the shear stress acts in the tangential direction. Also, since the ratio, $\tau_{zx}/\tau_{zy} = -y/x(I_y/I_x)$, is proportional to y/x, it does not vary along a radius 0D drawn from the origin to point D on the boundary as indicated in Figure 5.9b. The direction of the resultant shear stress along all points of 0D coincides with its direction at point D, which is tangential to the boundary of the cross section (Figure 5.9b). Hence, across concentric ellipses within the cross section, $(x/\lambda a)^2 + (y/\lambda b)^2 = 1$, λ is constant, the resultant shear stress is tangential to the boundary, and the stress function, $\phi = B(1 - \lambda^2)$, is constant. Such ellipses represent the stress lines of the cross section.

The warping of the cross section during elastic behavior is obtained by solving for the function ω from Equations 5.18. Substituting the expressions of the shear stresses from Equations 5.29 into 5.18 and making use of Equation 5.28 gives

$$\begin{aligned} \frac{\partial \omega}{\partial x} &= y\left(\frac{b^2 - a^2}{b^2 + a^2}\right) \\ \frac{\partial \omega}{\partial y} &= x\left(\frac{b^2 - a^2}{b^2 + a^2}\right) \end{aligned} \tag{d}$$

Performing integration on the derivatives, noting that the constants of integration in the first and second of Equations (d) are functions of y and x, respectively, which can be neglected because $\omega(x, y)$ satisfies Equation 5.20. The result is

$$\omega(x, y) = -\left(\frac{a^2 - b^2}{a^2 + b^2}\right)xy \tag{5.30}$$

The contour lines defined by $\omega = \text{constant}$, that is, $xy = \text{constant}$, are rectangular hyperbolas whose asymptotes are the principal axes of the ellipse, as shown in Figure 5.10. The solid

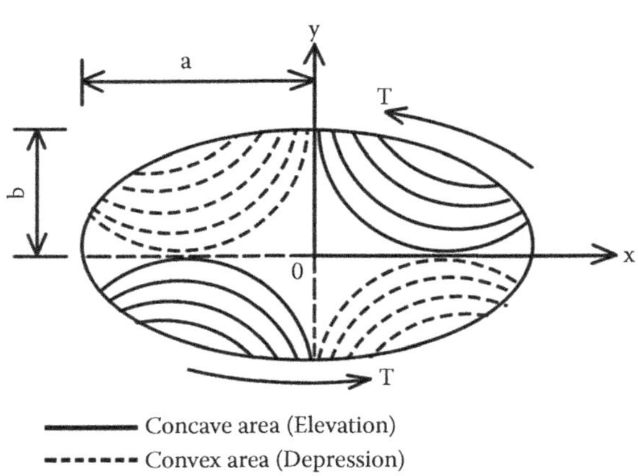

FIGURE 5.10
Contours of constant warping in an elliptical bar.

lines in the figure denote portions of the cross section that become convex (ω is negative) while the dotted lines denote concave portions (ω is positive).

The displacements in the bar are obtained from Equations 5.10 as

$$u = -\theta yz, \quad v = qxz, \quad \text{and} \quad w(x, y) = \frac{T(b^2 - a^2)xy}{\pi a^3 b^3 G} \tag{e}$$

When $a = b$, the above results reduce to those given in Section 5.2 for the circular bar.

Elliptical bar with a central elliptical hole: Suppose that the elliptical bar has a central elliptical hole of semiaxes λa and λb, λ is a constant, $\lambda < 1$. The boundary of the hole is the ellipse

$$\left(\frac{x}{\lambda a}\right)^2 + \left(\frac{y}{\lambda b}\right)^2 = 1 \tag{a}$$

In a solid bar with semiaxes a and b, the stress function (Equation 5.22) is constant along ellipse (a), and as discussed previously, the shear stress is tangential to concentric ellipses. It follows that if a cylinder of the material bounded by ellipse (a) is removed from the solid bar, the stresses in the outer material will not be affected. Thus, a similar stress function to that of the solid bar can be used for the hollow one.

The angle of twist of the solid bar is $G\theta = T(a^2 + b^2)/\pi a^3 b^3$, and the corresponding angle of the portion of the material removed from the hole (semiaxes $\lambda\theta$ and λb) is $G\theta = T(a^2 + b^2)/\lambda^4 \pi a^3 b^3$. It follows that, for the same angle of twist θ, the torque carried by the hollow shaft is reduced by the ratio, λ^4: 1 compared to the solid shaft. The torque of the hollow bar is then given by

$$T_{\text{hollow}} = (1 - \lambda^4)\frac{\pi G\theta a^3 b^3}{a^2 + b^2} \tag{b}$$

A suitable stress function for the hollow member is

$$\phi = \frac{T_{\text{hollow}}}{\pi ab(1 - \lambda^4)}\left(1 - \frac{x^2}{a^2} - \frac{y^2}{b^2}\right) \tag{c}$$

and the maximum stress is

$$\tau_{\max} = \frac{2T_{\text{hollow}}}{(1 - \lambda^4)\pi ab^2} \tag{d}$$

This is an example of a multiconnected cross section with the inner boundary coinciding with a stress line of the solid bar. When $a = b$, the above results reduce to those found in Example 5.2.

Note that if $\lambda = 0.4$, the maximum shear stress at the ends of the minor axis is $\tau_{\max} = 0.653$ T/ab^2.

Other polynomial solutions: In order to construct additional polynomial solutions, the stress function $\phi(x, y)$ is expressed as [5.3]

$$\phi(x, y) = -\frac{G\theta}{2}(x^2 + y^2) + F(x, y) \tag{a}$$

where F(x, y) is an arbitrary function. Inserting Equation (a) into Equation 5.22, it is found that the function F(x, y) is harmonic, that is, it is governed by Laplace's equation

$$\frac{\partial^2 F}{\partial x^2} + \frac{\partial^2 F}{\partial y^2} = 0 \tag{b}$$

and the associated boundary condition is

$$F(x, y) - \frac{G\theta}{2}(x^2 + y^2) = \text{constant on } C$$

To obtain solutions in the form of polynomials, we consider the complex variable function $z^n = (x + iy)^n$, n is an integer and

$$i = \sqrt{-1} \tag{c}$$

The real and complex parts of this function are harmonic functions and represent solutions of Equation (b). Thus, for $n = 2$, we have functions $x^2 - y^2$ and 2xy; for $n = 3$, functions $x^3 - 3xy^2$ and $3x^2y - y^3$; for $n = -1$, functions $x/(x^2 + y^2)$ and $-y/(x^2 + y^2)$, etc. Let us apply this procedure to bars of triangular cross sections.

Triangular cross section: Consider a prismatic bar whose cross section is an equilateral triangle of side = a. To simplify the mathematics, the origin of coordinates is shifted from center O to an apex A of the cross section, as shown in Figure 5.11. The analytical equations of the sides of the triangle are as follows: Line BC is represented by $x = h(= (\sqrt{3}/2)a)$, while the equations of lines AB and AC are $y = \pm(x/\sqrt{3})$. The product of three equations of the sides of the cross section is

$$(x - h)(x^2 - 3y^2) = x^3 - 3xy^2 - h(x^2 - 3y^2) \tag{d}$$

A suitable combination of harmonic functions is obtained by selecting the real parts of z^n with $n = 2$ and $n = 3$. In this manner, the stress function ϕ is sought in the form

$$\phi = C_1(x^2 + y^2) + C_2(x^2 - y^2) + x^3 - 3xy^2 \tag{e}$$

where C_j, $j = 1, 2$ are arbitrary constants. In order to satisfy the boundary condition $\phi = 0$ along the sides of the triangle, select $C_1 = h$ and $C_2 = -2h$ to ensure that Equations (d) and (e) are identical. Thus, the appropriate stress function is

$$\phi = B(x^3 - 3xy^2 - hx^2 + 3hy^2) \tag{5.31}$$

Here, B is an arbitrary constant. It is clear that ϕ vanishes on the boundary of the cross section and satisfies the governing Equation 5.22 provided that $B = -(G\theta/2h)$.

The applied torque is given by

$$T = 2\iint_A \phi\, dx\, dy = 4B\int_0^h dx \int_0^{x/\sqrt{3}} [x^3 - 3xy^2 - hx^2 + 3hy^2]dy$$

Evaluating the integral, it is found that

$$T = -\frac{3Ba^5}{80} = \frac{\sqrt{3}}{80}G\theta a^4 \tag{5.32}$$

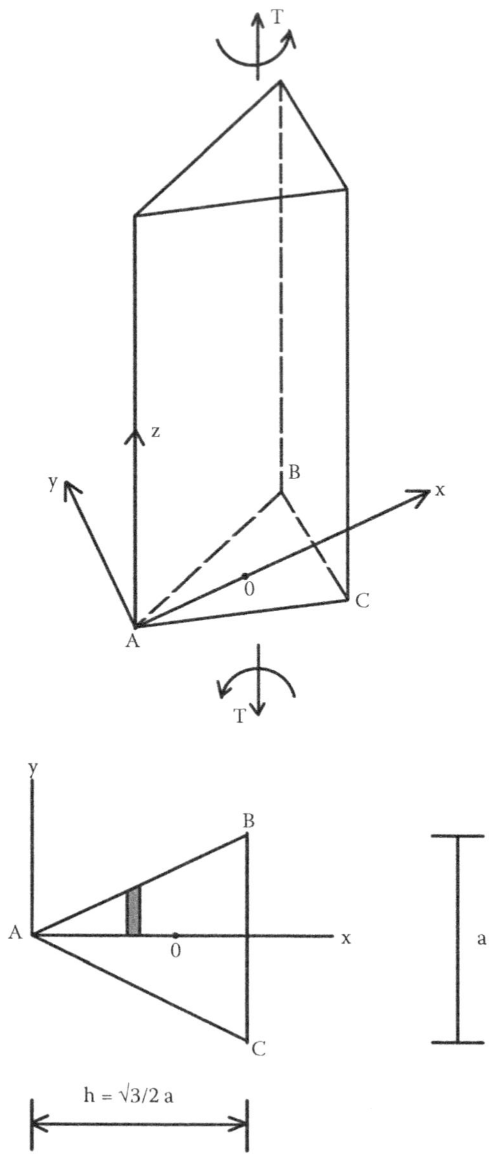

FIGURE 5.11
Elastic torsion of and equilateral triangular bar.

The torsional rigidity of the bar is $(T/\theta) = (\sqrt{3}/80)Ga^4 = 3/5GI_0$, where $I_0 = \sqrt{3}/48a^4$ is the polar moment of inertia of the cross section. (Note that, in the calculation of the expression for I_0, the origin of the coordinates is located at the centroid 0 of the cross section and not at the apex A, which was used in the derivation of the stress function.) The constant B is expressed in terms of the torque as $B = -(80T/3a^5)$, and the stress function becomes

$$\phi = -\frac{80T}{3a^5}(x^3 - 3xy^2 - hx^2 + 3hy^2) \tag{f}$$

The shear stresses are computed by differentiating the stress function

$$\tau_{xz} = \frac{\partial \phi}{\partial y} = \frac{160T}{a^5}(x - h)y$$
$$\tau_{yz} = -\frac{\partial \phi}{\partial x} = \frac{80T}{a^5}\left(x^2 - y^2 - \frac{2}{3}hx\right) \tag{5.33}$$

and $\tau_{zr} = (\tau_{xz}^2 + \tau_{yz}^2)^{1/2}$. At the corners of the triangle, point $A(x = y = 0)$ and points B and C $(x = h,\ y = \pm a/2)$, the stresses vanish while the maximum shear stress $\tau_{max} = \tau_{zr}$ occurs at midpoints of the sides, that is, at $x = (h/2)$, $y = \pm a/4$, and $x = h$, $y = 0$, and has the value

$$\tau_{max} = \frac{\sqrt{3}}{4}G\theta a = \frac{20T}{a^3} \tag{g}$$

The warping function is obtained by inserting the expressions of the stresses in Equation 5.33 into 5.18 and carrying out the integration. The result is

$$\omega = -2xy + \frac{\sqrt{3}}{a}\left(x^2 - \frac{y^2}{3}\right)y \tag{5.34}$$

Lines of constant warping are shown in Figure 5.12. The solid lines indicate concave areas, $\omega < 0$, while the dotted lines designate convex portions, $\omega > 0$.

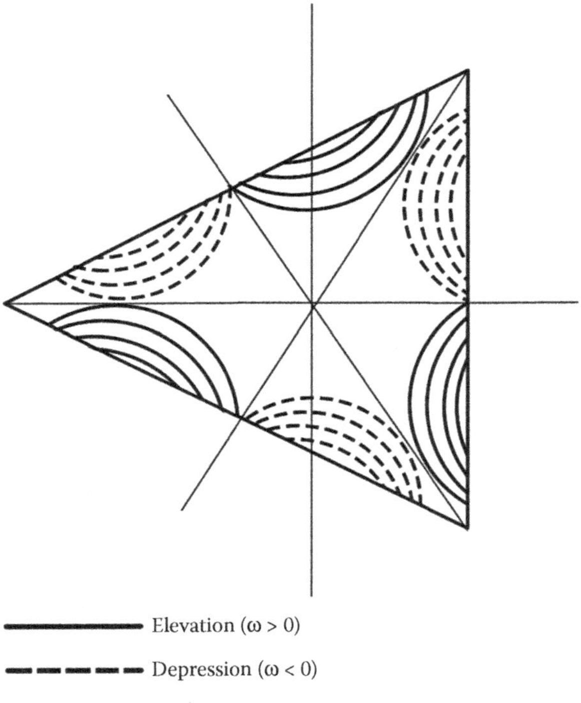

FIGURE 5.12
Contours of constant warping.

Polynomials of higher degrees were used by Saint-Venant to obtain, solutions of a number of similar singly connected cross sections.

EXAMPLE 5.3

A circular shaft of radius = b, has a semicircular notch or groove, radius = a, is subjected to a torque T. The cross section of the shaft is shown in Figure 5.13. Assuming the elastic material with shear modulus G, find the maximum shear stress developed in the shaft.

Solution

A closed-form solution of this problem can be obtained using a stress function consisting of equation of the boundary of the shaft in Cartesian or polar coordinates. Using the Cartesian coordinates shown in Figure 5.13, equations of the boundaries of notch and shaft are $x^2 + y^2 - a^2 = 0$ and $(x - b)^2 + y^2 - b^2 = 0$. In polar coordinates, the corresponding equations are $r^2 - a^2 = 0$ and $r^2 - 2br\cos\theta = 0$. Thus, a stress function that vanishes on the boundary of the shaft is the following:

$$\phi(r,\theta) = A(a^2 - r^2)\left(1 - \frac{2b\cos\theta}{r}\right) = A\left[a^2 - r^2 - 2b(r^2 - a^2)\frac{\cos\theta}{r}\right] \quad \text{(a)}$$

where A is constant. Inserting Equation (a) in 5.22, expressed in terms of polar coordinates, it is found that $A = G\,\theta^*/2$, where θ^* is the unit angle of twist.

Writing the stress function in rectangular coordinates,

$$\phi(x,y) = A\left[a^2 - x^2 - y^2 + 2bx - \frac{2ba^2x}{x^2 + y^2}\right] \quad \text{(b)}$$

Note that the function $x/(x^2 + y^2)$ is harmonic since it is obtained from $z = (x + iy)^n$, $n = -1$. In order to determine the shear stresses in the shaft, we need to differentiate the stress function (b).

$$\begin{aligned} \tau_{zx} &= \frac{\partial\phi}{\partial y} = A\left[-2y + \frac{4ba^2xy}{(x^2 + y^2)^2}\right] \\ \tau_{zy} &= -\frac{\partial\phi}{\partial x} = A\left[2x - 2b + \frac{2ba^2(y^2 - x^2)}{(x^2 + y^2)^2}\right] \end{aligned} \quad \text{(c)}$$

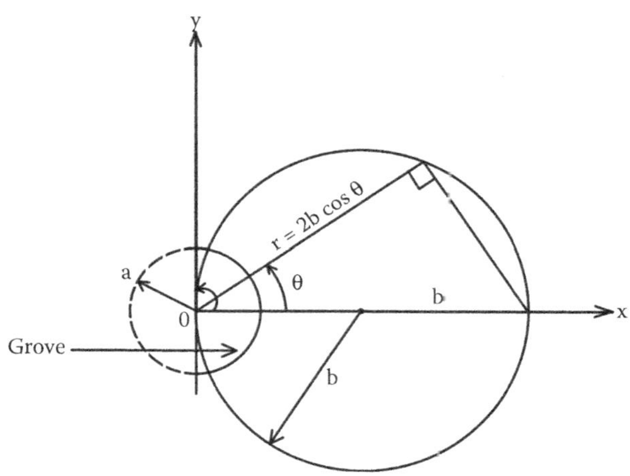

FIGURE 5.13
Example 5.3.

The maximum shear stress is obtained from Equation 5.21, $\tau_{max} = [(\partial\phi/\partial y)^2 + (-\partial\phi/\partial x)^2]^{1/2}$. It occurs at the center of the notch, x = a, y = 0, and

$$\tau_{max} = G\theta*(2b - a)$$

Note that when a = 0, the notch is infinitely small like a scratch in a solid circular bar and the maximum shear stress is twice that in the solid bar without the notch.

For the torque carried by the shaft, see problem P5.13.

5.4.2 Fourier Series Solutions

Fourier series techniques can be used to obtain the elastic stresses developed in a number of practically important cross sections when subjected to twisting moments. Few examples are discussed in this section.

Torsion of rectangular bars: Consider the torsion of a prismatic bar of rectangular cross section of sides a and b. The origin of rectangular coordinates x and y coincides with the center of the rectangle, as shown in Figure 5.14. It is required to find the stress function $\phi(x, y)$, which satisfies Equation 5.22 and the boundary conditions

$$\phi\left(\mp\frac{a}{2}, y\right) = \phi\left(x, \mp\frac{b}{2}\right) = 0 \tag{5.35}$$

which indicate that the stress function is symmetric with respect to the x and y axes. By the usual method of separation of the variables, the stress function is sought in the form

$$\phi = \sum_{n=1,3,5,..}^{\infty} f_n(y)\cos\frac{n\pi x}{a} \tag{5.36}$$

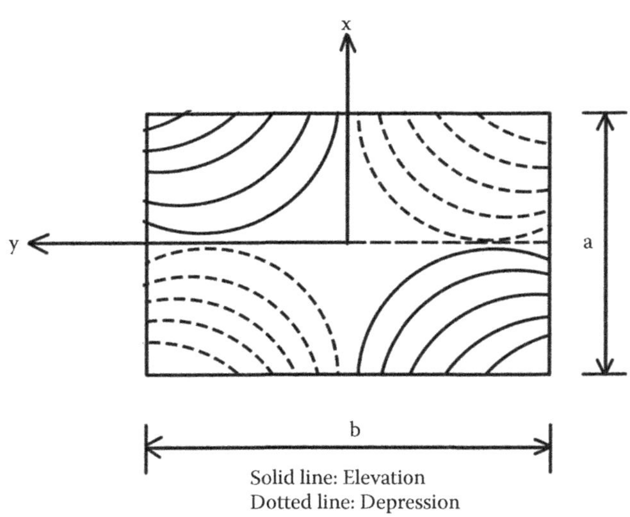

FIGURE 5.14
Torsion of a rectangular section bar.

where $f_n(y)$, $n = 1,3,5,\ldots$ are functions of the variable y only. To develop the stress function, we first expand the right-hand side of Equation 5.22 in a Fourier cosine series by writing

$$-2G\theta = \sum_{n=1,3,5}^{\infty} B_n \cos\frac{n\pi x}{a} \tag{a}$$

Multiplying both sides of Equation (a) by $\int_{-a/2}^{a/2} \cos(m\pi x/a)\,dx$, m = integer, and noting that

$$\int_{-a/2}^{a/2} \cos\frac{m\pi x}{a}\cos\frac{n\pi x}{a}\,dx = \begin{matrix} 0 & m \neq n \\ a/2 & m = n \end{matrix}$$

it follows that the constant $B_n = -8G\theta/n\pi(-1)^{n-1/2}$, and Equation 5.22 becomes

$$\frac{\partial^2\phi}{\partial x^2} + \frac{\partial^2\phi}{\partial y^2} = -\sum_{n=1,3,5}^{\infty} \frac{8G\theta}{n\pi}(-1)^{\frac{n-1}{2}}\cos\frac{n\pi x}{a} \tag{5.37}$$

Inserting Equation 5.36 into 5.37, it is found that the functions $f_n(y)$ satisfies the second-order differential equation

$$\frac{d^2 f_n}{dy^2} - \frac{n^2\pi^2}{a^2} f_n = -\frac{8G\theta}{n\pi}(-1)^{\frac{n-1}{2}}, \quad n = 1,3,5 \tag{b}$$

The solution is

$$f_n(y) = (-1)^{\frac{n-1}{2}}\frac{8G\theta a^2}{n^3\pi^3} + A_n \cosh\frac{n\pi y}{a} + C_n \sinh\frac{n\pi y}{a}$$

where A_n and C_n are constants. Because the function $f_n(y)$ is symmetric with respect to y, C_n must vanish, and A_n is determined from the boundary condition, $f_n(y = \mp b/2) = 0$, as

$$A_n = -\frac{8G\theta a^2}{\pi^3}\left[\frac{(-1)^{\frac{n-1}{2}}}{n^3\cosh(n\pi b/2a)}\right]$$

In this manner, the stress function is obtained as

$$\phi = \frac{8G\theta a^2}{\pi^3}\sum_{n=1,3,5}^{\infty}\frac{(-1)^{\frac{n-1}{2}}}{n^3}\left[1 - \frac{\cosh(n\pi y/a)}{\cosh(n\pi b/2a)}\right]\cos\frac{n\pi x}{a} \tag{5.38}$$

The stress components are obtained by differentiation of the stress function with respect to x and y. Thus,

$$\tau_{xz} = \frac{\partial\phi}{\partial y} = \frac{8G\theta a}{\pi^2}\sum_{n=1,3,5}^{\infty}\frac{(-1)^{\frac{n+1}{2}}}{n^2}\left[\frac{\sinh(n\pi y/a)}{\cosh(n\pi b/2a)}\right]\cos\frac{n\pi x}{a} \tag{5.39a}$$

$$\tau_{yz} = -\frac{\partial\phi}{\partial x} = \frac{8G\theta a}{\pi^2}\sum_{n=1,3,5}^{\infty}\frac{(-1)^{\frac{n-1}{2}}}{n^2}\left[1-\frac{\cosh(n\pi y/a)}{\cosh(n\pi b/2a)}\right]\sin\frac{n\pi x}{a} \tag{5.39b}$$

The resultant shear stress is $(\tau_{xz}^2 + \tau_{yz}^2)^{1/2}$. Assuming $b > a$, the maximum shear stress occurs at the midpoints of the long sides of the cross section, that is, at $y = 0$ and $x = \pm a$, where it is found that

$$\tau_{max} = \frac{8G\theta a}{\pi^2}\sum_{n=1,3,5}^{\infty}\frac{1}{n^2}\left[1-\frac{1}{\cosh\left(\frac{n\pi b}{2a}\right)}\right] \tag{c}$$

Since, $\sum_{n=1,3,5}^{\infty}(1/n^2) = (\pi^2/8)$, it follows that

$$\tau_{max} = G\theta a - \frac{8G\theta a}{\pi^2}\sum_{n=1,3,5}^{\infty}\frac{1}{n^2\cosh(n\pi b/2a)} = C_1 G\theta a \tag{5.40}$$

where C_1 is a constant, which can be easily computed for any value of the ratio b/a using the following expression (see Table 5.1):

$$C_1 = 1 - \frac{8}{\pi^2}\left[\frac{1}{\cosh(\pi b/2a)} + \frac{1}{3^2\cosh(3\pi b/2a)} + \cdots\right] \tag{d}$$

In case of a square cross section, $a = b$, $\tau_{max} = 0.675G\theta a$. Also, for a member with a narrow rectangular cross section, b/a is large, $\cosh(n\pi b/2a) \to \infty$ and the infinite series in Equation 5.40 can be neglected. Hence, $\tau_{max} = G\theta a$.

The torque required to induce the twisting action is computed from Equation 5.26. Because the cross section is symmetric with respect to x and y, Equation 5.26 is equivalent to

$$T = 8\int_0^{a/2}\int_0^{b/2}\phi(x,y)dxdy \tag{e}$$

TABLE 5.1

Torsion of a Rectangular Bar—Numerical Values of C_1 and C_2

b/a	$C_1 = \tau_{max}/G\theta a$	$C_2 = T/G\theta a^3 b$
1.0	0.675	0.141
1.5	0.848	0.196
2.0	0.930	0.229
2.5	0.968	0.249
3.0	0.985	0.263
3.5	0.993	0/272
4.0	0.997	0.281
4.5	0.998	0.285
5.0	0.999	0.291
10.0	1.000	0.312
∞	1.000	0.333

Substituting the expression for the stress function from Equation 5.38 and evaluating the resulting integrals, it is found that

$$T = \frac{32G\theta a^3 b}{\pi^4} \sum_{n=1,3,5}^{\infty} \left[\frac{1}{n^4} - \frac{2a \tanh(n\pi b/2a)}{b\pi n^5}\right] \tag{f}$$

Noting that the value of the infinite series $\sum_{n=1,3,5}^{\infty}(1/n^4) = (\pi^4/96)$, the expression of the torque simplifies to

$$T = \left[\frac{1}{3} - \frac{64a}{\pi^5 b}\sum_{n=1,3,5}^{\infty}\frac{1}{n^5}\tanh\left(\frac{n\pi b}{2a}\right)\right]G\theta a^3 b = C_2 G\theta a^3 b \tag{5.41}$$

where the constant C_2 is a numerical factor given by

$$C_2 = \frac{1}{3} - \frac{64a}{\pi^5 b}\left[\tanh\left(\frac{\pi b}{2a}\right) + \frac{1}{3^5}\tanh\left(\frac{3\pi b}{2a}\right) + \cdots\right] \tag{g}$$

Several values of the constant C_2 have been computed for different ratios b/a in Table 5.1. The torsional rigidity of the bar is given by

$$\frac{T}{\theta} = C_2 G a^3 b \tag{5.42}$$

and the maximum shear stress in terms of the torque is

$$\tau_{max} = \frac{C_1}{C_2}\left(\frac{T}{a^2 b}\right) \tag{5.43}$$

For a member with a square cross section, $a = b$, $C_2 = 0.141$, $C_1 = 0.675$, it follows that

$$T = 0.141G\theta a^4 \quad \text{and} \quad \tau_{max} = 4.8\left(\frac{T}{a^3}\right)$$

As b becomes large (narrow rectangular section), $\tanh(n\pi b/2a) \rightarrow 1$ and the expression for the torque (Equation 5.41) is reduced to

$$T = \left[\frac{1}{3} - \frac{64a}{\pi^5 b}\right]C_2 G\theta a^3 b \tag{5.44}$$

The warping of the cross sections is obtained by inserting the expressions of the stresses into Equation 5.18. Carrying out the integration results in the expression

$$\omega(x, y) = xy - \frac{8a^2}{\pi^3}\sum_{n=1,3,5,\ldots}^{\infty}\frac{(-1)^{\frac{n-1}{2}}}{n^3}\left[\frac{\sinh(n\pi y/a)}{\cosh(n\pi b/2a)}\right]\sin\frac{n\pi x}{a} \tag{5.45}$$

Contours of the warping function, $\omega(x, y)$, are shown in Figure 5.14. The solid lines represent $\omega > 0$ and the dotted lines $\omega < 0$.

Torsion of sectors of a circle: Another cross-sectional shape that can be solved by infinite series is a sector of a circle. Consider a prismatic bar whose cross section is the sector of a circle whose boundaries are given by the polar coordinates $r = 0$, $r = a$, and $\theta = \pm\beta$, as

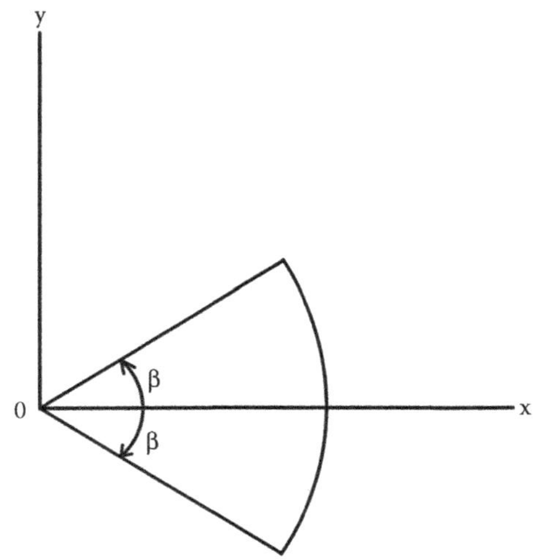

FIGURE 5.15
Section of a circle.

shown in Figure 5.15. The center of the circle coincides with the origin of coordinates and the bar is subject to end torques T.

It is convenient to use polar coordinates: $x = r\cos\theta$, $y = r\sin\theta$, $r^2 = x^2 + y^2$ and $\theta = \tan^{-1} y/x$. It follows that, $\partial r/\partial x = x/r = \cos\theta$, $\partial r/\partial y = y/r = \sin\theta$, $\partial\theta/\partial x = -y/r^2 = -\sin\theta/r$ and $\partial\theta/\partial y = x/r^2 = \cos\theta/r$.

The derivatives of the harmonic function F are developed as

$$\frac{\partial F}{\partial x} = \frac{\partial F}{\partial r}\frac{\partial r}{\partial x} + \frac{\partial F}{\partial \theta}\frac{\partial \theta}{\partial x} = \cos\theta\frac{\partial F}{\partial r} - \frac{\sin\theta}{r}\frac{\partial F}{\partial \theta}$$

$$\frac{\partial F}{\partial y} = \frac{\partial F}{\partial r}\frac{\partial r}{\partial y} + \frac{\partial F}{\partial \theta}\frac{\partial \theta}{\partial y} = \sin\theta\frac{\partial F}{\partial r} + \frac{\cos\theta}{r}\frac{\partial F}{\partial \theta}$$

and similarly for the second derivatives. It follows that the stress function to determine the state of stress in polar coordinates assumes the form

$$\phi(r,\theta) = -\frac{G^* r^2}{2} + F(r,\theta)$$

where G^* stands for the shear modulus of the material multiplied by the unit angle of twist $\theta = \phi/L$ (the symbol to designate the unit angle of twist should not be confused with the polar coordinate, θ), and the function F satisfies Laplace's equation in polar coordinates, that is,

$$\frac{\partial^2 F}{\partial r^2} + \frac{1}{r}\frac{\partial F}{\partial r} + \frac{1}{r^2}\frac{\partial^2 F}{\partial \theta^2} = 0$$

By the usual separation of variables, the function F can be sought as

$$F(r,\theta) = \sum_{m=1,2,3,..}^{\infty} A_m\left(\frac{r}{a}\right)^m \cos m\theta$$

TABLE 5.2

Numerical Values of the Constants C_j $j = 1, 2, 3$

β	$\pi/6$	$\pi/3$	$\pi/2$	π
C_1	0.035	0.148	0.298	0.878
C_2	0.452	0.622	0.728	–
C_3	0.490	0.652	0.849	–

Since the stress function must vanish on the straight line boundary, $\theta = \pm\beta$, it follows that by writing $m\beta = n\pi/2$, where n is an odd integer, the stress function becomes

$$\phi(r,\theta) = \frac{G^*}{2}\left[-r^2\left(1 - \frac{\cos 2\theta}{\cos 2\beta}\right) + a^2 \sum_{n=odd}^{\infty} A_n \left(\frac{r}{a}\right)^{\frac{n\pi}{2\beta}} \cos\frac{n\pi\theta}{2\beta}\right]$$

The stress function must also vanish on the curved boundary $r = a$ giving the following series to find A_n:

$$\sum_{n=1,3,5,..}^{\infty} A_n \cos\frac{n\pi\theta}{2\beta} = 1 - \frac{\cos 2\theta}{\cos 2\beta}$$

By the usual Fourier series analysis, the constants, A_n, are found as

$$A_n = (-1)^{\frac{n+1}{2}} \frac{64\beta^2}{\pi^3 n\left(n + \frac{4\beta}{\pi}\right)\left(n - \frac{4\beta}{\pi}\right)}$$

and the stress function is

$$\phi = \frac{G^*}{2}\left[-r^2\left(1 - \frac{\cos 2\theta}{\cos 2\beta}\right) + \frac{64\beta^2 a^2}{\pi^3} \sum_{n=odd}^{\infty} (-1)^{\frac{n+1}{2}} \left(\frac{r}{a}\right)^{\frac{n\pi}{2\beta}} \frac{\cos(n\pi\theta/2\beta)}{n(n + 4\beta/\pi)(n - 4\beta/\pi)}\right] \quad (5.46)$$

The stresses and displacements can be routinely computed using the expression for the stress function in Equation 5.46. In particular, the maximum shear stresses along the circular and straight boundaries, $\tau_{z\theta}$ and τ_{zr}, respectively, as well as the applied torque, $T = 2\iint \phi r\, dr d\theta$, are computed as

$$(\tau_{z\theta})_{max} = C_1 G^* a$$
$$(\tau_{zr})_{max} + C_2 G^* a$$
$$T = C_3 G^* a^3$$

where $G^* = G\theta$, θ is the unit angle of twist, and numerical values of the constants C_j, $j = 1, 2, 3$ are given in Table 5.2 [5.3].

5.4.3 Approximate Solution

In Section 5.4.1, it is shown that by choosing equation of the boundary of the cross section as stress function, an exact elastic solution of the torsion problem can be obtained because the stress function satisfies Equation 5.22 and the associated boundary condition. The examples used to illustrate this method include the solid elliptical cross section, the equilateral triangle cross section, and the circular shaft with a semicircular notch discussed in Example 5.3.

For members with rectangular cross sections, discussed in Section 5.4.2, trigonometric infinite series (Fourier series) solution is used because equation of the boundary of the cross section, Figure 5.14, namely

$$\phi(x, y) = G\theta\left(x^2 - \frac{a^2}{4}\right)\left(y^2 - \frac{b^2}{4}\right) \tag{a}$$

satisfies the boundary condition $\phi = 0$ on the boundary, but does not satisfy Poisson's equation, Equation 5.22, because derivatives of Equation (a) give

$$\nabla^2\phi(x, y) = -2G\theta\left(\frac{a^2 + b^2}{4} - x^2 - y^2\right) \tag{b}$$

which is obviously not a constant and cannot be used as stress function. However, there are few examples where such an approach produces *approximate* and reasonable solution to the elastic torsion problem under consideration. One such example is discussed next in Example 5.4.

EXAMPLE 5.4

The cross section of a member subjected to pure torsion consists of two parabolas of height b, width a at the midpoint of the cross section, as shown in Figure 5.16a. Find the maximum elastic torque supported by the member and compare it to that supported by a member with thin rectangular cross section of length b and width a (Figure 5.16b).

Solution

With the origin of rectangular coordinates chosen at the center of the cross section, the width of the cross section at arbitrary height is given by

$$c(y) = a\left(1 - \frac{4y^2}{b^2}\right) \tag{a}$$

A suitable stress function is

$$\phi(x, y) = G\theta\left[\frac{a^2}{4}\left(\frac{4y^2}{b^2} - 1\right)^2 - x^2\right] \tag{b}$$

The stress function (b) vanishes on the boundary of the cross section. However, the derivatives of the proposed stress function yield

$$\nabla^2\phi = -2G\theta\left[1 - \frac{24a^2y^2}{b^4}\right] + \text{Constant}$$

which indicates that the solution is at best approximate.

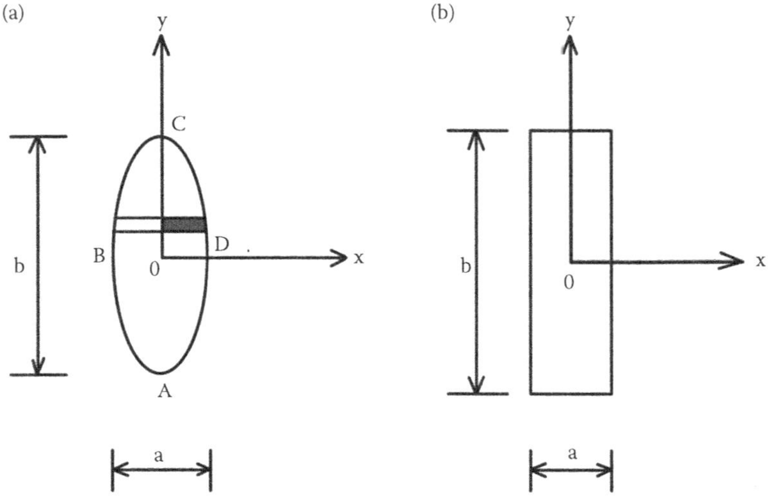

FIGURE 5.16
Example 5.4.

The shear stress is zero at points A and C and maximum at B and D. The applied torque is computed using Equation 5.26

$$T = 2\iint \phi dxdy = 8G\theta \int_0^{b/2} dy \int_0^{(a/2-2ay^2/b^2)} \left[\frac{a^2}{4}\left(\frac{4y^2}{b^2}-1\right)^2 - x^2\right] dx \qquad \text{(c)}$$

Performing the integration, the maximum elastic torque carried by the member is found to be

$$T = 0.152\ G\theta a^3 b$$

The corresponding elastic torque carried by a member with rectangular cross section, length b and width a, obtained using Fourier series is given in Equation 5.41, $T_{exact} = C_2$ $G\theta\ a^3$ b. Numerical values of the constant C_2 are listed in Table 5.1.

For $b/a = 1$, $C_2 = 0.141$ and the percent error = 7.8% ($T_{approx} > T_{exact}$)

For $b/a = 1.5$, $T_{approx} = 0.228\ G\theta\ a^4$ and $T_{exacr} = 0.294\ G\theta\ a^4$, % error = 22% ($T_{approx} < T_{exact}$)

5.5 Plastic Solutions

Consider the torsion of a prismatic bar made of an elastic–perfectly plastic material (Equation 4.34). A monotonic torque is applied to the ends of the bar. In the elastic range, the maximum shear stress has magnitude $|\text{grad}\phi|$ and attains its maximum value at one or more points of the boundary of the cross section. When the torque is increased beyond the elastic limit, the maximum shear stress becomes equal to the yield stress in simple shear τ_0. Thus, initiation of yield begins at some location on the boundary when

$$\tau_{xz}^2 + \tau_{yz}^2 = k^2$$

where k is the yield stress, $k = \sigma_0/2$, according to Tresca's yield criterion, and $k = (\sigma_0/\sqrt{3})$, according to Mises' criterion, σ_0 is the yield stress in tension, and $\sigma_0 = 2\tau_0$.

In terms of the stress function, $\phi(x, y)$, the governing equation for the plastic stress function is

$$\left(\frac{\partial \phi}{\partial x}\right)^2 + \left(\frac{\partial \phi}{\partial y}\right)^2 = k^2 \tag{5.47}$$

As the torque is increased further, plastic regions are formed near the boundary of the cross section and spread inward, one such region is shown in Figure 5.17a between the external boundary C and the elastic/plastic interface D. Since the material is nonhardening, the shear stress maintains its yield value throughout the plastic region. The stress function $\phi(x, y)$ satisfies Equation 5.47 in the plastic region, and because the lateral surface of the bar is free from stress, the corresponding boundary condition is $\phi = 0$, at C. It is recalled that in the elastic region, the stress function is governed by Equation 5.16, that is,

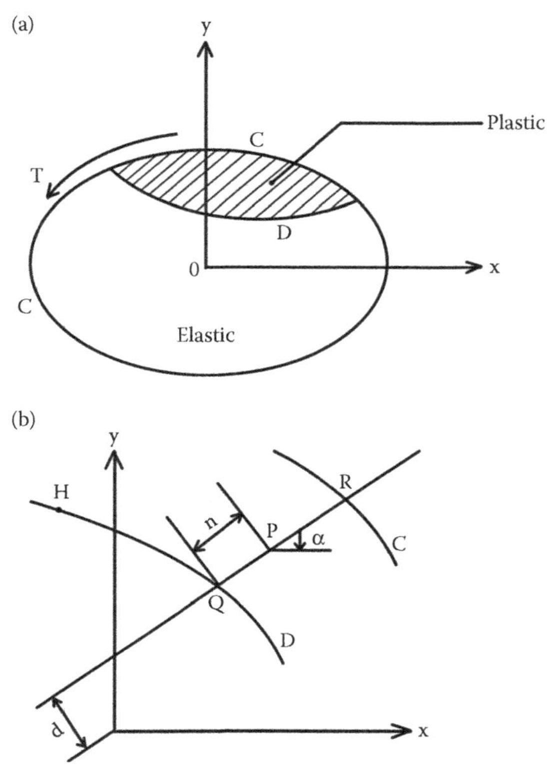

FIGURE 5.17
Elastic–plastic torsion of noncircular cross-section.

$\partial^2\phi/\partial x^2 + \partial^2\phi/\partial y^2 = -2G\theta$, subject to boundary condition $\phi = 0$ on C. Thus, identical boundary conditions for the stress function are valid in the elastic and plastic phases of the deformation. However, the stress function in the plastic region is independent of the angle of twist θ. At each stage of the deformation, the location of the elastic/plastic boundary is determined by requiring the shear stress to be continuous across the interface surface.

In the plastic zone, we have $|\text{grad}\phi| = k$ and the derivative of ϕ with respect to the inward normal n to the boundary is given by

$$\frac{\partial\phi}{\partial n} = k \tag{5.48}$$

It follows that the stress function at any point in the plastic region is k times the distance from the external boundary along the normal through that point, that is, $\phi = kn + \text{constant}$. The magnitude of the maximum shear stress is k and acts tangential to a curve drawn parallel to boundary C. The magnitude and direction of the shear stress do not change during subsequent deformation provided the geometry of cross section is not significantly altered. Also, because ϕ vanishes on C, contours of constant shear stress are parallel curves spaced at constant distance from boundary C.

5.5.1 Axial Displacement

In this section, we discuss the procedure necessary to determine the axial displacement of the cross section. Consider an arbitrary point, P, in the cross section (Figure 5.17b). In the elastic region, we have by Hooke's law,

$$\frac{\gamma_{xz}}{\gamma_{yz}} = \frac{\tau_{xz}}{\tau_{yz}} \tag{a}$$

and the axial displacement is determined from Equations 5.16 and 5.17. When the elastic/plastic boundary reaches point P, the elastic part of the displacement has disappeared, and it follows from that

$$\frac{d\gamma_{xz}}{d\gamma_{yz}} = \frac{\tau_{xz}}{\tau_{yz}} \tag{b}$$

Integration of Equation (b) yields

$$\gamma_{xz} = \frac{\tau_{xz}}{\tau_{yz}}\gamma_{yz} + \text{Constant} \tag{c}$$

The constant of integration is zero because Equation (c) is valid as soon as boundary D reaches point P, and at that instant of deformation, Equation (a) must hold. It follows that with increasing torque, the ratios of the strains remain constant in both the elastic and plastic zones, implying that

$$\frac{\gamma_{xz}}{\gamma_{yz}} = \frac{\tau_{xz}}{\tau_{yz}} = \frac{(\partial\omega/\partial x) - y}{(\partial\omega/\partial y) + x} \tag{5.49}$$

Hence, the displacement system introduced in Equations 5.16 is compatible with the state of stress in the plastic as well as the elastic phases of deformation.

Consider a typical normal, QR, to the external boundary intersecting the elastic/plastic boundary D at Q. The normal makes an angle α with the x-axis (Figure 5.17b). The shear stresses at a generic point P on the normal QR are

$$\tau_{xz} = -k\sin\alpha,\ \tau_{yz} = k\cos\alpha \tag{5.50}$$

Substituting Equation 5.50 into 5.49 gives

$$\frac{\partial\omega}{\partial x}\cos\alpha + \frac{\partial\omega}{\partial y}\sin\alpha = y\cos\alpha - x\sin\alpha \tag{5.51}$$

Equation 5.51 is a linear differential equation of the hyperbolic type whose characteristics are family of curves, $dx/\cos\alpha = dy/\sin\alpha$ or $dy/dx = \tan\alpha$. It follows that the characteristics coincide with the family of normal to the external boundary. Denoting the distance QP in Figure 5.17b by n, it is readily shown that

$$\frac{\partial\omega}{\partial n} = \frac{\partial\omega}{\partial x}\frac{\partial x}{\partial n} + \frac{\partial\omega}{\partial y}\frac{\partial y}{\partial n} = \frac{\partial\omega}{\partial x}\cos\alpha + \frac{\partial\omega}{\partial y}\sin\alpha$$

Hence,

$$\frac{\partial\omega}{\partial n} = y\cos\alpha - x\sin\alpha = d \tag{5.52}$$

where d is the perpendicular distance from the origin 0 to the characteristic QR. The sign of d is positive when the vector RQ has a clockwise moment about 0. Denoting the coordinates of R by ξ, η, and observing that $n\cos\alpha = x - \xi$, $n\sin\alpha = y - \eta$, it follows from Equation 5.52 that

$$\omega(x, y) - \omega(\xi, \eta) = x\eta - y\xi = nd \tag{5.53}$$

Since the warping function, ω, is continuous across the interface boundary, the value of the function $\omega(\xi, \eta)$ can be computed from integration of Equations 5.17 in the elastic region, and the warping $\omega(x, y)$ at any point is readily determined.

When the cross section has an axis of symmetry, the warping $\omega(x, y)$ vanishes on this axis in both the elastic and the plastic regions. The warping function is computed by direct integration along the plastic boundary as follows:

$$d\omega = \frac{\partial\omega}{\partial x}dx + \frac{\partial\omega}{\partial y}dy = ydx - xdy + \frac{1}{G\theta}(\tau_{xz}dx + \tau_{yz}dy)$$

Let the axis of symmetry intersect the plastic boundary at point H in Figure 5.17b. It follows that at H, $\omega = 0$, and integration of the previous equation along the plastic boundary yields

$$\omega(\xi, \eta) = \int_H^D (ydx - xdy) + \frac{1}{G\theta}\int_H^D (\tau_{xz}dx + \tau_{yz}dy) \tag{5.54}$$

If the twist angle is large, the contribution of the second integral can be neglected. Finite element and finite difference numerical techniques can be used to determine the displacement [5.4].

In the next section, the elastic/plastic deformation of an oval-shaped bar is considered.

5.5.2 Oval Cross Section

Sokolovsky [5.5] developed an inverse solution to determine the external boundary of a prismatic bar undergoing elastic/plastic deformation from a known solution in the elastic region. Consider an oval cross section with an elliptical elastic/plastic boundary of semiaxes a and b (Figure 5.18). In the elastic region, the shear stresses are known to be proportional to (−y) and x (Equations 5.18). Writing the stress function as

$$\phi(x, y) = A^* - \frac{k}{2}\left(\frac{x^2}{a} + \frac{y^2}{b}\right) \tag{a}$$

where A* is an arbitrary constant determined from the interface continuity condition. The shear stresses are

$$\begin{aligned} \tau_{xz} &= -k\frac{y}{b} \\ \tau_{yz} &= k\frac{x}{a} \end{aligned} \tag{5.55}$$

And, at the elastic/plastic boundary, $(x^2/a^2) + (y^2/b^2) = 1$, the yield criterion $\tau_{xz}^2 + \tau_{yz}^2 = k^2$ is recovered. The lines of constant shear stress in the elastic region are concentric ellipses with semimajor and minor axes proportional to $\sqrt{a}$ and $\sqrt{b}$, respectively. Inserting Equation 5.55 into 5.19 gives

$$\theta = \frac{k}{2G}\left(\frac{a+b}{ab}\right) \tag{b}$$

The warping displacement in the elastic region is readily computed from Equation 5.18 as

$$\omega = -xy\left(\frac{a-b}{a+b}\right) \tag{5.56}$$

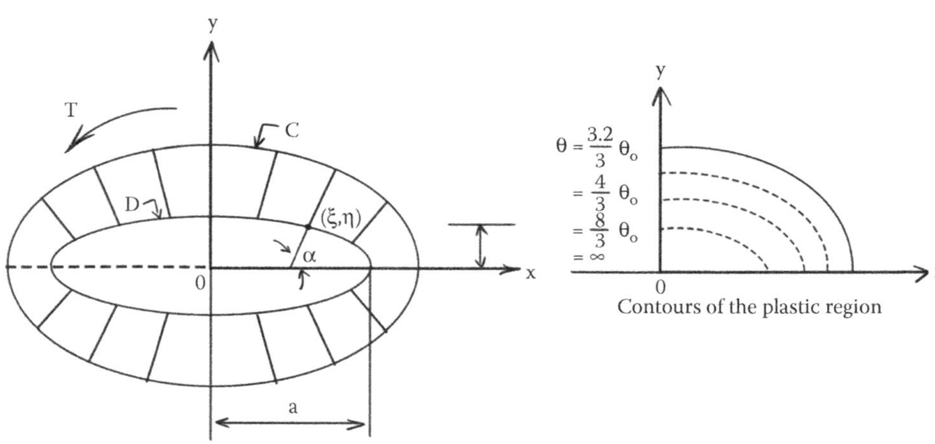

FIGURE 5.18
Elastic–plastic torsion of oval cross section.

The stresses at an arbitrary point in the plastic region are given by Equation 5.50, that is, $\tau_{xz} = -k\sin\alpha$ and $\tau_{yz} = k\cos\alpha$. Since the stresses must be continuous across the elastic/plastic boundary, the characteristic angle, α, at a point of the ellipse with coordinates (ξ, η), satisfies the relations $\xi = a\cos\alpha$, $\eta = b\sin\alpha$. Hence, α is the eccentric angle of the ellipse, and the equation of the characteristic through (ξ, η) is the straight line

$$y = x\tan\alpha - (a - b)\sin\alpha \tag{c}$$

Because the shearing contours in the plastic region are orthogonal trajectories to the family of straight lines (c), and $dy/dx = -\cot\alpha$ along such trajectories, differentiation of Equation (c) with respect to α gives

$$\frac{dx}{d\alpha} + x\tan\alpha = (a - b)\sin\alpha\cos^2\alpha \tag{d}$$

The solution of Equation (d) in parametric form gives the equation of the trajectory.

$$x = \cos\alpha\left[K - \frac{a-b}{4}\cos 2\alpha\right]$$

$$y = \sin\alpha\left[K - (a - b) - \frac{a-b}{4}\cos 2\alpha\right]$$

where K is an arbitrary constant of integration. Introducing the abbreviations
$K = A + 2B$ and $B = a–b/4$, A is a constant, it follows that

$$\begin{aligned} x &= \cos\alpha[A + B(2 - \cos 2\alpha)] \\ y &= \sin\alpha[A - B(2 + \cos 2\alpha)] \end{aligned} \tag{5.57}$$

For $A > 3B$, Equations 5.57 represents a closed oval—very nearly ellipse—with semiaxes $A + B$ $(\alpha = 0)$ and $A - B$ $(\alpha = \pi/2)$.

For given values of a, b, and A/B, Equations 5.57 determines the external boundary C of the cross section. Alternatively, when the external boundary is specified, the elastic/plastic boundary is obtained by solving for a and b from the following relations:

$$a - b = 4B, \quad \frac{a+b}{ab} = \frac{2G\theta}{k} \tag{e}$$

Thus, the elastic/plastic boundary corresponding to an angle of twist θ is determined by

$$\begin{aligned} a &= 2B + \frac{k}{2G\theta} + \sqrt{(2B)^2 + \left(\frac{k}{2G\theta}\right)^2} \\ b &= -2B + \frac{k}{2G\theta} + \sqrt{(2B)^2 + \left(\frac{k}{2G\theta}\right)^2} \end{aligned} \tag{5.58}$$

As the angle of twist is increased, a and b are reduced by the same amount, and when $\theta \to \infty$, the elastic core contracts toward the x-axis. Figure 5.18 shows the contours of the plastic regions with increasing angle of twist for external boundary corresponding to $A/B = 7$. The smallest angle of twist for which the elastic core is embedded in the cross section is obtained when $A + B = a$. Using the first of Equation 5.58, we have

$$\frac{k}{2G\theta} + 2B + \sqrt{(2B)^2 + \left(\frac{k}{2G\theta}\right)^2} < A + B \tag{f}$$

And the smallest angle of twist is found from Equation (f) as

$$\theta \geq \frac{(A - B)k}{(A + B)(A - 3B)G} \tag{5.59}$$

The shape of the elastic/plastic boundary cannot be determined analytically for smaller angles of twist.

Stress function: Consider a generic point along a characteristic of inclination α at distance n from the elastic/plastic boundary. The coordinates of the point are

$$x = (a + n)\cos\alpha \quad y = (b + n)\sin\alpha \tag{5.60}$$

where n and α are functions of x and y. It follows from Equations 5.57 that the distance along the characteristic between external and elastic/plastic boundaries is

$$n_0 = A - \frac{a + b}{2} - B\cos 2\alpha \tag{5.61}$$

Hence, the stress function in the plastic region is

$$\phi(x, y) = k(n_0 - n) = k\left[A - \frac{a + b}{2} - B\cos 2\alpha - n\right] \tag{5.62}$$

The stress function is continuous across the elastic/plastic boundary ($n = 0$) provided the constant A^* in Equation (a) is $A^* = (k/2)(1 + 2A - a - b)$.

Warping displacement: The warping displacement in the plastic region is determined from Equation 5.53, namely, $\omega(x, y) = \omega(\xi, \eta) + nd$. To determine the distance d, we can use Equation 5.52 in conjunction with Equation (c) to give $d = y\cos\alpha - x\sin\alpha = -2B\sin 2\alpha$ while the function, $\omega(\xi, \eta)$, can be determined from the expression of the warping displacement in the elastic region, Equation 5.50, making use of the relations $a + b = 2G\theta ab/k$ $x = a\cos\alpha$, and $y = b\sin\alpha$. The result is $\omega(\xi, \eta) = -(Bk/G\theta)\sin 2\alpha$. It follows that the warping displacement in the plastic region is

$$\omega(x, y) = -2B\left(n - \frac{k}{2G\theta}\right)\sin 2\alpha \quad 0 \leq n \leq n_0 \tag{5.63}$$

Contours of constant warping in the plastic region can be obtained from Equation 5.57 by calculating several values of n and α for specified values of the warping function, and the coordinates of the corresponding points on the contours are then computed from Equations 5.63. In the elastic region, Equation 5.50 gives the contours of constant warping. Figure 5.18 shows the lines of constant warping in one quadrant for a cross section with $A = 7B$ and $\theta = (4/3)\theta_0$ with θ_0 corresponding to an oval cross section containing an elastic core whose semi-axes are $a = A + B$ and $b = A - 3B$. On account of symmetry of the cross section, the axial displacement is positive in the second and fourth quadrants and negative in the first and third quadrants.

Applied torque: The torque applied to the cross section is computed by adding contributions from the stress distributions in the elastic and plastic regions. Thus,

$$T = \iint (x\tau_{yz} - y\tau_{xz})dxdy = T_e + T_p$$

The torque carried by the elastic region is

$$T_e = \iint_A (x\tau_{yz} - y\tau_{xz})\,dxdy = k\iint_A \left(\frac{x^2}{a^2} + \frac{y^2}{b^2}\right) dxdy$$

The integration is performed over an elliptical region of semiaxes a and b, and the result is $T_e = \pi/4kab(a + b)$. In view of Equation 5.50, the torque carried by the plastic region is

$$T_p = k\iint_A (x\cos\alpha + y\sin\alpha)dxdy \tag{g}$$

The integration is carried over the domain between the elastic/plastic boundary and external boundary. It is convenient to change the variables from x and y to n and α using the relations $x = (a + n)\cos\alpha$ and $y = (b + n)\sin\alpha$. The Jacobian of the transformation is the determinant

$$\begin{vmatrix} \dfrac{\partial x}{\partial n} & \dfrac{\partial x}{\partial \alpha} \\ \dfrac{\partial y}{\partial n} & \dfrac{\partial y}{\partial \alpha} \end{vmatrix} = a\sin^2\alpha + b\cos^2\alpha + n$$

and Equation (g) becomes

$$T_p = 4k\int_0^{\frac{\pi}{2}}\int_0^{n_0} (a\cos^2\alpha + b\sin^2\alpha + n)(a\sin^2\alpha + b\cos^2\alpha + n)dnd\alpha$$

Carrying out the integration, the result is

$$T_p = \pi k\left[\frac{2}{3}A^3 - 3AB^2 + \left(\frac{a+b}{2}\right)\left(\frac{a^2+b^2}{2} - 5ab\right)\right] \tag{h}$$

and the total torque is

$$T = \pi k\left[\frac{2}{3}A^3 - 3AB^2 + \frac{a^3 + b^3 - 9ab(a+b)}{24}\right] \tag{5.64}$$

where a and b are given in Equation 5.58. The torque angle of twist result (Equation 5.64) is valid for $\theta \geq \theta_0$ where ϑ_0 is the angle of twist corresponding to a torque obtained from Equation 5.64 by setting $a = A + B$ and $b = A - 3B$. For nonhardening material, as the applied torque is continuously increased, the elastic core in the limit shrinks to a line of discontinuity along the x-axis. A fully plastic cross section is realized and the corresponding torque is obtained from Equation 5.64 by setting $b = 0$ and $a = 4B$. Further details are provided by Chakrabarty [2.2].

5.5.3 Fully Plastic Torque

The fully plastic or ultimate torque in a member made of elastic–perfectly plastic material is obtained as the asymptotic limit of the magnitude of the torque when the angle of twist increases beyond the elastic limit. The magnitude of the fully plastic torque has some physical significance because the accompanying deformation is of the same order of magnitude as the elastic deformation. The fully plastic torque for a bar with solid circular cross section, radius $= a$, is $T_p = (2/3\pi a^3)\tau_0$, where $\tau_0 =$ half the yield stress in tension.

Nadai [4.12] proposed a "sand hill" analogy to determine the fully plastic torque. Dry sand is piled on a horizontal base, whose shape is identical to that of the cross section of the twisted bar. The slope of the sand hill is constant and is controlled by internal friction of the sand. Using known solutions, for example, cylindrical bars, a factor of proportionality is introduced to make the slope of the sand hill equal to τ_0/G. The fully plastic torque applied to the member is

$$T_p = 2GV \tag{5.65}$$

where V is the volume of the sand hill. The analogy is illustrated in the following examples:

Circular member: Consider a solid bar of circular cross section of radius $= a$. Volume of the sand hill spread over the base is $V = (1/3)\pi a^2 h$. Since $h/a = \tau_0/G$, it follows that the fully plastic torque is

$$T_p = \frac{2}{3}\pi a^3 \tau_0 \tag{5.66a}$$

Similarly, for a hollow bar with external and internal radii, b and a, respectively, $b > a$, the fully plastic torque is

$$T_p = \frac{2}{3}\pi(b^3 - a^3)\tau_0 \tag{5.66b}$$

The results in Equations 5.66 coincide with those in Section 5.2.1. Note that $T_p/T_e = 4/3$ for the solid bar and

$$\frac{T_p}{T_e} = \frac{4b(b^3 - a^3)}{3(b^4 - a^3)}$$

Triangular cross section: Consider a solid bar whose cross section is an equilateral triangle of side $= a$. The elastic analysis in Section 5.4 indicates that maximum shear stress occurs at the midpoint of each side and has the value $(\sqrt{3}/4)G\theta a$. Thus, plastic yielding begins at those locations when the angle of twist reaches the value $\theta_e = 4\tau_0/\sqrt{3}Ga$. The corresponding torque from Equation 5.32 is

$$T_e = \frac{\sqrt{3}G\theta_e a^4}{80} = \frac{\sqrt{3}Ga^4}{80}\left(\frac{4\tau_0}{\sqrt{3}Ga}\right) = \frac{\tau_0 a^3}{20} \tag{a}$$

In the plastic zone,

$$\tau_{xz}^2 + \tau_{yz}^2 = \left(\frac{\partial\phi}{\partial x}\right)^2 + \left(\frac{\partial\phi}{\partial y}\right)^2 = \tau_0^2 \tag{b}$$

While in the elastic region, Equation 5.22 governs the stress function, and across the interface the shear stress is continuous.

Further increase in the torque causes the plastic zone to spread inward until the cross section becomes fully plastic for nonhardening material. To determine the fully plastic torque:

The volume of a sand heap spread over the base is (Figure 5.11)

$$V = \frac{1}{3}\left(\frac{\sqrt{3}a^2}{4}\right)h$$

where

$$h = \left(\frac{a}{2\sqrt{3}}\right)\frac{\tau_0}{G}$$

It follows that the fully plastic torque is

$$T_p = 2G\left[\frac{1}{3}\left(\frac{\sqrt{3}a^3}{4}\right)h\right] = \frac{1}{12}a^3\tau_0 \tag{5.66c}$$

and $T_p/T_e = 1.67$.

Rectangular cross section: Consider a solid bar of rectangular cross section, sides a and b, $b > a$ (Figure 5.14). In the elastic range, the maximum shear stress occurs at the midpoint of the longer side, that is, at $x = \pm, y = 0$, its magnitude from Equation 5.40 is $C_1G\theta a$. Plastic yielding begins when

$$\theta_e = \frac{\tau_0}{C_1G\theta}$$

and using Equation 5.42, the torque at initiation of yield is

$$T_e = \frac{C_2}{C_1}a^2b\tau_0$$

where C_j, $j = 1, 2$ are constants listed in Table 5.1.

A sketch of the piled sand over the cross section is shown in Figure 5.19. Within a proportionality scale factor, the height of the central ridge is $h = \tau_0 a/2G$, and the volume of the sand consists of the volume of a square pyramid of height h and base a^2, that is, $(1/3)a^2h$, and the volume of a triangular prism of length (b–a) and cross-sectional area ah/2, that is, (b–a) ah/2. Hence, the fully plastic torque is

$$T_p = \frac{\tau_0 a^2}{2}\left(b - \frac{a}{3}\right) \tag{5.66d}$$

and $T_p/T_e = C_1(b - a/3)/2C_2b$.

Square cross section: When $b = a$

$$T_p = \frac{1}{3}\tau_0 a^3 \tag{5.66e}$$

and $T_p/T_e = C_1/3C_2$.

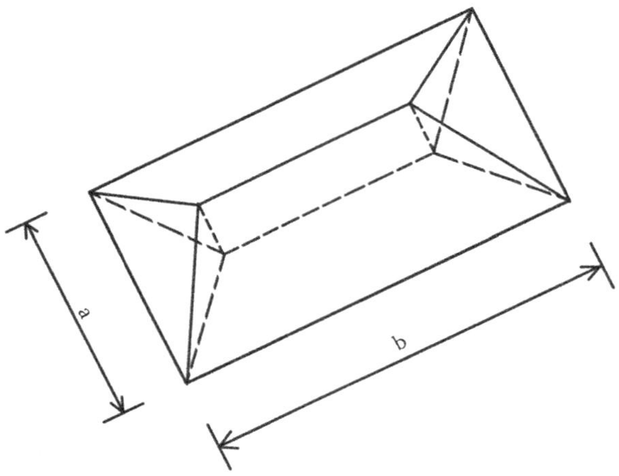

FIGURE 5.19
Sand hill with rectangular base.

The magnitude of T_p/T_e varies between 0.333 C_1/C_2 for $b/a = 1$ to 0.05 C_1/C_2 for $b/a = \infty$.

NOTE: In the sand hill analogy instead of measuring the volume of a sand hill, its weight can be recorded and compared to that of a circular cross section to find the factor of proportionality. The ratio of the two weights is proportional to the ratio of the two fully plastic torques.

5.6 Membrane Analogy

Prandtl presented an experimental technique to obtain solutions of the torsion problem using a mathematical analogy between the torsion equation of a twisted bar and equation of the deflection of a uniformly stretched membrane subject to lateral pressure [5.6].

Consider a membrane covering a hole of the same shape as the cross section of the twisted bar (Figure 5.20). The membrane is under the action of lateral pressure p and is stretched by a uniform force F per unit length of the boundary.

The deflection of the membrane is represented by z. The inclination of F on side ab of an element of the membrane is $\partial z/\partial x$, because z varies with x and y, the inclination of z on side cd of the element is

$$\frac{\partial z}{\partial x} + \frac{\partial^2 z}{\partial x^2} dx$$

Similar angles of inclination occur on sides, ac and bd of the element. Vertical equilibrium of the forces acting on the membrane yields

$$\frac{\partial}{\partial x}\left(F\frac{\partial z}{\partial x}\delta y\right)\delta x + \frac{\partial}{\partial y}\left(F\frac{\partial z}{\partial y}x\delta\right)y\delta + p\delta x\delta y = 0$$

On simplifying, the previous equation is reduced to

$$\frac{\partial^2 z}{\partial x^2} + \frac{\partial^2 z}{\partial y^2} = -\frac{p}{F} \tag{5.67}$$

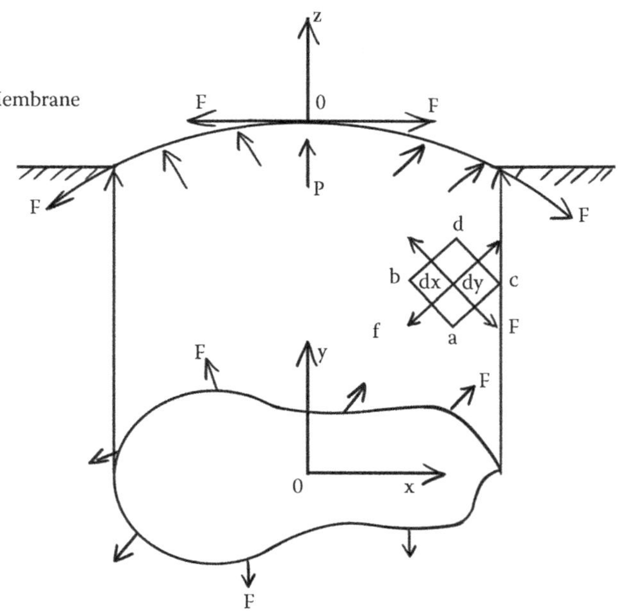

FIGURE 5.20
Analogy of torsion problem with equilibrium of membrane.

Equation 5.67 is of Poisson's type and is analogues to Equation 5.22 provided that z is replaced by ϕ and $2G\theta$ by p/F.

The volume under the membrane is $V = \iint z dxdy$, which is comparable to $\iint \phi dxdy$.

Thus, a measure of the torque required to cause a unit angle of twist θ can be obtained.

Since $z = 0$ at the boundary of the membrane, the boundary condition of the torsion problem, that is, $\phi = 0$, is also satisfied. On any contour of constant deflection of the membrane, $dz/ds = 0$, where s is the tangential direction. Thus,

$$\frac{d\phi}{ds} = \frac{\partial\varphi}{\partial x}\frac{dx}{ds} + \frac{\partial\phi}{\partial y}\frac{dy}{ds} = -\tau_{yz}\frac{dx}{ds} + \tau_{xz}\frac{dy}{ds} = 0$$

The above equation implies that contours of constant deflection correspond to lines of shearing of the twisted member. Since the resultant shear stress at any point is given by (Figure 5.8)

$$\tau_{max} = \tau_{yz}\frac{dx}{dn} - \tau_{xz}\frac{dy}{dn} = -\left(\frac{\partial\phi}{\partial y}\frac{dy}{dn} + \frac{\partial\phi}{\partial x}\frac{dx}{dn}\right) = -\frac{d\phi}{dn} \sim -\frac{dz}{dn}$$

It follows that the maximum shearing stress at any point is represented by the maximum slope at the corresponding point of the membrane, which is $|\text{grad}\phi|$.

The shear stresses at arbitrary points of the bar can be obtained from knowledge of the slopes of the deflected membrane at those points because $\tau_{xz} = \partial\phi/\partial y \sim G\, \partial z/\partial y$ and $\tau_{yz} = -\partial\phi/\partial x \sim -G\partial z/\partial x$.

In sum, the membrane analogy provides ratios between pressure and angle of twist, slope and stress, and volume and torque. The ratios are found by performing an experiment on a circular membrane subject to given pressure, measuring the maximum slope and volume, and relating the results to the corresponding quantities in the known solution of a circular bar subject to twisting moment.

Prandtl also extended the membrane analogy to deal with elastic/plastic torsion problems for nonhardening materials. A roof of constant slope equal to τ_0/G, τ_0 is the yield stress of the bar, is erected over the stretched membrane. As the lateral pressure on the membrane is increased, it begins to touch the roof indicating the initiation of yield in the twisted bar. With further increase in the lateral pressure, certain parts of the membrane are in contact with the roof and the remaining parts are free to expand, which correspond to the plastic and elastic zones of the cross section, respectively. The corresponding applied torque is equal to 2G times the volume under the entire membrane. With further increase in pressure, the total membrane is in contact with the roof, indicating that the cross section is fully plastic.

Various experimental techniques for using membrane analogy have been discussed in Hetenyi [5.7].

Next, a simple example is used to illustrate the basic features of membrane analogy. Also, the analogy is applied in Sections 5.6.1 and 5.6.2 to obtain approximate analytical solutions to bars of common cross sections.

EXAMPLE 5.5

A membrane is anchored around a circle of radius "a," and is subjected to lateral pressure p. By using membrane analogy, derive the well-known formulas for the twisting of a solid round bar by torque T.

Solution

As p is increased, the membrane forms a segment of a paraboloid with the top section having radius r, as shown in Figure 5.21.

Equating the vertical forces acting on the membrane, we have

$$p\pi r^2 = -F\frac{\partial z}{\partial r} \cdot 2r\pi$$

$$\frac{\partial z}{\partial r} = -\frac{p}{2F}r$$

Hence,

$$z = -\frac{pr^2}{4F} + C$$

The constant C is found from the condition when $z = 0, r = a$, so $z = p/4F(a^2 - r^2)$. Using membrane analogy, $\phi = G\theta/2(a^2 - r^2)$. Maximum shear stress, $|\text{grad}\phi|$ which is $-\partial\phi/\partial r = G\theta r$ occur at $r = a$. Hence, $\tau_{max} = G\theta a$.

The volume of paraboloid between $z = 0$ and the membrane is

$$V = \iint_A z\,dA = \int_0^a \frac{p}{4F}(a^2 - r^2)(2r\pi)dr = \frac{p\pi a^4}{8F}$$

It follows that the torque $T = 2\,V$, and using the membrane analogy, we find

$$T = \frac{G\theta\pi a^4}{2} = G\theta J$$

where $J = (\pi a^4)/2$ is the polar moment of inertia of the circular cross section about the normal axis through the centroid.

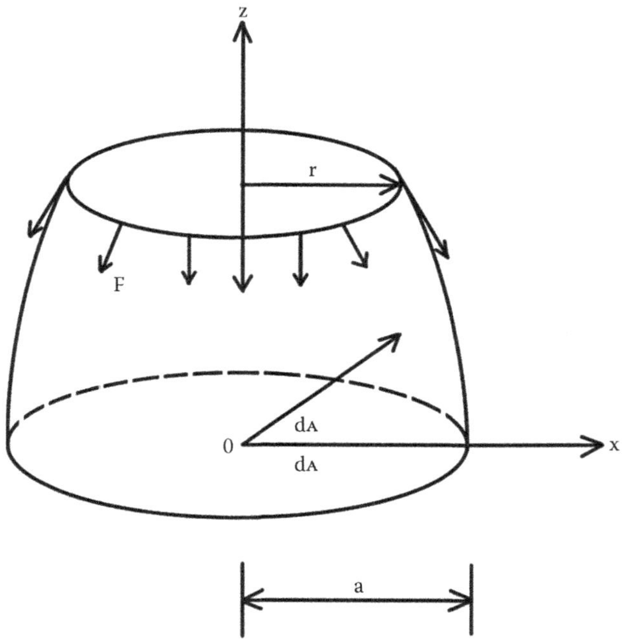

FIGURE 5.21
Example 5.5.

5.6.1 Torsion of Thin-Walled Members

The membrane analogy can be applied to thin-walled members like I = beams, channels, tubes, and other multiconnected open sections.

Consider a bar whose cross section is a narrow rectangle of length L and width a ($a \ll L$). The deflected membrane maintains somewhat constant cylindrical shape over the entire domain except near the shorter sides $y = \pm L/2$ (Figure 5.22). Elastic and plastic behaviors are discussed.

Elastic behavior: Neglecting the stress concentration at the corners of the cross section, it is reasonable to assume that $\partial z/\partial y = 0$. Hence, Equation 5.67 is reduced to $d^2z/dx^2 = -p/F$. Integrating twice and using boundary conditions $dz/dx = 0$ at $x = 0$, and $z = 0$ at $x = \pm a/2$, it follows that

$$z = \frac{p}{2F}\left(\frac{a^2}{4} - x^2\right) \tag{a}$$

Making use of membrane analogy, z is replaced by ϕ and p/F by $2G\theta$. It follows that the maximum deflection at $x = 0$ is $\delta = z_{max} = a^2G\theta/4$, and the stress function is

$$\phi = G\theta\left(\frac{a^2}{4} - x^2\right) \tag{5.68}$$

The stresses in the member are $\tau_{xz} = \partial\phi/\partial y = 0$ and $\tau_{yz} = -\partial\phi/\partial x = 2G\theta x$, which are valid everywhere in the member except near the ends.

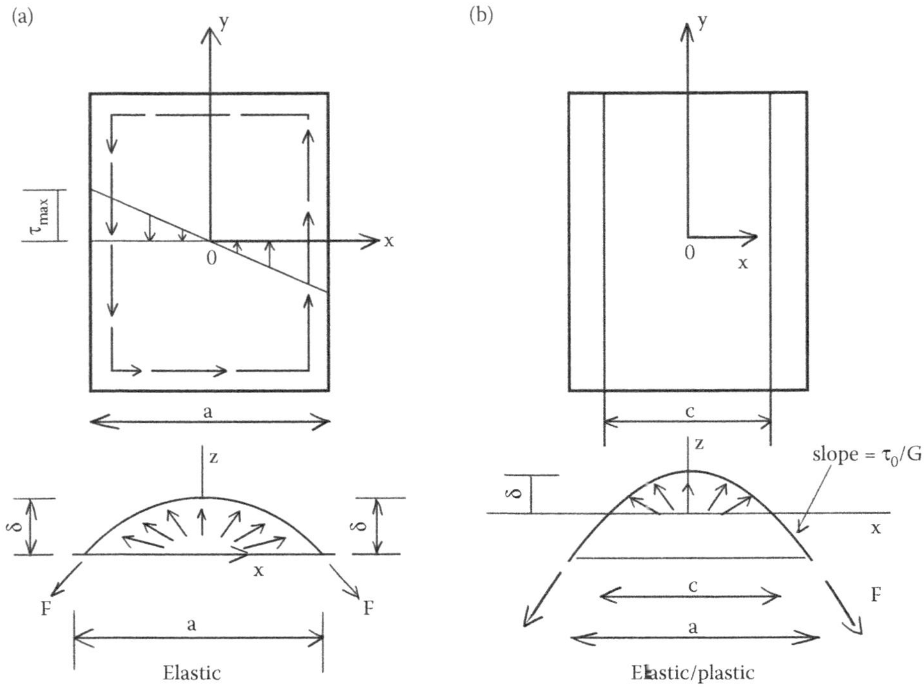

FIGURE 5.22
Elastic–plastic torsion of a bar of narrow rectangular cross section.

The volume enclosed by the membrane and the xy-plane is

$$V = \frac{2}{3}(aL)\delta = \frac{1}{6}G\theta a^3 L \tag{b}$$

And the applied torque producing the angle of twist is twice the volume

$$T = \frac{1}{3}G\theta a^3 L = GJ_{eff}\theta \tag{5.69}$$

where $J_{eff} = 1/3a^3L$ is the effective polar moment of inertia.

The torsional rigidity is

$$C = \frac{T}{\theta} = GJ_{eff} \tag{c}$$

and the unit angle of twist from Equation 5.69 is

$$\theta = \frac{3T}{a^3 LG} \tag{d}$$

The maximum shear stress that occurs along the long sides. $x = \pm a/2$, is obtained as

$$\tau_{max} = G\theta a = \frac{3T}{a^2 L} \tag{5.70}$$

However, examining the shear flow in the member, Figure 5.22a, the torque produced by the maximum shear stress is

$$T = 2\left(\frac{1}{2}\tau_{max}\right)\left(\frac{a}{3}\right)L = \frac{1}{6}a^2L\tau_{max} \tag{e}$$

which is half the amount given in Equation 5.70. Apparently, the remaining amount of the applied torque is resisted by the shear stresses τ_{xz}, which is neglected in the above analysis.

The maximum elastic torque is reached when the maximum shear stress is equal to the yield stress, τ_0, that is,

$$T_e = \frac{1}{3}a^2L\tau_0 \tag{5.71}$$

The corresponding unit angle of twist at initiation of yield is $\theta_e = \tau_0/Ga$.

As an example, consider a member with I—section as shown in Figure 5.23. When the member is subjected to torque T, the angle of twist and maximum shear stress are determined as follows:

The effective polar moment of inertia, Equation 5.69, is

$$J_e = \sum \frac{1}{3}Lt^3 = \frac{1}{3}(ht_1^3 + 2bt_2^3)$$

It follows that the unit angle of twist and maximum shear stress are

$$\theta = \frac{T}{GJ_e} = \frac{3T}{G(ht_1^3 + 2bt_2^3)}$$

$$\tau_{max} = G\theta t_i = \frac{3Tt_i}{ht_1^3 + 2bt_2^3}$$

where t_i is the larger of t_1 and t_2.

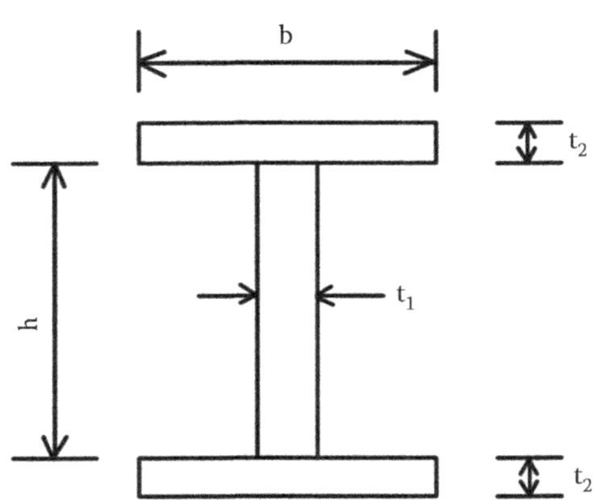

FIGURE 5.23
I Sectem.

Plastic behavior: When the applied torque is increased further, $T > T_e$, the plastic zone spreads inward and encloses a rectangular elastic core of length L and width c (Figure 5.22b). The shear stress τ_{xz} is still zero and

$$\begin{aligned} \tau_{yz} &= 2\frac{\tau_0}{c}x, \quad 0 \le x \le \frac{c}{2} \\ \tau_{yz} &= \tau_0, \quad \frac{c}{2} \le x \le \frac{a}{2} \end{aligned} \tag{f}$$

The membrane now is in contact with the erected roof over the stretched membrane in the plastic region and has a cylindrical shape in the elastic core (Figure 5.22b). Equilibrium of the membrane requires that $\theta = \tau_0/Gc$ in the plastic region while the maximum deflection of the parabolic cylinder in the free region is $\delta = \tau_0 c/4G$.

The volume under the membrane is computed as

$$V = \frac{2}{3}Lc\delta + \frac{1}{4}\frac{\tau_0}{G}L(a^2 - c^2) \tag{g}$$

Substituting for δ from the previous expression, the torque is obtained by multiplying the volume by 2G and the result is

$$T = \frac{1}{2}\tau_0 L\left(a^2 - \frac{1}{3}c^2\right) \tag{h}$$

The torque–twist relation for the member becomes

$$T_p = \frac{1}{2}\tau_0 L\left(a^2 - \frac{\tau_0^2}{3G^2\theta^2}\right) \tag{5.72}$$

It should be noted that the analysis neglects significant stress concentration at the reentrant corners where two members meet.

5.6.2 Torsion of Thin-Walled Tubes

In this section, we use membrane analogy to determine the torsional deformation of thin-walled hollow members. Consider a tubular member of arbitrary cross section, wall thickness t(s), where t is small and s is the distance along the mean perimeter (Figure 5.24). Because there are no stresses in the hollow region, a horizontal rigid plate is attached to the outer boundary of the membrane to provide constant stress function. In addition, a weightless horizontal rigid plate is attached to the inner boundary of the membrane, thus connecting the outer and inner contours over a distance t. The inner plate is constrained to move vertically under lateral pressure and is rendered weightless by using counterbalance devices. Its movement is constrained so that it does not experience sideward motion. Because the thickness of the tube, t, is small, the slope of the deflected membrane does not vary significantly with the thickness. The resultant shear stress τ(s) is constant and given by

$$\tau = h/t \tag{a}$$

where h denotes the height of the deflected membrane, its slope is h/t and the volume enclosed by the membrane and horizontal plate is Ah, where A represents the area enclosed

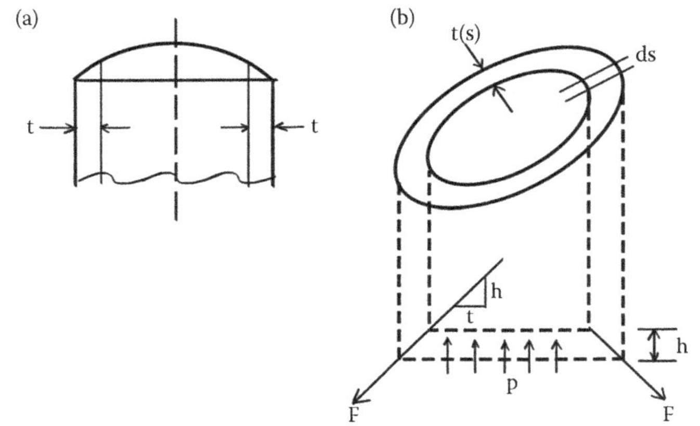

FIGURE 5.24
Membrane for thin-walled tubular member.

by the mean perimeter of the cross section (shown dotted in Figure 5.24). The membrane analogy gives the torque T as twice the volume under the membrane. Thus,

$$T = 2\,Ah$$

and excluding corners and abrupt variations in tube thickness, the shear stress is

$$\tau = \frac{T}{2At} \tag{5.73}$$

The shear stress is maximum at the thinnest part of the tube.

It is noted that the classical torsion formula, Equation 5.3, for thin-walled cylindrical bar with outer radius b, inner radius a, thickness $b - a = t$, gives, $\tau = 2Tr/\pi(b^4 - a^4)$, where $r = a + t/2 = b - t/2$, and $A = \pi r^2$. Inserting r in the classical formula and expanding for small t, formula (5.73) is recovered.

The torque angle of twist relation is obtained from consideration of vertical equilibrium of the membrane (Figure 5.24). Observing that h is small compared to t, it follows that $\mathrm{Tan}(h/t) \approx h/t$, and equilibrium of the vertical forces acting on the plate implies that

$$pA = F\oint \left(\frac{h}{t}F\right) ds = \oint \tau F ds \tag{b}$$

The integral in Equation (b) is evaluated over the mean perimeter of the cross section. Using membrane analogy, $p/F = 2G\theta$, (b) provides

$$\theta = \frac{1}{2GA}\oint \tau ds \tag{5.74}$$

Formulas (5.73) and (5.74) are valid in the elastic region. In the literature, they are referred to as Bredt's formula [5.8].

As the applied torque is increased, the shear stress attains its yield value τ_0; first at the location where the thickness t is minimum, and then the plastic region spreads and creates an arc of length s* along the mean perimeter of the tube. The corresponding height of the rigid plate is $h = \tau_0(t^*/G)$, where t* is the thickness of the plastic arc. The corresponding torque is

$$T = 2\tau_0 A t^* \tag{5.75}$$

and the corresponding angle of twist per unit length is

$$\theta = \frac{T}{4GA^2}\left(\frac{s^*}{t^*} + \int \frac{ds}{s}\right) \tag{5.76}$$

EXAMPLE 5.6

A torque T is applied to a thin-walled tube with an octagonal cross section of side length a and thickness t. The modulus of rigidity of the material is G. It is required to derive the formulas for the shear stress τ and unit angle of twist θ of the tube.

Solution

Consider the cross section of an octagonal tube of side length = a (Figure 5.25). The magnitude of the interior angles is equal to $(n-2)\ \pi$, $n = 8$, so each angle = 135°. It is readily confirmed that

$$A = (5 + 4\sqrt{2})a^2 = 10.657\, a^2$$

Using Equation 5.73, we have

$$\tau = \frac{T}{2At} = \frac{T}{21.3137a^2t} \tag{a}$$

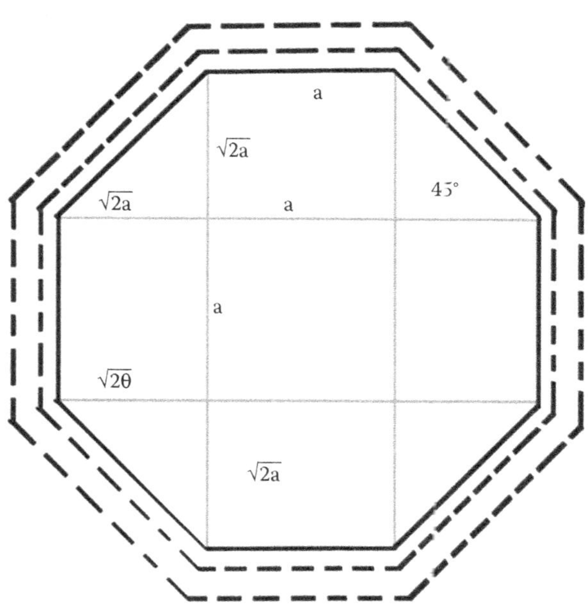

FIGURE 5.25
Octagonal tube.

The length of the mean perimeter is 8a, and Equations 5.74 and (a) furnish

$$\theta = \frac{\tau(8a)}{2GA} = \frac{2T}{(5+4\sqrt{2})^2 a^3 tG} = \frac{T}{10.657a^3 tG} \tag{b}$$

Note that if T = 50 kN, G = 28 GPa, a = 80 mm, and t – 10 mm, then

$$\tau = 36.7\,\text{MPa}, \quad \text{and} \quad \theta = 32.73 \times 10^{-3}\,\text{rad/m}$$

5.7 Torsion of Restrained Structural Sections

In this section, we consider the response of thin-walled structural members such as I-beam, channel, angle, etc. to twisting moments. A mechanics of material's approach is used because analytical solutions are not available.

We note that if both ends of the member are free to rotate, the torsional rigidity is $C = GJ_e$ (J_e is the polar moment of inertia of the cross section) and the angle of twist is $\phi = TL/C$ throughout the member. However, if one cross section of the member is fixed, the angle of twist and the warping will vary along the axis and there will be tension and compression in the longitudinal fibers. The resulting additional stresses give rise to torsional resistance, which cannot be neglected in regions away from the restrained section as in the case of members made of solid sections like cylinders and rectangular bars. In the case of an I-section member, for example, bending of the flanges will give rise to a twisting moment, which needs to be included in the analysis to determine the torsional resistance.

Consider an I-section cantilever of length L subjected to torque T at the free end (Figure 5.26a). The applied torque T is balanced by two parts, $T = T_1 + T_2$, where T_1 is due to shear stresses developed in both the web and flanges assuming all cross sections are free to wrap. The additional part, T_2, is contributed by the moment of equal and opposite shearing forces due to bending of the flanges.

Assuming that the origin of coordinates is at the fixed end, then T_1 is proportional to the rate of change of the angle of twist along the axis

$$T_1 = C\theta = C\frac{d\phi}{dz} \tag{a}$$

The usual right-hand rule of sign is used to determine the positive sign of the angle of twist ϕ. The shearing stresses due to bending of the upper and lower flanges give rise to a pair of lateral shear forces V_f at distance h (Figure 5.26b), which in turn leads to

$$T_2 = V_f h \tag{b}$$

An expression for V_f can be derived in terms of displacement u along the x-axis of the flange. Because the deformation is small and the cross section is symmetrical, it follows that $\tan\phi = \phi = u/(h/2)$ and the variation of u along the axis is

$$\frac{du}{dz} = \frac{h}{2}\frac{d\phi}{dz} \tag{c}$$

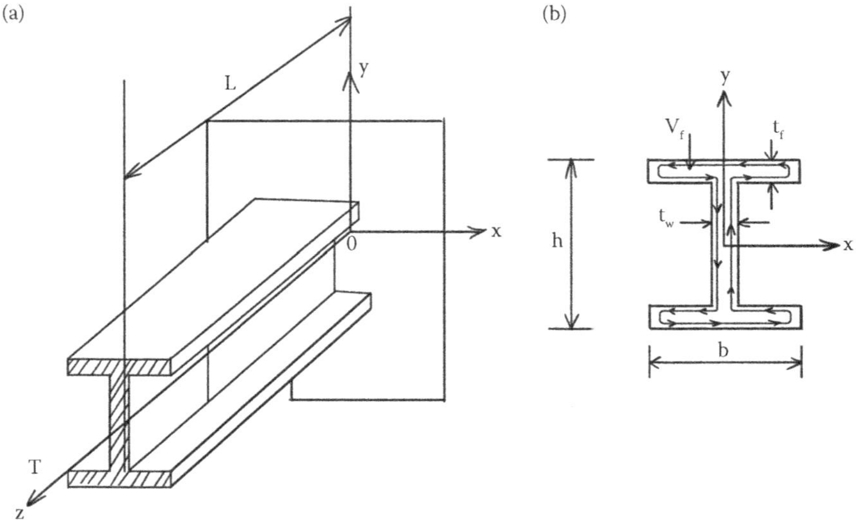

FIGURE 5.26
Torsion of a restrained I-section member.

Using beam theory, the moment and shear force in the flange are

$$\begin{aligned} M_f &= EI_f \frac{d^2u}{dz^2} = \frac{EI_f h}{2}\frac{d^2\phi}{dz^2} \\ V_f &= -EI_f \frac{d^3u}{dz^3} = -\frac{EI_f h}{2}\frac{d^3\phi}{dz^3} \end{aligned} \tag{d}$$

where I_f is the moment of inertia of one flange about the y-axis. It follows that the total torque carried by the member from Equations (a) and (b) is

$$T = C\frac{d\phi}{dz} - EI_f \frac{h^2}{2}\frac{d^3\phi}{dz^3} \tag{5.77}$$

Introducing the abbreviation, $\eta^2 = 2C/Eh^2I_f$, Equation 5 77 yields the differential equation governing the angle of twist in the cantilever

$$\frac{d^3\phi}{dz^3} - \eta^2\frac{d\phi}{dz} + \frac{\eta^2}{C}T = 0 \tag{5.78}$$

Since the slope and moment are zero at the fixed and free ends of the cantilever, respectively, the corresponding boundary conditions required to solve Equation 5.78 are

$$\begin{aligned} \frac{d\phi}{dz} &= 0, z = 0 \\ \frac{d^2\phi}{dz^2} &= 0, z = L \end{aligned} \tag{5.79}$$

The solution of Equation 5.78 consists of a homogeneous solution, $d\phi/dz = Ae^{\eta z} + Be^{-\eta z}$, and a particular solution, $d\phi/dz = T/C$. The constants A and B are found from boundary conditions (5.79). The result is

$$\frac{d\phi}{dz} = \frac{T}{C}\left[1 - \frac{e^{\eta(z-L)} + e^{-\eta(z-L)}}{2\cosh \eta L}\right] = \frac{T}{C}\left[1 - \frac{\cosh(z-L)}{\cosh L}\right] \tag{e}$$

When L is large, $e^{-\eta L} \rightarrow 0$, $2\cosh \eta L = e^{\eta L}$, and Equation (e) is reduced to

$$\frac{d\phi}{dz} = \frac{T}{C}(1 - e^{-\eta z}) \tag{f}$$

Using Equations (a) and (b), it follows that the contributions of T_1 and T_2 are

$$\begin{aligned} T_1 &= T(1 - e^{-\eta z}) \\ T_2 &= Te^{-\eta z} \end{aligned} \tag{5.80}$$

At $z = 0$, $T_1 = 0$, and $T_2 = T$, implying that the applied torque is balanced by the effect of shearing forces V_f in the flange, $T_2 = V_f\, h$.

The maximum bending moment in the flange at $z = 0$ is obtained from Equations (d) and (f). The result is

$$(M_f)_{max} = \frac{T}{\eta h} \tag{g}$$

The angle of twist is found by integrating Equation (f), which yields

$$\phi = \frac{T}{C}\left(z + \frac{e^{-\eta z}}{\eta} + A\right)$$

The constant A is determined from boundary condition $z = 0$, $\phi = 0$, and $A = -\frac{1}{\eta}$. Hence,

$$\phi = \frac{T}{C}\left[z + \frac{1}{\eta}(e^{-\eta z} - 1)\right] \tag{5.81a}$$

For relatively long members, the term $e^{-\eta z}$ can be neglected and the angle of twist is given by

$$\phi = \frac{T}{C}\left[L - \frac{1}{\eta}\right] \tag{5.81b}$$

For comparison, the angle of twist in an unrestrained member is $\phi = (TL/C)$. Hence, the stiffness of a restrained member has increased due to the restraints of the flanges (effect of the term $-1/\eta$ in Equations 5.81).

EXAMPLE 5.7

A wide flange cantilever beam (WF 96 × 10) of length 14 ft. is subjected to a torque of 30 kip-in. The geometric properties of a typical cross section shown in Figure 5.26b are $h = 16.32$ in., $t_w = 0.535$ in., $b = 11.533$ in., and $t_f = 0.875$ in. Assume $E = 30 \times 10^6$ psi and $G = 12 \times 10^6$ psi.

Find the maximum bending stress in the flange and the angle of twist at the free end.

Solution

The polar moment of inertia of the cross section from Equation 5.69 is

$$J_e = \frac{1}{3}\left[h_w t_w^3 + 2bt_f^3\right] = \frac{1}{3}\left[14.57 \times (0.535)^3 + 2(11.533)(0.875)^3\right] = 5.895\,\text{in.}^4$$

Hence, C G $J_0 = 5.895$ G.
The flexural rigidity of one flange is

$$EI_f = E\frac{t_f b^3}{12} = E\frac{0.875(11.533)^3}{12} = 111\,855E$$

It follows that the parameter η introduced in Equation 5.78 is

$$\frac{1}{\eta} = h\left[\frac{EI_f}{2C}\right]^{1/2} = 4.87\,h$$

The maximum bending moment in the flange is $(M_f)_{max} = T/\eta h = 4.87T$, and the maximum longitudinal bending stress is

$$(\sigma_f)_{max} = \frac{(M_f)_{max}(x)}{I_f} = \left(\frac{4.87T}{111.855}\right)\left(\frac{11.533}{2}\right) = 7.53\,\text{ksi}$$

The angle of twist for a constrained member using Equation 5.81b is $\phi = (T/C)(L - 1/\eta) = 0.0375$ radians.
For a member with free ends, the angle of twist is $\phi_{free} = (TL/C) = 0.0713$ radians.

Problems

5.1 A cantilever solid cylindrical shaft is subjected to a torque of 1.0 kN m at the free end. The length of the shaft = 1.0 m and its diameter is 30 mm. The material is elastic–perfectly plastic with $\tau_0 = 160$ MPa and $G = 80$ GPa. Determine (a) the angle of twist at the free end; (b) the residual angle of twist if the torque is completely unloaded.

Ans: $\phi = 9.9°$, $\phi^R = 0.9°$.

5.2 A hollow circular shaft, length 20 in., external and internal radii, 3 and 1 in., respectively, is fixed at one end and subjected to a torque $T = 1.2\,T_e$ at the free end. The material is elastic–perfectly plastic with yield stress in shear = 15 ksi and modulus of rigidity 12×10^3 ksi. Find the extent of the plastic zone.

Ans: 2.22 in.

5.3 The material of a hollow cylindrical shaft is elastic–perfectly plastic and its inner to outer radius ratio is α. The shaft is subjected to pure torsion until it is completely plastic and then the torque is released. Find the values of the shear stress at the outer radius for $\alpha = 0, 1/3, 2/3$.

Ans: $\tau^R/\tau = 1, 0.57, 0.78$.

5.4 A solid cylindrical bar of radius a is made of a work-hardening material with yield stress in shear τ_0 and constant plastic modulus $G_p = 0.3$ G (Figure P5.4). The bar is subjected to pure torsion until the torque reaches the value of the fully plastic torque for no work hardening. Find the ratio of the unit angle of twist θ/θ_e.

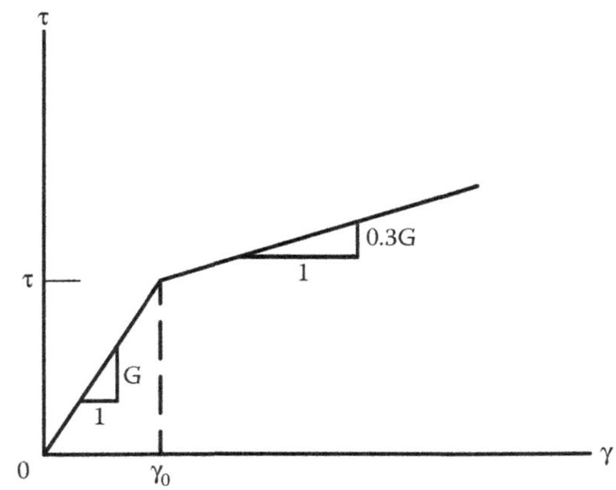

FIGURE P5.4

Ans: 1.86.

5.5 In problem 5.4, if $a = 12$ mm, $G = 80$ GPa, $\tau_0 = 160$ MPa, and the bar is experiencing pure torsion, find the torque and the corresponding unit angle of twist if extent of the plastic zone $c = 4$ mm.

Ans: 551 N m, 0.5 N/m.

5.6 Show that when the stress function in a shaft subjected to pure torsion has a constant value, K, on the boundary instead of zero, like that for multiconnected cross sections, the value of the torque in Equation 5.26 is

$$T = 2\iint (\phi - K)dxdy$$

Confirm the result by considering an elliptical shaft with a central elliptical hole.

5.7 Two bars undergoing pure torsion. One bar has a solid circular section with radius b, the other is elliptical with semiaxes a and b, $a = 1.5$ b. Determine (a) τ_e/τ_c, when the angles of twist in both bars are equal; (b) T_e/T_c, when the allowable shearing stresses in both bars are equal; (c) C_e/Cc, C is the torsional rigidity and both bars have equal cross-sectional area.

Ans: 1.39, 1.5, 0.92.

5.8 Consider problem 5.7 for square shaft with side length = a and an equilateral triangle shaft with side length = 2a.

Ans: 0.78, 0.52, 1.22.

5.9 Repeat problem 5.7 for an elliptical bar with semiaxes a and b, $a = 2b$, and a bar with rectangular cross section, side lengths $2a \times a$.

Ans: 0.86, 0.8. 0.34.

5.10 Show that the stress function

$$\phi(x, y) = \frac{G\theta}{2}\left[\frac{a^2}{9} - x^2 - y^2 - \frac{2x}{a\sqrt{3}}(x^2 - 3y^2)\right]$$

Vanish on the boundary of a bar with equilateral triangle cross section of side length = a (Figure P5.10). Determine the shear stresses and compare the results with those in Section 5.4.1.

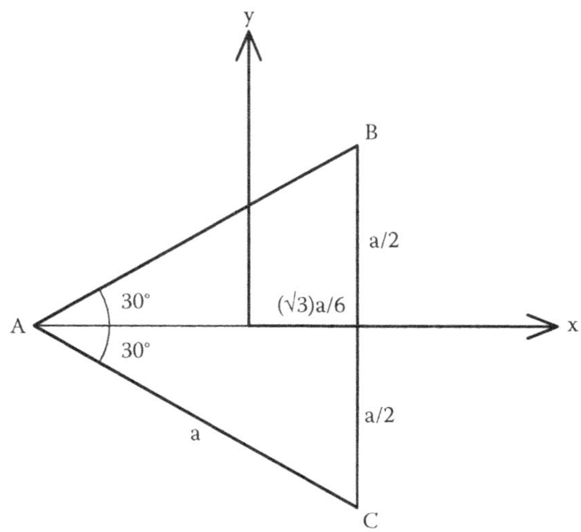

FIGURE P5.10

5.11 The following polynomial of the fourth degree was used by Saint-Venant to solve the torsion problem of a shaft subjected to pure torsion [5.1]:

$$\phi(x, y) = K[x^2 + y^2 - 0.5(x^4 - 6x^2y^2 + y^4 + 1)]$$

(a) Find the constant K; (b) sketch the cross section of the shaft circumscribed by a Rhombus of side length = a.

Ans: $K = -G\theta/2$, θ is the unit angle of twist.

5.12 Find the constants K and c so that the following polynomial is a valid stress function for a bar experiencing pure torsion:

$$\phi = K[c(x^4 + y^4) + (c^2 + 1)x^2y^2 + x^2 + y^2]$$

Ans: $c = -5.83$ or -0.175.

5.13 Determine the torsional rigidity of the bar with the cross-section shown in Figure 5.13. If the angle of the notch $\alpha = 10°$, compare the torsional rigidity with that of the complete circular cross section.

Ans: $C = T/\theta = -G\, b^4[1.9 + 1.72(a/b)^2 + \cdots]$.

5.14 Show that the function

$$\phi(r, \theta) = -\frac{G\theta^*}{2}r^2 + A_0 + A_3r^3\cos 3\theta$$

where A_0, A_3 are constants and θ^* is the unit angle of twist, is a suitable stress function for solving the elastic torsion problem of a bar with a cross section obtained by drawing circular arc of radius = 2a from the vertices A, B, and C of an equilateral triangle of side length = 2a, as shown in Figure P5.14. Find the values of A_0 and A_3.

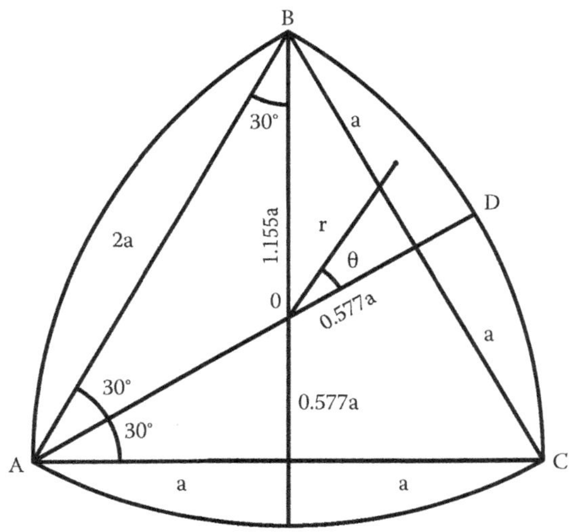

FIGURE P5.14

Ans: $A_0 = 0.444\ Ga^2\ \theta^*$, $A_3 = -0.144\ Ga^2\theta^*$.

5.15 Use the stress function given in Equation 5.68 to find an approximate expression of the torsional rigidity of a shaft with the narrow triangular cross section shown in Figure P5.15.

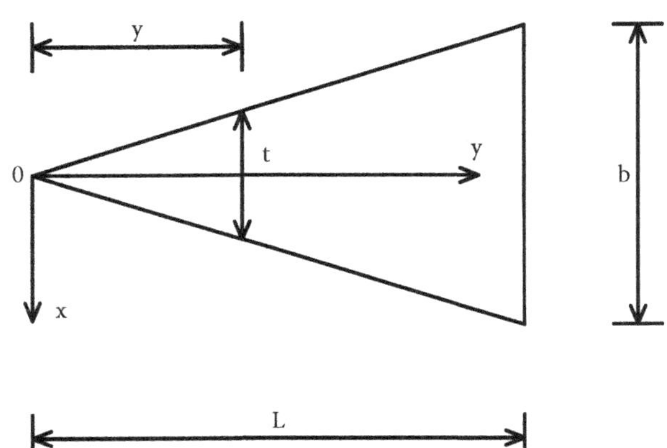

FIGURE P5.15

Ans: $(1/12)\ b^3\ GL$.

5.16 The cross section of a thin-walled tube is in the form of a hexagon of side length a = 100 mm and constant thickness t = 8 mm. The tube is subjected to a torque T = 600 N m, find the shear stress and unit angle of twist. Assume G = 28 GPa.

Ans: 144 MPa, 5.95×10^{-5} rad/m.

5.17 An angle steel shaft whose cross section is shown in Figure P5.17 is subjected to a torque T = 600 N.m. Assume G = 200 GPa, determine the maximum shear stress and the unit angle of twist.

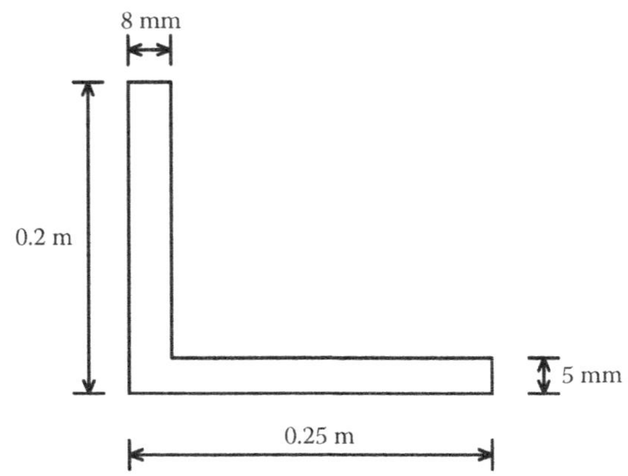

FIGURE P5.17

Ans: 67 MPa, 3.9°.

5.18 A wide flange cantilever beam is subjected to a torque T = 2 kN m. The geometric properties of the cross section (Figure P5.18) are L = 2.5 m, b = 0.15 m, h = 0.22 m, t_{flange} = 12 mm, and t_{web} = 10 mm.

Assume E = 200 GPa and G = 12 GPa, find the maximum longitudinal bending stress and the ratio of the angles of twist of restrained beams to that of free beams at both ends.

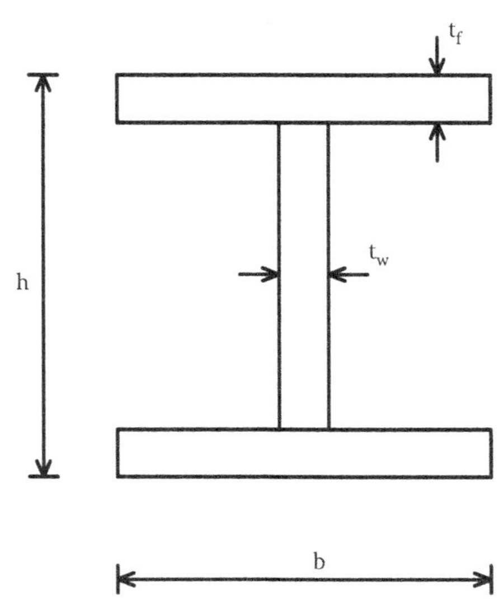

FIGURE P5.18

Ans: 186 MPa, 1.58.

References

5.1. B. Saint-Venant, *Savant etranger Science Math. Physics*, vol. 14, p. 383, 1855.
5.2. B. D. Anin and G. P. Chereoanov, *Elastic–Plastic Problems*, ASME Press, New York, 1988.
5.3. S. P. Timoshenko and J. N. Goodier, *Theory of Elasticity*, 3rd edition, McGraw-Hill Book Company, New York, 1970.
5.4. P. G. Hodge, Jr., C. J. Herakovich and R. R., Stout, *J. Appl. Mech., ASME*, vol. 35, p. 454, 1968.
5.5. W. W. Sokolovsky, *Prikl. Mat. Mech.*, vol. 6, p. 741, 1942.
5.6. L. Prandtl, *Physics 2*, vol. 4, p. 758, 1913.
5.7. H. Hetenyl, ed., *Handbook of Experimental Stress Analysis*, Wiley, New York, 1950.
5.8. R. Bredt, *ver Dtsch. Ing.*, vol. 40, p. 815, 1896.

6

Two-Dimensional Problems

A 2D problem in elasticity is defined as one that can be modeled in the xy-plane. To exclude the torsion problem discussed in Chapter 5, the shear strains in the z-plane are assumed to vanish, that is, $\gamma_{xz} = \gamma_{yz} = 0$ for 2D problems.

2D problems are of two types: *plane stress* and *plane strain*. Plane stress problems are concerned with thin, sheet-like or plate component loaded at the edges by coplanar forces parallel to the xy-plane and uniformly distributed over the thickness, while plane strain problems deal with long member symmetrically loaded on the boundary at each transverse cross section. The formulation of both types contains certain approximations needed to simplify the 3D problem and reduce it to a 2D one. Although these are two distinct physical problems, the mathematical formulation and methods of solutions to both problems are identical.

6.1 Plane Stress

Considering an element of a thin member, the midplane of the element is taken as the xy-plane and the z-axis is the normal as shown in Figure 6.1. The loads are applied at the boundary and uniformly distributed across its thickness. Because the top and bottom faces of the member are free from stress, it follows that $\sigma_z = \tau_{zx} = \tau_{zy} = 0$ at the top and bottom surfaces, and because the member is thin, it can be *assumed* that the same stresses vanish throughout the member. The nonvanishing stresses, σ_x, σ_y, and τ_{xy}, are *assumed* to be functions of x and y throughout the thickness of the member. As we shall see later, those assumptions imply that the stresses and deformations generated do not satisfy all the conditions of compatibility derived in Chapter 3. Thus, purely from the mathematical view point, plane stress solutions are not exact. However, for practical applications they are realistic as long as the members are relatively thin. To summarize, the stress distribution in a plane stress component is determined by determining σ_x, σ_y, τ_{xy}, as functions of x and y and $\sigma_z = \tau_{zx} = \tau_{zy} = 0$.

The corresponding elastic strains are also independent of the variable z and are obtained from Hooke's law as follows:

$$
\begin{aligned}
\varepsilon_x &= \frac{1}{E}(\sigma_x - \nu\sigma_y) \\
\varepsilon_y &= \frac{1}{E}(\sigma_y - \nu\sigma_x) \\
\gamma_{xy} &= \frac{2(1+\nu)}{E}\tau_{xy}
\end{aligned}
\tag{6.1}
$$

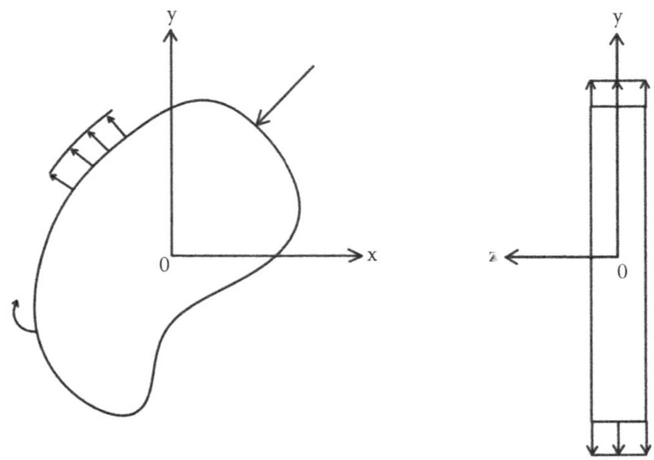

FIGURE 6.1
Thin sheet in plane stress.

$$\varepsilon_z = -\frac{\nu}{E}(\sigma_x + \sigma_y)$$

$$\gamma_{xz} = \gamma_{yz} = 0$$

Adding the first two equations in Equation 6.1 gives $\varepsilon_x + \varepsilon_y = ((1 - \upsilon)/E)(\sigma_x + \sigma_y)$ and it follows that the normal strain, ε_z, is also equivalent to

$$\varepsilon_z = -\frac{\nu}{1-\nu}(\varepsilon_x + \varepsilon_y) \tag{a}$$

Once the stresses and strains are determined, the displacement components, u, v, and w, which are also functions of x and y only, are determined by integrating the usual strain–displacement relations

$$\varepsilon_x = \frac{\partial u}{\partial x}$$

$$\varepsilon_y = \frac{\partial v}{\partial y}$$

$$\gamma_{xy} = \frac{\partial u}{\partial y} + \frac{\partial v}{\partial x}$$

It should be noted that the normal displacement w is obtained from the basic relation $\varepsilon_z = (\partial w/\partial z)$, which indicates that displacement w varies linearly with the depth z of the member, and usually is of no interest in thin members.

Substituting Equations 6.1 into the first relation of the compatibility requirements given in Equations 3.22 is repeated here

$$\frac{\partial^2 \varepsilon_x}{\partial y^2} + \frac{\partial^2 \varepsilon_y}{\partial x^2} = \frac{\partial^2 \gamma_{xy}}{\partial x \partial y}$$

The following relation is obtained:

$$\frac{\partial^2}{\partial y^2}(\sigma_x - \nu\sigma_y) + \frac{\partial^2}{\partial x^2}(\sigma_y - \nu\sigma_x) = 2(1+\nu)\frac{\partial^2 \tau_{xy}}{\partial x \partial y} \quad \text{(b)}$$

Differentiating the first and second equilibrium equations given in Equation 1.6 with respect to x and y, respectively, the following relation is obtained:

$$\frac{\partial^2 \sigma_x}{\partial x^2} + \frac{\partial^2 \sigma_y}{\partial y^2} = -2\frac{\partial^2 \tau_{xy}}{\partial x \partial y} \quad \text{(c)}$$

Inserting Equation (c) into (b), cancelling the terms containing Poisson's ratio, the following compatibility equation is obtained:

$$\left(\frac{\partial^2}{\partial x^2} + \frac{\partial^2}{\partial y^2}\right)(\sigma_x + \sigma_y) = 0 \quad \text{(6.2a)}$$

Relation (6.2a) is independent of the material properties. Similarly, inserting Equation 6.1 in the other compatibility equations given in Equation 3.22 results in

$$\frac{\partial^2 \varepsilon_z}{\partial x^2} = 0, \quad \frac{\partial^2 \varepsilon_z}{\partial y^2} = 0, \quad \frac{\partial^2 \varepsilon_z}{\partial x \partial y} = 0 \quad \text{(d)}$$

An alternative way of writing the previous relations is to substitute Equations 6.1 in (d) to give

$$\left(\frac{\partial^2}{\partial x^2}, \frac{\partial^2}{\partial y^2}, \frac{\partial^2}{\partial x \partial y}\right)(\sigma_x + \sigma_y) = 0 \quad \text{(6.2b)}$$

It is noted that because the compatibility equation does not contain elastic constants of the material, the solution is valid for any isotropic material. This conclusion is important in experimental stress science because actual stresses in structures can be measured in the laboratory using special materials to build model of the structure.

When body forces are included in the equilibrium equations then Equation 1.7 is used to arrive at the compatibility condition for plane stress problems in the presence of body forces.

$$\left(\frac{\partial^2}{\partial x^2} + \frac{\partial^2}{\partial y^2}\right)(\sigma_x + \sigma_y) = -(1+\upsilon)\left(\frac{\partial F_x}{\partial x} + \frac{\partial F_y}{\partial y}\right) \quad \text{(6.3)}$$

And Equations (d) is also applicable. Note that because $\sigma_z = \tau_{xz} = \tau_{yz} = 0$, the equilibrium equations in the presence of body forces imply that the body force F_z must vanish.

6.2 Plane Strain

A plane strain state of stress exists when a long prismatic member is subject to constant forces acting normal to the surface such that the axial strain is zero. If the z-axis is oriented

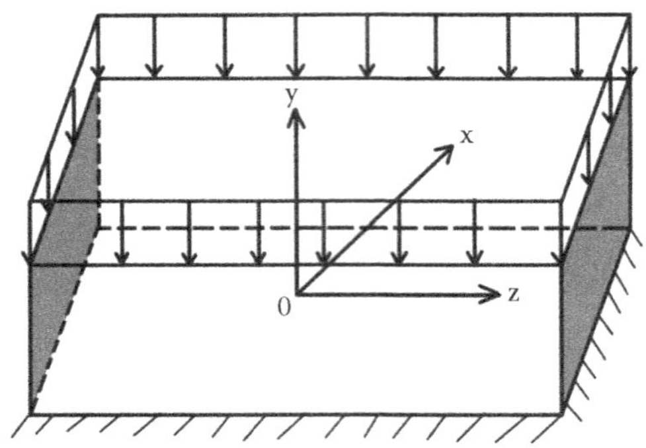

FIGURE 6.2
Member in plane strain condition.

along the length of the member, the applied forces are functions of x and y and the end sections are constrained such that there is no axial displacement. Typical examples include cylindrical tubes with internal pressures, retaining walls subject to lateral pressures, tunnels under the action of overburden forces and the like. It suffices to model a slice of unit thickness in the xy-plane as a 2D problem subject to prescribed surface loads at the boundary (Figure 6.2). The basic assumptions, which are used to reduce the 3D problem into a plane one, consist of specifying the displacement components, u, v, and w, in the form

$$u = u(x, y), \quad v = v(x, y), \quad w = 0 \tag{a}$$

It follows that the strains become

$$\gamma_{xy}(x, y), \quad \varepsilon_z = \gamma_{xz} = \gamma_{yz} = 0 \tag{b}$$

Using Hooke's law, the stresses are obtained as

$$\begin{aligned}
\sigma_x &= \lambda(\varepsilon_x + \varepsilon_y) + 2G\varepsilon_x \\
\sigma_y &= \lambda(\varepsilon_x + \varepsilon_y) + 2G\varepsilon_y \\
\tau_{xy} &= G\gamma_{xy} \\
\sigma_z &= \nu(\sigma_x + \sigma_y) \\
\tau_{xz} &= \tau_{yz} = 0
\end{aligned} \tag{6.4a}$$

The nonvanishing strains are given by

$$\begin{aligned}
\varepsilon_x &= \frac{1+\nu}{E}[(1-\nu)\sigma_x - \nu\sigma_y] \\
\varepsilon_y &= \frac{1+\nu}{E}[(1-\nu)\sigma_y - \nu\sigma_x] \\
\gamma_{xy} &= \frac{2(1+\nu)}{E}\tau_{xy}
\end{aligned} \tag{6.4b}$$

Problems involving plane strain members subjected to *uniform axial strain* at all points of the cross section, for example, open-ended pressurized cylinders, can be reduced to the above case by the addition of uniform axial stress, σ_z. The resulting strains and displacements are readily calculated.

Inserting Equations 6.4 into the compatibility conditions (3.22) and proceeding as in the plane stress case, the compatibility requirement for plane strain problems in terms of stresses is

$$\nabla^2(\sigma_x + \sigma_y) = 0 \tag{c}$$

Here, $\nabla^2 = (\partial^2/\partial x^2) + (\partial^2/\partial y^2)$ stands for the harmonic operator in two dimensions.

When body forces are present, Equations 1.7 is used instead of Equations 1.6, and the compatibility equation in terms of stresses becomes

$$\nabla^2(\sigma_x + \sigma_y) = -\frac{1}{1-\nu}\left(\frac{\partial F_x}{\partial x} + \frac{\partial F_y}{\partial y}\right) \tag{6.5}$$

Here, $\nabla^2 = (\partial^2/\partial x^2) + (\partial^2/\partial y^2)$ stands for the harmonic operator in two dimensions.

To summarize, the stresses in plane stress and plane strain problems are governed by identical compatibility equations. In order to determine the stresses in plane stress and plane strain problems, one needs to solve the differential equations of equilibrium together with the compatibility equation subject to the proper boundary conditions. This will be done in the next section. Once the stresses are determined, the strains are obtained from Equations 6.1 and 6.4b and the displacement components are then obtained from the strain–displacement relations. Comparisons of the strain expressions in Equations 6.1 and 6.4b indicate that to convert the strains from plane stress to plane strain, one needs to replace E and ν in plane stress by $E/(1-\nu^2)$ and $\nu/(1-\nu)$ for plane strain, respectively. Conversely, to convert from plane strain to plane stress, one needs to substitute for the constants E and ν in plane strain by $((1+2\nu)E)/(1+\nu)^2$ and $\nu/(1+\nu)$ for plane stress, respectively.

6.3 Stress Function

In Chapter 5, we have shown that in the absence of body forces, plane stress and plane strain problems are governed by two equilibrium equations

$$\begin{aligned}\frac{\partial \sigma_x}{\partial x} + \frac{\partial \tau_{xy}}{\partial y} &= 0\\ \frac{\partial \tau_{xy}}{\partial x} + \frac{\partial \sigma_y}{\partial y} &= 0\end{aligned} \tag{a}$$

The resulting stresses are determined subject to satisfying the compatibility relation

$$\left(\frac{\partial^2}{\partial x^2} + \frac{\partial^2}{\partial y^2}\right)(\sigma_x + \sigma_y) = 0 \tag{b}$$

and the appropriate boundary conditions. To facilitate the finding of the stresses, we introduce a new function of x and y, $\phi(x, y)$, such that the equilibrium equations are automatically satisfied. Such function was first introduced by Airy [6.1] in 1863

$$\sigma_x = \frac{\partial^2 \phi}{\partial y^2}$$
$$\sigma_y = \frac{\partial^2 \phi}{\partial x^2} \tag{6.6}$$
$$\tau_{xy} = -\frac{\partial^2 \phi}{\partial x \partial y}$$

It is readily confirmed that Equations (a) is automatically satisfied. Inserting expression (6.6) into the compatibility equation, it is found that the stress function is governed by a fourth-order partial differential equation

$$\frac{\partial^4 \phi}{\partial x^4} + 2\frac{\partial^4 \phi}{\partial x^2 \partial y^2} + \frac{\partial^4 \phi}{\partial y^4} = 0, \quad \text{or} \quad \nabla^4 \phi = 0 \tag{6.7}$$

where ∇^4 the bi-harmonic operator stands for the abbreviation

$$\nabla^4 = \frac{\partial^4}{\partial x^4} + 2\frac{\partial^4}{\partial x^2 \partial y^2} + \frac{\partial^4}{\partial y^4} \tag{c}$$

The stress function, ϕ, is determined by finding the solution of Equation 6.7 subject to satisfying appropriate boundary conditions of the problem.

Boundary conditions: The induced stresses vary throughout the element as functions of x and y. At the boundary, the stresses must be in equilibrium with the external forces present at that locality. Consider a small triangular prism at the boundary of a 2D element, as shown in Figure 6.3. Denote the components of the external load per unit area by p_x and p_y and the direction cosines of the normal to the boundary surface by l and m. Equilibrium of the element yields

$$\begin{aligned} p_x &= \sigma_x l + \tau_{xy} m \\ p_y &= \tau_{xy} l + \sigma_y m \end{aligned} \tag{d}$$

When the normal to the boundary is the x-axis, then $\alpha = 0$, $l = 1$, and $m = 0$. Similarly, when the normal is the y-axis, $\alpha = 90°$, $l = 0$, and $m = 1$.

Body force: When body forces are present, similar procedures can be used. Suppose that the body force is conservative, that is, it has a potential, denoted by V(x, y). Then, the components of the body force are given by the equations

$$F_x = -\frac{\partial V}{\partial x}, \quad F_y = -\frac{\partial V}{\partial y} \tag{e}$$

The equilibrium equations now become

$$\begin{aligned} \frac{\partial}{\partial x}(\sigma_x - V) + \frac{\partial \tau_{xy}}{\partial y} &= 0 \\ \frac{\partial \tau_{xy}}{\partial x} + \frac{\partial}{\partial y}(\sigma_y - V) &= 0 \end{aligned} \tag{f}$$

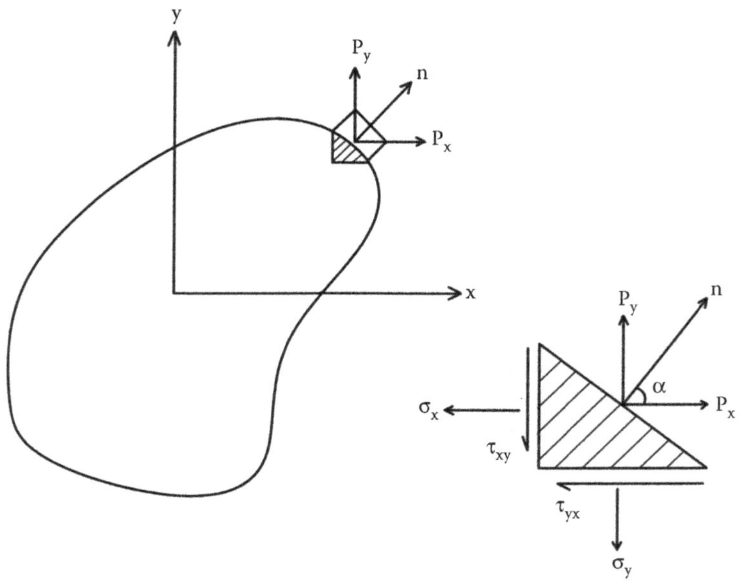

FIGURE 6.3
Boundary conditions in two-dimensional problems.

These equations can be satisfied by taking

$$\sigma_x - V = \frac{\partial^2 \phi}{\partial y^2}$$
$$\sigma_y - V = \frac{\partial^2 \phi}{\partial x^2} \quad (6.8)$$
$$\tau_{xy} = -\frac{\partial^2 \phi}{\partial x \partial y}$$

And the compatibility equation for plane stress problems is

$$\nabla^4 \phi = -(1+\nu)\nabla^2 V \quad (6.9a)$$

A similar equation for plane strain problems is derived by replacing ν in Equation 6.9a by $\upsilon/(1-\upsilon)$, which yields

$$\nabla^4 \phi = -\left(\frac{1}{1-\nu}\right)\nabla^2 V \quad (6.9b)$$

Solutions of elasticity problems: It has been shown in the previous sections that the problem of determining the stresses in plane elastic bodies requires developing solutions of the biharmonic equation such that the stress components satisfy the appropriate boundary conditions.

In the next few sections, bi-harmonic functions in rectangular as well as polar coordinates are used to provide analytical solutions to problems of practical interest. The most general solution of 2D elasticity problems involves the use of complex variables [6.2].

6.4 Polynomial Solutions

It can be shown by simple differentiation that a number of algebraic polynomials satisfy the bi-harmonic equation and may be used as stress functions to solve simple problems in rectangular coordinates. The stress functions of interest are polynomials consisting of linear combination of quadratic, cubic, and higher terms. Let us begin with a quadratic stress function

$$\phi = Ax^2 + Bxy + Cy^2 \tag{a}$$

Here, A, B, and C are constants. Substituting in the bi-harmonic Equation 6.7 yields constant stresses

$$\sigma_x = \frac{\partial^2\phi}{\partial y^2} = 2C, \quad \sigma_y = \frac{\partial^2\phi}{\partial x^2} = 2A, \quad \text{and} \quad \tau_{xy} = -\frac{\partial^2\phi}{\partial x\partial y} = -B$$

Similarly, using the cubic function

$$\phi = Ax^3 + Bx^2y + Cxy^2 + Dy^3 \tag{b}$$

Equation 6.7 is automatically satisfied and the resulting stresses are linear functions of x and y.

$$\sigma_x = \frac{\partial^2\phi}{\partial y^2} = 2Cx + 6Dy$$

$$\sigma_y = \frac{\partial^2\phi}{\partial x^2} = 6Ax + 2By$$

$$\tau_{xy} = -\frac{\partial^2\phi}{\partial x\partial y} = -2Bx - 2Cy$$

The fourth-order stress function

$$\phi = Ax^4 + Bx^3y + Cx^2y^2 + Dxy^3 + Ey^4 \tag{c}$$

satisfies Equation 6.7 provided that $3A + C + 3E = 0$ and furnishes quadratic stresses

$$\sigma_x = 2Cx^2 + 6Dxy + 12Ey^2$$

$$\sigma_y = 12Ax^2 + 6Bxy + 2Cy^2$$

$$\tau_{xy} = -3Bx^2 - 4Cxy - 3Dy^2$$

Additional stress functions can be constructed by using higher-order polynomials to solve problems of practical interest. Since Equation 6.7 is a linear differential equation, several solutions can be summed and used to satisfy the boundary conditions. Next, we apply this methodology to obtain elasticity solutions to simple beam problems.

6.4.1 Cantilever Beam Loaded by a Concentrated Force

A cantilever of length L, depth 2h, and unit width is loaded by a concentrated force at the free end as shown in Figure 6.4. It is required to determine the stresses and displacements in the cantilever.

The end $x = 0$ is free from normal stress and is loaded by shearing forces having a resultant P while the surfaces $y = \mp h$ are free from load. It follows that the boundary conditions are

$$x = 0, -h \le y \le h : \sigma_x = 0 \quad \text{and} \quad \iint_A \tau_{xy}\, dA = P$$

$$0 \le x \le L, y = \mp h : \sigma_y = \tau_{xy} = 0 \qquad \text{(a)}$$

We can proceed to construct the proper stress function by combining polynomials of second, third, and fourth degrees as follows:

$$\phi(x,y) = C_1x^2 + C_2xy + C_3y^2 + C_4x^3 + C_5x^2y + C_6xy^2 + C_7y^3 + C_8x^4 + C_9x^3y + \cdots$$

where C_j, $j = 1, 2, 3, \ldots, 12$ are constants. Inserting the proposed 12-term stress function in the bi-harmonic equation and applying the boundary conditions that the normal stress at the free end $x = 0$ and the normal stresses across the top and bottom surfaces $y = \mp h$ vanish, it is readily confirmed that the stress function reduces to

$$\phi = Axy + Bxy^3 \qquad \text{(b)}$$

where A and B are arbitrary constants. The corresponding stresses are

$$\sigma_x = 6Bxy, \quad \sigma_y = 0, \quad \tau_{xy} = -A - 3By^2$$

Since on the surfaces $y = \mp h$, $0 \le x \le L$, $\tau_{xy} = 0$, it follows that $A = -3h^2B$ and $\tau_{xy} = 3B(h^2 - y^2)$.

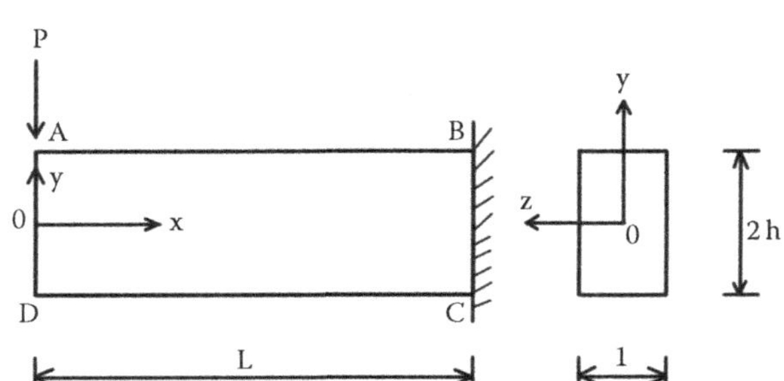

FIGURE 6.4
Cantilever subjected to a concentrated load.

The requirement that $\iint_A \tau_{xy}\, dA = P$ on the surface $x = 0$, $-h \le y \le h$ gives

$$\int_{-h}^{h} 3B(h^2 - y^2)dy = P$$

and evaluating the integral one finds $B = (P/4h^3) = (P/6I)$, where $I = (2h^3/3)$ is the moment of inertia of the cross section. It follows that

$$\sigma_x = \frac{Pxy}{I}, \quad \sigma_y = 0$$
$$\tau_{xy} = \frac{P}{2I}(h^2 - y^2) \tag{6.10}$$

The solution (6.10) coincides with the elementary solution given in elementary courses on strength of materials; see Equation 1.4.

In sum, the previous solution is an exact solution provided that the distribution of the shear forces on the loaded end coincides with the parabolic distribution of the shear stress τ_{xy}, and the normal forces at the built-in end vary linearly with the depth as indicated in Equation 6.10.

The displacements are determined by integrating the strain–displacement relations and making use of Hooke's law for plane stress. Thus,

$$\frac{\partial u}{\partial x} = \frac{1}{E}(\sigma_x - \nu\sigma_y) = \frac{Pxy}{EI}$$

Integrating with respect to x yields

$$u = \frac{Px^2y}{2EI} + f(y)$$

where f(y) is the constant of integration. Similarly,

$$\frac{\partial V}{\partial y} = \frac{1}{E}(\sigma_y - \nu\sigma_x) = -\frac{\nu Pxy}{EI}$$

giving

$$V = -\frac{\nu Pxy^2}{2EI} + g(x)$$

Since $(\partial u/\partial y) + (\partial V/\partial x) = (1 + \nu)/EI(h^2 - y^2)P$, it follows by differentiating the previous expressions for u(x, y) and v(x, y) and rearranging the terms that

$$\frac{Px^2}{2EI} + \frac{dg}{dx} + \frac{(2+\nu)Py^2}{2EI} + \frac{df}{dy} = \frac{(1+\nu)Ph^2}{EI}$$

This equation can be satisfied if

$$\frac{dg}{dx} + \frac{Px^2}{2EI} = C_1$$

and

$$\frac{df}{dy} + \frac{(2+\nu)Py^2}{2EI} = C_2$$

where the constants C_1 and C_2 are related by

$$C_1 + C_2 = \frac{(1+\nu)Ph^2}{EI} \tag{c}$$

The functions g(x) and f(y) are obtained by simple integration

$$g(x) = -\frac{Px^3}{6EI} + C_1x + C_3$$

$$f(y) = -\frac{(2+\nu)Py^3}{6EI} + C_2y + C_4$$

where C_3 and C_4 are two more constants, and the expressions of the displacements become

$$u(x, y) = \frac{Px^2y}{2EI} - \frac{(2+\nu)Py^3}{6EI} + C_2y + C_4$$

$$V(x, y) = -\frac{\nu Pxy^2}{2EI} - \frac{Px^3}{6EI} + C_1x + C_3$$

The constants C_i, $i = 1, 2, 3$, and 4 are found from relation (c) and the conditions of constraint at the built-in end $x = L$.

Suppose that at $x = L$, $y = 0$: $u = v = 0$ and, $(\partial V/\partial x) = 0$, that is, an element of the axis of the beam is fixed at $x = L$, $y = 0$. It follows that

$$C_1 = \frac{PL^2}{2EI}$$

$$C_2 = \frac{(1+\nu)Ph^2}{EI} - \frac{PL^2}{2EI}$$

$$C_3 = -\frac{PL^3}{3EI}$$

$$C_4 = 0$$

The final expressions for the displacements are

$$u = \frac{Px^2y}{2EI} - \frac{(2+\nu)Py^3}{6EI} + \left[\frac{(1+\nu)Ph^2}{EI} - \frac{PL^2}{2EI}\right]y$$

$$V = -\frac{Px^3}{6EI} + \frac{PL^2x}{2EI} - \frac{PL^3}{3EI} - \frac{\nu Pxy^2}{2EI} \tag{6.11}$$

The first three terms in the expression of the vertical displacement agree with the deflection of a cantilever derived in the elementary courses on strength of materials. At the free end, $x = 0$, the vertical displacement is $V = -(PL^3/3EI)$, which coincides with the elementary solution. Further discussion on types of constraints at the fixed end is available in Reference 5.3; see Problem 6.8 at the end of this chapter.

Plastic behavior: The maximum bending moment occurs at the built-in end. Assuming elastic–perfectly plastic behavior, yield is initiated at the top and bottom fibers of the built-in end. The corresponding moment is the elastic limit moment, $M_e = -P_eL$. As the applied moment increases beyond the elastic limit, plastic zones are formed at corners B and C of the built-in ends and spread along the sides BA, CD, and depth BC of the cantilever, as shown in Figure 6.5a. Suppose that at a certain stage of the loading, the elastic limit moment M_e spreads to a distance, a, from the free end, then if M is the bending moment at any distance x in the plastic zone, we have $(M/M_e) = (x/a)$ and $(P/P_e) = (a/L)$. Also, if b is the semi-height of the plastic zone at distance x, we have $(b/h) = (\rho/\rho_e)$ where ρ and ρ_e are the radii of curvature of the neutral axis. Following the procedure used in deriving relations (1.16) and (1.19) of Chapter 1, we arrive at

$$\left(\frac{b}{h}\right)^2 = 3 - 2\frac{M}{M_e} = 3 - 2\frac{x}{a} \tag{d}$$

It is readily confirmed that relation (d) is equivalent to

$$\left(\frac{b}{h}\right)^2 = 3 - 2\frac{x}{L}\left(\frac{P}{P_e}\right) \tag{6.12}$$

Thus, the plastic/elastic boundary is part of a parabola having its vertex at the x-axis at distance 1.5a from the free end. Further increase in the load causes the bending moment at the built-in end to reach the fully plastic value of $M_p = 1.5\ M_e$. At this stage, $P_u = 1.5\ P$ and $a = (2/3)L$. A hinge is assumed to have formed at the built-in end, which causes the cantilever to collapse (Figure 6.5b).

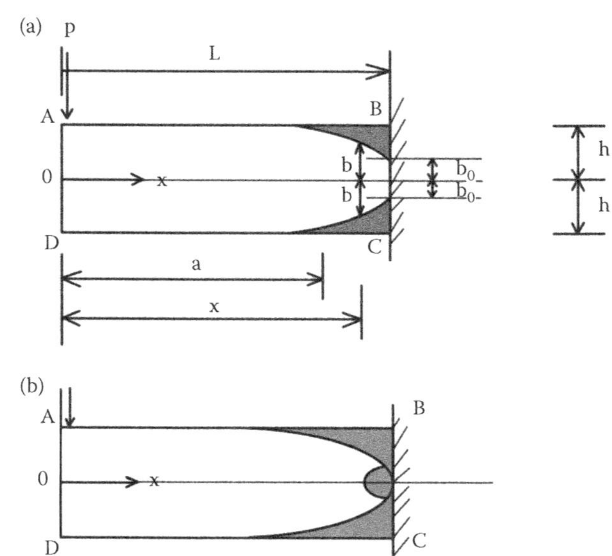

FIGURE 6.5
Beam loaded beyond the elastic limit. (a) Elastic/plastic stage; (b) collapse stage.

6.4.2 Bending of a Simply Supported Beam by Uniform Load

In this example, the distribution of stress in a simply supported beam of length 2L, depth 2h, and unit thickness is determined. The beam is loaded by a uniform load of intensity q, as shown in Figure 6.6. With the origin of the x and y coordinates located at the center of the beam, the boundary conditions of this plane stress problem are

$$\begin{aligned} y &= h, -L \le x \le L : \tau_{yx} = 0, \sigma_y = -q \\ y &= -h, -L \le x \le L : \tau_{yx} = \sigma_y = 0 \end{aligned} \tag{a}$$

In addition, the longitudinal force and bending moment vanish at the ends of the beam. This implies that

$$\begin{aligned} &x = \mp L, -h \le y \le h : \\ &\iint_A \sigma_x \, dA = 0, \quad \iint_A (\sigma_x \, dA) y = 0 \end{aligned} \tag{b}$$

and the distribution of shearing forces gives

$$\int_{-h}^{h} |\tau_{xy}|_{x=\mp L} dy = \mp qL \tag{c}$$

A proper stress function is constructed by combining polynomials of the second, third, and fifth degrees. Inserting the resulting stress function in the bi-harmonic equation and utilizing the boundary conditions, it is readily confirmed that the nonvanishing terms in the stress function are

$$\phi = Ax^2 + Bx^2y + Cy^3 + D(y^5 - 5x^2y^3) \tag{d}$$

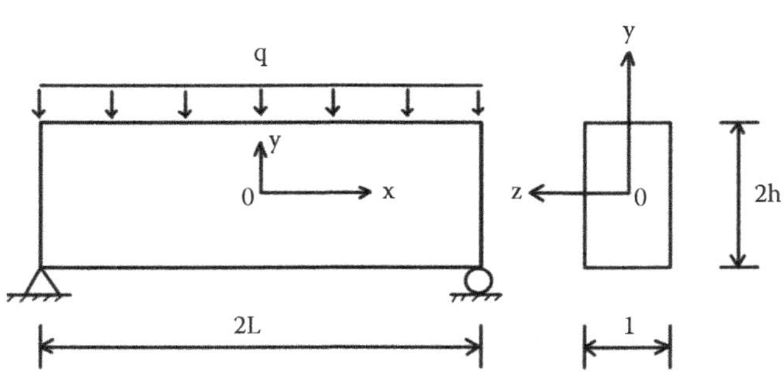

FIGURE 6.6
Simply supported beam subjected to uniformly distributed load.

where A, B, ... are arbitrary constants. The corresponding stresses are

$$\sigma_x = \frac{\partial^2\phi}{\partial y^2} = 6Cy + D(20y^3 - 30x^2y)$$

$$\sigma_y = \frac{\partial^2\phi}{\partial x^2} = 2A + 2By - 10Dy^3$$

$$\tau_{xy} = -\frac{\partial^2\phi}{\partial x\partial y} = -2Bx + 30Dxy^2$$

The conditions that $\tau_{xy} = 0$ at $y = \mp h$, $\sigma_y = 0$ at $y = -h$, and $\sigma_y = -q$ at $y = h$ provide

$$A = -\frac{qh^3}{6I}$$

$$B = -\frac{qh^2}{4I}$$

$$D = -\frac{q}{60I}$$

where $I = (2h^3/3)$ is the moment of inertia of the cross section about the z-axis. The remaining constant, C, is determined from the condition that the bending moment at the end sections is zero, that is

$$\int_{-h}^{h} \sigma_x(\mp L, y)y\,dy = 0$$

This gives

$$C = -\left(L^2 - \frac{2h^2}{5}\right)\frac{q}{12I}$$

Finally, the stress distribution in the beam is

$$\begin{aligned}
\sigma_x &= -\frac{q}{2I}(L^2 - x^2)y + \frac{q}{I}\left(\frac{h^2y}{5} - \frac{y^3}{3}\right) \\
\sigma_y &= \frac{q}{2I}\left(-\frac{2h^3}{3} - h^2y + \frac{y^3}{3}\right) \\
\tau_{xy} &= \frac{q}{2I}(h^2 - y^2)x
\end{aligned} \tag{6.13}$$

The expression for the shear stress and the first term in the expression for σ_x coincides with the solution obtained from elementary analysis. The remaining terms are corrections on the elementary solutions. The corrections do not depend on x, and in a beam with large span/depth ratio, are usually small in comparison with the maximum stress. In the elementary solution, the normal stress σ_y is assumed to vanish while the above expression indicates that there are compressive stresses between the longitudinal fibers. The distribution of the compressive stress σ_y varies between $-q$ at the top fiber and zero at the bottom.

At the centerline $y = 0$, $\sigma_y = -(q/2)$, $\sigma_x = 0$, and it follows from Hooke's law that the centerline experiences a tensile strain $\varepsilon_x = (\nu q/2E)$.

The displacements for this plane stress problem are derived as follows:

$$\frac{\partial u}{\partial x} = \frac{1}{E}(\sigma_x - \nu\sigma_y) = \frac{q}{2EI}\left[-L^2y + x^2y + \frac{2h^2y}{5} - \frac{2y^3}{3} + \nu\left(\frac{2h^3}{3} + h^2y - \frac{y^3}{3}\right)\right]$$

$$\frac{\partial V}{\partial y} = \frac{1}{E}(\sigma_y - \nu\sigma_x) = \frac{q}{2EI}\left[-\frac{2h^3}{3} - h^2y + \frac{y^3}{3} + \nu\left(L^2y - x^2y - \frac{2h^2y}{5} + \frac{2y^3}{3}\right)\right]$$

After integration, one obtains

$$u = \frac{q}{2EI}\left[-L^2yx + \frac{x^3y}{3} + \frac{2h^2xy}{5} - \frac{2y^3x}{3} + \nu x\left(\frac{2h^3}{3} + h^2y - \frac{y^3}{3}\right)\right] + f(y)$$

$$V = \frac{q}{2EI}\left[-\frac{2h^3y}{3} - \frac{h^2y^2}{2} + \frac{y^4}{12} + \nu\left(\frac{L^2y^2}{2} - \frac{x^2y^2}{2} - \frac{2h^2y^2}{10} + \frac{y^4}{6}\right)\right] + g(x)$$

where f(y) and g(x) are arbitrary functions. Hence

$$\gamma_{xy} = \frac{\partial u}{\partial y} + \frac{\partial V}{\partial x} = \frac{2(1+\nu)q}{2EI}(h^2 - y^2)x$$

Differentiating the displacement u with respect to y and V with respect to x and inserting the results in the last expression yields

$$\frac{q}{2EI}\left[-L^2x + \frac{x^3}{3} - \frac{8h^2x}{5} - \nu h^2x\right] + \frac{dg}{dx} = -\frac{df}{dy} = C_1$$

Hence,

$$f(y) = -C_1y + C_2$$

$$g(x) = C_1x + \frac{q}{2EI}\left[\frac{L^2x^2}{2} - \frac{x^4}{12} + \frac{4h^2x^2}{5} + \nu\frac{h^2x^2}{2}\right] + C_3$$

where C_j, $j = 1, 2$, and 3 are constants determined by the nature of the constraints on the beam. Assuming that the deflection of the centerline is zero at the supports, that is, $V = 0$ at $x = \mp L$, $y = 0$, it is easily confirmed that

$$C_1 = 0 \quad \text{and} \quad C_3 = -\frac{q}{2EI}\left(\frac{5L^4}{12} + \frac{4h^2L^2}{5} + \frac{\nu h^2L^2}{2}\right)$$

Also, if $u=0$ at the center of the beam ($x=0$, $y=0$), it follows that $C_2=0$, and the displacement components become

$$u=\frac{qx}{2EI}\left[\left(-L^2+\frac{x^2}{3}+\frac{2h^2}{5}-\frac{2y^2}{3}\right)y+\nu\left(\frac{2h^3}{3}+h^2y-\frac{y^3}{3}\right)\right]$$

$$V=\frac{q}{2EI}\left[\frac{y^4}{12}-\frac{h^2y^2}{2}-\frac{2h^3y}{3}+\nu\left(\frac{L^2y^2}{2}-\frac{x^2y^2}{2}+\frac{y^4}{6}-\frac{h^2y^2}{5}\right)\right.$$
$$\left.+\frac{L^2x^2}{2}-\frac{x^4}{12}+\frac{4h^2x^2}{5}+\nu\frac{h^2x^2}{2}-\frac{5L^4}{12}-\frac{4h^2L^2}{5}-\nu\frac{h^2L^2}{2}\right] \tag{6.14}$$

The deflection of the centerline of the beam is obtained by setting $y=0$ in the expression for the vertical displacement. The result is

$$V|_{y=0}=\frac{q}{24EI}\left[-x^4-6L^2x^2-5L^4+12\left(\frac{4}{5}+\frac{\nu}{2}\right)(x^2h^2-L^2h^2)\right]$$

The first three terms coincide with the deflection curve determined in elementary analysis assuming cross sections of the beam remain plane and the remaining terms represent correction due to the effect of shearing force on the deflection. The deflection at the center point ($x=0$, $y=0$) is $V|_{\substack{x=0\\y=0}}=-(5qL^4/24EI)$.

Plastic behavior: The maximum bending moment occurs at points on the midsection, $x=0$, $y=\pm h$, of the top and bottom fibers of the beam. Thus, yield is initiated at those locations where $M_e=(1/2)q_e\,L^2$. As the load increases beyond q_e, two symmetrical plastic zones are formed covering a distance 2 a on the top and bottom surfaces and spreading inward, as shown in the shaded regions of Figure 6.7. The moment at an arbitrary cross section in the plastic zone is $M_x=(1/2)q\,(1-(x^2/L^2))$ while the moment at $x=\pm a$ is the elastic limit moment M_e. It follows that

$$\frac{M_x}{M_e}=\frac{q}{q_e}\left(1-\frac{x^2}{L^2}\right)\quad\text{and}\quad\frac{a}{L}=\sqrt{1-\frac{q_e}{q}} \tag{e}$$

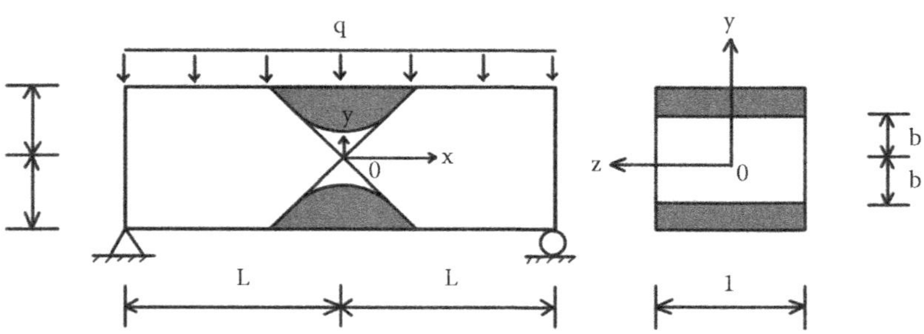

FIGURE 6.7
Beam loaded in the plastic range.

Following the procedures used in deriving relations (1.16) and (1.10), the boundary of the plastic/elastic interface is found to be given by

$$\left(\frac{b}{h}\right)^2 - 2\frac{q}{q_e}\left(\frac{x}{L}\right)y^2 = 3 - 2\frac{q}{q_e} \tag{6.15}$$

where b stands for the extent of the elastic core at $x = 0$. Equation 6.30 represents hyperbolas with asymptotes $y = \pm\sqrt{(3h/L)}x$. Further increase in the load causes the two plastic zones to enlarge and join at the origin. For elastic–perfectly material, a hinge is formed at the midcross section of the beam and the beam is assumed to collapse when $q_u = 3/2\ q$. See Reference 2.2.

6.4.3 Cantilever Beam Subjected to Hydrostatic Pressure

A cantilever of depth 2a, unit width, and length L is loaded by linearly varying load such that the load at distance x from one end is qx, as shown in Figure 6.8. The boundary conditions on the longitudinal sides are

$$\begin{aligned} y &= a : \sigma_y = -qx, \tau_{yx} = 0 \\ y &= -a : \sigma_y = \tau_{yx} = 0 \end{aligned} \tag{a}$$

and on the top end of the cantilever ($x = 0$), the stresses σ_x and τ_{xy} vanish.

By combining polynomials of second, third, fourth, fifth, and sixth degrees (a total of 25 terms) and considering the boundary conditions stated in Equation (a), it is not difficult to confirm that the following stress function will satisfy the stress conditions on the sides of the cantilever other than the fixed end:

$$\phi = Axy + Bx^3 + Cx^3y + Dxy^3 + Ex^3y^3 + Fxy^5 \tag{b}$$

where A, … , F are arbitrary constants. It is not difficult to confirm that the bi-harmonic equation, $\nabla^4\phi = 0$, is satisfied provided that $F = -3/5\ E$, and the stress function reduces to

$$\phi = Axy + Bx^3 + 6Cxy + Dxy^3 + E\left(x^3y^3 - \frac{3}{5}xy^5\right) \tag{c}$$

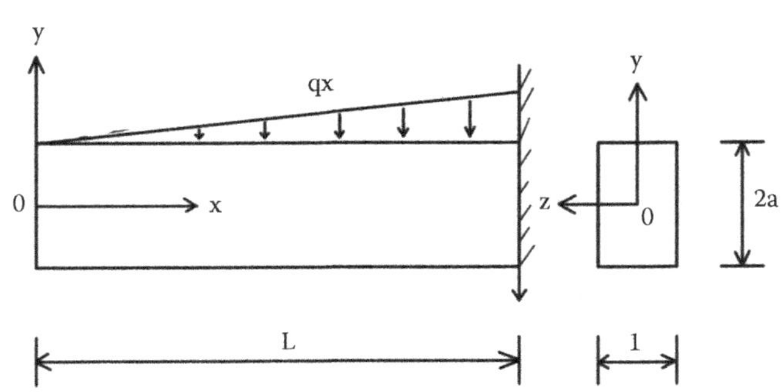

FIGURE 6.8
Cantilever subjected to hydrostatic pressure.

The stresses are

$$\sigma_x = \frac{\partial^2 \phi}{\partial y^2} = 6Dxy + 6E(x^3y - 2xy^3)$$

$$\sigma_y = \frac{\partial^2 \phi}{\partial x^2} = 6Bx + 6Cxy + 6Exy^3 \qquad \text{(d)}$$

$$\tau_{xy} = -\frac{\partial^2 \phi}{\partial x \partial y} = -[A + 3Cx^2 + 3Dy^2 + E(9x^2y^2 - 3y^4)]$$

On $y = \pm a$, the conditions on σ_y from Equations (a) imply

$$B + aC + a^3E = -\frac{q}{6}$$

$$B - aC - a^3E = 0$$

giving

$$B = -\frac{q}{12} \qquad \text{(e)}$$

$$C + a^2E = -\frac{q}{12a} \qquad \text{(f)}$$

The boundary conditions on the shear stress τ_{xy} in relation (a) yield

$$\begin{aligned} 3C + 9a^2E &= 0 \\ A + 3a^2D - 3a^4E &= 0 \end{aligned} \qquad \text{(g)}$$

Using Equations (e) and (f), it is readily found that

$$C = -\frac{q}{8a}, \quad E = \frac{q}{24a^3}, \quad \text{and} \quad A + 3a^2D = \frac{qa}{8} \qquad \text{(h)}$$

In order to solve for the constants A and D, a second relation is obtained by requiring the total shear force on the top face of the beam to vanish, that is

$$\iint_A \tau_{xy}(0, y)dA = 0 \quad \text{or} \quad \int_{-a}^{+a} \tau_{xy}(0. y)(1)dy = 0$$

Noting that $\tau_{xy}(0, y) = -[A + 3Dy^2 - (qy^4/8a^3)]$ and evaluating the integral, the second relation between the constants A and D is

$$A + a^2D = \frac{qa}{40} \qquad \text{(i)}$$

Using Equations (h) and (i), it is found that $A = -(qa/40)$ and $D = (q/20a)$, and the stresses generated by the stress function are readily obtained from Equations (d)

$$\begin{aligned}
\sigma_x &= \frac{qx^3y}{4a^3} + \frac{3q}{10a}xy\left(1 - \frac{5y^2}{3a^2}\right) \\
\sigma_y &= -\frac{qx}{2}\left(1 + \frac{3y}{2a} - \frac{y^3}{2a^3}\right) \\
\tau_{xy} &= \frac{3qx^2}{8a^3}(a^2 - y^2) + \frac{qa}{8}\left(1 - \frac{y^2}{a^2}\right)\left(\frac{1}{5} - \frac{y^2}{a^2}\right)
\end{aligned} \tag{6.16}$$

It is observed that the first term in the expressions for σ_x and τ_{xy} are identical to those obtained using the elementary formulas of strength of materials, namely,

$$\sigma_x = \frac{My}{I} = \frac{qx^3y}{6}\left(\frac{3}{2a^3}\right) \quad \text{and} \quad \tau_{xy} = \frac{VQ}{Ib} = \frac{qx^2}{2}\left(\frac{3}{2a^3}\right)\left(\frac{a^2 - y^2}{2}\right)$$

6.5 Saint-Venant's Principle

In Sections 6.4.1 through 6.4.3, exact solutions of elementary beam problems were presented using polynomial stress functions. In each example, the support forces at the ends of the beam must have the same distribution as required by the solution. Thus, in the case of the cantilever subjected to concentrated load at the free end, the normal traction at end $x = L$ provided by the support must be proportional to y in order for the solution given in Equation 6.10 to be valid. If the moment provided by the support is applied in any different manner, then the boundary condition will be different and the solution in Equation 6.10 is no longer correct. However, if the support provides a statically equivalent moment to the one required by Equation 6.10, then the stress distribution throughout the beam except the region near the ends is the same. This observation was first noted by Saint-Venant in 1855 and is commonly referred to as Saint-Venant principle [6.3]. The principle can be stated as follows: "If any load distribution at the boundary of a member is replaced by a statically equivalent system (same resultant force and moment), the induced stress and strain throughout the member are altered only near the region of the load application." To illustrate the principle, consider a member subject to axial load P, as shown in Figure 6.9a. Figure 6.9b shows the same member subject to an equivalent load but uniformly distributed at the end with intensity q, $P = qA$ where A is the area of the cross section. The axial stress in both members is $\sigma_x = q$ except in the region where the concentrated load P is applied. The maximum stress for member (b) near the point of application of the load P is much higher than the average stress q. The load diminishes and approaches the average stress q at a distance approximately equal to the width b.

EXAMPLE 6.1

A cantilever of length L, width b, and depth h is loaded at the free end by two forces and a bending moment, as shown in Figure 6.10. It is required to determine the stresses using polynomial stress function approach.

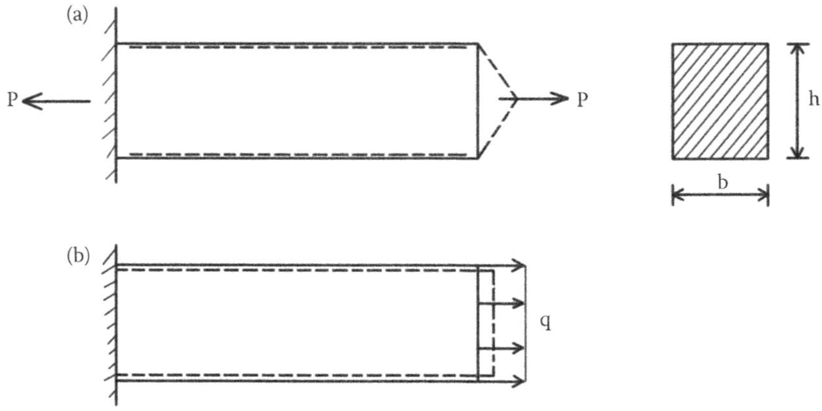

FIGURE 6.9
Illustration of Saint-Venant's principle. (a) Concentrated load; (b) uniformly distributed load.

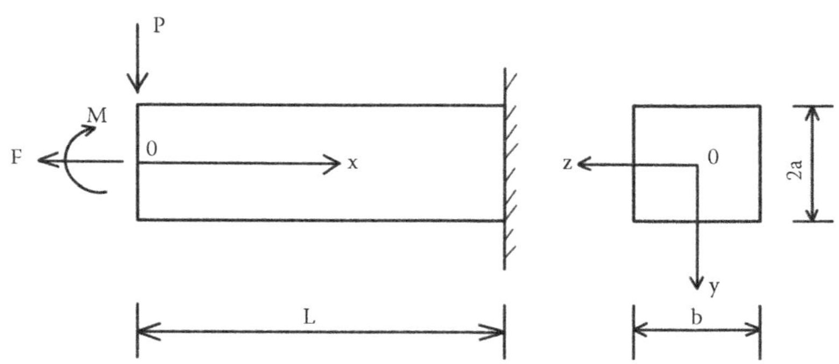

FIGURE 6.10
Example 6.1.

Solution

As explained in Sections 6.4.1 and 6.4.2, the stress function that satisfies the loading is the following polynomial:

$$\phi(x, y) = C_1xy + C_2xy^3 + C_3y^2 + C_4y^3$$

where C_j, $j = 1, 2, \ldots, 4$ are constants. Using Equations 6.6, the stresses are

$$\sigma_x = \frac{\partial^2\phi}{\partial y^2} = 6C_2xy + 2C_3 + 6C_4y$$

$$\sigma_y = \frac{\partial^2\phi}{\partial x^2} = 0$$

$$\tau_{xy} = -\frac{\partial^2\phi}{\partial x\partial y} = -C_1 - 3C_2y^2$$

The boundary conditions are

a. At $y = d$ and $y = -d$, the shear stress is zero, which gives $C_1 = -3C_2\,d^2$

b. At any section distance x from the free end, we have

$$\iint_A \sigma_x b\,dy = F$$

$$\iint_A (\sigma_x b\,dy)y = M - Px \tag{a}$$

$$\iint_A \tau_{xy} b\,dy = -P$$

The first condition in Equation (a) is

$$\int_{-d}^{d} (2C_3 + 6C_2xy + 6C_4y)b\,dy = F$$

Evaluating the integral provides $C_3 = F/(4bd)$.

The second condition in Equation (a), $\int_{-d}^{d}(2C_3y + 6C_2xy^2 + 6C_4y^2)$ $b\,dy = M - Px$, gives $C_2 = -P/(4bd^3)$ and $C_4 = M/(4bd^3)$. Hence, the stresses induced in the cantilever are

$$\begin{aligned}\sigma_x &= \frac{F}{2bd} + \frac{3My}{2bd^3} - \frac{3Pxy}{2bd^3}\\ \tau_{xy} &= -\frac{3P}{4bd} + \frac{3Py^2}{4bd^3}\end{aligned} \tag{b}$$

Note that the above results are identical to those derived using strength of materials approach (the moment of inertia of the cross section is $I = (2/3)$ bd^3). Also, the third condition in Equation (a) is automatically satisfied.

EXAMPLE 6.2

The plane stress distribution in a plate of length L, width w, and unit thickness (Figure 6.11) is given by

$$\sigma_x = x^3y - 2y^3x$$

$$\sigma_y = C_1x - 2C_2xy + y^3x$$

$$\tau_{xy} = C_3 + C_2x^2 - \frac{3}{2}x^2y^2 + \frac{y^4}{2}$$

where C_j, $j = 1, 2, 3$ are constants.

a. Neglecting body forces prove that equilibrium exists.
b. If the following boundary conditions apply:

$$\tau_{xy} = 0 \quad \text{at, } y = \pm\frac{w}{2}$$

$$\sigma_y = 0 \quad \text{at, } y = -\frac{w}{2}$$

determine the constants C_j and the total load on edge $y = w/2$.

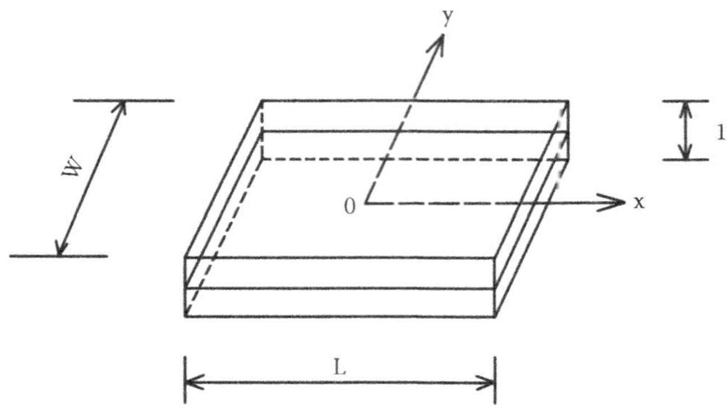

FIGURE 6.11
Example 6.2.

Solution

a. Equations (a) in Section 6.3 implies, after performing the differentiation, that

$$3x^2y - 2y^3 - 3x^2y + 2y^3 = 0$$
$$2C_2x - 3xy^2 - 2C_2x + 3y^2x = 0$$

Hence, equilibrium is satisfied.

b. Applying the first boundary condition, we get $C_3 - C_2\ x^2 - (3/2)(w^2/4)\ x^2 + (w^4/32) = 0$, which yields $C_2 = (3/8)w^2$ and $C_3 = -(w^4/32)$.

Similarly, the second boundary condition, $C_1\ x - 2(3/8)\ w^2\ (-w/2)x + (-w^{3/8})x = 0$, gives $C_1 = -w^3/4$.

And the total load on edge $y = w/2$ is given by

$$\text{Load} = \iint_A (\sigma_y)_{y=w/2} dA = \int_0^L \left(-\frac{w^3}{2}x\right)(1)dx = -\frac{w^3L^2}{4}$$

6.6 Solutions in Polar Coordinates

The use of polar coordinates enables us to solve a wide range of practical problems including those with axial symmetry like cylinders, disks, curved beams, wedges, etc. The following equations relate polar coordinates (r,θ) to the rectangular coordinates (x, y) (Figure 6.12):

$$x = r\cos\theta, \quad y = r\sin\theta$$
$$r^2 = x^2 + y^2, \quad \theta = \tan^{-1} y/x \tag{a}$$

The derivatives of r and θ with respect to x and y are obtained from Equation (a)

$$\frac{\partial r}{\partial x} = \frac{x}{r} = \cos\theta, \quad \frac{\partial r}{\partial y} = \frac{y}{r} = \sin\theta$$
$$\frac{\partial \theta}{\partial x} = -\frac{y}{r^2} = -\frac{\sin\theta}{r}, \quad \frac{\partial \theta}{\partial y} = \frac{y}{r^2} = \frac{\cos\theta}{r} \tag{b}$$

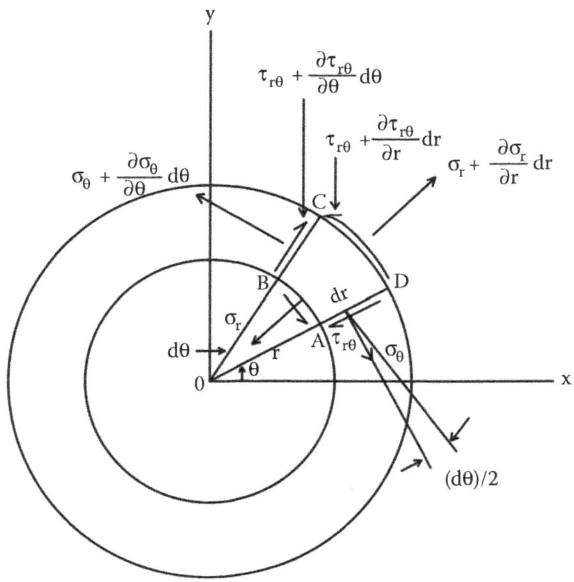

FIGURE 6.12
Element ABCD in polar coordinates.

Applying the chain rule, the derivatives with respect to the Cartesian coordinates are

$$\begin{aligned}\frac{\partial}{\partial x} &= \frac{\partial r}{\partial x}\frac{\partial}{\partial r} + \frac{\partial \theta}{\partial x}\frac{\partial}{\partial \theta} = \cos\theta\frac{\partial}{\partial r} - \frac{\sin\theta}{r}\frac{\partial}{\partial \theta} \\ \frac{\partial}{\partial y} &= \frac{\partial r}{\partial y}\frac{\partial}{\partial r} + \frac{\partial \theta}{\partial y}\frac{\partial}{\partial \theta} = \sin\theta\frac{\partial}{\partial r} + \frac{\cos\theta}{r}\frac{\partial}{\partial \theta}\end{aligned} \tag{c}$$

Consider an infinitesimal element ABCD in the xy-plane with unit thickness along the normal axis (Figure 6.12). The lengths of the sides are $AB = rd\theta$, $CD = (r + dr)d\theta$, and $AD = BC = dr$.

The stresses acting on the sides of the element in the positive directions are as follows: in the radial direction $(\sigma_r, \tau_{r\theta})$ on side AB, and $(\sigma_r + (\partial\sigma_r/\partial r)dr, \tau_{r\theta} + (\partial\tau_{r\theta}/\partial r)dr)$ on side CD. In the tangential direction, the stresses acting on sides AB and BC are $(\sigma_\theta, \tau_{r\theta})$ and $(\sigma_\theta + (\partial\sigma_\theta/\partial\theta)d\theta, \tau_{r\theta} + (\partial\tau_{r\theta}/\partial\theta)d\theta)$, respectively. The corresponding forces are obtained by multiplying the stresses times the areas on which they act.

Equilibrium of the forces in the radial direction including body force, F_r, per unit volume gives

$$\begin{aligned}&\left(\sigma_r + \frac{\partial\sigma_r}{\partial r}dr\right)(r + dr)d\theta - \sigma_r(rd\theta) + \left(\tau_{r\theta} + \frac{\partial\tau_{r\theta}}{\partial\theta}d\theta - \tau_{r\theta}\right)\left(dr\cos\frac{d\theta}{2}\right) \\ &\quad - \left(\sigma_\theta + \frac{\partial\sigma_\theta}{\partial\theta}d\theta - \sigma_\theta\right)\left(dr\sin\frac{d\theta}{2}\right) + F_r\, r dr\, d\theta = 0\end{aligned}$$

Letting dr and dθ approach zero, noting that in the limit sin dθ/2 = dθ/2 and cos dθ/2 = 1, and cancelling r dr dθ from each term, the result is

$$\frac{\partial \sigma_r}{\partial r} + \frac{1}{r}\frac{\partial \tau_{r\theta}}{\partial \theta} + \frac{\sigma_r - \sigma_\theta}{r} + F_r = 0 \tag{6.17a}$$

Similarly, equilibrium of the forces in the θ direction gives

$$\frac{1}{r}\frac{\partial \sigma_\theta}{\partial \theta} + \frac{\partial \tau_{r\theta}}{\partial r} + \frac{2\tau_{r\theta}}{r} + F_\theta = 0 \tag{6.17b}$$

Stress function: In the absence of body forces, $F_r = F_\theta = 0$, it is convenient to introduce a stress function $\phi(r, \theta)$ such that equilibrium in the radial and tangential directions is automatically satisfied. This can be accomplished by introducing the following relations:

$$\begin{aligned} \sigma_r &= \frac{1}{r}\frac{\partial \phi}{\partial r} + \frac{1}{r^2}\frac{\partial^2 \phi}{\partial \theta^2} \\ \sigma_\theta &= \frac{\partial^2 \phi}{\partial r^2} \\ \tau_{r\theta} &= \frac{1}{r^2}\frac{\partial \phi}{\partial \theta} - \frac{1}{r}\frac{\partial^2 \phi}{\partial r \partial \theta} = -\frac{\partial}{\partial r}\left(\frac{1}{r}\frac{\partial \phi}{\partial \theta}\right) \end{aligned} \tag{6.18}$$

It is readily confirmed that Equations 6.17 are satisfied. The stress function ϕ is governed by the bi-harmonic equation $\nabla^4\phi = 0$ in polar coordinates (Equation 6.21).

Strain–displacement relations: In polar coordinates, a structural element encounters deformations in the radial and tangential directions as well as rotations and changes of angles. The deformation of element ABCD is shown as element A′B′C′D′ in Figure 6.13a.

Denote the displacements of the element in the radial and tangential directions by $u(r, \theta)$ and $v(r, \theta)$, respectively. The radial displacement of side AB is u and that of side CD is $u + (\partial u/\partial r)dr$ (Figure 6.13b). Hence, the strain in the radial direction is given by

$$\varepsilon_r = \frac{u + \frac{\partial u}{\partial r}dr - u}{dr} = \frac{\partial u}{\partial r} \tag{6.19a}$$

The tangential strain ε_θ consists of two components. The first component arises due to the radial deformation. Because the length A′B′ in the deformed element is $(r + u)d\theta$, the radial contribution to the tangential strain is $((r + u)d\theta - rd\theta)/rd\theta = (u/r)$. The second component exists because the deformation in the tangential direction is $(v + (\partial v/\partial \theta)d\theta - v)/rd\theta$ (Figure 6.13c). Hence, the tangential strain is

$$\varepsilon_\theta = \frac{u}{r} + \frac{1}{r}\frac{\partial v}{\partial \theta} \tag{6.19b}$$

The shearing strain $\gamma_{r\theta}$ is the change in the right angle BAD to angle B′A′D′. The displacements u and v of the element contribute to the change in this angle. The side AB rotates into A′B′ and the contribution of the rotation due to the radial displacement u is $(1/r)(\partial u/\partial \theta)$. Also, the rotation of side AD due to the displacement v is $\partial v/\partial r$. Because an initial rotation of magnitude v/r has occurred due to rigid body motion of the element about an axis

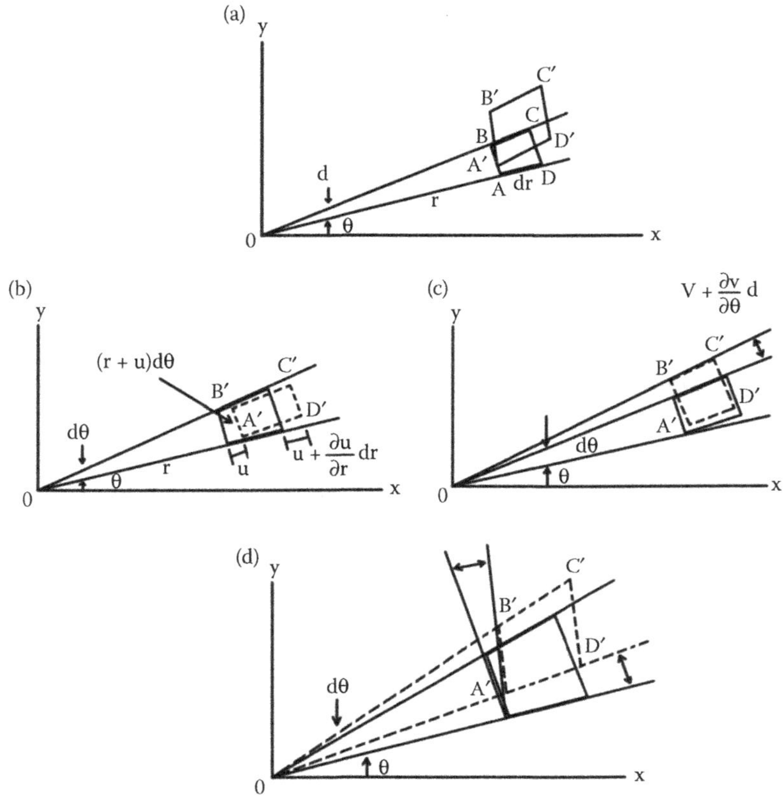

FIGURE 6.13
(a,d) Deformation of element ABCD in polar coordinates; (b) radial deformation; (c) tangential deformation.

through point D, the net contribution to the shearing strain due to the tangential displacement is $(\partial v/\partial r) - (v/r)$ (Figure 6.13d). Hence, the shearing strain is

$$\gamma_{r\theta} = \frac{1}{r}\frac{\partial u}{\partial \theta} + \frac{\partial v}{\partial r} - \frac{v}{r} \tag{6.19c}$$

Compatibility equation: Equations 6.19 relate three strains, two normal and one shearing strain, in terms of two independent variables, namely, displacements u and v. It follows that there is a relation between the strains known as the compatibility equation. It can be confirmed that the compatibility equation in terms of strains is (see Equation 3.21)

$$\frac{\partial^2 \varepsilon_\theta}{\partial r^2} + \frac{1}{r^2}\frac{\partial^2 \varepsilon_r}{\partial \theta^2} + \frac{2}{r}\frac{\partial \varepsilon_\theta}{\partial r} - \frac{1}{r}\frac{\partial \varepsilon_r}{\partial r} = \frac{1}{r}\frac{\partial^2 \gamma_{r\theta}}{\partial r \partial \theta} + \frac{1}{r^2}\frac{\partial \gamma_{r\theta}}{\partial \theta} \tag{6.20}$$

In order to express the compatibility equation in terms of the stresses, it is necessary to determine the operator, $\nabla^2 = (\partial^2/\partial x^2) + (\partial^2/\partial y^2)$, in terms of polar coordinates. Following the chain rule procedure used to determine relation (c), it can be confirmed that $\nabla^2\phi = ((\partial^2/\partial r^2) + (1/r)(\partial/\partial r) + (1/r^2)(\partial^2/\partial \theta^2))\phi(r,\theta)$, and similarly the bi-harmonic equation is

$$\nabla^4\phi = \left(\frac{\partial^2}{\partial r^2} + \frac{1}{r}\frac{\partial}{\partial r} + \frac{1}{r^2}\frac{\partial^2}{\partial \theta^2}\right)\nabla^2\phi = 0 \tag{6.21}$$

Performing the lengthy derivatives indicated in Equation 6.21, the compatibility equation in polar coordinates assumes the form

$$\frac{\partial^4\phi}{\partial r^4} + \frac{2}{r}\frac{\partial^3\phi}{\partial r^3} - \frac{1}{r^2}\frac{\partial^2\phi}{\partial r^2} + \frac{1}{r^3}\frac{\partial\phi}{\partial r} + \frac{2}{r^2}\frac{\partial^4\phi}{\partial r^2\partial\theta^2} - \frac{2}{r^3}\frac{\partial^3\phi}{\partial r\partial\theta^2} + \frac{4}{r^4}\frac{\partial^2\phi}{\partial\theta^2} + \frac{1}{r^4}\frac{\partial^4\phi}{\partial\theta^4} = 0 \tag{6.22}$$

Stress–strain relations: For elastic behavior, Hooke's law in polar coordinates is obtained by replacing the coordinates x and y in the corresponding relations in rectangular coordinates by r and θ, respectively. Thus, for plane stress problems, we have

$$\begin{aligned}
\varepsilon_r &= \frac{1}{E}(\sigma_r - \upsilon\sigma_\theta)\\
\varepsilon_\theta &= \frac{1}{E}(\sigma_\theta - \upsilon\sigma_r)\\
\gamma_{r\theta} &= \frac{1}{G}\tau_{r\theta}\\
\varepsilon_z &= -\frac{\upsilon}{E}(\sigma_r + \sigma_\theta)\\
\gamma_{rz} &= \gamma_{\theta z} = 0
\end{aligned} \tag{6.23a}$$

while for plane strain problems, the relations are

$$\begin{aligned}
\varepsilon_r &= \frac{1+\upsilon}{E}[(1-\upsilon)\sigma_r - \upsilon\sigma_\theta]\\
\varepsilon_\theta &= \frac{1+\upsilon}{E}[(1-\upsilon)\sigma_\theta - \upsilon\sigma_r]\\
\gamma_{r\theta} &= \frac{\tau_{r\theta}}{G}\\
\varepsilon_z &= \gamma_{rz} = \gamma_{\theta z} = 0
\end{aligned} \tag{6.23b}$$

6.7 Solutions of the Compatibility Equation

This section contains few mathematical solutions of the compatibility equation in polar coordinates (Equation 6.22), which can be used to find the stress distribution in problems of practical interest. We begin with a summary of the solutions and then proceed to determine the state of stress and deformation.

a. *Axially symmetric problems*: For axially symmetric problems with the normal axis coinciding with the z-axis, the stress function is $\phi(r, \theta) = \phi(r)$, and Equation 6.22

reduces to an ordinary differential equation of the fourth order.

$$r^3 \frac{d^4\phi}{dr^4} + 2r^2 \frac{d^3\phi}{dr^3} - r\frac{d^2\phi}{dr^2} + \frac{d\phi}{dr} = 0 \tag{a}$$

whose solution, containing four constants, can be found by standard methods.

$$\phi(r) = A \ln r + Br^2 \ln r + Cr^2 + D \tag{6.24}$$

A, B, C, and D are constants.

b. General forms of the stress function include the following relations:

$$\phi(r, \theta) = f(r) \cos n\theta \quad \text{or} \quad \phi = f(r) \sin n\theta \tag{b}$$

where n is an integer. Inserting Equation (b) in the compatibility equation, the following ordinary differential equation governing the function f(r) is obtained:

$$r^4 f^{iv} + 2r^3 i^{iii} - (1 + 2n^2)r^2 f^{ii} + (1 + 2n^2)rf^{i} + n^2(n^2 - 4)f = 0 \tag{c}$$

In Equation (c), the first derivative of f(r) with respect to r is denoted by f^{i} and so on.

When $n = 0$, the axially symmetric case discussed above is recovered. For higher values of n, there are two cases to consider. When $n = 1$, the following equation is obtained:

$$r^4 \frac{d^4 f}{dr^4} + 2r^3 \frac{d^3 f}{dr^3} - 3r^2 \frac{d^2 f}{dr^2} + 3r\frac{df}{dr} - 3f = 0 \tag{d}$$

Solving Equation (d), the stress function becomes

$$\phi(r, \theta) = \left[Ar + Br\,\mathrm{Ln}\,r + Cr^3 + \frac{D}{r} \right] [\cos\theta, \text{ or } \sin\theta] \tag{6.25a}$$

When $n \geq 2$, the differential equation governing f(r) is

$$r^4 \frac{d^4 f}{dr^4} + 2r^3 \frac{d^3 f}{dr^3} - (1 + 2n^2)r^2 \frac{d^2 f}{dr^2} + (1 + 2n^2)r\frac{df}{dr} + n^2(n^2 - 4)f = 0 \tag{e}$$

Solving Equation (e), the corresponding stress function is

$$\phi(r, \theta) = \left[A_n r^n + \frac{B_n}{r^n} + C_n r^{n+2} + D_n r^{-n+2} \right] [\cos n\theta, \text{ or } \sin n\theta], \quad n \geq 2 \tag{6.25b}$$

In cases where it is not possible to represent the force(s) acting on the member by a single cosine or sine term, then a Fourier series representation can be used in relation (b).

c. *Stress function*: $\phi(r, \phi) = \phi(\theta)$: When the stress function is a function of the polar angle only, Equation 6.22 reduces to

$$\frac{d^4\phi}{d\theta^4} + 4\frac{d^2\phi}{d\theta^2} = 0$$

and the stress function is

$$\phi(\theta) = A\sin 2\theta + B\cos 2\theta + C\theta + D \tag{6.26}$$

Here, A, B, C, and D are constants.

d. *Stress function* $\phi(r, \theta) = r\ g(\theta)$: Inserting the stress function in the compatibility Equation 6.22, it is found that $g(\theta)$ satisfies the ordinary differential equation

$$\frac{d^4g}{d\theta^4} + 2\frac{d^2g}{d\theta^2} + g = 0 \tag{f}$$

The solution of Equation (f) is readily found as

$$g(\theta) = A\theta\cos\theta + B\theta\sin\theta + C\cos\theta + D\sin\theta \tag{g}$$

where A, B, C, and D are constants. It follows that the applicable stress function is

$$\phi(r, \theta) = Ar\theta\cos\theta + Br\theta\sin\theta \tag{6.27}$$

The terms containing constants C and D in Equation (g) represent rigid body motion and do not contribute to the stresses.

In the next few sections, illustrative examples are given to explain the use of stress functions; Equations 6.24 through 6.27.

6.7.1 Axially Symmetric Problems

For this class of problems, we use the stress function (6.24). The corresponding stresses are readily found as follows:

$$\begin{aligned}
\sigma_r &= \frac{A}{r^2} + 2B[1 + 2\,\mathrm{Ln}\,r] + 2C \\
\sigma_\theta &= -\frac{A}{r^2} + 2B[3 + 2\,\mathrm{Ln}\,r] + 2C \\
\tau_{r\theta} &= 0
\end{aligned} \tag{6.28}$$

It is clear that if the point $r = 0$ is situated within the component, for example circular disks and solid shafts, then the terms containing constants A and B cannot be allowed because the stresses become unbounded. Moreover, it turns out that, for certain domains in which the angle θ is less than 2π, the displacements are multivalued, that is, they contain at least one term that gives different displacement value each time the angle θ is increased by 2π, and such term cannot be allowed. To explore this point further, let us examine the displacements resulting from expressions (6.28).

The displacements u and v for plane stress elements are computed from the strain–displacement relations in Equations 6.23a as follows:

$$E\frac{\partial u}{\partial r} = \sigma_r - \upsilon\sigma_\theta = (1+\upsilon)\frac{A}{r^2} + 2B[1 - 3\upsilon + 2(1-\upsilon)\mathrm{Ln}\, r] + 2(1-\upsilon)C$$

Performing the integration, it is found that

$$u = \frac{1}{E}\left[-\frac{(1+\upsilon)A}{r} + 4B(1-\upsilon)Br\,\mathrm{Ln}\, r - 2(1+\upsilon)Br + 2(1-\upsilon)Cr\right] + f(\theta) \qquad \text{(a)}$$

where f(θ) is constant of integration. The tangential displacement v is computed using the relation $(1/r)(\partial v/\partial\theta) = (\sigma_r - \sigma_\theta)/E - (u/r)$. Substituting the stress expressions in Equations 6.28 and the radial displacement in Equation (a) results in the relation

$$\frac{\partial v}{\partial \theta} = \frac{8Br}{E} + f(\theta)$$

Carrying out the integration, the tangential displacement is

$$v = \frac{8Br}{E}\theta + \int f(\theta)d\theta + g(r) \qquad \text{(b)}$$

where the functions f(θ) and g(r) are determined from the boundary conditions describing the constraints of the member. Since the first term in Equation (b) predicts different displacement values because the angle θ can take different values at the same point of the element simply by adding multiple values of 2π, such term must vanish and for such members Equations 6.28 reduces to

$$\sigma_r = \frac{A}{r^2} + 2C$$
$$\sigma_\theta = -\frac{A}{r^2} + 2C \qquad (6.29)$$
$$\tau_{r\theta} = 0$$

It is noted that the sum of the normal stresses is constant throughout the element. Also, if r = 0 is located in the element, the term containing A must vanish and the stresses are either constant tension or compression throughout the element.

For application to elements where the angle θ is restricted to values less than 360°, the B term in the displacement is single-valued and expressions (6.28) are applicable. Next, we use examples of axially symmetric problems to illustrate the procedure. Additional examples are provided in Chapter 7.

a. *Expansion of a circular hole*: Consider an infinite sheet of uniform thickness containing a circular of radius, a, subjected to radial pressure p. The material is assumed to be elastic–perfectly plastic with yield stress σ_0. Assuming von Mises' yield criterion, it is required to find the ultimate pressure and extent of the plastic zone around the hole (Figure 6.14).

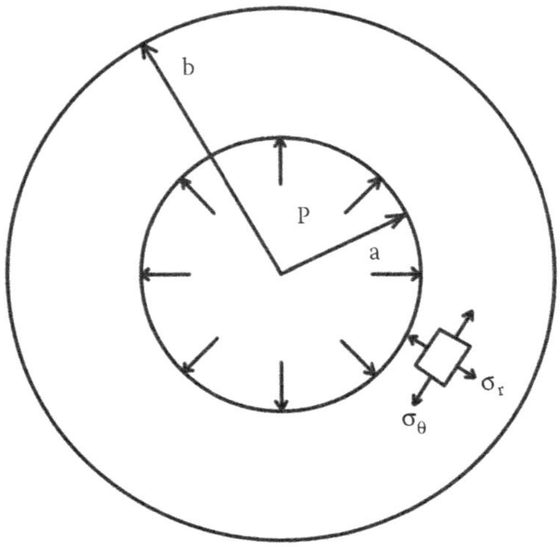

FIGURE 6.14
Large plate containing radially pressurized circular hole of radius a.

Solution: In the elastic range, the proper solution of the governing equation is provided by Equations 6.29. Equation 6.29 indicates that the sum of the normal stresses, $\sigma_r + \sigma_\theta$, is constant throughout the sheet. The boundary conditions are $r = a$: $\sigma_r = -p$ and at infinity ($r = \infty$) σ_r and σ_θ must vanish.

Hence, Equations 6.29 gives $C = 0$ and $A = -p\,a^2$. The resulting stresses are

$$\sigma_r = -p\frac{a^2}{r^2}, \quad \sigma_\theta = p\frac{a^2}{r^2} \tag{6.30}$$

Plastic analysis: Beyond the elastic range the material, which is assumed to be elastic–perfectly plastic with yield stress σ_0, yields according to von Mises's criterion if the stresses satisfy the following equation:

$$\sigma_r^2 - \sigma_r\sigma_\theta + \sigma_\theta^2 = \sigma_0^2 \tag{a}$$

Inserting the values of the stresses from Equation 6.30 at $r = a$ into (a), the pressure that initiates yield is

$$p_e = \frac{\sigma_0}{\sqrt{3}} \tag{6.31}$$

When p is increased beyond p_e, yielding will spread from the edge of the hole and a circular plastic zone is established, as shown in Figure 6.14. Assume that at a certain pressure p the plastic zone extends to a distance b from the center of the hole. In the region $r \geq b$, the behavior is elastic, and using Equations 6.30 and (a), the stresses are $\sigma_\theta = -\sigma_r = \left(\sigma_0 b^2/\sqrt{3}r^2\right)$.

In order to obtain the stresses in the plastic region, $a \leq r \leq b$, we need to solve the equilibrium equation

$$\frac{\partial\sigma_r}{\partial r} = \frac{\sigma_\theta - \sigma_r}{r} \tag{b}$$

Subject to satisfying the yield criterion (a) and the following boundary conditions

$$r = a, \quad \sigma_r = -p; \quad r = b, \quad \sigma_\theta = -\sigma_r = \frac{\sigma_0}{\sqrt{3}} \tag{c}$$

We use a solution introduced by Nadai [4.12]. The yield criterion is readily shown to be satisfied by the following parametric representation of the stresses:

$$\begin{aligned} \sigma_r &= -\frac{2\sigma_0}{\sqrt{3}}\sin(30+\beta) \\ \sigma_\theta &= \frac{2\sigma_0}{\sqrt{3}}\sin(30-\beta) \end{aligned} \tag{6.32}$$

where β is arbitrary. Note that when $\beta = 0$, the values of the stresses proposed in Equation 6.32 are $\sigma_\theta = -\sigma_r = (\sigma_0/\sqrt{3})$, which are valid in the elastic region $r \geq b$.

Inserting stresses (6.32) into Equation (b) and simplifying the algebra, the following differential equation is obtained for β:

$$\left(\frac{\sqrt{3}}{2} - \frac{1}{2}\tan\beta\right)\frac{d\beta}{dr} = -\frac{1}{r} \tag{d}$$

The integral of Equation (d) is $(\sqrt{3}/2)\beta + (1/2)\text{Ln}\cos\beta = -\text{Ln}\, r + C$, where the constant C is found from the boundary condition that $\beta = 0$ at $r = b$, or $C = \text{Ln}\, b$. Simplifying the algebra, the values of β are given by the relation

$$\frac{b^2}{r^2} = e^{\beta\sqrt{3}}\cos\beta \tag{e}$$

The stresses in the plastic zone are calculated from Equation 6.30. In particular, at $r = a$, the applied pressure at the hole is

$$p = -\sigma_r = \frac{2\sigma_0}{\sqrt{3}}\sin(30+\alpha) \tag{f}$$

where $\beta = \alpha$ at $r = a$. and $(b^2/a^2) = e^{\alpha\sqrt{3}}\cos\alpha$. Maximum pressure occurs when $\alpha = 60°$.

As the pressure increases from its yield initiation value, $(\sigma_0/\sqrt{3})$, α is increased from zero to 60°. Hence, the extent of the plastic zone is $(b/a) = \left[(1/2)e^{(\pi/\sqrt{3})}\right]^{(1/2)} = 1.751$ and the ultimate pressure is

$$P_u = \frac{2\sigma_0}{\sqrt{3}} \tag{6.33}$$

The tangential stress at the edge of the hole is compression of magnitude, $(\sigma_0/\sqrt{3})$, and any further increase in the pressure causes the material surrounding the hole to thicken. It is observed that the sum of the normal stresses in the sheet vanish at all stages of the loading.

b. *Bending of curved beams*: We consider a curved beam (curved plate) of narrow rectangular cross section and circular axis subjected to end moments M per unit width in plane strain condition. The radii of the inner and outer surfaces are denoted by a, b, respectively (Figure 6.15). Although the geometry of the curved beam is not symmetric about the normal axis to its plane, the stresses and strains derived in Equations 6.28 can be used to determine the in-plane stresses, σ_r and σ_θ, and the longitudinal stress is given by $\sigma_z = \upsilon(\sigma_r + \sigma_\theta)$.

Elastic solution: The stress state derived in Equations 6.28 is applicable because the point r = 0 is outside the member and the displacements are not multivalued.

To simplify the formulation of the problem, the constants A and 2C in Equations 6.28 are replaced by b^2A and C – B – 2B Ln b, respectively. The stress expressions become

$$\begin{aligned}\sigma_r &= \frac{b^2A}{r^2} + 2B\,\mathrm{Ln}\frac{r}{b} + C\\ \sigma_\theta &= -\frac{b^2A}{r^2} + 2B\left(1 + \mathrm{Ln}\frac{r}{b}\right) + C\\ \tau_{r\theta} &= 0\end{aligned} \tag{a}$$

The boundary conditions required to determine constants A, B, and C are as follows: the radial stress σ_r must vanish at the curved surfaces r = a, r = b, and the tangential stress σ_θ across any radial cross section must give rise to zero normal force and moment M, that is,

$$\sigma_r = 0 \text{ at } r = b \quad \text{and} \quad r = a \tag{b}$$

$$\iint_A \sigma_\theta\, dA = \int_a^b \sigma_\theta(1)dr = 0 \tag{c}$$

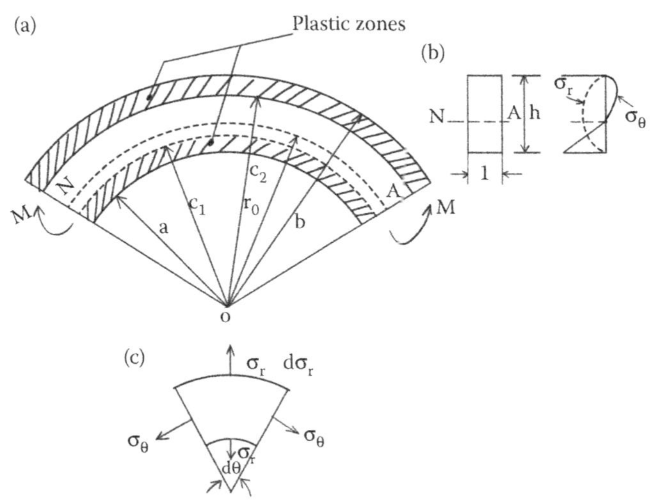

FIGURE 6.15
Bending of a curved beam. (a) Geometery of the beam; (b) distribution of stress; and (c) stress on an element.

$$\iint_A \sigma_\theta r dA = \int_a^b \sigma_\theta r(1)dr = -M \tag{d}$$

The first boundary condition in Equation (b) gives $C = -A$ and the second boundary condition furnishes the relation

$$A(b^2 - a^2) = 2a^2 B \,\text{Ln}\frac{b}{a} \tag{6.34}$$

Integrating the tangential stress as required by condition (c) and using relation (6.34), it is confirmed that the sum of normal forces on any cross section vanish for all values of the constants A and B. The stresses in Equation (a) become

$$\begin{aligned}\sigma_r &= A\left(\frac{b^2}{r^2} - 1\right) + 2B\,\text{Ln}\frac{r}{b}\\ \sigma_\theta &= -A\left(\frac{b^2}{r^2} + 1\right) + 2B\left(1 + \text{Ln}\frac{r}{b}\right)\end{aligned} \tag{6.35}$$

The constants A and B are determined in terms of the applied moment by applying boundary condition (d). To simplify the integration, it can be shown (see Problem 6.13) that condition (d) is equivalent to

$$M = \frac{1}{2}\int_a^b (\sigma_r - \sigma_\theta) r dr = \int_a^b \left(\frac{A}{r} - Br\right)dr = A\,\text{Ln}\frac{b}{a} - \frac{B}{2}(b^2 - a^2) \tag{e}$$

The constants A and B are obtained from Equations 6.34 and (e) as follows:

$$A = \frac{4M}{a^2 D}\text{Ln}\frac{a}{b} \quad \text{and} \quad B = \frac{2\left(1 - \dfrac{b^2}{a^2}\right)M}{a^2 D} \tag{f}$$

where D stands for the abbreviation

$$D = \left(\frac{b^2}{a^2} - 1\right)^2 - 4\frac{b^2}{a^2}\left(\text{Ln}\frac{b}{a}\right)^2 \tag{g}$$

and the stresses in the beam are

$$\begin{aligned}\sigma_r &= \frac{4M}{a^2 D}\left[\left(\frac{b^2}{r^2} - 1\right)\text{Ln}\frac{a}{b} + \left(1 - \frac{b^2}{a^2}\right)\text{Ln}\frac{r}{b}\right]\\ \sigma_\theta &= \frac{4M}{a^2 D}\left[-\left(\frac{b^2}{r^2} + 1\right)\text{Ln}\frac{a}{b} + \left(1 - \frac{b^2}{a^2}\right)\left(1 + \text{Ln}\frac{r}{b}\right)\right]\\ \tau_{r\theta} &= 0\end{aligned} \tag{6.36}$$

The radial stress is positive throughout the cross section of the beam and has a maximum value at $r = b\sqrt{A/B}$. The tangential stress is positive at $r = a$, and changes to negative at $r = b$. The greatest value of the tangential stress occurs at the inner boundary $r = a$. The neutral surface, determined from the condition $\sigma_\theta = 0$, is independent of the applied moment.

The radial and tangential displacements, u and v, are determined by using the strain–displacement relations in Equations 6.19 after inserting Equation 6.36 into Hooke's law for plain strain (Equations 6.23b). The results are

$$2Gu = r\left\{-A\left[1 - 2\upsilon + \frac{b^2}{r^2}\right] + 2B\left[-(1-\upsilon) + (1-2\upsilon)\mathrm{Ln}\frac{r}{b}\right]\right\} + \frac{df(\vartheta)}{d\theta} \qquad \text{(h)}$$

$$2Gv = 4(1-\upsilon)Br\theta - f(\theta) + g(r)$$

where $f(\theta)$ and $g(r)$ are arbitrary constants of integration to be found from the specified constraint of the beam. Assuming that the middle cross section of the beam is constrained during bending, then

$$u = v = \frac{\partial v}{\partial r} = 0 \text{ at } \theta = 0, \quad r = r_0 = \frac{a+b}{2} \qquad \text{(i)}$$

Using relations (h) and (i), it is readily confirmed that $f(\theta) = C_1 \sin\theta$ and $g(r) = 0$, and the elastic displacements are

$$2Gu = r\left\{-A\left(1 - 2\upsilon + \frac{b^2}{r^2}\right) + 2B\left[-(1-\upsilon) + (1-2\upsilon)\mathrm{Ln}\frac{r}{b}\right]\right\} + C_1\cos\theta \qquad \text{(6.37a)}$$

$$2Gv = 4(1-\upsilon)Br\theta - C_1\sin\theta \qquad \text{(6.37b)}$$

where the constant C_1 stands for

$$C_1 = r_0\left\{A\left(1 - 2\upsilon + \frac{b^2}{r_0^2}\right) + 2B\left[1 - \upsilon - (1-2\upsilon)\mathrm{Ln}\frac{r_0}{b}\right]\right\} \qquad \text{(6.37c)}$$

The tangential displacement, v, is single valued because the magnitude of the polar angle θ cannot exceed 360°. Also, because v is a linear function of r, all radial cross sections of the beam remain plane after bending.

It is of interest to compare the above elasticity solution with the strength of materials solutions of curved beams [6.4–6.6]. If the depth, $b - a$, is small relative to the central axis $(a + b)/2$ then the elementary solution, Equation 1.15, can be used to compute the bending stress. If the depth is not small, then Winkler's theory of the bending of initially curved beams can be used to obtain the stress distribution [6.7]. In all cases, the maximum and minimum values of the tangential stress can be expressed in the form

$$\sigma_\theta = m\frac{M}{a^2} \qquad \text{(6.38)}$$

where m is a numerical factor whose values are given in Table 6.1.

TABLE 6.1

Values of Coefficient m in Equation 6.36

	Linear Stress Distribution	Winkler's Theory (Hyperbolic Stress Distribution)		Elasticity Theory (Equation 6.36)	
(b/a)	($\sigma_\theta = (My/I)$)	r = a	r = b	r = a	r = b
1.3	±66.67	+72.98	−61.27	+73.05	−61.35
2.0	±6.000	+7.725	−4.863	+7.755	−4.917
3.0	±1.500	+2.285	−1.095	+2.292	−1.130

It is clear from the table that the strength of material solution based on the hypothesis of plane sections remains plane after bending, especially Winkler's formula gives good results. Any inaccuracy in the strength of material solution arises from the fact that the radial stress is neglected and the longitudinal fibers are assumed either in tension or in compression. In the elasticity solution, the tangential stresses give rise to tensile forces in the radial direction, which tends to separate the longitudinal fibers and produce tensile stresses in the radial direction [5.3].

Plastic analysis: When the elastic moment reaches the proportional limit, M_e, plastic yielding begins at the inner surface r = a. Using Tresca's yield criterion, $\sigma_\theta - \sigma_r = \sigma_0$, where σ_0 denotes the yield stress in tension and assuming that $\sigma_0 = 2\tau_0$, where τ_0 is the yield stress in shear, the moment per unit width, which initiates yield, is obtained from the second of Equations 6.36 as

$$M_e = \frac{a^2 D}{2\left[1 - \frac{b^2}{a^2}\left(1 + \mathrm{Ln}\frac{b^2}{a^2}\right)\right]}(\tau_0) \tag{6.39}$$

As the applied moment is increased beyond M_e, a plastic zone is formed around the inner surface and spread upward. Because of symmetry, the elastic/plastic boundary is a circular arc of radius, say C_1, as shown in Figure 6.15.

In the plastic zone, $a \le r \le C_1$, the stresses are required to satisfy Tresca's criterion mentioned above and the equilibrium equation

$$\frac{\partial \sigma_r}{\partial r} = \frac{\sigma_\theta - \sigma_r}{r} \qquad r = a,\ \sigma_r = 0 \tag{a}$$

It is readily confirmed that the stresses that satisfy Equation (a) and Tresca's criterion are $\sigma_r = 2\tau_0\,\mathrm{Ln}(r/a)$ and $\sigma_\theta = 2\tau_0(1 + \mathrm{Ln}(r/a))$. In the elastic region, the stresses are given by Equations 6.35 with the constants A and B to be determined from the conditions of continuity of σ_r and σ_θ at $r = C_1$. Further increase in the applied moment causes the radius C_1 to increase and the neutral axis r_0 to change its location from the elastic to the plastic region. A stage is reached when the outside fiber, r = b, begins to yield and a second plastic zone of radius C_2 is formed (Figure 6.15). Further increase in the applied moment causes the plastic zones to expand and the elastic layer in between to shrink in thickness. Eventually, the elastic layer reduces to a line and the whole cross section becomes plastic.

Plastic moment (M_p): The fully plastic moment is computed as follows:
For $a \leq r \leq r_0$,

$$\sigma_r = 2\tau_0 \operatorname{Ln}\frac{r}{a} \quad \text{and} \quad \sigma_\theta = 2\tau_0\left(1 + \operatorname{Ln}\frac{r}{a}\right)$$

For $r_0 \leq r \leq b$,

$$\sigma_r = -2\tau_0 \operatorname{Ln}\frac{r}{b} = 2\tau_0 \ln\frac{b}{r} \quad \text{and} \quad \sigma_\theta = -2\tau_0\left(1 - \ln\frac{b}{r}\right)$$

Since the radial stress must be continuous across $r = r_0$, we have Ln $(r_0/a) = \ln(b/r_0)$ and the radius of the neutral axis is given by

$$r_0 = \sqrt{ab} \tag{6.40}$$

Hence, the fully plastic moment is given by

$$M_p = \frac{1}{2}\int_a^{r_0} (\sigma_r - \sigma_\theta) r dr + \frac{1}{2}\int_{r_0}^{b} (\sigma_r - \sigma_\theta) r dr$$

Inserting the values of the stresses in the zones above and below the neutral axis and evaluating the integrals, the plastic moment is

$$M_p = \frac{1}{2}\left(\frac{b}{a} - 1\right)^2 \tau_0 a^2 \tag{6.41}$$

For a beam having $b/a = 2$, the moment that initiates yield in the beam at the inner curved surface, computed using Equation 6.53, is $M_e = 0.258\ \tau_0 a^2$. At this stage of loading, the tangential stress at $r = b$ is $\sigma_\theta = -1.267\tau_0$. The moment required to make the whole beam plastic is found from Equation 6.55 as $M_p = 0.5\ \tau_0 a^2$ ($M_p = 1.94\ M_e$). The neutral axis is located at $r_0 = 1.444$ a. The radial stress has a maximum value $(\sigma_r)_{max} = 1.07\ M_e/a^2 = 0.276\ \tau_0$ and is located at $r = 1.36$ a, which is slightly lower than the radius of the neutral axis.

Additional results are available in Reference 6.6. In particular, the variations of σ_r/τ_0 and σ_θ/τ_0 with r/a as the ratio M/M_p varies between 0.516 (initiation of yield) and 1:00 (fully plastic cross section) are provided. The oscillations in the position of the neutral axis are also given. These authors also showed that changes in the radii of the surfaces of the beam during bending are small, less than 2%, and may be neglected.

c. *Other axially symmetric problems*: The stresses in Equations 6.28 can be used to solve other problems like hollow cylinders and thick-wall pressure vessels subjected to internal and external loads. Such problems are discussed in Chapter 7.

EXAMPLE 6.3

An annular plate is fixed at the inner boundary $r = a$, and subjected to uniform shear stress q at the outer boundary $r = b$, the thickness of the plate is unity (Figure 6.16). Assuming elastic behavior of the material, it is required to determine the stresses and displacements in the plate.

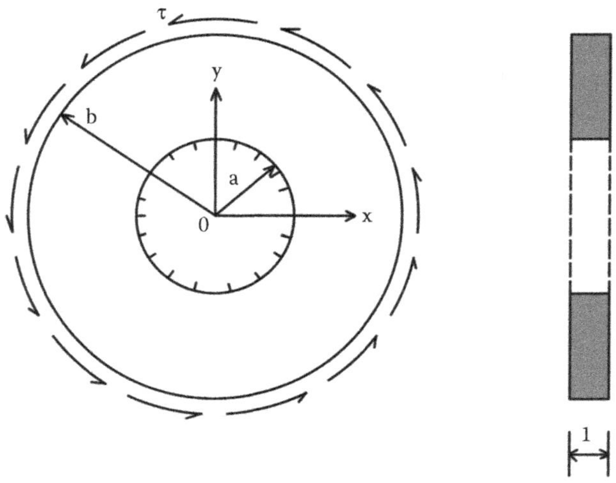

FIGURE 6.16
Example 6.3.

Solution

The applied load is equivalent to a twisting moment $M = q\,(2\pi b \times 1)\,b = 2\pi b^2 q$ about an axis through the center of the plate normal to its surface. The boundary conditions are at $r = a$, the normal and tangential displacements, $u = v = 0$ and at $r = b$, the stresses $\sigma_r = 0$, $\tau_{r\theta} = q$.

The stress function that satisfies the loading is $\phi = C\,\theta$ (Equation 6.26). The corresponding stresses are $\tau_{r\theta} = C/r^2$ and zero radial and tangential stresses. Using the boundary condition, $C = b^2 q$. Thus, the stresses in the plate are

$$\tau_{r\theta}\frac{b^2}{r}q, \quad \sigma_r = \sigma_\theta = \sigma_z = 0$$

The radial and tangential displacements vanish at $r = a$. It follows that $\varepsilon_r = (\partial u(r,\theta))/\partial r = 0$. Hence, $u = f(\theta) = 0$, and $u(r,\theta) = 0$.

$\varepsilon_\theta = (u/r) + (1/r)(\partial V(r,\theta))/\partial\theta = 0$. It follows that V is a function of r only, making use of Hooke's law

$$\gamma_{r\theta} = \frac{\tau_{r\theta}}{G} = \frac{1}{r}\frac{\partial u}{\partial\theta} + \frac{\partial V(r,\theta)}{\partial r} - \frac{V(r,\theta)}{r}$$

we have

$$r\frac{dV(r)}{dr} - V(r) = \frac{b^2 q}{Gr} = \frac{M}{2\pi Gr}$$

$$V(r) = Ar - \frac{M}{4\pi Gr}$$

The constant A is found from the condition that $V = 0$ at $r = b$. Hence,

$$V(r) = \frac{b^2 q}{2Gr}\left(\frac{r^2}{a^2} - 1\right) = \frac{M}{4\pi Gr}\left(\frac{r^2}{a^2} - 1\right)$$

EXAMPLE 6.4

At the outside of a solid circular disk of radius $r = a$, unit thickness, the following radial and tangential displacements are imposed: $u = u_0 \sin 2\theta$, $V = 0$, where $u_0 =$ constant. Assuming elastic behavior, Poisson's ratio $= 0.3$, find the stresses induced in the disk.

Solution

The stress function that satisfies the loading of the disk is obtained from Equation 6.25b with $n = 2$. Note that the disk is in a plane stress condition.

$$\phi(r, \theta) = \left(Ar^2 + \frac{B}{r^2} + Cr^4 + D\right) \sin 2\theta$$

where A, B, C, and D are constants. The resulting stresses that satisfy the regularity condition at $r = 0$ (stresses are finite) are as follows:

$$\sigma_r = -2A \sin 2\theta$$
$$\sigma_\theta = 2(A + 6Cr^2) \sin 2\theta$$
$$\tau_{r\theta} = -2(A + 3Cr^2) \cos 2\theta$$

The corresponding radial and tangential displacements are obtained using Hooke's law, eliminating the routine details, they are

$$\begin{aligned} u &= -\frac{2}{E}[(1 + \upsilon)Ar + 2\upsilon Cr^3] \sin 2\theta + f(\theta) \\ V &= -\frac{2}{E}[(1 + \upsilon)Ar + (3 + \upsilon)Cr^3] \cos 2\theta + g(r) \end{aligned} \tag{a}$$

where $f(\theta)$ and $g(r)$ are constants of integration. They can be determined from the relation

$$\gamma_{r\theta} = \frac{1}{r}\frac{\partial u}{\partial \theta} + \frac{\partial V}{\partial r} - \frac{V}{r} = \frac{2(1 + \upsilon)}{E}\tau_{r\theta}$$

which yields the following simple differential equations and their solutions:

$$\frac{d^2 f}{d\theta^2} + f(\theta) = 0, \quad f(\theta) = C_1 \sin\theta + C_2 \cos\theta$$
$$\frac{d^2 g}{dr^2} - g(r) = \text{constant}, \quad g(r) = C_3 + C_4 r$$

Since the displacement vanishes at $r = 0$, $u = V = 0$, it follows that the constants C_j, $j = 1, \ldots, 4$ are zero.

The constants A and C are determined from the boundary conditions at $r = a$, namely, $u = u_0$ and $V = 0$. Thus,

$$C = -\left(\frac{1 + \upsilon}{3 + \upsilon}\right)\frac{A}{a^2} = -0.394\frac{A}{a^2}$$
$$A = -0.47\frac{Eu_0}{a}, \quad C = 0.185\frac{Eu_0}{a}$$

And the stresses follow as

$$\sigma_r = 0.94\frac{Eu_0}{a}\sin 2\theta$$

$$\sigma_\theta = 2\left(-0.47 + 1.11\frac{r^2}{a^2}\right)\frac{Eu_0}{a}\cos 2\theta$$

$$\tau_{r\theta} = 2\left(0.47 - 0.555\frac{r^2}{a^2}\right)\frac{Eu_0}{a}\cos 2\theta$$

The maximum stress is

$$(\sigma_\theta)_{r=a,\theta=\pi/4} = 1.28\frac{Eu_0}{a}$$

6.7.2 Wedge Problems

This section is concerned with determining the stress distribution in wedges subjected to concentrated line loads at the apex and bending couples at the edges.

a. *Compressive line load at the apex*: Consider the problem of a symmetrical wedge compressed by a concentrated force P/unit length at the apex, as shown in Figure 6.17. The length of the wedge is along the z-axis and the cross section is in the xy-plane. The stress function, $\phi(r, \theta) = BP\ r\ \theta \sin\theta$, B = constant, derived in Equation 6.27, gives rise to stresses

$$\sigma_r = 2\frac{BP}{r}\cos\theta, \quad \sigma_\theta = \tau_{r\theta} = 0 \qquad \text{(a)}$$

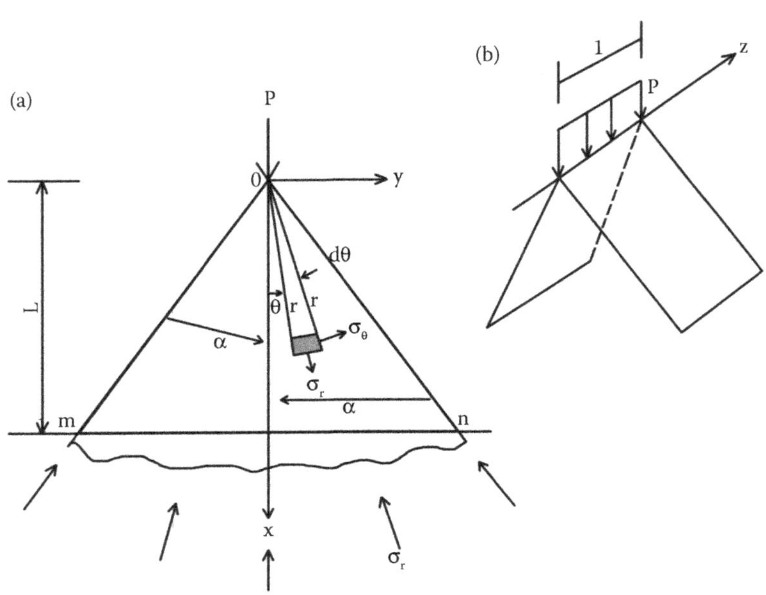

FIGURE 6.17
Load P at apex of a wedge. (a) Vertical section; (b) load on wedge.

which automatically satisfies the requirements that the faces $\theta = \pm\alpha$, $0 \leq r < \infty$ of the wedge are stress-free or $\sigma_\theta = \tau_{r\theta} = 0$; and at large distances, $r \to \infty$, $-\alpha \leq \theta \leq \alpha$, the radial stress must vanish. The constant B is found from the boundary condition that equilibrium is satisfied at point 0 (Figure 6.17a), which requires the following:

$$P + 2\int_0^\alpha (\sigma_r r d\theta)\cos\theta = 0 \tag{b}$$

Inserting the radial stress from Equation (a) into (b) and evaluating the integral, the constant B is found to be given

$$B = -\frac{1}{(2\alpha + \sin 2\alpha)}$$

Hence, the radial stress is

$$\sigma_r = -\frac{2P\cos\theta}{r(2\alpha + \sin 2\alpha)} \tag{6.42}$$

This solution is due to J. H. Michell [6.7]. The radial stress is maximum at $\theta = 0$ and minimum at $\theta = \pm\alpha$. At distant $x = L$ along the axis of symmetry of the wedge (section m–n in Figure 6.17a), we have $L = r\cos\theta$ and applying the transformation rule, the normal stress is obtained as

$$\sigma_x = \sigma_r \cos^2\theta = -\frac{P\cos^4\theta}{L(\alpha + \sin\alpha\cos\alpha)} \tag{c}$$

At section m–n, the distribution of the normal stresses is not uniform. The normal stress is maximum at $\theta = 0$ and minimum at $\theta = \pm\alpha$. The ratio of minimum to maximum normal stress is $\cos^4\alpha$.

The strength of material solution is $\sigma_x = -(P/2L\tan\alpha) = -(P/A)$, where A stands for the cross-sectional area.

For $\alpha = 15^\circ$ we have $(\sigma_x)_{strength} = -1.866(P/L)$ and $(\sigma_x)_{elasticity} = -1.954$ $(P\cos^4\theta/L)$.

The elasticity solution varies between 1.954 P/L for $\theta = 0$ and 0.871 P/L for $\theta = 15^\circ$. For small values of the angle α, the strength of material solution gives good results.

b. *Bending of a wedge by line load*: The previous solution can be modified to obtain the stresses in a wedge subjected to bending by line load F/unit length applied normal to the x-axis, as shown in Figure 6.18. We use the angle $\theta_1 = (\pi/2) - \theta$ and stress function $\phi(r, \theta) = BF\, r\, \theta_1 \sin\theta_1$ to obtain the stresses

$$\sigma_r = 2\frac{BF}{r}\cos\theta_1, \quad \sigma_\theta = \tau_{r\theta} = 0 \tag{a}$$

The constant B is found from satisfying equilibrium requirements at point 0 of the apex, namely, summation of forces in the vertical and horizontal directions must vanish.

$$\int_{\frac{\pi}{2}-\alpha}^{\frac{\pi}{2}+\alpha} (\sigma_r r d\theta_1)\cos\theta_1 + F = 0, \qquad \int_{\frac{\pi}{2}-\alpha}^{\frac{\pi}{2}+\alpha} (\sigma_r r d\theta_1)\sin\theta_1 = 0 \tag{b}$$

Inserting the radial stress in Equation (a) and evaluating the integrals, it is found that the second integral vanishes and the first integral yields the value of B as

$$B - \frac{1}{2(\alpha - \sin\alpha\cos\alpha)}$$

The stresses in the wedge in terms of the angle θ are

$$\sigma_r = -\frac{2F\sin\theta}{r(2\alpha - \sin 2\alpha)}, \quad \sigma_\theta = \tau_{r\theta} = 0 \tag{6.43}$$

Equation 6.43 indicates that if θ is negative the radial stress is tensile in that region as indicated in Figure 6.18. In order to compare the elasticity results in Equation 6.43 with the strength of materials solution, the stresses in rectangular coordinates at section m–n, distance x from the origin, are computed using the usual transformation rules and noting that $x = r\cos\theta$, $y = r\sin\theta$, and $r = (x^2 + y^2)^{1/2}$,

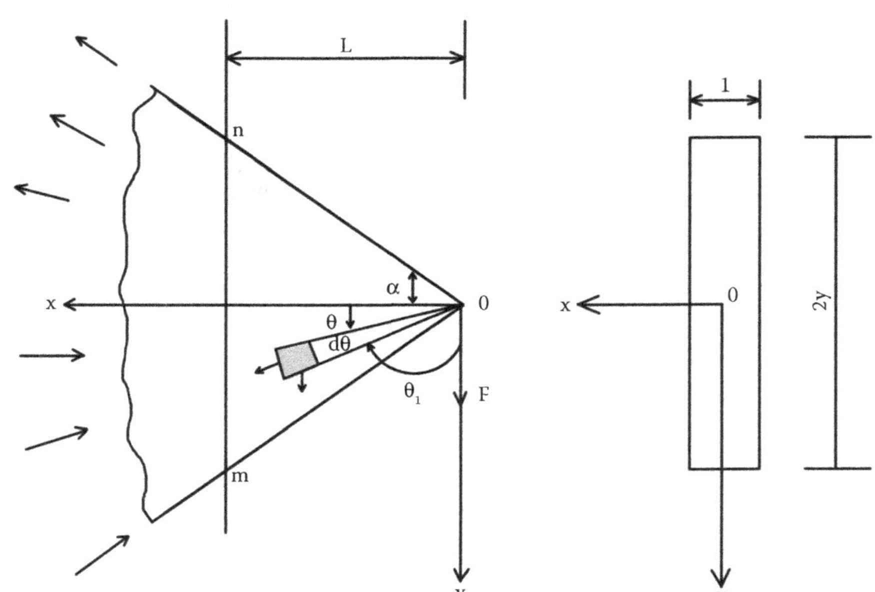

FIGURE 6.18
Bending of a wedge by force F.

the results are

$$\sigma_x = \sigma_r \cos^2\theta = -\left[\frac{2F}{(2\alpha - \sin 2\alpha)}\right]\frac{x^2y}{r^4}$$

$$\sigma_y = \sigma_r \sin^2\theta = -\left[\frac{2F}{(2\alpha - \sin 2\alpha)}\right]\frac{y^3}{r^4} \quad \text{(c)}$$

$$\tau_{xy} = \sigma_r \sin\theta\cos\theta = -\left[\frac{2F}{(2\alpha - \sin 2\alpha)}\right]\frac{xy^2}{r^4}$$

The moment of inertia of a rectangular cross section of the wedge at section m–n of unit thickness is $I = (2/3)y^3 = (2/3)x^3\tan^3\alpha$. For wedges with small angle 2α, the Sine term in the denominator of (f) can be expanded in series $\sin 2\alpha = 2\alpha - ((2\alpha)^3/3!) + ((2\alpha)^5/5!) - \cdots$, and keeping two terms of the series, we have $2\alpha - \sin 2\alpha = (4/3)\alpha^3$. The stresses σ_x and τ_{xy} in Equation (c) can then be written as

$$\sigma_x = -\frac{Fxy}{I}\left[\left(\frac{\tan\alpha}{\alpha}\right)^3\cos^4\theta\right]$$

$$\tau_{xy} = -\frac{Fy^2}{I}\left[\left(\frac{\tan\alpha}{\alpha}\right)^3\cos^4\theta\right] \quad \text{(d)}$$

For a wedge with small angle, the magnitude of the term in brackets is approximately unity and it appears that σ_x coincides with the elementary strength of material solution, σ_y is neglected in the strength of material solution, and the magnitude of the shear stress is twice the maximum value computed using the formula, $VQ/Ib = Fy^2/2I$. Also, the maximum shear stress occurs at points m and n of the cross section while in the elementary approach the maximum shear stress occurs at the neutral axis.

c. *Bending of a wedge by end couples*: Consider a wedge subjected to two equal and opposite moments M applied at the ends (Figure 6.19). Using the stress function,

$$\phi = A\sin 2\theta + C\theta, \quad \text{A and C are constants} \quad \text{(a)}$$

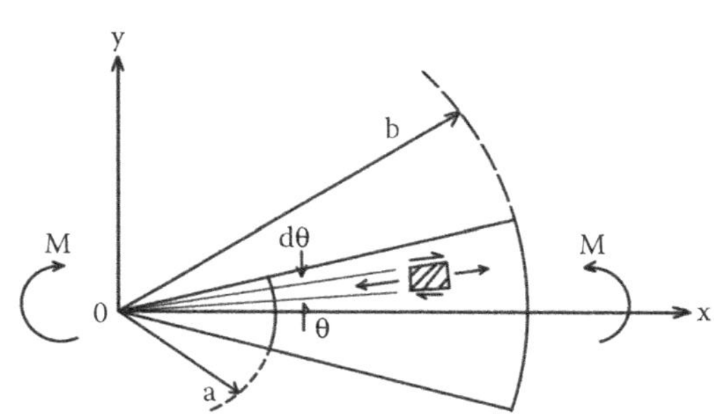

FIGURE 6.19
Wedge subjected to end moments.

The corresponding stresses are

$$\sigma_r = -\frac{4A}{r^2}\sin 2\theta, \quad \sigma_\theta = 0, \quad \text{and} \quad \tau_{r\theta} = (2A\cos 2\theta + C)/r^2 \tag{b}$$

The boundary conditions on the straight surfaces of the wedge, $\theta = \pm\alpha$, are

$$\tau_{r\theta} = \sigma_\theta = 0 \tag{c}$$

which gives, $C = -2\,A\cos 2\alpha$, and

$$\begin{aligned} \sigma_r &= -\frac{4A}{r^2}\sin 2\theta \\ \sigma_\theta &= 0 \\ \tau_{r\theta} &= \frac{2A}{r^2}(\cos 2\theta - \cos 2\alpha) \end{aligned} \tag{6.44a}$$

On the cylindrical surface of the wedge, the sum of the forces formed by the radial and shear stresses in Equation 6.44a along the x- and y-directions must vanish. This can be confirmed by observing that in the x-direction we have along any cylindrical surface of radius r

$$\int_{-\alpha}^{\alpha} (\sigma_r\cos\theta - \tau_{r\theta}\sin\theta)r d\theta = -\frac{A}{r}\int_{-\alpha}^{\alpha} \sin\theta(6\cos^2\theta - 1 - \cos 2\theta)d\theta \tag{d}$$

The integrand in Equation (d) is an odd function of θ, and hence it vanishes. Similarly, in the vertical direction, we have

$$\int_{-\alpha}^{\alpha} (\sigma_r\sin\theta + \tau_{r\theta}\cos\theta)r\, d\theta = \frac{2A}{r}\int_{0}^{\alpha} \cos\theta(1 - 6\sin^2\theta - \cos 2\alpha)d\theta = 0 \tag{e}$$

The remaining boundary condition, namely, the moment of the shear stresses at radius r of a cylindrical surface about the origin, determines the constant A in terms of the applied moment by means of the integral

$$M = \int_{-\alpha}^{\alpha} (\tau_{r\theta} r\, d\theta)r \tag{f}$$

Evaluating the integral using the shear stress in Equation 6.44a provides

$$A = \frac{M}{2(\sin 2\alpha - 2\alpha\cos 2\alpha)} \tag{6.44b}$$

And the stress distribution in the wedge is obtained from Equations 6.44a. Equation 6.44b indicates that the moment vanishes when $\sin 2\alpha - 2\alpha\cos 2\alpha = 0$, or $2\alpha = 257.4°$, implying that the contributions of the A and C terms to M are equal and opposite (see Equation (b) for the radial stress). In order for the solution to be appropriate, the denominator in Equation 6.44b cannot vanish for any value of angle α, otherwise, constant A becomes undefined. Thus, the usefulness of this solution is restricted to wedges of relatively small angles in which the effect of change of load distribution on surfaces r = a and r = b is localized. This issue has been discussed in detail in Reference 6.8.

6.7.3 Semi-Infinite Solid

In this section, we apply the solutions derived in the previous section to determine the stresses in a semi-infinite solid subjected to concentrated forces applied at a point of the straight boundary. We consider the three cases of wedge loading separately.

a. *Concentrated vertical load P per unit length*: Consider a line load consisting of a force P per unit length acting at point 0 of the horizontal boundary of a semi-infinite solid, as shown in Figure 6.20. The solution derived in Equation 6.42 with $\alpha = 90^\circ$ gives

$$\phi(r, \theta) = \frac{2\,Pr\cos\theta}{\pi}$$

and the corresponding stresses are

$$\sigma_r = -\frac{2P\cos\theta}{\pi r}, \quad \sigma_\theta = \tau_{r\theta} = 0 \tag{6.45}$$

These stresses satisfy the boundary conditions that the stresses diminish at large distances from point 0 and the horizontal surface ($\theta = \pm\alpha$) is stress-free. The stresses in Equation 6.45 are valid throughout the solid except at r = 0 at which the radial stress is infinity. Practically speaking, the line load occupies an area of small width and in that semicylindrical region of small radius shown dotted in Figure 6.20b plastic deformation will occur.

Consider a circle of diameter d on the x-axis and tangent to the y-axis then d $\cos\theta = 0A$ (Figure 6.17a). Using Equation 6.45, the radial stress at an element at point A, distance r from the origin, $\sigma_r = -(2P/\pi d)$, is constant at all points of the circle. This solution was first obtained by Alfred A. Flamant in 1892 [6.9].

The rectangular stresses at any point in the solid are

$$\sigma_x = \sigma_r\cos^2\theta = -\frac{2P\cos^3\theta}{\pi r} = -\frac{2P}{\pi}\frac{x^3}{(x^2 + y^2)^2}$$

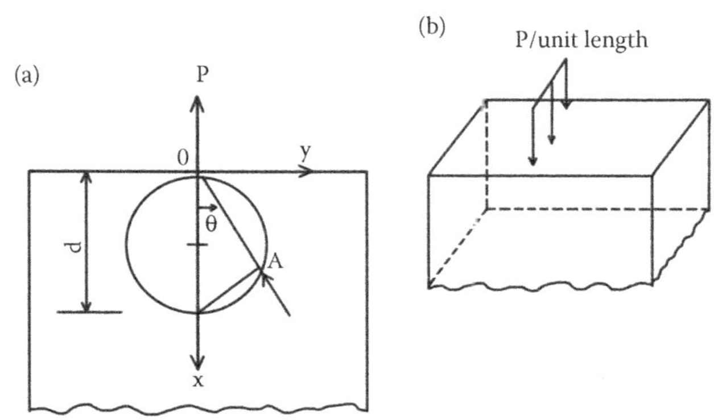

FIGURE 6.20
Concentrated force on a semi-infinite body. (a) Vertical section; (b) load on semi-infinite solid.

$$\sigma_y = \sigma_r \sin^2\theta = -\frac{2P}{\pi r}\cos\theta\sin^2\theta = -\frac{2P}{\pi}\frac{xy^2}{(x^2+y^2)^2} \tag{a}$$

$$\tau_{xy} = \sigma_r \sin\theta\cos\theta = -\frac{2P}{\pi r}\cos^2\theta\sin\theta = -\frac{2P}{\pi}\frac{x^2y}{(x^2+y^2)^2}$$

Distributed load: The above solution can be used to obtain the stresses due to a uniform load, q per unit length, distributed over a region of the horizontal surface of the semi-infinite solid bounded by angles θ_1 and θ_2 (Figure 6.21). Consider an element of the load q dy acting at distance y from the origin, as shown in Figure 6.21a, then $dP = q\ dy$ and from geometry of the element shown in Figure 6.21b, $dy\cos\theta = r\ d\theta$.

Using the result (Equation 6.45), it follows that

$$d\sigma_r = -\frac{2q\,dy}{\pi r}\cos\theta = -\frac{2q\,d\theta}{\pi}$$

Hence,

$$\sigma_r = \int_{\theta_1}^{\theta_2} -\frac{2q}{\pi}d\theta = -\frac{2q}{\pi}(\theta_2-\theta_1) \tag{6.46}$$

and the rectangular stresses are

$$\sigma_x = -\frac{2q}{\pi}\int_{\theta_1}^{\theta_2}\cos^2\theta\,d\theta = -\frac{q}{2\pi}[2(\theta_2-\theta_1)+\sin 2\theta_2 - \sin 2\theta_1]$$

$$\sigma_y = -\frac{2q}{\pi}\int_{\theta_1}^{\theta_2}\sin^2\theta\,d\theta = -\frac{q}{2\pi}[2(\theta_2-\theta_1)-(\sin 2\theta_2 - \sin 2\theta_1)] \tag{b}$$

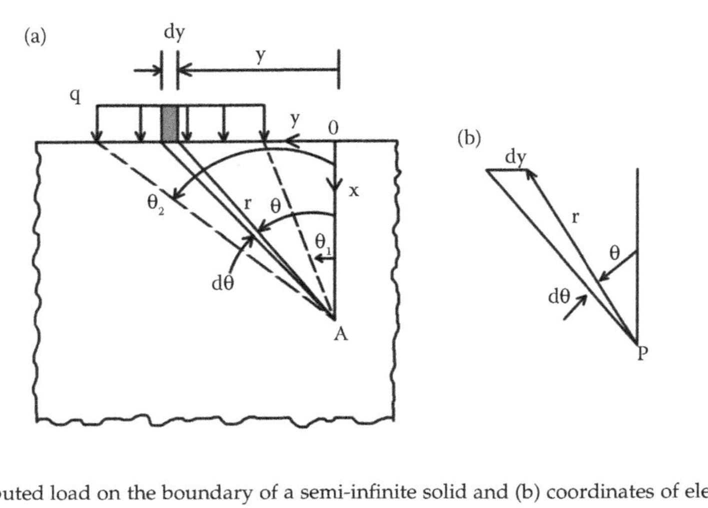

FIGURE 6.21
(a) Uniformly distributed load on the boundary of a semi-infinite solid and (b) coordinates of element dy.

$$\tau_{xy} = -\frac{2q}{\pi}\int_{\theta_1}^{\theta_2} \sin\theta\cos\theta\, d\theta = -\frac{q}{2\pi}[\cos 2\theta_2 - \cos 2\theta_1]$$

b. *Concentrated horizontal force F per unit length*: When the concentrated force is applied parallel to the horizontal surface at point 0, shown in Figure 6.22, the solution derived in Equation 6.43 with the wedge angle $\alpha = 90°$ and angle $\theta = (\pi/2) - \vartheta_1$ yields

$$\sigma_r = -\frac{2F}{\pi r}\cos\theta_1, \quad \sigma_\theta = \tau_{r\theta} = 0 \tag{6.47}$$

The tangential force F is balanced by the resultant forces acting on any cylindrical surface distance r from the origin, see the dotted line in Figure 6.22, as indicated by the integral

$$\int_0^\pi (\sigma_r \cos\theta_1) r\, d\theta_1 = -\frac{2F}{\pi}\int_0^\pi \cos^2\theta_1\, d\theta_1 = -F \tag{c}$$

The boundary conditions of the semi-infinite solid, namely, the stresses vanish at large distance from the origin and stress-free horizontal surface, are satisfied.

c. *Concentrated moment on the horizontal surface of semi-infinite solid*: Consider a semi-infinite solid subjected to concentrated moment M at the horizontal boundary of a semi-infinite solid, as shown in Figure 6.23. The moment can be thought of as the limiting action of two equal and opposite vertical forces, P, at a small distance apart.

Seeking a stress function similar to the one used in Section 6.7.2 with $\alpha = \pi/2$, the appropriate stress function for this case is as follows:

$$\phi = \frac{M}{\pi}(\theta + \sin\theta\cos\theta) \tag{a}$$

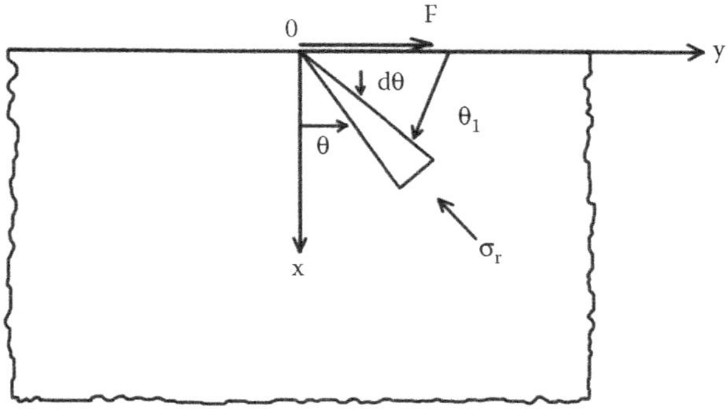

FIGURE 6.22
Horizontal force at the surface of a semi-infinite solid.

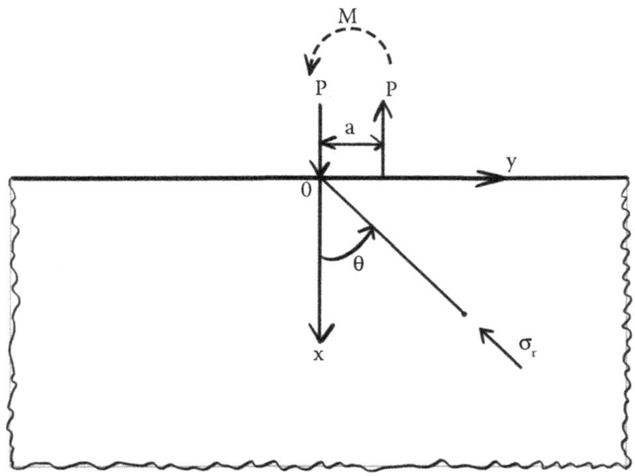

FIGURE 6.23
Moment acting on a semi-infinite solid.

And the corresponding stresses are

$$\sigma_r = -\frac{2M}{\pi}\frac{\sin 2\theta}{r^2}, \quad \sigma_\theta = 0, \quad \tau_{r\theta} = \frac{2M}{\pi}\frac{\cos^2\theta}{r^2} \tag{6.48}$$

An alternative way of determining the stress state is to consider the stress function of two equal and opposite vertical forces, P, at distance a apart, as shown in Figure 6.23. As was discussed in previous article, the stress function of load P at point 0 is $\phi = (P/\pi)r\theta\sin\theta = \phi(x, y)$ in rectangular coordinates, and the stress function of the load at point 0_1 is $\phi = -(P/\pi)r\theta\sin\theta$, which in rectangular coordinates is $-\phi(x, y + a)$.

Consider a new stress function

$$\phi_1 = -\phi(x, y + a) + \phi(x, y) \tag{b}$$

which in the limit as $a \to 0$, the combined stress function is

$$\phi_1 = -a\frac{\partial\phi}{\partial y} = -a\left[\frac{\partial\phi}{\partial r}\sin\theta + \frac{\cos\theta}{r}\frac{\partial\phi}{\partial\theta}\right] \tag{c}$$

Performing the derivatives in Equation (c) using the stress function of the tensile load ($-P$), the stress function (a) is recovered.

EXAMPLE 6.5

A tapered cantilever of unit thickness is supporting a uniformly distributed load w per unit length, as shown in Figure 6.24. The stress function that satisfies the boundary conditions on the upper and lower edges is

$$\phi(r, \theta) = Cr^2[\alpha - \theta + \sin\theta\cos\theta - \tan^2\alpha\cos^2\theta]$$

where C is a constant.

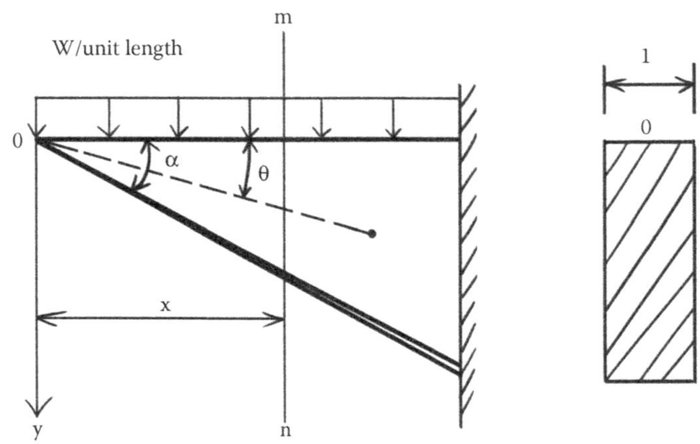

FIGURE 6.24
Example 6.5.

a. Find the stresses in the cantilever.
b. Compare the magnitude of the normal stress σ_x ($\theta = 0$, x) at a section distance x from the origin with that obtained using strength of materials solution when $\alpha = 10°$, $30°$, and $45°$.
   Determine the % error in each case.

Solution

First, we need to confirm that the given stress function satisfies the compatibility Equation 6.21. Differentiating the stress function, we have

$$\nabla^2\phi(r,\theta) = \frac{\partial^2\phi}{\partial r^2} + \frac{1}{r}\frac{\partial\phi}{\partial r} + \frac{1}{r^2}\frac{\partial^2\phi}{\partial\theta^2} = 2C[2(\alpha - \theta) - \tan\alpha]$$

It follows that

$$\nabla^2\nabla^2\phi = \nabla^4\phi = 0$$

a. The stresses in the cantilever are obtained using Equation 6.18 as follows:

$$\sigma_r = 2C[\alpha - \theta - \sin\theta\cos\theta - \tan\alpha\sin^2\theta]$$

$$\sigma_\theta = 2C[\alpha - \vartheta + \sin\theta\cos\theta - \tan\alpha\cos^2\theta]$$

$$\tau_{r\theta} = -2C[\sin^2\theta - \tan\alpha\sin\theta\cos\theta]$$

It is readily confirmed that the shear stress vanishes at the upper and lower edges of the cantilever ($\theta = 0$ and $\theta = \alpha$). The tangential stress vanishes at the lower edge, and at the upper edge, $\theta = 0$, we have $-w = 2C\ [\alpha - \tan\alpha]$, or

$$C = \frac{W}{2(\tan\alpha - \alpha)}$$

b. In order to determine σ_x in the cantilever at a section distance x from the origin, it is convenient to express the stress function in terms of rectangular coordinates x and y. The result is

$$\phi(x,y) = C\left[\alpha(x^2+y^2) - (x^2+y^2)\tan^{-1}\frac{y}{x} + xy - x^2\tan\alpha\right]$$

And the normal stress using elasticity theory is

$$\sigma_x = \frac{\partial^2\phi}{\partial y^2} = \frac{w}{\tan\alpha - \alpha}[\alpha - \theta - \sin\theta\cos\theta] \qquad \text{(a)}$$

The normal stress using the strength of materials solution is obtained from the formula $\sigma_x = (My)/I$, where $M = -wx^2/2$, $I = (1/12)\ x^3\tan^3\alpha$, and y is replaced by $y - (x/2)\tan\alpha$. Thus,

$$\sigma_x = \frac{3w}{\cos\theta\tan^3\alpha}(\cos\theta\tan\alpha - 2\sin\theta) \qquad \text{(b)}$$

For $\theta = 0$, the normal stresses in Equations (a) and (b) reduce to

$$\frac{\sigma_x}{w}(\text{elasticity}) = \frac{\alpha}{\tan\alpha - \alpha}$$

$$\alpha = 10^\circ : \frac{\sigma_x}{w} = 96.961$$

$$\alpha = 30^\circ : \frac{\sigma_x}{w} = 9.741$$

$$\alpha = 45^\circ : \frac{\sigma_x}{w} = 3.66$$

and

$$\frac{\sigma_x}{w}(\text{strength}) = \frac{3}{\tan^2\alpha}$$

$$\alpha = 10^\circ : \frac{\sigma_x}{w} = 96.49$$

$$\alpha = 30^\circ : \frac{\sigma_x}{w} = 9$$

$$\alpha = 45^\circ : \frac{\sigma_x}{w} = 3$$

The percentage errors are 0.49, 8.22, 22 (strength of materials magnitudes are lower).

6.8 Rotating Circular Disks

In this section, we examine several stress distributions in circular disks rotating with constant angular speed ω about a normal axis through the center. The angular speed is usually measured by radian per second or by revolution per minute (RPM). Solid and annular disks of constant and variable thicknesses are considered. The thickness of the disk is assumed small compared to its radius so that any variation in the radial and tangential stresses across the thickness is negligible. The stress state in such disks is a plane stress state with axial symmetry such that $\tau_{r\theta} = \sigma_z = 0$, and σ_r and σ_θ are independent of θ. Neglecting the weight of the disk, the rotation induces centrifugal inertia force, which is included in the equilibrium equation as body force per unit volume with components

$$F_r = \rho\omega^2 r, \quad F_\theta = 0 \qquad \text{(a)}$$

Here, ρ is the mass density of the disk. Because of axial symmetry, equilibrium of the disk is reduced to the following equation:

$$\frac{d\sigma_r}{dr} + \frac{\sigma_r - \sigma_\theta}{r} + \rho\omega^2 r = 0 \tag{6.49}$$

The elastic stress–strain relations for plane stress are

$$\begin{aligned} \varepsilon_r &= \frac{du}{dr} = \frac{1}{E}(\sigma_r - \upsilon\sigma_\theta) \\ \varepsilon_\theta &= \frac{u}{r} = \frac{1}{E}(\sigma_\theta - \upsilon\sigma_r) \end{aligned} \tag{b}$$

Eliminating the radial displacement from Equations (b), the compatibility equation in terms of stresses is obtained

$$\frac{d}{dr}(\sigma_r + \sigma_\theta) + (1 + \upsilon)\rho\omega^2 r = 0 \tag{6.50}$$

Integrating Equation 6.50 yields

$$\sigma_r + \sigma_\theta = 2A - \frac{1 + \upsilon}{2}\rho\omega^2 r^2 \tag{c}$$

where 2A is constant. Multiplying Equation 6.49 by r and adding the result to Equation (c), the tangential stress is cancelled. Multiplying the resulting equation by r furnishes the following differential equation:

$$\frac{d}{dr}(r^2\sigma_r) = 2Ar - \frac{3 + \upsilon}{2}\rho\omega^2 r^3 \tag{d}$$

Integrating Equation (d) gives the radial stress, and relation (c) provides the tangential stress. The results are

$$\begin{aligned} \sigma_r &= A + \frac{B}{r^2} - \frac{3 + \upsilon}{8}\rho\omega^2 r^2 \\ \sigma_\theta &= A - \frac{B}{r^2} - \frac{1 + 3\upsilon}{8}\rho\omega^2 r^2 \end{aligned} \tag{6.51}$$

where A, B are arbitrary constants to be determined from the boundary conditions.

General expression for u can be obtained by solving for the radial and tangential stresses in Equation (b). The result is

$$\begin{aligned} \sigma_r &= \frac{E}{1 - \upsilon^2}\left(\frac{du}{dr} + \upsilon\frac{u}{r}\right) \\ \sigma_\theta &= \frac{E}{1 - \upsilon^2}\left(\upsilon\frac{du}{dr} + \frac{u}{r}\right) \end{aligned} \tag{e}$$

Inserting Equation (e) into 6.49 and simplifying the algebra, the following equation is obtained for the radial displacement u:

$$r^2\frac{d^2u}{dr^2} + r\frac{du}{dr} - u = -\frac{1-\upsilon^2}{E}\rho\omega^2 r^3 \tag{f}$$

Equation (f) is an equidimensional ordinary differential equation whose solution consists of homogeneous and particular parts as follows:

$$u(r) = C_1 r + \frac{C_2}{r} - \frac{1-\upsilon^2}{8E}\rho\omega^2 r^3 \tag{g}$$

The constants C_1 and C_2 are determined from the boundary conditions at the edges of the disk. Inserting u(r) into Equations (e), Equations 6.51 are basically recovered.

6.8.1 Solid Disk

A solid disk of radius a, thickness t is rotating about a normal axis with gradually increasing angular speed ω (Figure 6.25). In order to exclude infinite stresses at $r = 0$ (regulatory condition), the constant B in Equation 6.51 must vanish, and A is found from the boundary condition that $\sigma_r = 0$ at $r = b$ giving $A = ((3+\upsilon)/8)\rho\omega^2 a^2$. Hence, the elastic stresses are

$$\begin{aligned}
\sigma_r &= \frac{3+\upsilon}{8}\rho\omega^2(a^2 - r^2) \\
\sigma_\theta &= \frac{\rho\omega^2}{8}[(3+\upsilon)a^2 - (1+3\upsilon)r^2], \quad 0 \le r \le a, \\
\sigma_z &= 0
\end{aligned} \tag{6.52}$$

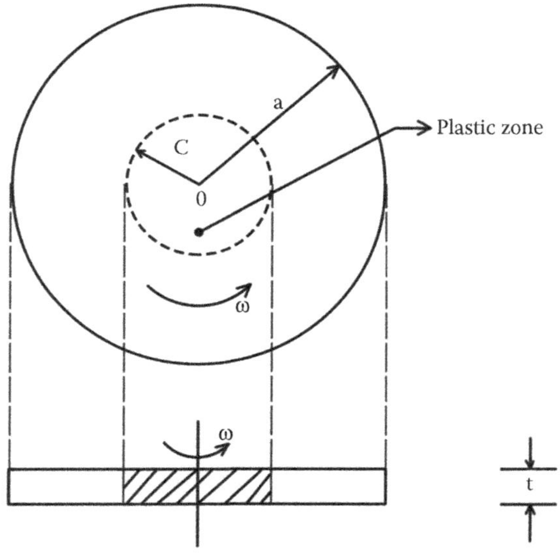

FIGURE 6.25
Thin solid disk rotating with angular velocity ω.

and the radial displacement is

$$u = \frac{1-\upsilon}{8E}\rho\omega^2 r[(3+\upsilon)a^2 - (1+\upsilon)r^2] \tag{6.53}$$

The radial and tangential stresses in Equation 6.52 are both positive (tensile) with $\sigma_\theta \geq \sigma_r$. The maximum stress occurs at $r = 0$, where

$$(\sigma_r)_{max} = (\sigma_\theta)_{max} = \frac{3+\upsilon}{8}\rho\omega^2 a^2 = \frac{3+\upsilon}{8}\rho V^2 \tag{a}$$

where $V = a\omega$ is the tangential velocity of the disk.

Assuming elastic–perfectly plastic material with yield stress σ_0, applying either maximum shear stress or maximum energy of distortion yield criterion, we get $\sigma_\theta = \sigma_0$, and the onset of yielding occurs when the speed of rotation, ω_e, is given by

$$\rho\omega_e^2 a^2 = \frac{8\sigma_0}{3+\upsilon} \tag{6.54}$$

Plastic analysis: When the speed of rotation is increased beyond the maximum elastic value given in Equation 6.54, a circular plastic zone, say of radius $r = c$, is formed around the origin. It is surrounded by an elastic region $c \leq r \leq a$ (Figure 6.25).

In the plastic zone, $0 \leq r \leq c$, the stresses are required to satisfy the equilibrium, Equation 6.49, and Tresca's yield criterion, $\sigma_\theta = \sigma_0$. Equation 6.49 is equivalent to

$$\frac{d}{dr}(r\sigma_r) = \sigma_0 - \rho\omega^2 r^2$$

Integrating this equation with respect to r leads to

$$\sigma_r = \sigma_0 - \frac{\rho\omega^2 r^2}{3} + \frac{A}{r} \tag{b}$$

Here, A is a constant of integration. Because the radial stress has to be finite at $r = 0$, A must vanish in case of solid disk, and the stresses in the plastic region are

$$\begin{aligned} \sigma_r &= \sigma_0 - \frac{\rho\omega^2 r^2}{3}, \quad 0 \leq r \leq c \\ \sigma_\theta &= \sigma_0 \end{aligned} \tag{6.55}$$

In the elastic region, $c \leq r \leq a$, the stresses are obtained from Equations 6.51 subject to satisfying the boundary conditions, namely, the radial stress vanishes at $r = a$ and is continuous at the elastic/plastic interface, $r = c$. The condition at $r = a$ gives $A = -(B/a^2) + ((3 + \upsilon)/8)\rho\omega^2 a^2$, and the stresses in the elastic region in terms of the constant B assume the form

$$\begin{aligned} \sigma_r &= \frac{B}{a^2}\left(\frac{a^2}{r^2} - 1\right) + \frac{3+\upsilon}{8}\rho\omega^2 a^2\left(1 - \frac{r^2}{a^2}\right) \\ \sigma_\theta &= -\frac{B}{a^2}\left(\frac{a^2}{r^2} + 1\right) + \frac{\rho\omega^2 a^2}{8}\left[3 + \upsilon - (1 + 3\upsilon)\frac{r^2}{a^2}\right] \end{aligned}, \quad c \leq r \leq a \tag{c}$$

The continuity of the radial stress at $r = c$ requires equating the radial stresses from Equations 6.55 and (c) at $r = c$. Hence,

$$\sigma_0 - \frac{\rho\omega^2 c^2}{3} = B\left(\frac{1}{c^2} - \frac{1}{a^2}\right) + \frac{3+\upsilon}{8}\rho\omega^2(a^2 - c^2) \tag{d}$$

Equating the tangential stresses from Equations 6.55 and (c) at $r = c$ provides

$$\sigma_0 = -B\left(\frac{1}{c^2} + \frac{1}{a^2}\right) + \frac{\rho\omega^2}{8}[(3+\upsilon)a^2 + (1+3\upsilon)c^2] \tag{e}$$

Solving Equations (d) and (e) simultaneously gives the magnitude of constant $B = -((1+3\upsilon)/24)\rho\omega^2 c^4$ and the following relation between the angular speed and extent of the plastic zone:

$$\rho\omega^2 a^2 = \frac{3\sigma_0}{1 + \frac{1+3\upsilon}{8}\left(1 - \frac{c^2}{a^2}\right)^2} \tag{6.56}$$

The stresses in the elastic region, from Equations (c), become

$$\left.\begin{aligned} \sigma_r &= \frac{\rho\omega^2 a^2}{8}\left[3 + \upsilon - \frac{1+3\upsilon}{3}\left(\frac{c^4}{a^2 r^2}\right)\right]\left(1 - \frac{r^2}{a^2}\right) \\ \sigma_\theta &= \frac{\rho\omega^2 a^2}{8}\left\{3 + \upsilon + (1+3\upsilon)\left[-\frac{r^2}{a^2} + \frac{c^4}{3a^4}\left(1 + \frac{a^2}{r^2}\right)\right]\right\} \end{aligned}\right\}, \quad c \le r \le a \tag{6.57}$$

The displacement in the elastic region is readily computed from the relation $Eu = r(\sigma_\theta - \upsilon\sigma_r)$, where the stresses are given by Equation 6.57. In the plastic region, numerical methods are usually needed to find the displacement; further information is available in Reference 2.2. Note that when $c = 0$, Equation 6.56 gives the speed that initiates yield derived in Equation 6.54 and the stresses in Equation 6.57 reduce to those in Equation 6.52.

The critical value of the rotational speed, ω_u, is reached when the boundary of the plastic region reaches the edge of the disk, $c = a$, and Equation 6.56 gives

$$\rho\omega_u^2 a^2 = 3\sigma_0 \tag{6.58}$$

The critical speed in Equation 6.58 could have been derived from Equation 6.55 by setting $r = c = a$, and making use of the condition $\sigma_r = 0$. It is worth observing that the critical rotational speed does not depend on Poisson's ratio of the material. For $\upsilon = 0.3$, the ratio $\omega_u/\omega_e = 1.11$.

When $\omega = \omega_u$, using Equations 6.55 and 6.58, the stresses in the disk at the critical speed are

$$\left.\begin{aligned} \sigma_r &= \sigma_0\left(1 - \frac{r^2}{a^2}\right) \\ \sigma_\theta &= \sigma_0 \\ \sigma_z &= 0 \end{aligned}\right\}, \quad 0 \le r \le a \tag{6.59}$$

EXAMPLE 6.6

A steel solid disk of radius 150 mm is rotating at a speed equal to $(1/2)(\omega_e + \omega_u)$, find the tangential velocity of the disk and extent of the plastic zone. For steel, mass density = 7850 kg/m^3, yield stress = 250 MPa and Poisson's ration = 0.3.

Solution

The rotational speed that initiates yield in the disk, given by Equation 6.54, is $\rho\omega^2 a^2 = 8\sigma_0/(3 + \upsilon)$, that is

$$\omega_e^2 = \frac{8\left(250 \times 10^6 \dfrac{\text{N}}{\text{m}^2}\right)}{3.3\left(7850 \dfrac{\text{kg}}{\text{m}^3}\right)(0.15\,\text{m})^2} = (3.431) \times 10^6 \frac{\text{rad}}{\text{s}^2}$$

Or ω_e = 1852 rad/s (17,785 RPM).
Similarly, the critical speed is obtained from Equation 6.58

$$\omega_u^2 = \frac{3 \times 250 \times 10^6 \dfrac{\text{N}}{\text{m}^2}}{\left(7850 \dfrac{\text{kg}}{\text{m}^3}\right)(0.15\,\text{m})^2} = 4.227 \times 10^6 \frac{\text{rad}}{\text{s}^2}$$

giving ω_u = 2061 rad/s (19,689 RPM).
Hence, the disk is rotating with a speed = 0.5(1852 + 2061) = 1954 rad/s.
The tangential velocity of the disk is $V = a\omega$ = (0.15 m) (1954 rad/s) = 293 m/s.
The extent of the plastic zone is determined from Equation 6.56.

$$\left[1 - \frac{c^2}{(0.15\,\text{m})^2}\right]^2 = \frac{8}{1.9}\left[\frac{3\left(250 \times 10^6 \dfrac{\text{N}}{\text{m}^2}\right)}{\left(7850 \dfrac{\text{kg}}{\text{m}^3}\right)\left(1954 \dfrac{\text{rad}}{\text{s}}\right)^2 (0.15\,\text{m})^2} - 1\right]$$

which gives extent of the plastic zone, c = 85 mm.

6.8.2 Annular Disk

Consider an annular disk of internal and external radii, a, b, respectively, rotating about the normal axis (Figure 6.26). The stresses are obtained from Equations 6.51 in which the constants A and B are determined subject to satisfying the boundary conditions, $\sigma_r = 0$ at $r = a$ and $r = b$, giving

$$A = \frac{3 + \upsilon}{8}\rho\omega^2(a^2 + b^2), \quad B = -\frac{3 + \upsilon}{8}\rho\omega^2 a^2 b^2$$

It follows that the elastic stresses are obtained as follows:

$$\begin{aligned} \sigma_r &= \frac{3 + \upsilon}{8}\rho\omega^2\left(a^2 + b^2 - \frac{a^2b^2}{r^2} - r^2\right) \\ \sigma_\theta &= \frac{3 + \upsilon}{8}\rho\omega^2\left(a^2 + b^2 + \frac{a^2b^2}{r^2} - \frac{1 + 3\upsilon}{3 + \upsilon}r^2\right) \\ \sigma_z &= 0 \end{aligned} \tag{6.60}$$

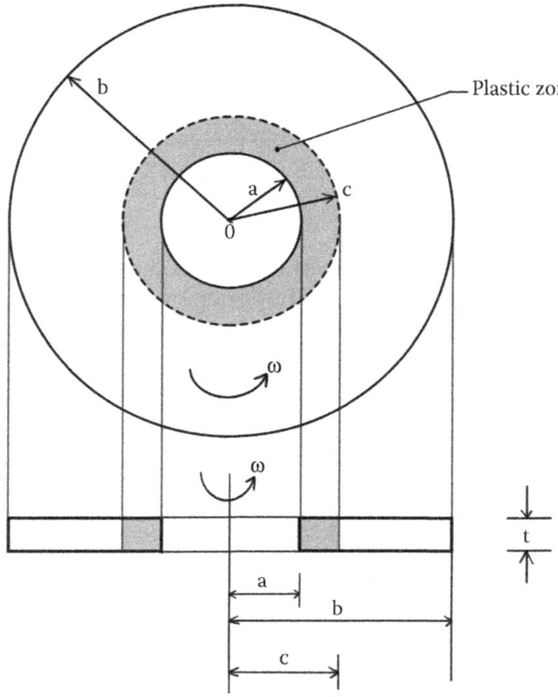

FIGURE 6.26
Annular disk rotating with angular velocity ω.

Since the disk is in plane stress condition, the radial displacement, $u = (r/E)(\sigma_\theta - \upsilon\sigma_r)$, is given by

$$u = \frac{(3+\upsilon)(1-\upsilon)}{8E}\rho\omega^2 r\left[a^2 + b^2 - \left(\frac{1+\upsilon}{3+\upsilon}\right)r^2 + \left(\frac{1+\upsilon}{1-\upsilon}\right)\frac{a^2b^2}{r^2}\right] \tag{6.61}$$

Examining the variations of the stresses with r, the radial stress becomes maximum when $d\sigma_r/dr = 0$, giving $r = (ab)^{1/2}$, and

$$(\sigma_r)_{max} = \frac{3+\upsilon}{8}\rho\omega^2(b-a)^2 \tag{a}$$

The maximum hoop stress occurs at the inner boundary, $r = a$, where

$$(\sigma_\theta)_{max} = \frac{3+\upsilon}{4}\rho\omega^2\left(b^2 + \frac{1-\upsilon}{3+\upsilon}a^2\right) \tag{b}$$

It is observed that by letting $a = 0$ in Equation (b), the maximum hoop stress is double the amount in the solid disk. This is because of the presence of stress concentration at the boundary of the hole. The phenomenon of stress concentration around holes will be discussed in detail in Chapter 7. The maximum tangential stress is greater than that of the radial stress.

Using Tresca's yield criterion, the onset of yield occurs at $r = a$, and the corresponding rotational speed is

$$\rho\omega_e^2 b^2 = \frac{4\sigma_0}{\left[3 + \upsilon + (1 - \upsilon)\frac{a^2}{b^2}\right]} \tag{6.62}$$

Plastic analysis: When the rotational speed is increased beyond the elastic limit, a plastic region, say of radius c, is formed around the inner boundary (Figure 6.26). The stresses in the plastic region are computed in a similar manner to that used in case of the solid disk, that is, the stresses must satisfy equilibrium, Equation 6.49, and Tresca's yield criterion. Following the procedure used in the solid disk, in this case the constant of integration is found from the condition that $\sigma_r = 0$ at $r = a$, the stresses in the plastic region of the annular disk become

$$\begin{aligned} \sigma_r &= \left(1 - \frac{a}{r}\right)\sigma_0 + \frac{1}{3}\rho\omega^2 r^2\left(\frac{a^3}{r^3} - 1\right), \quad a \leq r \leq c \\ \sigma_\theta &= \sigma_0 \end{aligned} \tag{6.63}$$

In the elastic region, $c \leq r \leq b$, the basic stress expressions in Equations 6.51 are applicable. Applying condition $\sigma_r = 0$ at $r = b$ to Equations 6.51, furnish

$$\begin{aligned} \sigma_r &= \frac{B}{r^2} - \frac{B}{b^2} + \frac{3 + \upsilon}{8}\rho\omega^2(b^2 - r^2) \\ \sigma_\theta &= -\left(\frac{B}{r^2} + \frac{B}{b^2}\right) + \frac{\rho\omega^2}{8}[(3 + \upsilon)b^2 - (1 + 3\upsilon)r^2] \end{aligned}, \quad c < r < b \tag{c}$$

The constant B is determined by requiring the radial stress to be continuous across the elastic/plastic boundary at $r = c$, that is

$$\left(1 - \frac{a}{c}\right)\sigma_0 + \frac{\rho\omega^2 c^2}{3}\left(\frac{a^2}{c^2} - 1\right) = B\left(\frac{1}{c^2} - \frac{1}{b^2}\right) + \frac{3 + \upsilon}{8}\rho\omega^2(b^2 - c^2) \tag{d}$$

Also, equating the tangential stresses at $r = c$ from Equations 6.63 and (c) gives

$$\sigma_0 = -B\left(\frac{1}{c^2} + \frac{1}{b^2}\right) + \frac{\rho\omega^2}{8}[(3 + \upsilon)b^2 - (1 + 3\upsilon)c^2] \tag{e}$$

Equations (d) and (e) provide the value of constant B

$$B = -\frac{ac\sigma_0}{2} - \frac{\rho\omega^2 c^4}{24}\left(1 + 3\upsilon - 4\frac{a^3}{c^3}\right) \tag{f}$$

and the following relation between the rotational speed ω and c:

$$\frac{\rho\omega^2 b^2}{3\sigma_0} = \frac{2 - \frac{a}{c}\left(1 + \frac{c^2}{b^2}\right)}{2 + \frac{1+3\upsilon}{4}\left(1 - \frac{c^2}{b^2}\right)^2 - \frac{a^3}{b^2 c}\left(1 + \frac{c^2}{b^2}\right)} \tag{6.64}$$

Notice that we used identical procedure to derive the corresponding relation between ω and c for the solid disk (Equation 6.56). When c approaches the inner boundary, $c = a$, the rotational speed in Equation 6.64 reduces to the value given in Equation 6.62 for the speed causing the initiation of yield. Also, when $a = 0$, Equation 6.64 reduces to the corresponding relation between ω and c for the solid disk (Equation 6.56).

As c approaches the outer boundary, $c = b$, the ultimate rotational speed is obtained

$$\frac{\rho\omega_u^2 b^2}{\sigma_0} = \frac{3}{1 + \frac{a}{b} + \frac{a^2}{b^2}} \tag{6.65}$$

It is worth noting that the critical speed does not depend on Poisson's ratio. Note that formula (6.65) for the critical rotational speed could have been obtained from relation (6.64) by setting $c = b$, and from Equation 6.63 by letting $r = c = b$ and $\sigma_r = 0$. In general, Equation 6.64 has to be solved numerically to determine the extent of the plastic region for a specified rotational speed in the range $\omega_e \le \omega \le \omega_u$.

In a fully plastic annular disk, the stresses are obtained from relations (6.63) by setting $c = b$ and $\omega = \omega_u$

$$\sigma_r = \left(1 - \frac{a}{r}\right)\left[1 - \frac{\left(1 - \frac{a}{b}\right)\left(1 + \frac{a}{r} + \frac{a^2}{r^2}\right)\frac{r^2}{b^2}}{1 - \frac{a^3}{b^3}}\right]\sigma_0, \quad a \le r \le b \tag{6.66}$$

$$\sigma_\theta = \sigma_0; \sigma_z = 0$$

Methods of determining the radial displacement in the elastic and plastic stages are available in Reference 2.2.

EXAMPLE 6.7

Consider an annular disk of internal radius $a = 0.06$ m and external radius $b = 0.24$ m rotating about its normal axis. The disk is made of elastic–perfectly plastic steel with Poisson's ratio $\upsilon = 0.3$, yield stress = 250 MPa, and mass density $\rho = 7800$ kg/m^3.

a. If the disk rotates at an angular velocity of 7500 RPM, find the largest stress.
b. Find the extent of the plastic zone when the rotational speed of the disk is equal to $(1/2)(\omega_e + \omega_u)$.

Solution

First, we determine ω_e and ω_u. We have $a/b = 0.25$, $\upsilon = 0.3$, and mass density = 7800 kg/m^3. Using Equations 6.62 and 6.65, it is readily confirmed that $\omega_e = 815.9$ rad/s (7791 RPM) and $\omega_u = 1127.8$ rad/s (10,769 RPM).

a. At $\omega = 7500$ RPM $= 785.4$ rad/s, the material is still elastic, and Equations (a) and (b) in Section 6.8.2 provide the following results:

$$(\sigma_r)_{max} = 13.365 \times 10^{-3}\ \rho\omega^2 = 13.365 \times 10^{-3}\ (7800)\ (785.4)^2 = 64.31\ \text{MPa}$$
$$(\sigma_\theta)_{max} = 4.815 \times 10^{-3}\ \rho\omega^2 = 4.815 \times 10^{-3}\ (7800)\ (785.4)^2 = 231.67\ \text{MPa}$$

b. In order to find the extent of the plastic zone when the disk rotates at a speed of $(815.9 + 1127.8)/2 = 971.83$ rad/s, Equation 6.64 furnishes the following relation:

$$565.77 \times 10^{-3} = \frac{2 - \frac{a}{c} - 0.25\frac{c}{b}}{2 + 0.475\left(1 - \frac{c^2}{b^2}\right)^2 - 62.5 \times 10^{-3}\left(\frac{a}{c}\right) - 15.6 \times 10^{-3}\left(\frac{c}{b}\right)}$$

which needs to be programmed and solved numerically. The result is approximately $c = 0.10$ m.

6.8.3 Angular Rotation of a Thin Ring or Cylinder

A thin ring or cylinder of radius r is rotating about a central axis with constant angular velocity ω, as shown in Figure 6.27a. The centrifugal effect of the mass causes radial pressure p per unit length of the circumference given by

$$P = m\omega^2 r \tag{6.67}$$

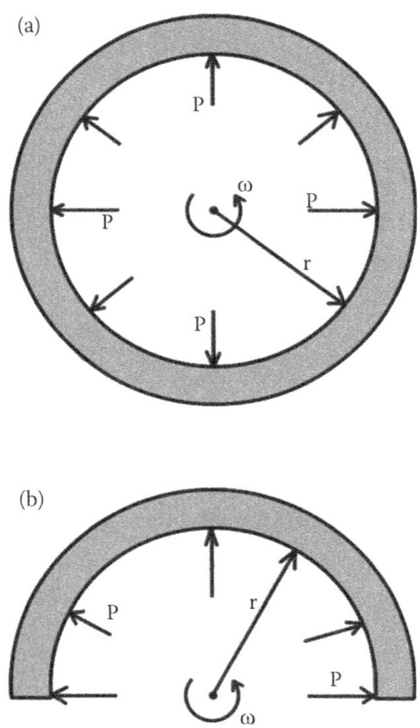

FIGURE 6.27
Thin ring (cylinder) rotating with angular velocity ω. (a) Thin ring (cylinder); (b) semicircular ring.

Equilibrium of half of the ring, shown in Figure 6.27b, gives 2F = p (2r), that is the hoop tension due to the rotation is given by

$$F = pr = m\omega^2 r^2 \tag{a}$$

The thickness of the ring is assumed small so that the centrifugal effect is constant across the wall thickness. Also, F = mass × acceleration = $m\omega^2 r \times r$. This tension is transmitted through the complete circumference of the ring and is resisted by the complete cross-sectional area, so if the cross-sectional area is A, then we have

$$\text{Hoop stress} = \frac{F}{A} = \frac{m\omega^2 r^2}{A} \tag{b}$$

Since m/A is the mass of the material per unit volume or density ρ, and the above calculations are based on unit length, the hoop stress in Equation (b) is equivalent to $\rho\omega^2 r^2$.

6.8.4 Disks of Variable Thickness

In previous sections, it was observed that the onset of yield in rotating disks occurs at fibers close to the axis of rotation. Thus, in order to reduce the weight of the disk and obtain lower rotational velocities, disks with variable thicknesses are used, usually the thickness of the disk decreases gradually from the periphery near the axis of rotation to the outer boundary. Disks with variable thicknesses are also used to achieve the constant stress conditions in the disk.

Consider an annular disk with internal and external radii, a, b, respectively. The thickness of the disk is a function of r, say t(r), assumed to vary from t_0 at the inner boundary to t_1 at the outside (Figure 6.28). Assuming that the stresses do not vary across the thickness of the

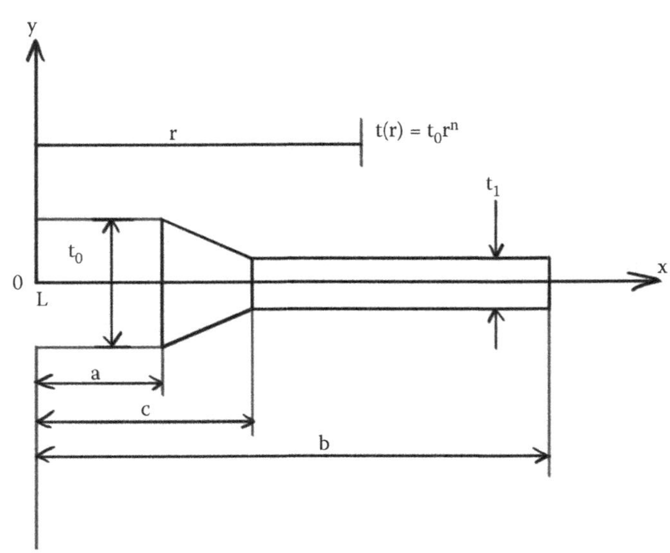

FIGURE 6.28
Disk of variable thickness t(r).

disk, the equation of equilibrium of an element of the disk takes the form (see Figure 6.12 and Equation 6.49a)

$$\frac{d}{dr}(t r \sigma_r) - t\sigma_\theta + t\rho\omega^2 r^2 = 0 \tag{6.68}$$

The compatibility equation of axially symmetric strains is

$$r\frac{d\varepsilon_\theta}{dr} + \varepsilon_\theta - \varepsilon_r = 0 \tag{a}$$

In term of stresses, Equation 6.23a, the compatibility equation becomes

$$r\frac{d\sigma_\theta}{dr} - \upsilon r\frac{d\sigma_r}{dr} + (1+\upsilon)(\sigma_\theta - \sigma_r) = 0 \tag{b}$$

Using Equation 6.68 to eliminate the tangential stress in Equation (b) results in the following differential equation:

$$r^2\frac{d^2\sigma_r}{dr^2} + \left(3 + \frac{r}{t}\frac{dt}{dr}\right) r\frac{d\sigma_r}{dr} + \left[r\frac{d}{dt}\left(\frac{r}{t}\frac{dt}{dr}\right) + (1+\upsilon)\left(\frac{r}{t}\frac{dt}{dr}\right)\right]\sigma_r = -(3+\upsilon)\rho\omega^2 r^2 \tag{6.69}$$

Equation 6.69 can be easily integrated when the thickness is

$$t(r) = t_0 r^n \tag{6.70}$$

Here, n is a constant, usually taken as negative to provide thicker profile near the axis of rotation, and t_0, also constant, is the thickness at $r = \text{unity}$.

Inserting Equation 6.70 into Equation 6.69, the following equation is obtained:

$$r^2\frac{d^2\sigma_r}{dr^2} + (3+n) r\frac{d\sigma_r}{dr} + (1+\upsilon) n\sigma_r = -(3+\upsilon)\rho\omega^2 r^2 \tag{6.71}$$

Equation 6.71 is an equidimensional differential equation (which can be reduced to an ordinary differential equation with constant coefficients by introducing a new variable, $r = e^z$). It turns out that the homogenous part of Equation 6.71 has a solution of the form, $\sigma_r = r^k$, where k is a constant giving the two roots of the characteristic equation as follows:

$$\begin{aligned}
&k^2 + (2+n)k + (1+\upsilon)n = 0 \\
&k_1 = -1 - n/2 + \left[\left(\frac{n}{2}\right)^2 + 1 - n\upsilon\right]^{\frac{1}{2}} \\
&k_2 = -1 - n/2 - \left[\left(\frac{n}{2}\right)^2 + 1 - n\upsilon\right]^{\frac{1}{2}}
\end{aligned} \tag{6.72}$$

Thus, the solution of the homogeneous part of Equation 6.71 is $\sigma_r = A\, r_1^k + B\, r_2^k$, where A and B are constants. The particular solution of Equation 6.71 is

$$(\sigma_r)_{part} = -\frac{(3+\upsilon)\rho\omega^2 r^2}{8+(3+\upsilon)n} \tag{c}$$

Hence, the stresses in the disk are

$$\begin{aligned} \sigma_r &= Ar^{k_1} + Br^{k_2} - \frac{(3+\upsilon)\rho\omega^2 r^2}{8+(3+\upsilon)n} \\ \sigma_\theta &= (1+n+k_1)Ar^{k_1} + (1+n+k_2)Br^{k_2} - \frac{(1+3\upsilon)\rho\omega^2 r^2}{8+(3+\upsilon)n} \end{aligned} \tag{6.73}$$

where the constants A and B are determined from the conditions that $\sigma_r = 0$ at $r = a$, and at $r = b$. The results are

$$\begin{aligned} A &= \frac{(3+\upsilon)\rho\omega^2}{8+(3+\upsilon)n} a^{2-k_1} - Ba^{k_2-k_1} \\ B &= \frac{(3+\upsilon)\rho\omega^2}{8+(3+\upsilon)n} a^{2-k_2} \left[\frac{1-\left(\frac{b}{a}\right)^{2-k_1}}{1-\left(\frac{b}{a}\right)^{k_2-k_1}} \right] \end{aligned} \tag{6.74}$$

Equation 6.73 reveals that the tangential stress is greater than the radial stress. The maximum tangential stress occurs at $r = a$. When the maximum tangential stress reaches the yield stress in tension, σ_0, the elastic rotational speed is maximum. Its magnitude, ω_e, can be obtained by setting in the second equation of Equation 6.73, $\sigma_\theta = \sigma_0$, $r = a$, and using the values of A, B given in Equation 6.74 to solve for the rotational speed.

Note that if $n = 0$, $t(r) = t_0$, $k_1 = 0$, $k_2 = -2$ and the above results are reduced to those derived in Section 6.8.2 for constant thickness disk.

Plastic analysis: If the material is elastic–perfectly plastic with yield stress σ_0 the rotational speed that will initiate yield according to Tresca's criterion is obtained from Equation 6.73, as mentioned previously.

When the rotational speed is increased beyond the elastic limit, $\omega > \omega_e$, an annular plastic zone spreads from the inner boundary to a radius = c. In this zone, the stresses are required to be in equilibrium and satisfy Tresca's yield criterion.

Equilibrium is maintained if the stresses satisfy the equation

$$\frac{d}{dr}(tr\sigma_r) = t\sigma_0 - t\rho\omega^2 r^2, \quad \text{or} \quad r\frac{d\sigma_r}{dr} + \left(1 + \frac{r}{t}\frac{dt}{dr}\right)\sigma_r = \sigma_0 - \rho\omega^2 r^2 \tag{a}$$

Substituting for t(r) from Equation 6.70 in (a), the stresses in the plastic zone are obtained by solving the differential equation

$$\frac{d\sigma_r}{dr} + \frac{1+n}{r}\sigma_r = \frac{\sigma_0}{r} - \rho\omega^2 r, \quad a \le r \le c \tag{6.75}$$

There are two cases to consider: case (1), where n takes all negative values except $n = -1$, and case (2), where $n = -1$ only.

Case (1): $n \neq -1$—Equation 6.75 is a first-order ordinary differential equation whose solution is readily obtained by multiplying both sides by the integrating factor

$$e^{\int \frac{1+n}{r} dr} = r^{1+n}$$

Hence, we have

$$\begin{aligned} \frac{d}{dr}(r^{1+n}\sigma_r) &= \sigma_0 r^n - \rho\omega^2 r^{n+2} \\ r^{1+n}\sigma_r &= \frac{\sigma_0 r^{n+1}}{n+1} - \frac{\rho\omega^2 r^{n+3}}{n+3} + D \end{aligned} \tag{b}$$

where D is a constant of integration to be determined from the condition that the radial stress vanishes when $r = a$. It follows that the stresses in the plastic region are obtained as follows:

$$\left.\begin{aligned} \sigma_r &= \frac{\sigma_0}{n+1}\left[1 - \left(\frac{a}{r}\right)^{n+1}\right] + \frac{\rho\omega^2}{3+n} r^2 \left[\left(\frac{a}{r}\right)^{3+n} - 1\right] \\ \sigma_\theta &= \sigma_0 \\ \sigma_z &= 0 \end{aligned}\right\} \quad a < r < c \tag{6.76}$$

In the elastic region, $c \leq r \leq b$, the stresses in Equations 6.73 and 6.76 are used. The boundary conditions on the radial stress are $\sigma_r = 0$ at $r = b$, which determine one constant, say, A in terms of B, and the second condition is continuity of σ_r across the interface $r = c$.

The first condition gives

$$A = Kb^{2-k_1} - Bb^{k_2-k_1}, \quad \text{where } K = \frac{(3+\upsilon)\rho\omega^2}{8+(3+\upsilon)n} \tag{c}$$

and the stresses in the elastic region, $c < r < b$, are

$$\begin{aligned} \sigma_r &= B\left[r^{k_2} - b^{k_2-k_1} r^{k_1}\right] + K\left[b^{2-k_1} r^{k_1} - r^2\right] \\ \sigma_\theta &= B\left[(1+n+k_2)r^{k_2} - (1+n+k_1)b^{k_2-k_1} r^{k_1}\right] + K\left[(1-n+k_1)b^{2-k_1} r^{k_1} - \frac{1+3\upsilon}{3+\upsilon} r^2\right] \end{aligned} \tag{6.77}$$

The second condition on the radial stress determines the constant B by equating the radial stresses in Equations 6.76 and 6.77 at the interface $r = c$, that is,

$$\frac{\sigma_0}{1+n}\left[1 - \left(\frac{a}{c}\right)^{1+n}\right] + \frac{\rho\omega^2 c^2}{3+n}\left[\left(\frac{a}{c}\right)^{3+n} - 1\right] = B\left[c^{k_2} - b^{k_2-k_1} c^{k_1}\right] + K\left[b^{2-k_1} c^{k_1} - c^2\right] \tag{6.78}$$

In addition, the tangential stress in the plastic and elastic regions must be equal at $r = c$, that is,

$$\sigma_0 = B\left[(1+n+k_2)c^{k_2} - (1+n+k_1)b^{k_2-k_1} c^{k_1}\right] + K\left[(1+n+k_1)b^{2-k_1} c^{k_1} - \frac{1+3\upsilon}{3+\upsilon} c^2\right] \tag{6.79}$$

Substituting for B from Equations 6.78 and 6.79 gives the values of the rotational speed ω for any value of c in the elastic region. By letting $c = b$, the critical speed is determined.

Alternatively, the critical speed can be determined by using the first equation of Equation 6.76 and setting $\sigma_r = 0$ at $r = c = b$. This procedure yields the critical rotational speed in which the whole disk is plastic.

$$\frac{\rho\omega_u^2 b^2}{\sigma_0} = \frac{3+n}{1+n}\left[\frac{1-\left(\frac{a}{b}\right)^{1+n}}{1-\left(\frac{a}{b}\right)^{3+n}}\right] \tag{6.80}$$

Note that when $n = 0$, Equations 6.74 and 6.80 reduce to the results given in Equations 6.62 and 6.65 for the annular disk with constant thickness.

Case (2): $n = -1$—Equation 6.75 becomes $(d\sigma_r/dr) = (\sigma_0/r) - \rho\omega^2 r$, that is,

$$\sigma_r = \sigma_0 \log r - \frac{\rho\omega^2 r^2}{2} + D \tag{d}$$

The constant D is found from the condition that $\sigma_r = 0$ at $r = a$. Hence, in the plastic region $a < r < c$, we have

$$\left.\begin{aligned} \sigma_r &= \sigma_0 \log\frac{r}{a} + \frac{\rho\omega^2 r^2}{2}\left(1-\frac{r^2}{a^2}\right) \\ \sigma_\theta &= \sigma_0 \\ \sigma_z &= 0 \end{aligned}\right\} \quad a < r < c \tag{6.81}$$

In the elastic region, $c < r < b$, Equations 6.77 are valid with $n = -1$. Equating the radial stresses in Equations 6.77 and 6.81 at $r = c$, the constant B is determined from the equation

$$\sigma_0 \log\frac{c}{a} + \frac{\rho\omega^2 a^2}{2}\left(1-\frac{c^2}{a^2}\right) = B\left(c^{k_2} - b^{k_2-k_1}\right) + K\left(b^{2-k_1} - c^2\right) \tag{6.82}$$

In this case, $n = -1$ and

$$K = \frac{(3+\upsilon)\rho\omega^2}{5-\upsilon} \tag{e}$$

The second equation involving B is obtained from Equation 6.79 by setting $n = -1$, that is,

$$\sigma_0 = B\left[k_2 c^{k_2} - k_1 b^{k_2-k_1} c^{k_1}\right] + K\left[k_1 b^{2-k_1} c^{k_1} - \frac{1+3\upsilon}{3+\upsilon} c^2\right] \tag{6.83}$$

As in the previous case, Equations 6.82 and 6.83 determine B and a relation which gives the extent of the plastic zone for a given rotational speed. By letting $r = c = b$, the critical rotational speed is obtained, or we can proceed as follows: The critical rotational speed for this case is obtained by setting the radial stress in Equation 6.81 to be zero at $r = b$.

Hence, for case $n = -1$, we have

$$\frac{\rho\omega_u^2 a^2}{\sigma_0} = \frac{2\log(b/a)}{(b/a)^2 - 1} \tag{6.84}$$

6.8.5 Rotating Disk of Uniform Strength

Disks rotating at high speed, for example, turbine blades, are subject to very high-centrifugal forces. For such disks, it is often economically desirable to design for constant stress condition so that the use of the material is most efficient.

Consider a disk rotating about its central axis with angular velocity ω such that the induced radial and tangential stresses are equal

$$\sigma_r = \sigma_\theta = \sigma(\text{constant}) \tag{6.85}$$

In order to achieve such stress condition, disks with variable thickness, t(r), are needed. Equilibrium of a disk, Equation 6.68, is reduced to

$$\sigma\frac{d}{dr}(rt) - tr + \rho\omega^2 tr^2 = 0 \tag{a}$$

which can be simplified to the following:

$$\frac{dt}{t} = -\frac{\rho\omega^2}{\sigma} r\,dr \tag{6.86}$$

Integrating Equation 6.86, we have

$$\log_e t = -\frac{\rho\omega^2 r^2}{2\sigma} + \log_e C \tag{b}$$

where C is a constant, and

$$t(r) = Ce^{-\frac{\rho\omega^2 r^2}{2\sigma}} \tag{c}$$

At $r = 0$, $t(r) = t_0$, so that $C = t_0$, and in order to achieve uniform stress condition, the thickness profile of the disk is

$$t(r) = t_0 e^{-\frac{\rho\omega^2 r^2}{2\sigma}} \tag{6.87}$$

EXAMPLE 6.8

A steam turbine disk has a hyperbolic profile $t(r) = t_0/r$ with radii $a = 0.06$ m and $b = 0.24$ m. The material is elastic–perfectly plastic with $\sigma_0 = 250$ MPa, $\nu = 0.3$, $E = 200$ GPa, and density $\rho = 7800\ \text{kg/m}^3$.

a. If the disk is rotating at a speed of $\omega = 7500$ RPM, find the maximum elastic stress.
b. Find the rotational speed that initiates yield and the critical speed of the disk.

Solution

Using Equation 6.72 with $n = -1$, $(b/a) = 4$, and Poisson's ratio = 0.3, we find $k_1 = 0.745$ and $k_2 = -1.745$. The constants A and B are computed using Equation 6.74

$$A = 120.25245 \times 10^{-3}\ \rho\omega^2$$

$$B = -0.09042 \times 10^{-3}\ \rho\omega^2$$

a. The maximum elastic stress occurs in the tangential stress at $r = 0.06$ m. Equation 6.73 gives when the speed is 7500 RPM = 785.4 rad/s

$$(\sigma_\theta)_{max} = 0.0309\ \rho\omega^2 = 148.7\ \text{MPa}$$

b. The rotational speed that initiates yield is given by $\sigma_0 = 0.0309\ \rho\omega^2$, which gives

$$\omega_e = 1018\ \text{rad/s} = 9726\ \text{RPM}$$

The ultimate or critical speed is obtained using Equation 6.80, which gives

$$\omega_u = 1282.8\ \text{rad/s} = 12{,}250\ \text{RPM}$$

Note that the corresponding values for the constant thickness disk in Example 6.7 are

$$\omega_e = 7791\ \text{RPM} \quad \text{and} \quad \omega_u = 10{,}769\ \text{RPM}$$

Problems

6.1 In strength of materials, the bending stress in prismatic beams is determined from the formula $\sigma_x = My/I$ (positive moment is sagging and y coordinate is taken downward). Use equilibrium equations of the theory of elasticity to determine the stresses σ_y and τ_{xy} in a simply supported rectangular beam of length L, depth 2h, and unit width, subjected to uniformly distributed load q/unit length.

Ans:

$$\sigma_y = \frac{q}{2I}\left(h^2 y - \frac{y^3}{3} - \frac{2}{3}h^3\right)$$

$$\tau_{xy} = \frac{V}{2I}(h^2 - y^2)$$

I is moment of inertia and V = shear force.

6.2 The cross section of a simply supported I-beam loaded by equal and opposite moments M is shown in Figure P6.2. Show that the stress function $\phi(x, y) = Ay^3$ provides an elasticity solution, which satisfies the boundary conditions and find the constant A in terms of M, b, and h.

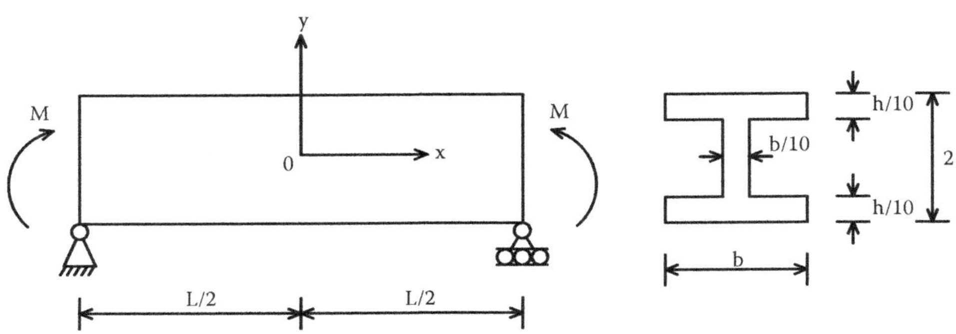

FIGURE P6.2

Ans: $A = 0.73M/bh^3$.

6.3 Consider a plane stress rectangular plate, unit thickness and sides L and h, as shown in Figure P6.3. Investigate what problem is solved by the stress function

$$\phi(x, y) = \frac{Fy^2}{2h} + \frac{2My^3}{h^3} - \frac{2Wxy^3}{h^3} + \frac{3Wxy}{2h}$$

where F, M, W, and h are constants. Sketch the stress resultants on the boundary of the plate and check equilibrium.

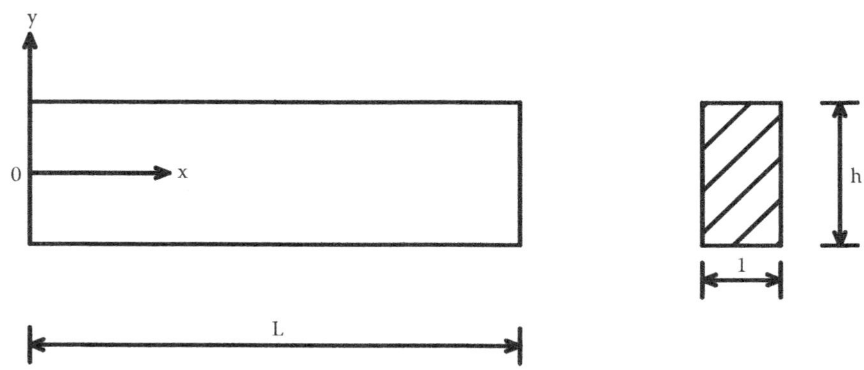

FIGURE P6.3

6.4 The plane stress sheet shown in Figure P6.4 is subjected to stresses governed by the stress function

$$\phi(x, y) = \frac{\sigma x^2 y^2}{2a^2}$$

If $a = 20$ cm, $t = 1.5$ cm, $E = 200$ GPa, $\nu = 0.3$, and $\sigma = 200$ MPa, find (a) stress resultants F_x, F_y, V_x, and V_y; (b) deformed length of diagonal AC.

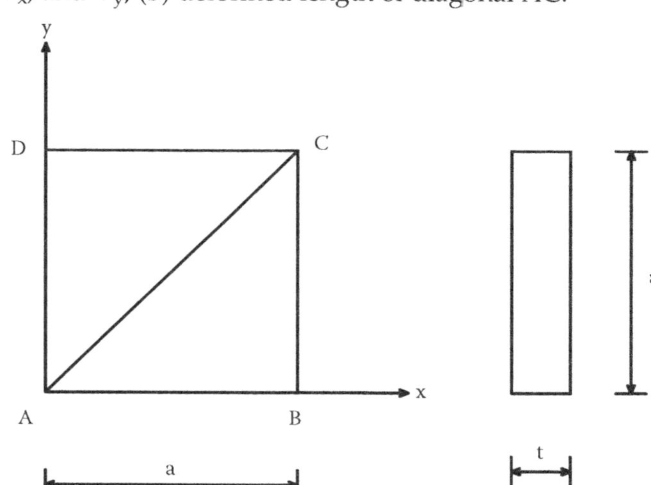

FIGURE P6.4

Ans: AC = 282.6 mm.

6.5 In the problem of the cantilever beam loaded by a concentrated force P discussed in Section 6.4.1, the support conditions at $x = L$ are hinges specified as follows:

At $y = 0$, $u = V = 0$, at $y = +h$ (point B), and at $y = -h$ (point C) in Figure 6.4, $u = 0$ only.

Prove that the vertical deflection of the free end $x = 0$ is given by

$$V = -\frac{PL^3}{3EI}\left[1 + \frac{1}{2}(4 + 5\upsilon)\frac{h^2}{L^2}\right]$$

where ν is Poisson's ratio.

6.6 Given that the following expression:

$$\phi(x, y) = \frac{q}{20h^3}(20x^2y^3 - 30hx^2y^2 - 4y^5 + 10hy^4 - 8h^2y^3 + 2h^3y^2)$$

where q and h are constants which represent a stress function. Check its validity, find the stresses σ_x, σ_y, τ_{xy}, and find the resultant shear force V_y on edge $x = 2h$ of an element $0 < x < 2h$, $0 < y\ h$, $0 < z < 1$.

Ans: $V_y = -qh$.

6.7 The plane stress distribution in a flat plate of unit thickness, Figure P6.7, is given by

$$\sigma_x = x^3y - 2y^3x$$

$$\sigma_y = y^3x - 2c_1xy + c_2x$$

$$\tau_{xy} = 0.5y^4 - 1.5x^2y^2 + c_1x^2 + c_3$$

where c_j, $j = 1, 2$, and 3 are constants. (a) Show that if body forces are neglected, equilibrium of the plate exists; (b) If the boundary conditions on the element are $\tau_{xy} = 0$ at $y = \pm(b/2)$ and $\sigma_y = 0$ at $y = -b/2$, determine the three constants c_j and the total load on edge $y = b/2$.

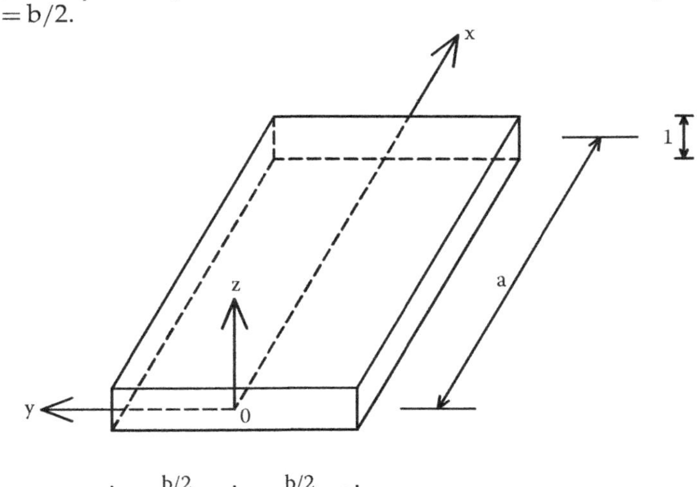

FIGURE P6.7

Ans: $c_1 = (3/8)b^2$, $c_2 = -(b^3/8)$, $c_3 = -(b^4/32)$.

6.8 A cantilever $0<x<L$, $y=\pm d$, $z=\pm t/2$ is loaded by a uniform shear stress τ along the upper edge, the lower edge, and end $x=L$ are free from stress (Figure P6.8).

It is proposed to use the following stress function to solve the problem:

$$\phi(x, y) = Axy + Bxy^2 + Cxy^3 + Dy^2 + Ey^3$$

where, A, B, C, D, and E are constants. Use the boundary conditions to find the constants and indicate what is wrong with the proposed stress function.

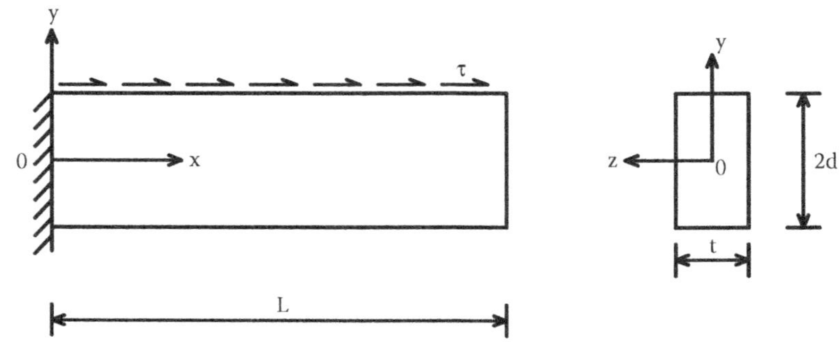

FIGURE P6.8

Ans: Boundary conditions on $x=L$ are not satisfied.

6.9 In the previous problem if the cantilever is subjected to a linearly increasing shearing stress ($\tau\, x/L$) on the upper edge, the following polynomial is proposed to solve the problem:

$$\phi(x, y) = Ax^2 + Bx^2y + Cx^2y^2 + Dx^2y^3 + Ey^2 + Fy^3 + Gy^4 + Hy^5$$

where A, B, C, D, E, F, and H are constants. (a) How many independent terms of the polynomial are needed to be a proper stress function? (b) Determine the constants in term of L, d, and τ. Indicate any deficiency of the solution.

Ans: Constants G and H are not independent.

6.10 Show that if $V(x, y)$ is a plane harmonic function, that is, it satisfies Laplace equation in two dimensions, $(\partial^2V/\partial x^2) + (\partial^2V/\partial y^2) = 0$ then, each of the functions xV, yV, and $(x^2+y^2)V$ is a stress function.

6.11 Show that the function $V(x, y) = \tan^{-1}(y/x)$ is a plane harmonic function and determine the stresses σ_x, σ_y, and τ_{xy} induced by the stress function

$$\phi(x, y) = -\frac{P}{\pi} yV(x, y)$$

where P is a constant (Section 6.7.3).

6.12 If the strain components of a plane strain problem in cylindrical coordinates are

$$\varepsilon_r = \frac{\partial u}{\partial r}$$

$$\varepsilon_\theta = \frac{u}{r}$$

$$\varepsilon_z = \gamma_{r\theta} = \gamma_{\theta z} = \gamma_{zr} = 0$$

prove that the compatibility equation in term of stresses is

$$\upsilon r \frac{\partial \sigma_r}{\partial r} - (1 - r) r \frac{\partial \sigma_\theta}{\partial r} + \sigma_r - \sigma_\theta = 0$$

where ν is Poisson's ratio, and find the corresponding equation for a plane stress problem.

6.13 Show that the moment boundary condition needed to solve the problem of bending of a curved beam by end moment, Equation (d) in section (b) of Section 6.7.1, which is

$$\int_a^b \sigma_\theta r(1)dr = -M$$

is equivalent to relation (e), namely, $M = \frac{1}{2}\int_a^b (\sigma_r - \sigma_\theta) r\, dr$, which is more convenient for finding constants A and B.

6.14 A curved bar of radii a and b and unit thickness is fixed at the lower end and subjected to tangential force F at the other end, as shown in Figure P6.14. Find the stresses developed in the bar (Hint: use $\phi(\theta, r) = f(r) \sin \theta$).

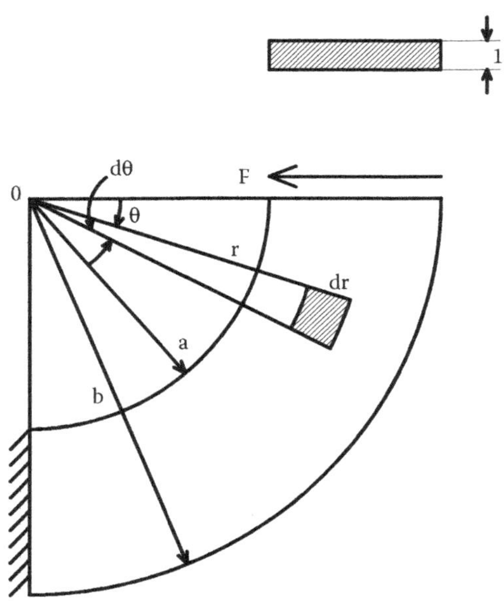

FIGURE P6.14

6.15 Instead of the force F, the curved bar of the previous problem is subjected to a normal force Q at the free end, as shown in Figure P6.15. Find the stresses in the bar (Hint: use $\phi(r, \theta) = f(r) \cos \theta$).

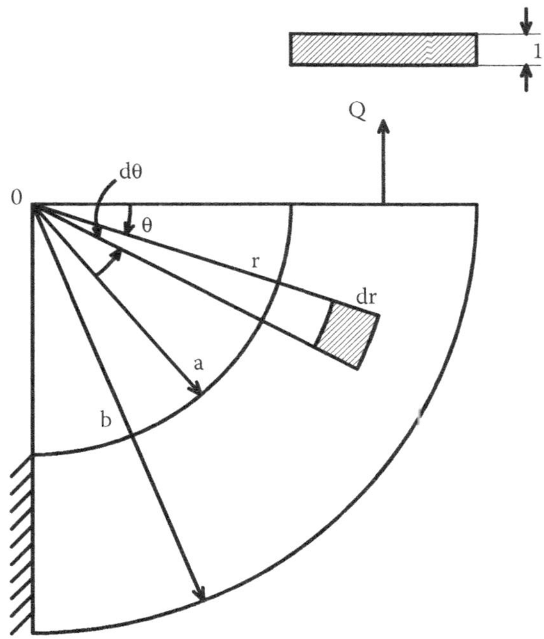

FIGURE P6.15

6.16 Find the stress distribution in a quadrant of a circular plate of radius b and unit thickness generated by the stress function $\phi(r, \theta) = p\, r^2\, \theta$, where p is constant. Use separate diagrams to sketch the normal and shearing stresses on the boundary of the plate.

Ans: $2p\,\theta$, $2p\,\theta$, $-p$.

6.17 Show that the expression $\phi(r, \theta) = (\cos^3 \theta)/r$ is a stress function. Determine the stresses and sketch the radial and shear stresses (σ_r, $\tau_{r\theta}$) on the boundary of a circular plate of radius a.

Ans: $\sigma_r = (2/r^3)\cos\theta(3 - 5\cos^2 \theta)$.

6.18 Consider the stress function $\phi(r, \theta) = (\sin^3 \theta)/r$ and repeat Problem 6.17.

Ans: $\sigma_r = (2\sin\theta/r^3)(3 - 5\sin^2 \theta)$.

6.19 A solid circular disk of radius b, unit thickness, is subjected to stress $\sigma_r = q \sin \theta$ at $r = b$, $q =$ constant. Find expressions for the radial and tangential elastic displacements, u and v in the disk. If $b = 25$ cm, $q = 200$ MPa, $E = 200$ GPa, and $\nu = 0.3$, find numerical values of the displacements at the outer edge of the disk.

Ans: $u = 0.013 \sin \theta$ mm, $v = -0.663 \cos \theta$ mm.

6.20 Determine the stresses generated by the stress function

$$\phi(r, \theta) = -\frac{p}{\pi} r^2(\theta - \sin \theta \cos \theta)$$

in the semi-infinite plate shown in Figure P6.20, $p =$ constant.

Sketch the loading on axis $y = 0$ and examine the behavior of the shearing stress $\tau_{r\theta}$ as r approaches the origin from the vertical and horizontal directions in the plate.

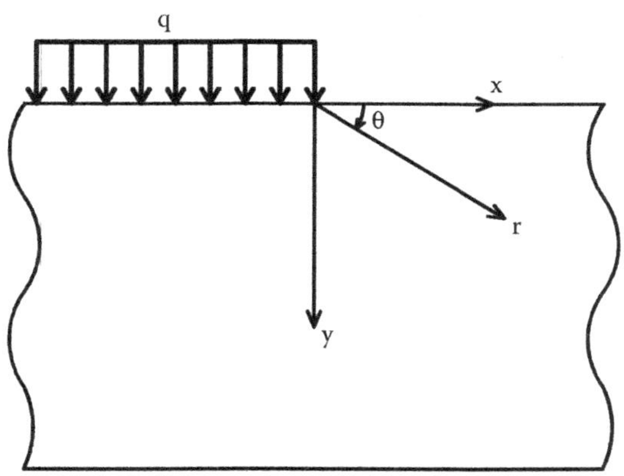

FIGURE P6.20

Ans: $\tau_{r\theta} = (p/\pi)\sin^2\theta$, the loading creates discontinuity.

6.21 Determine the rectangular stresses, σ_x, σ_y, and τ_{xy}, generated by the stress function

$$\phi(r, \theta) = -\frac{P}{\pi} r\theta \sin\theta$$

where P is a constant in the semi-infinite plate shown in Figure P6.20. Compare your results with those given in Equation (a) of Section 6.7.3. Show that as x and y become large the stresses tend to zero.

6.22 Determine the loads acting along the boundary of a semi-infinite plate, Figure P6.20, $-\infty < x < \infty$, $y \geq 0$, derived from the following stress functions:

$$\text{(a)}\ \phi(r, \theta) = \frac{pr^2}{\pi}\theta$$

$$\text{(b)}\ \phi(r, \theta) = -\frac{pr^2}{\pi}\theta$$

where p is constant. Show your results on separate sketches of the plate.

Ans:

$$\sigma_\theta = +2p \quad \text{and} \quad -2p$$
$$\tau_{r\theta} = -p/\pi \quad \text{and} \quad \tau_{r\theta} = +p/\pi$$

6.23 A line load of intensity 40 kip/in. acts at point 0 on the surface of a plane strain semi-infinite plate. The yield stress of the material is $\sigma_0 = 36$ ksi. Sketch the region in which yield is initiated using the maximum distortion energy theory.

Ans: Circle of diameter 1.57 in. touching the boundary at 0.

6.24 Determine the stresses in a wedge subjected to uniform shear loading of intensity q on each edge as shown in Figure P6.24 using the stress function

$$\phi(r, \theta) = Ar^2(\cos 2\theta - \cos 2\alpha)$$

where A is a constant. Verify that no concentrated force or couple acts on the vertex of the wedge.

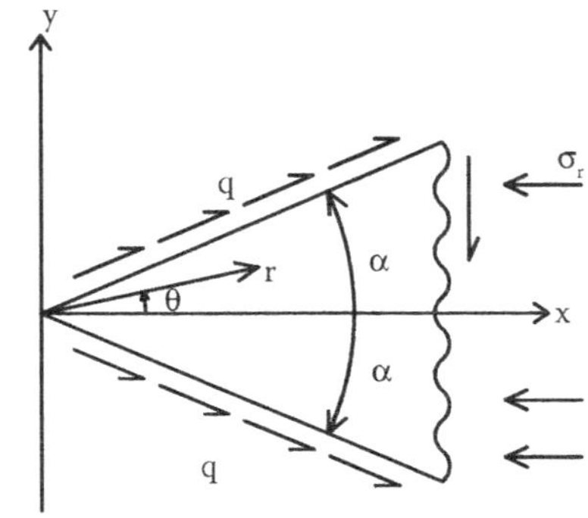

FIGURE P6.24

Ans:

$$\tau_{r\theta} = q\frac{\sin 2\theta}{\sin 2\alpha}$$

6.25 In the previous problem if the wedge is subjected to shearing load $\tau_{r\theta} = qr$ on each edge (q = constant), find the stresses using the function

$$\phi(r, \theta) = Ar^3 \cos 3\theta + Br^3 \cos \theta$$

where A, B are constants. Verify that the above function is a stress function.
Ans:

$$\tau_{r\theta} = 6Ar \sin 3\theta + 2Br \sin \theta; \quad A = \frac{q}{16(\sin \alpha \cos^2 \alpha)}$$

6.26 A ring plate of 30 cm outer radius and 20 cm internal radius is shrunk on a solid shaft. The material is arranged such that the radial pressure between the mating surfaces will not fall below 40 MPa while the assembly rotates. The material is assumed to be elastic–perfectly plastic with $\rho = 7800\,kg/m^3$, E = 200 GPa, and $\nu =$ 0.3 yield stress = 250 MPa. If the maximum stress in the plate is limited to the yield stress, determine the maximum angular velocity.

Ans: 4800 RPM.

6.27 An annular disk of internal and external radii, a, b, respectively, (b/a) = 2 is rotating about its normal axis. The material is assumed to be elastic-perfectly plastic with yield stress σ_0 and $\nu = 1/3$. Find the ratio of maximum tangential to radial stress at the yield initiation stage and at the fully plastic stage using Tresca's yield criterion.

Ans: σ_0, 0.19 σ_0.

6.28 An annular steel disk of outer and inner radii 200 and 100 mm, respectively, rotates about its normal axis with angular speed ω. If the outer radius is not to increase by more than 1.5%, find the maximum permissible rotational speed. Assume elastic behavior, E = 200 GPa, Poisson's ratio = 0.3, and the mass density = 7800 kg/m^3.

Ans: 5300 RPM.

6.29 A ring plate of outer diameter 600 mm is shrunk onto a steel shaft of 100 mm diameter. The interface in the diameters is 0.05 mm. The material properties of the ring are $E = 200$ GPa, $\nu = 0.3$, and $\rho = 7800\ kg/m^3$. Assuming the elastic behavior, find (a) the maximum tangential stress in the ring at rest in terms of the pressure p between ring and shaft; (b) the angular speed ω at which the contact pressure between the plate and shaft is zero.

Ans: 1.06 p, 1613 RPM.

6.30 An annular steel disk of uniform thickness rotates about its normal axis with a speed of 110 revolution/s. The outside diameter of the disk is 800 mm. If the maximum allowable stress in the disk did not to exceed its yield strength, find the diameter of a central hole in the disk. For steel $\rho = 7450\ kg/m^3$, $\nu = 1/3$, and $\sigma_0 = 207$ MPa.

Ans: 53 mm.

6.31 A solid circular disk of radius b is made of an elastic–perfectly material having yield stress σ_0, Poisson's ratio $= 1/3$, and density ρ. The disk is rotating at an angular speed of magnitude equal to half the sum of the speed needed to initiate yield and the ultimate speed, that is $(1/2)\ (\omega_e + \omega_u)$. Find the extent of the plastic zone, distance c from the origin, when (a) the disk has no central hole; (b) the disk is annular with external and internal radii b and b/3, respectively.

Ans: (i) 0.57 b, (ii) 0.72 b.

References

6.1. G. B. Airy, *Phil. Trans. R. Soc.*, vol. 153, pp. 49–80, 1863.
6.2. N. I. Muskhleishelishvil, *Some Basic Problems of the Mathematical Theory of Elasticity*, translated by I. R. M. Radok, P. Noordhoff Ltd., Gronigen, The Netherland, 1963.
6.3. B. de Saint-Venant, *Memeires des Savants Etrangers*, vol. 14, pp. 233–250, 1855.
6.4. E. P. Popov, *Mechanics of Materials*, Prentice-Hall Inc., Englewood Cliffs, NJ, 1976.
6.5. A. C. Ugural and S. K. Fenster, *Advanced Strength and Applied Elasticity*, PTR Prentice-Hall, Englewood Cliffs, NJ, 1995.
6.6. B. W. Shaffer and R. N. House, The elastic–plastic stress distribution within a wide curved bar subjected to pure bending, *J. Appl. Mech., ASME*, vol. 24, 1955.
6.7. J. H. Michell, *The Collected Mathematical Works*, of J. H. and A. G. M. Michell, P. Noordhoff Publishing, Groningen, The Netherlands, 1964. Also, *Proc. London Math. Soc.*, vol. 34, 1902.
6.8. E. Stenberg and W. Koiter, *J. Appl. Mech., ASME*, vol. 25, pp. 575–581, 1958.
6.9. A. Flamant, *Compt. Rend.*, vol. 114, p. 1465, 1802.

7

Other Elastoplastic Problems

In addition to the problems considered in Chapter 6, other practically important problems that possess axial or spherical symmetry include cylindrical and spherical pressure vessels, stress concentration around holes and notches, and contact stresses between components of structures and machines.

7.1 Thick-Walled Cylindrical Tubes

Consider a thick-walled cylinder subjected to constant pressures p and p_1 on the inner and outer surfaces, respectively, and loaded at both ends by equal and opposite forces acting along the axis of the cylinder. The z-axis of cylindrical coordinates, r, θ, z, is taken along the longitudinal axis of the tube, and r and θ in the transverse direction. The ends of the cylinder are either closed or allowed to deform; however, the cylinder is considered long compared to its diameter so that the stresses at a distance at least a diameter away from the ends do not vary along the length of the tube. This implies that transverse sections remain plane during deformation and the longitudinal strain is independent of the variable r and remains constant throughout the cylinder at each stage of the loading. Figure 7.1 shows a typical cross section of the cylinder where the inner radius is denoted by a, and the outer radius by b.

Hooke's law in cylindrical coordinates is given by

$$\begin{aligned} \varepsilon_r &= \frac{1}{E}[\sigma_r - \upsilon(\sigma_\theta + \sigma_z)] \\ \varepsilon_\theta &= \frac{1}{E}[\sigma_\theta - \upsilon(\sigma_z + \sigma_r)] \\ \varepsilon_z &= \frac{1}{E}[\sigma_z - \upsilon(\sigma_r + \sigma_\theta)] \end{aligned} \tag{a}$$

Because of the axial symmetry of the problem, the equation of equilibrium in the absence of body forces reduces to (see Section 6.8)

$$\frac{d\sigma_r}{dr} = \frac{\sigma_\theta - \sigma_r}{r} \tag{7.1}$$

The nonvanishing displacement is u (r) and the corresponding 2D strains are $\varepsilon_r = (du/dr)$, $\varepsilon_\theta = (u/r)$, and $\gamma_{r\theta} = 0$.

It follows that the compatibility equation of strains is

$$r\frac{d\varepsilon_\theta}{dr} + \varepsilon_\theta - \varepsilon_r = 0 \tag{7.2}$$

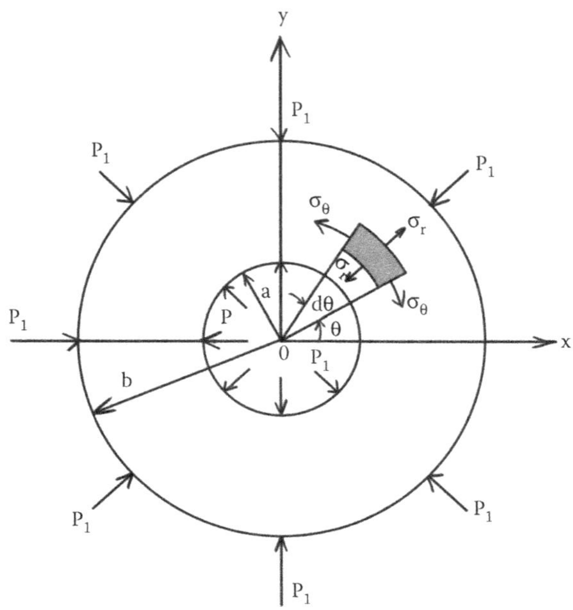

FIGURE 7.1
Thick-walled cylinder subjected to internal and external pressure.

Substituting $\sigma_z = E\varepsilon_z + \upsilon(\sigma_r + \sigma_\theta)$ in the first two equations of Equation (a) and using equilibrium, the compatibility Equation 7.2, in terms of stresses, assumes the form

$$\frac{d}{dr}(\sigma_r + \sigma_\theta) = 0 \quad \text{or} \quad \sigma_r + \sigma_\theta = 2A \tag{b}$$

The stress components that satisfy Equations (b) and 7.1 are

$$\sigma_r = A + \frac{B}{r^2}, \quad \sigma_\theta = A - \frac{B}{r^2} \tag{7.3}$$

where A and B are constants to be determined from the boundary conditions. For a cylinder subjected to uniform internal and external pressures p and p_1, respectively, the boundary conditions are $\sigma_r = -P$ at $r = a$, and $\sigma_r = -P_1$ at $r = b$. Using the first equation in Equation 7.3, the constants A and B are determined

$$A = \frac{a^2p - b^2p_1}{b^2 - a^2}, \quad B = \frac{a^2b^2(p_1 - p)}{b^2 - a^2} \tag{a}$$

And the stress components follow from Equations 7.3 as

$$\begin{aligned} \sigma_r &= \frac{1}{b^2 - a^2}\left[a^2p - b^2p_1 - (p - p_1)\frac{a^2b^2}{r^2}\right] \\ \sigma_\theta &= \frac{1}{b^2 - a^2}\left[a^2p - b^2p_1 + (p - p_1)\frac{a^2b^2}{r^2}\right] \end{aligned} \tag{7.4a}$$

The axial stress is determined from Hooke's law

$$\sigma_z = E\varepsilon_z + \upsilon(\sigma_r + \sigma_\theta) = E\varepsilon_z + \frac{2\upsilon(a^2p - b^2p_1)}{b^2 - a^2} \tag{7.4b}$$

The radial displacement is determined from the relation $u = r\,\varepsilon_\theta$

$$u = -\upsilon\varepsilon_z r + \frac{(1+\upsilon)r}{E(b^2 - a^2)}\left[(1 - 2\upsilon)(a^2p - b^2p_1) + (p - p_1)\frac{a^2b^2}{r^2}\right] \tag{7.5}$$

These expressions were first derived by G. Lame in 1833 and commonly referred to in the literature as Lame's solution [7.1]. Observe that the sum $\sigma_r + \sigma_\theta$ and σ_z are constants throughout the tube. The magnitude of the axial stress depends on the end condition of the cylinder. For open ends, $\sigma_z = 0$, for closed ends, if the resultant axial force is denoted by F, then $F = \pi a^2 p$, p is the gauge pressure, and overall equilibrium requires

$$\pi a^2 p = \int_a^b 2\pi\sigma_z r\,dr \tag{7.6}$$

A third condition that is sometimes used because of its mathematical simplicity is the plane strain condition, $\varepsilon_z = 0$ (zero extension) or $\varepsilon_z = \text{constant}$ indicating that the cylinder extends without bending.

7.1.1 Cylindrical Tube Subjected to Internal Pressure

In this case, $p_1 = 0$ and the stresses in Equations 7.4 reduce to

$$\begin{aligned}
\sigma_r &= -\frac{a^2p}{(b^2 - a^2)}\left(\frac{b^2}{r^2} - 1\right)\\
\sigma_\theta &= \frac{a^2p}{(b^2 - a^2)}\left(\frac{b^2}{r^2} + 1\right)\\
\sigma_z &= E\varepsilon_z + \frac{2\upsilon a^2p}{(b^2 - a^2)}
\end{aligned} \tag{7.7}$$

The radial stress is compressive (negative) for all r except at $r = b$, in which it is zero. The maximum radial stress occurs at $r = a$, and has the value $-p$. The tangential stress σ_θ is tensile (positive) and has its maximum value also at $r = a$, $(\sigma_\theta)_{max} = p\,(b^2 + a^2)/(b^2 - a^2)$.

The magnitude of the axial stress varies with the end condition, $\sigma_z = 0$ for open ends, for closed ends, Equation 7.6 gives $\sigma_z = (a^2p)/(b^2 - a^2) = (1/2)\,(\sigma_r + \sigma_\theta)$, and for the plane strain condition, $\varepsilon_z = 0$, we get, from the third equation in Equation 7.7, $\sigma_z = \nu(\sigma_r + \sigma_\theta) = (2\nu a^2p)/(b^2 - a^2)$.

The values of the corresponding axial strain can be summarized as follows:

$$\begin{aligned}
&\frac{(1 - 2\upsilon)a^2p}{b^2 - a^2} && \text{Closed-End}\\
&E\varepsilon_z = 0 && \text{Plane-Strain}\\
&-\frac{2\upsilon a^2p}{b^2 - a^2} && \text{Open-End}
\end{aligned} \tag{7.8}$$

It is observed that the axial strain in the closed-end condition is tensile (positive) while that in the open end is contraction (negative) and the plane strain one is intermediate between the two. The magnitude of the axial stress in the plane strain case is identical to that in the closed ends if the material is incompressible ($\upsilon = 1/2$). Note that all stresses are principal stresses and the magnitude of the axial stress is always intermediate between the other two.

The radial displacement is given by Equation 7.5 with $p_1 = 0$

$$\frac{u}{r} = -\upsilon\varepsilon_z + \frac{(1+\upsilon)a^2p}{(b^2-a^2)E}\left(\frac{b^2}{r^2} + 1 - 2\upsilon\right) \tag{7.9a}$$

Using the axial strains in Equation 7.8, the radial displacement for the three end conditions can be summarized as follows:

$$u = \frac{pa^2r}{E(b^2-a^2)}\left[\beta + (1+\upsilon)\frac{b^2}{r^2}\right] \tag{7.9b}$$

where β stands for $(1-2\upsilon)$ in case of closed ends, $(1-\upsilon)$ for open ends, and $(1+\upsilon)(1-2\upsilon)$ for plane strain condition. In particular, the radial displacement at the outer boundary is given by

$$\frac{u_b}{b} = (1+\upsilon+\beta)\frac{pa^2}{E(b^2-a^2)}$$

Since $\sigma_\theta > \sigma_z > \sigma_r$ for the three end conditions, the maximum shearing stress occurs at $r = a$, and has the value

$$\tau_{max} = \frac{1}{2}(\sigma_\theta - \sigma_r) = \frac{b^2p}{b^2-a^2} \tag{a}$$

Yield initiation: The material is assumed to be elastic–perfectly plastic with yield stress in tension $= \sigma_0$ (yield stress in shear is approximately $= \sigma_0/2$). As the pressure p is increased, a stage is reached where yield will initiate in the tube at the inner boundary and spread outward (Figure 7.1). The pressure that will initiate yield depends on the principal stresses and yielding hypothesis used in the analysis. Two yield criteria are used, Tresca's maximum shearing stress criterion (Equation 4.40) and Von Mises' maximum energy criterion (Equation 4.42).

Tresca's criterion: If the material yields according to Tresca's criterion, $\sigma_\theta - \sigma_r = \sigma_0$, then using relation (a), the pressure that initiates yield regardless of the end conditions of the cylinder is

$$p_e = \frac{\sigma_0}{2}\left(1 - \frac{a^2}{b^2}\right) \tag{7.10}$$

Von Mises' criterion: This criterion assumes that yielding occurs when the strain energy of distortion reaches a critical value. The pressure that initiates yield is obtained from Equation 4.42a, which assumes the form

$$(\sigma_\theta - \sigma_r)^2 + (\sigma_z - \sigma_\theta)^2 + (\sigma_r - \sigma_z)^2 = 2\sigma_0^2 \tag{b}$$

Instead of using Equation (b), it is more convenient to use the following identical relation in which the second term is independent of r:

$$\frac{3}{2}(\sigma_\theta - \sigma_r)^2 + 2\left(\sigma_z - \frac{\sigma_\theta + \sigma_r}{2}\right)^2 = 2\sigma_0^2 \tag{c}$$

Observing that the maximum principal stresses occur at $r = a$, making use of the stress expressions in Equation 7.7, relation (c) gives

$$3\left(\frac{b^2 p_e}{b^2 - a^2}\right) + \left(\sigma_z - \frac{a^2 p_e}{b^2 - a^2}\right)^2 = \sigma_0^2 \tag{d}$$

The magnitude of the axial stress and consequently the yielding pressure depends on the end conditions, which are summarized in the discussion leading to Equation 7.8. Thus for:

Closed ends: $\sigma_z = (a^2 p_e)/(b^2 - a^2)$ and yield is initiated in the cylinder when the applied pressure is given by

$$p_e = \frac{\sigma_0}{\sqrt{3}}\left(1 - \frac{a^2}{b^2}\right) \tag{7.11a}$$

Open ends: In this case, the axial stress is zero and the pressure is

$$p_e = \frac{\sigma_0}{\sqrt{3}}\left[\frac{1 - a^2/b^2}{\left(1 + \frac{a^4}{3b^4}\right)^{\frac{1}{2}}}\right] \tag{7.11b}$$

Plane strain condition: Using the third relation in Equation 7.8, yield is initiated when the pressure becomes

$$p_e = \frac{\sigma_0}{\sqrt{3}}\left\{\frac{1 - a^2/b^2}{\left[1 + (1 - 2\upsilon)^2 \frac{a^4}{3b^4}\right]^{\frac{1}{2}}}\right\} \tag{7.11c}$$

It is interesting to observe that the formulas for p_e, except for the plane strain condition, are functions of the ratio b/a, and do not include Poisson's ratio. As an example, consider a cylinder with $b/a = 2$, $\upsilon = (1/3)$, and $\sigma_0 = 36$ ksi (250 MPa), then Tresca's criterion gives $p_e = 13.5$ ksi (93.75 MPa) while the following values are obtained using Mises' criterion:

Closed ends—$p_e = 15.59$ ksi (108.25 MPa)
Open ends—$p_e = 15.43$ ksi (107.13 MPa)
Plane strain condition—$p_e = 15.57$ ksi (108.07 MPa)

The differences in the numerical values between the two criteria and the various end conditions are not significant. As b increases to infinity, the pressure required to initiate yield is $p_e = \sigma_0/\sqrt{3}$.

7.1.2 External Pressure Only

When the cylinder is subjected to external pressure only, p_1, the stress distribution can be obtained from Equations 7.4 and 7.5 by putting $p = 0$, or by interchanging a and b in Equation 7.7. The result is

$$\sigma_r = -\frac{b^2 p_1}{b^2 - a^2}\left(1 - \frac{a^2}{r^2}\right)$$

$$\sigma_\theta = -\frac{b^2 p_1}{b^2 - a^2}\left(1 + \frac{a^2}{r^2}\right) \tag{7.12}$$

$$\sigma_z = E\varepsilon_z - \frac{2\upsilon b^2 p_1}{b^2 - a^2}$$

In this case, both radial and tangential stresses are compressive, maximum radial stress occurs at $r = b$, $(\sigma_r)_{max} = -p_1$, maximum tangential stress is at $r = a$, $(\sigma_\theta)_{max} = -(2b^2 p_1)/(b^2 - a^2)$, which is greater than $(\sigma_r)_{max}$. The magnitude of maximum axial stress is in-between the other two stresses as in the case of internal pressure.

The radial displacement is

$$\frac{u}{r} = -\upsilon\varepsilon_z + \frac{(1+\upsilon)b^2 p_1}{E(b^2 - a^2)}\left[\frac{a^2}{r^2} - (1 - 2\upsilon)\right] \tag{7.13}$$

For a cylinder with open ends, $\sigma_z = 0$, $\varepsilon_z = (2\upsilon\, b^2 p_1)/[(b^2 - a^2)E]$, and Equation 7.13 gives

$$\frac{u}{r} = \frac{b^2 p_1}{(b^2 - a^2)E}\left[1 - \upsilon + (1+\upsilon)\frac{a^2}{r^2}\right] \tag{7.14}$$

As the external pressure is increased, yielding begins at $r = a$, and according to Tresca's criterion, $(\sigma_\theta - \sigma_r)_{max} = -\sigma_0$, or $(2b^2 p_1)/(b^2 - a^2) = \sigma_0$, the pressure that initiates yield is

$$(p_1)_e = \frac{\sigma_0}{2}\left(1 - \frac{a^2}{b^2}\right) \tag{7.15}$$

which is identical to Equations 7.10 for the case of internal pressure. As the pressure increases further, a circular plastic zone forms around the inner boundary and spreads inward.

When a thick tube already under external pressure less than the pressure necessary to initiate yield is subjected to gradually increasing internal pressure p, Equations 7.4 give

$$\sigma_\theta - \sigma_r = \frac{2(p - p_1)}{b^2 - a^2}\left(\frac{a^2 b^2}{r^2}\right)$$

The magnitude of $\sigma_\theta - \sigma_r$ at any radius r $(a \le r \le b)$ increases when $p > p_1$, and initiation of yield according to Tresca's criterion begins at $r = a$ when

$$p - p_1 = \frac{\sigma_0}{2}\left(1 - \frac{a^2}{b^2}\right) \tag{7.16}$$

Thus, the internal pressure is $p = p_1 + p_e$, where p_e is given by Equations 7.10.

7.1.3 Elastic–Plastic Expansion of a Cylinder Subjected to Internal Pressure

As the pressure is increased beyond initiating yield, $p > p_e$, a plastic zone is formed around the inner surface and spreads inward. At a certain stage of the pressurization, let $r = c$ denote the radius of the plastic zone, as shown shaded in Figure 7.2, the remaining area of the cross section is still elastic. In the remaining discussion, Tresca's criterion is used for the initiation of yield.

In the elastic region, $c \le r \le b$, Lame's solution in Equations 7.3 is applicable, the constants A and B are now determined from the boundary conditions, $\sigma_r = 0$ at $r = b$, and at $r = c$, $\sigma_\theta - \sigma_r = \sigma_0$. These conditions give $A = -(\sigma_0 c^2/2)$, $B = (\sigma_0 c^2/2b^2)$. It follows that the stresses in the elastic region are

$$\sigma_r = -\frac{\sigma_0 c^2}{2b^2}\left(\frac{b^2}{r^2} - 1\right)$$

$$\sigma_\theta = \frac{\sigma_0 c^2}{2b^2}\left(\frac{b^2}{r^2} + 1\right) \tag{7.17}$$

$$\sigma_z = E\varepsilon_z + \upsilon\sigma_0\frac{c^2}{b^2}, \quad c \le r \le b$$

The axial strain, $E\varepsilon_z$, depends on the end conditions of the cylinder, as indicated in Equation 7.8, its value cannot be determined unless the deformation of the plastic region is considered except, of course, in case of plane strain condition where $\varepsilon_z = 0$.

The radial displacement in the elastic region is determined from the expression, $u = r\varepsilon_\theta = (r/E)[\sigma_\theta - \upsilon\,(\sigma_r + \sigma_z)]$, making use of Equation 7.17. The result is

$$u = -\upsilon r\varepsilon_z + \frac{\sigma_0 c^2}{2b^2 E}(1+\upsilon)\left[\frac{b^2}{r} + (1 - 2\upsilon)r\right], \quad c \le r \le b \tag{7.18}$$

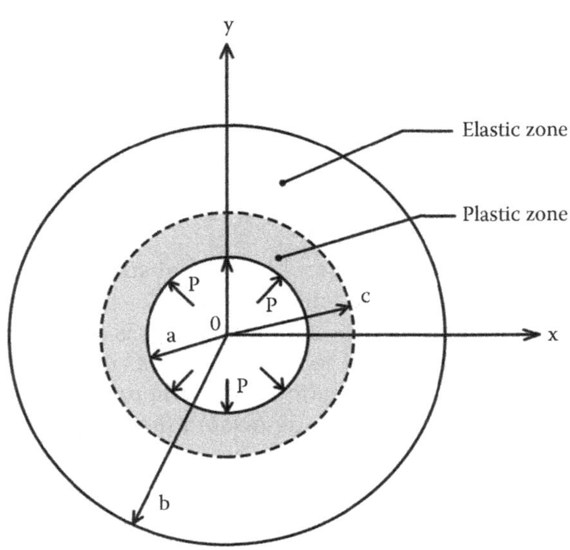

FIGURE 7.2
Plastic zone around inner boundary.

In particular, the displacement of the outer boundary is $(u_b/b) = -\upsilon\varepsilon_z + (1-\upsilon^2)(\sigma_0 c^2/Eb^2)$.

In the plastic region, $a \le r \le c$, the axial stress is *assumed* to be the intermediate principal stress between the tangential and radial stresses as in the elastic region. As will be shown later, this assumption will impose certain restrictions on the wall thickness of the cylinder for the solution to be valid. Since the material is assumed to be elastic–perfectly plastic (no hardening occurs) and Tresca's criterion is used for yielding, the applicable equilibrium equation, $r(d\sigma_r/dr) = \sigma_\theta - \sigma_r = \sigma_0$, enable us to find the stresses in the plastic region. Integrating the equilibrium equation and determining the constant of integration from the boundary condition: $\sigma_r = -p$ at $r = a$, the radial and tangential stresses are

$$\begin{aligned} \sigma_r &= -p + \sigma_0 \,\mathrm{Ln}\frac{r}{a} \\ \sigma_\theta &= \sigma_0 - p + \sigma_0 \,\mathrm{Ln}\frac{r}{a}, \quad a \le r \le c \end{aligned} \tag{e}$$

At the elastic/plastic boundary, $r = c$, the radial stress must be continuous. This implies that $-(\sigma_0 c^2/2b^2)((b^2/c^2) - 1) = -p + \sigma_0 \mathrm{Ln}(c/a)$, which gives the relation between the applied pressure and the elastic/plastic boundary as

$$p = \frac{\sigma_0}{2}\left[1 - \frac{c^2}{b^2} + \mathrm{Ln}\frac{c^2}{a^2}\right] \tag{7.19a}$$

When $c = b$, the ultimate pressure is reached

$$p_u = \sigma_0 \,\mathrm{Ln}\frac{b}{a} \tag{7.19b}$$

and Equations (e) upon using Equation 7.19a furnishes

$$\sigma_r = -\frac{\sigma_0}{2}\left[1 - \frac{c^2}{b^2} + \mathrm{Ln}\frac{c^2}{r^2}\right], \quad a \le r \le c \tag{7.20a}$$

$$\sigma_\theta = \frac{\sigma_0}{2}\left[1 + \frac{c^2}{b^2} - \mathrm{Ln}\frac{c^2}{r^2}\right], \quad a \le r \le c \tag{7.20b}$$

Note that the expressions of the radial and tangential stresses do not depend on the end condition of the cylinder. A plot of the variations of σ_r/σ_0 and σ_θ/σ_0 with r/b is shown in Figure 7.3 for $b/a = 2$; see Reference 2.2. The radial stress is always negative while the tangential stress is positive and its peak values occur at points on the elastic/plastic boundary. To determine the axial stress in the plastic region where the material is incompressible ($\varepsilon_r^p + \varepsilon_\theta^p + \varepsilon_z^p = 0$), we make use of Tresca's associated flow rule to conclude that the plastic strain increments are

$$d\varepsilon_\theta^p = -d\varepsilon_r^p > 0, \quad d\varepsilon_z^p = 0 \tag{f}$$

Because the plastic part of the axial strain is zero, the total axial strain is entirely elastic and is determined using Hooke's law, $\sigma_z = E\varepsilon_z + \upsilon\,(\sigma_r + \sigma_\theta)$. Making use of Equations 7.20a and

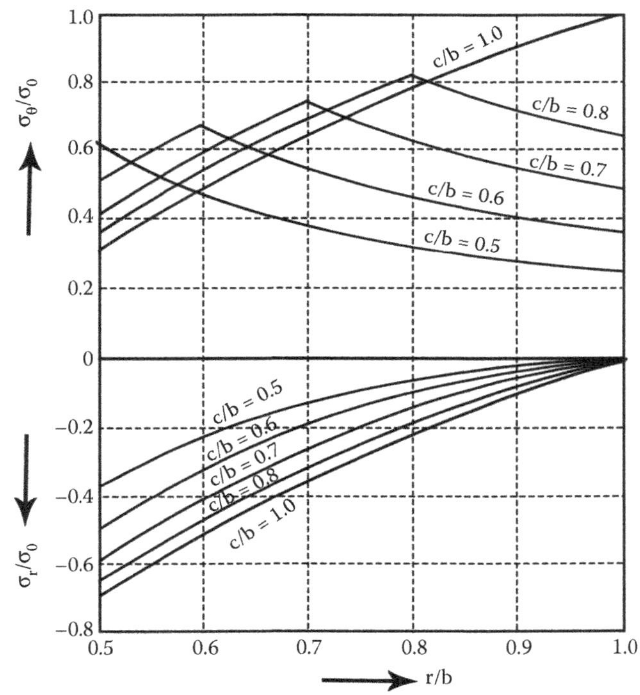

FIGURE 7.3
Distribution of radial and tangential stresses for a cylinder with $b/a = 2$ [2.2].

7.20b, the axial stress in the plastic zone is

$$\sigma_z = E\varepsilon_z + \upsilon\sigma_0\left(\frac{c^2}{b^2} - \text{Ln}\frac{c^2}{r^2}\right), \quad a \le r \le c \tag{7.20c}$$

In order to find the axial strain in the plastic region, it is necessary to consider overall vertical equilibrium of the cylinder. The resultant axial force is given by

$$F = 2\pi\int_a^b r\sigma_z dr = \pi(b^2 - a^2)E\varepsilon_z + 2\pi\upsilon\int_a^b r(\sigma_r + \sigma_\theta)dr$$

Instead of using expression (7.15) to perform the integration, Equation 7.1 can be used to show that $r(\sigma_r + \sigma_\theta) = (d/dr)(r^2\sigma_r)$, performing the integration and using the conditions $\sigma_r = 0$ at $r = b$ and $\sigma_r = -p$ at $r = a$, the resultant axial force is given by $F = \pi(b^2 - a^2)E\varepsilon_z + 2\pi\upsilon\, a^2p$, or $E\varepsilon_z = \sigma_z - (2\upsilon\, a^2p)/(b^2 - a^2)$, as in the elastic case (Equation 7.7). Hence, the axial strain in the plastic region is identical to the elastic case and is given by Equation 7.8 for the various end conditions.

The radial displacement in the plastic region is obtained by noting that since the plastic strains in partially plastic cylinder are negligible, $\varepsilon_r^p + \varepsilon_\theta^p = 0$, it follows that the sum of the radial and tangential strains consists of the elastic component only, that is, $\varepsilon_r + \varepsilon_\theta = (1/E)[(1 - \upsilon)(\sigma_r + \sigma_\theta) - 2\upsilon\sigma_z]$, or equivalently

$$\frac{du}{dr} + \frac{u}{r} = -2\upsilon\varepsilon_z + \frac{1 - 2\upsilon}{2G}(\sigma_r + \sigma_\theta) \tag{g}$$

In Equation (g), G denotes the shearing modulus of the material. Since $\sigma_r + \sigma_\theta = (1/r)(d/dr)(r^2\sigma_r)$, Equation (g) can be written as follows:

$$\frac{d}{dr}\left[ru - \frac{1 - 2\upsilon}{2G}r^2\sigma_r\right] = -2\upsilon r\varepsilon_z \tag{h}$$

Integrating Equation (h) with respect to r, finding the constant of integration using the condition $\sigma_r = 0$ at $r = b$, and comparing the result with the value of u given in Equation 7.13, it follows that the radial displacement in the plastic region, $a \leq r \leq c$, is

$$\frac{u}{r} = -\upsilon\varepsilon_z + \frac{(1 - 2\upsilon)\sigma_r}{2G} + \frac{(1 - \upsilon)\sigma_0 c^2}{2Gr^2} \tag{7.21}$$

which is identical to expression (7.13) of the displacement in the elastic region (replace σ_r by its value from the first equation in Equation 7.12 and simplify the algebra). In particular, the displacement of the outer boundary, obtained from either equation, is

$$\frac{u_b}{b} = -\upsilon\varepsilon_z + (1 - \upsilon^2)\frac{\sigma_0 c^2}{Eb^2}$$

The solution given in Equations 7.17 through 7.21 hinges on the assumption that the axial stress is the intermediate principal stress between the tangential and radial stresses throughout the plastic region. This assumption requires the principal stresses to satisfy the inequalities, $\sigma_r \leq E\varepsilon_z + \upsilon(\sigma_r + \sigma_\theta) \leq \sigma_\theta$, or using Tresca's yield criterion to substitute for σ_θ, the following inequalities must be satisfied:

$$\sigma_r \leq E\varepsilon_z + \upsilon(2\sigma_r + \sigma_0) \leq \sigma_r + \sigma_0 \tag{7.22a}$$

The first inequality requires that $(1 - 2\upsilon)\sigma_r \leq E\varepsilon_z + \sigma_0$, which is satisfied because the right-hand side is positive for all end conditions of the cylinder.

The second inequality requires that

$$E\varepsilon_z - (1 - 2\upsilon)\sigma_r \leq (1 - \upsilon)\sigma_0 \quad \text{or} \quad E\varepsilon_z + (1 - 2\upsilon)p \leq (1 - \upsilon)\sigma_0 \tag{7.22b}$$

And using Equation 7.8 to substitute the values of the axial strain for the three end conditions, the inequality (7.22b) gives the critical values of the pressure in the cylinder [7.2].

$$\frac{p_u}{\sigma_0} \leq \begin{cases} \dfrac{1 - \upsilon}{1 - 2\upsilon}\left(1 - \dfrac{a^2}{b^2}\right), & \text{closed-end} \\[2ex] \dfrac{(1 - \upsilon)\left(1 - \dfrac{a^2}{b^2}\right)}{\left(1 - 2\upsilon - \dfrac{a^2}{b^2}\right)}, & \text{open-end} \\[2ex] \dfrac{1 - \upsilon}{1 - 2\upsilon}, & \text{plane-strain} \end{cases} \tag{7.23}$$

Hence the maximum value of the pressure is obtained from Equations 7.19 by setting $c = b$, that is, $(p_u/\sigma_0) = \text{Ln}(b/a)$. Thus, the values of the wall ratios for which the solution is valid are obtained from

$$\text{Ln}\frac{b}{a} \le \beta \quad \begin{cases} \dfrac{1-\upsilon}{1-2\upsilon}\left(1-\dfrac{a^2}{b^2}\right), & \text{closed-end} \\[2ex] \dfrac{(1-\upsilon)\left(1-\dfrac{a^2}{b^2}\right)}{\left(1-2\upsilon-\dfrac{a^2}{b^2}\right)}, & \text{open-end} \\[2ex] \dfrac{1-\upsilon}{1-2\upsilon}, & \text{plane-strain} \end{cases} \tag{7.24}$$

For $\upsilon = 1/3$, the limiting values of the wall ratio (b/a) are found as follows:

Closed ends: $b/a = 7.10$, open ends: $b/a = 7.40$, and for plane strain condition: $b/a = 7.38$.

EXAMPLE 7.1

It is known that when a thick-walled cylinder is subjected to either internal pressure p or external pressure p_1, the maximum elastic stress is the tangential stress at the inner boundary $r = a$ (Equations 7.7 and 7.12).

Show that if the cylinder is subjected to both pressures p and p_1 at the same time there are certain ratios of p_1/p where the maximum elastic stress is the tangential stress at the outer boundary $r = b$.

Solution

In Equations 7.4a denote the ratios b/a by R and p_1/p by p^*, then we have

$$(\sigma_\theta)_{r=a} = \frac{p}{R^2-1}[1 + R^2(1-2p^*)]$$

$$(\sigma_\theta)_{r=b} = \frac{p}{R^2-1}[2 - (R^2+1)p^*]$$

Consider the following two cases:

1. $p^* = 1$: The magnitude of the ratio, $(\sigma_\theta)\ r = a/(\sigma_\theta)r = b$, obtained from the above equations, is equal to one.
2. $p^* = (1+R^2)/(2R^2)$: The magnitude of the same ratio of the tangential stresses is equal to zero.

This implies that for values of p^* in the range, $(1+R^2)/(2R^2) < p^* < 1$, or equivalently $(b^2+a^2)/(2b^2) < p^* < 1$, the following expression is valid:

$$0 < \frac{(\sigma_\theta)_{r=a}}{(\sigma_\theta)_{r=b}} < 1$$

which proves that the maximum elastic stress in the cylinder is the tangential stress at $r = b$ for all values of p^* given by the above inequality range; Reference 7.3.

EXAMPLE 7.2

A closed-ended thick-walled tube, radii a and b, $b > a$, is subjected to an internal pressure p and a torque T about the longitudinal axis (Figure 7.4). Find the ratio T/p to initiate simultaneous yielding at both internal and external surfaces. Use Mises' criterion.

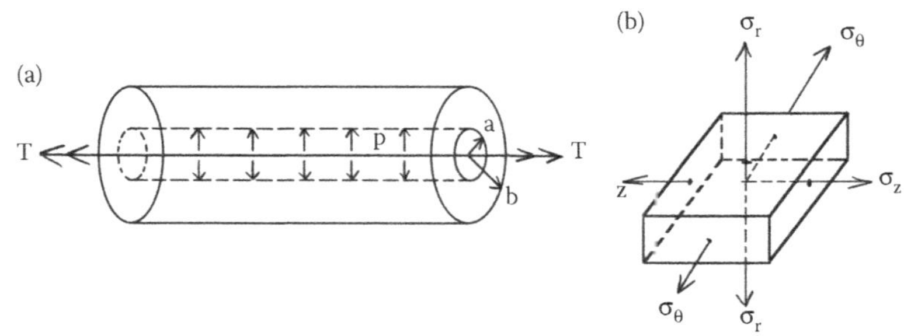

FIGURE 7.4
(a) Cylindrical tube subjected to internal pressure and torque; (b) stresses on tube element.

Solution

Mises' yield criterion in terms of general cylindrical stress is (Chapter 4 Equation (a) in Section 4.4.2)

$$[(\sigma_\theta - \sigma_r)^2 + (\sigma_r - \sigma_z)^2 + (\sigma_z - \sigma_\theta)^2 + 6\tau_{r\theta} + 6\tau_{\theta z} + 6\tau_{rz}]^{\frac{1}{2}} = \tau_0 \quad \text{(a)}$$

where τ_0 is the shearing stress of the material. The stresses acting on an element of the tube are shown in Figure 7.4b. Writing $b/a = R$, we have for a thick-walled cylinder with closed ends, $\sigma_z = p/(R^2 - 1)$, and using Equations 7.12, we obtain

$$\sigma_\theta - \sigma_r = \frac{2b^2p}{(R^2-1)r^2}$$

$$\sigma_r - \sigma_z = \sigma_z - \sigma_\theta = -\frac{b^2p}{(R^2-1)r^2}$$

Noting that the shear stress due to the torque is obtained from the well-known formula

$$\tau = \frac{Tr}{J} = \frac{Tr}{(\pi/32)(b^4 - a^4)} = \frac{32Tr}{\pi(R^4-1)a^4}$$

Inserting the above stresses in Equation (a) gives

$$\frac{6p^2b^4}{(R^2-1)^2r^4} + 6\left[\frac{32Tr}{\pi(R^4-1)a^4}\right]^2 = \tau_0^2$$

The condition that yield occurs at the same time at $r = a$ and at $r = b$ is satisfied by the relation

$$\frac{6p^2b^4}{(R^2-1)^2b^4} + 6\left[\frac{32Tb}{\pi(R^4-1)a^4}\right]^2 = \frac{6p^2b^4}{(R^2-1)^2a^4} + 6\left[\frac{32Ta}{\pi(R^4-1)a^4}\right]^2$$

Simplifying the algebra, the ratio T/p is determined as follows:

$$\frac{T}{p} = \frac{\pi b^3(R^2+1)^{\frac{3}{2}}}{32R^6}$$

EXAMPLE 7.3

A closed-ended thick-walled cylinder of internal and external radii a and b, $b > a$, respectively, is subjected to internal pressure p during elastic–plastic expansion, $p_e < p < p_u$. The yield stress of the material in shear is $\tau_0 = \sigma_0/2$ where σ_0 is the yield stress in tension.

a. Find the maximum shearing strain γ in the elastic region and the radial displacement u_b at the outer surface $r = b$.
b. Assuming the same expression for γ to hold in the plastic region, and the expression for the engineering shear strain to be given by $\tau = \tau_0\,(G\,\gamma/\tau_0)^n$, where n is constant and G is the shear modulus of the material, prove that the pressure applied at $r = a$ is given by

$$p = \tau_0\left\{1 - \frac{c^2}{b^2} + \frac{1}{n}\left[\left(\frac{c}{a}\right)^{2n} - 1\right]\right\}$$

Solution

a. The maximum shearing strain is $\gamma_{max} = \dfrac{\tau_{max}}{G} = \dfrac{1}{2G}(\sigma_\theta - \sigma_r)$

   Using Equations 7.17, the maximum shearing strain in the elastic region becomes

$$\gamma_{max} = \frac{1}{2G}\frac{\sigma_0 c^2}{2b^2}\left[\frac{b^2}{r^2} + 1 + \frac{b^2}{r^2} - 1\right] = \frac{\sigma_0 c^2}{2Gr^2}$$

   The radial displacement at $r = b$ is obtained from Equation 7.18

$$u_b = -\frac{\upsilon b}{E}[\sigma_z - \upsilon(\sigma_r + \sigma_\theta)] + (1 - \upsilon^2)\frac{\sigma_0 c^2}{bE}$$

   For closed-ended cylinder $\sigma_z = (1/2)\,(\sigma_r + \sigma_\theta)$, and using Equations 7.17, the displacement assumes the form

$$u_b = \left(\frac{2 - \upsilon}{1 + \upsilon}\right)\frac{\sigma_0 c^2}{4bG}$$

b. In the plastic region, we have $\tau_{r\theta} = \tau_0(G\gamma/\tau_0)^n = (1/2)(\sigma_\theta - \sigma_r)$, and inserting this expression in the equation of equilibrium, $r(d\sigma_r/dr) = \sigma_\theta - \sigma_r$, after using Equations 7.17 for the tangential and radial stresses, we obtain the following relations:

$$\frac{d\sigma_r}{dr} = 2\tau_0 c^{2n} r^{-2n-1}$$

   or

$$\sigma_r = -\tau_0 c^{2n}\left(\frac{r^{-2n}}{n}\right) + K$$

   where the constant K is determined from the condition $\sigma_r = -p$ at $r = a$. Hence,

$$\sigma_r = -p - \frac{\tau_0}{n}\left(\frac{c}{r}\right)^{2n} + \frac{\tau_0}{n}\left(\frac{c}{a}\right)^{2n}$$

   At $r = c$, the radial stress is continuous, which requires equating the above equation to the first equation of Equation 7.17, that is

$$-p - \frac{\tau_0}{n} + \frac{\tau_0}{n}\left(\frac{c}{a}\right)^{2n} = -\tau_0\left(1 - \frac{c^2}{b^2}\right)$$

Hence,

$$p = \left\{1 - \frac{c^2}{b^2} + \frac{1}{n}\left[\left(\frac{c}{a}\right)^{2n} - 1\right]\right\}$$

7.1.4 Residual Stresses

When the pressure in a partially plastic cylinder, $p_e < p < p_u$, is completely released, $p \to 0$, the deformation during unloading is entirely *elastic* and the resulting stresses are obtained by subtracting the stresses determined in Equations 7.7 from those in Equations 7.17 and 7.20. The results in the outer and inner regions are as follows:

In the *elastic region*, $c \leq r \leq b$,

$$\sigma_r = -\frac{\sigma_0}{2}\left(\frac{a^2}{r^2} - \frac{a^2}{b^2}\right)\left(\frac{c^2}{a^2} - \frac{p}{p_e}\right)$$
$$\sigma_\theta = \frac{\sigma_0}{2}\left(\frac{a^2}{r^2} + \frac{a^2}{b^2}\right)\left(\frac{c^2}{a^2} - \frac{p}{p_e}\right) \tag{7.25a}$$
$$\sigma_z = \upsilon\sigma_0\frac{a^2}{b^2}\left(\frac{c^2}{a^2} - \frac{p}{p_e}\right)$$

And in the plastic region, $a \leq r \leq c$,

$$\sigma_r = -\frac{\sigma_0}{2}\left[\left(1 - \frac{a^2}{r^2}\right)\frac{p}{p_e} - \text{Ln}\frac{r^2}{a^2}\right]$$
$$\sigma_\theta = -\frac{\sigma_0}{2}\left[\left(1 + \frac{a^2}{r^2}\right)\frac{p}{p_e} - 2 - \text{Ln}\frac{r^2}{a^2}\right] \tag{7.25b}$$
$$\sigma_z = \upsilon\sigma_0\left[1 - \frac{p}{p_e} + \text{Ln}\frac{r^2}{a^2}\right]$$

Here, p_e and p are given by Equations 7.10 and 7.19a, respectively. Note that Equations 7.25 indicate that the residual axial stress throughout the tube satisfies the relation, $\sigma_z = \upsilon(\sigma_r + \sigma_\theta)$. This is expected because the axial strain is basically elastic and is removed upon unloading.

The residual radial stress is compressive everywhere in the cylinder, its maximum value is obtained by requiring $(d\sigma_r/dr) = 0$ in the region $a \leq r \leq c$ and solving for r, the result is $r = a\sqrt{(p/p_e)}$.

The residual tangential stress is tensile in the outer part of the cylinder and compressive in the inner part vanishing at a radius somewhere in the plastic region, $a \leq r \leq c$.

The third of Equations 7.25b indicates that the residual axial stress is compressive in the plastic region because $(1/2)((p/p_e) - 1) \geq \text{Ln}(r/a)$, that is, within a radius, $r = a\, e^{1/2(p/p_e - 1)}$.

If the pressure is applied again, yield will initiate for the second time when the emerging stresses satisfy the yield criterion. Equations 7.25 provide the following relations:

$$\sigma_\theta - \sigma_r = \sigma_0\frac{a^2}{r^2}\left(\frac{c^2}{a^2} - \frac{p}{p_e}\right), \quad c \leq r \leq b$$
$$\sigma_\theta - \sigma_r = \sigma_0\left(1 - \frac{a^2}{r^2}\frac{p}{p_e}\right), \quad a \leq r \leq c \tag{7.26}$$

The greatest magnitude of $\sigma_\theta - \sigma_r$ occurs at $r = a$ in the plastic region. Application of Tresca's criterion results in

$$\sigma_0\left(1 - \frac{p}{p_e}\right) = -\sigma_0$$

Hence, yielding will restart if $p = 2p_e$. Since $p_e = (\sigma_0/2)(1 - (a^2/b^2))$ and if we let $p = p_u = \sigma_0 \,\mathrm{Ln}(b/a)$, it follows that the critical wall ratio is governed by the relation

$$\mathrm{Ln}\frac{b}{a} = 1 - \frac{a^2}{b^2} \tag{7.27}$$

Equation 7.27 gives $b/a = 2.22$ and the corresponding critical pressure is $p = 0.8\sigma_0$.

Any further reloading gives rise to elastic stresses, which prestress and strengthen the material at the surface of the inner boundary. Such process is known as *autofrettage*. The process has many practical applications in the manufacture of cylindrical tubes and gun barrels, the initial overstrain due to the process strengthens the finish surface of the cylinder and maintains the initial diameter, which is required for adequate performance. For autofrettage tube under closed-end condition, the tube is completely closed by plugs attached to the tube while in the open-end condition floating pistons are used, which allow free expansion.

7.1.5 Compound Cylinders

Compound or multilayer cylinders are designed to resist large internal pressure more efficiently than a single cylinder [2.1,2.2,6.5]. The cylinders are initially prestressed and thus require less material to resist the applied pressure. A typical method of assembling the cylinders is by the "shrink-fit" process; the external radius of an inner layer is made larger than the internal radius of the surrounding layer and so on. The cylinders are assembled by heating the outer layer in order to expand and enclose the inner one. After cooling, contact between the cylinders is achieved with contact stresses at the interfaces.

Consider a compound cylinder consisting of two layers, the inner layer has radii a, b and the radii of the outer layer are b, c, as shown in Figure 7.5. The external radius of the inner

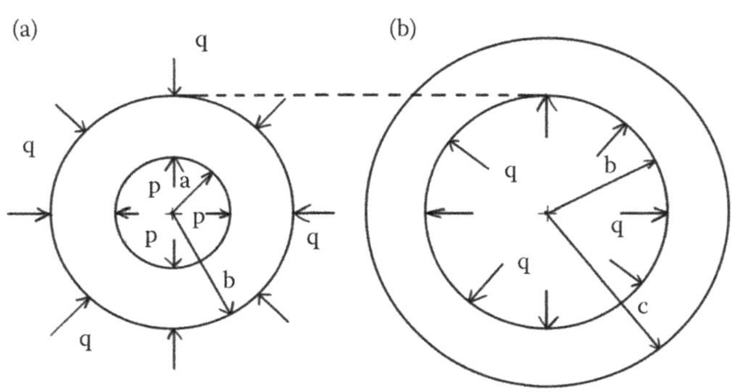

FIGURE 7.5
Compound cylinder, (a) inner layer; (b) outer layer.

layer is made larger than the internal radius of the outer layer by an amount, δ, commonly known as the *shrinkage allowance*. The outer layer or jacket is heated so that it can be fitted around the inner layer. After cooling, contact is achieved and a compound cylinder is formed with interface pressure, say q at $r = b$. Denoting the radial displacements of the outer and inner layers by u_2 and u_1, respectively, the shrinkage allowance is equivalent to the sum of the increase in the radius of the outer layer found from Equation 7.9b and decrease in the outer radius of the inner layer found from Equation 7.14 at $r = b$, that is,

$$u_2 + (-u_1) = \delta \tag{a}$$

Assuming open-ended cylinder and noting that Equation 7.9b gives the displacement of the outer cylinder

$$u_2 = \frac{qb^3}{E(c^2 - b^2)}[1 - \upsilon + (1 + \upsilon)(c^2/b^2)]$$

and Equation 7.14 gives the displacement of the inner cylinder

$$u_1 = \frac{-qb^3}{E(b^2 - a^2)}[1 - \upsilon + (1 + \upsilon)(a^2/b^2)]$$

it follows that the relation between the shrinkage allowance and the corresponding contact pressure is obtained from relation (a) as follows:

$$q = \frac{E\delta}{2b}\left[\frac{(b^2 - a^2)(c^2 - b^2)}{b^2(c^2 - a^2)}\right] \tag{7.28}$$

The stresses in the outer and inner layers of the cylinder due to the pressures p and q can now be determined.

EXAMPLE 7.4

A two-layer thick cylinder is made of two different materials. The inner layer has internal radius a, yield stress $k^2\sigma_0$ (k is constant), and the outer layer has external radius b and yield stress σ_0. The cylinder is subjected to internal pressure p. Find

1. The optimum pressure needed to cause simultaneous yielding in both layers using Tresca's yield criterion
2. The initial shrinkage allowance

Solution

Suppose that r is the common radius between the two layers, $a < r < b$ (Figure 7.6a).

1. As the pressure p gradually increases, the radial stress induces a pressure $\sigma_r = -q$ at distance r, which acts internally on the outer cylinder and externally on the inner one (Figure 7.6b and c).

   The pressure that causes yield in the outer layer according to Tresca's criterion, obtained from Equation 7.10, is

$$q_e = \frac{\sigma_0}{2}\left(1 - \frac{r^2}{b^2}\right) \tag{a}$$

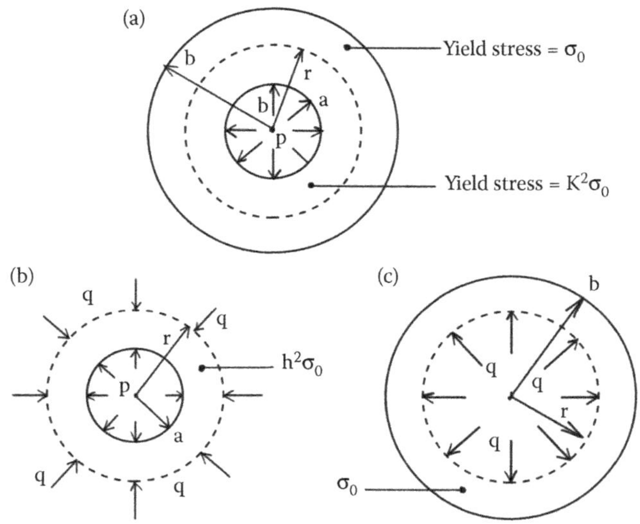

FIGURE 7.6
(a) Compound cylinder; (b) inner cylinder; and (c) outer cylinder.

The inner cylinder is subjected to internal pressure p and external pressure q (Figure 7.6c). Using Equation 7.16, the inner cylinder will yield at radius r when

$$p = \frac{k^2\sigma_0}{2}\left(1 - \frac{a^2}{r^2}\right) + \frac{\sigma_0}{2}\left(1 - \frac{r^2}{b^2}\right) \tag{b}$$

The maximum pressure in Equation (b) is found by requiring $dp/dr = 0$, which gives $r = \sqrt{kab}$, and the optimum pressure is

$$p_{opt} = \frac{\sigma_0}{2}\left(1 - 2k\frac{a}{b} + k^2\right) \tag{c}$$

2. The shrinkage allowance required for assembling the cylinders is found from Equation 7.28 by replacing b by $\sqrt{kab}$, c by b, and making use of the relation $q = (\sigma_0/2b)(b - ka)$. The result is

$$\delta = (b + a)\sqrt{\frac{a}{b}}\frac{\sigma_0 k^{\frac{3}{2}}}{E} \tag{d}$$

For $k = 1$, expressions (c) and (d) reduce to $p_{opt} = \sigma_0(1 - (a/b))$ and $\delta = (b + a)(a/b)^{1/2}(\sigma_0/E)$.

Multilayer cylinders: Consider a compound cylinder consisting of n layers made of the same material, which is assumed to be elastic–perfectly plastic with yield stress, σ_0. The internal and external radii of the various layers are (a, r_1), (r_1, r_2), (r_2, r_3), ..., (r_{n-1}, b). The cylinder is subjected to internal pressure p at its internal boundary $r = a$ while the external boundary $r = b$ is free from stress. Denoting the interface pressures between the various layers by $p_1, p_2, p_3, \ldots, p_{n-1}$, the pressure that causes simultaneous yield in all the layers is governed by the following equations obtained using Equation 7.16.

$$p - p_1 = \frac{\sigma_0}{2}\left[1 - \left(\frac{a}{r_1}\right)^2\right]$$
$$p_1 - p_2 = \frac{\sigma_0}{2}\left[1 - \left(\frac{r_1}{r_2}\right)^2\right] \quad \text{(e)}$$
$$\cdots$$
$$\cdots$$
$$p_{n-1} - p_n = \frac{\sigma_0}{2}\left[1 - \left(\frac{r_{n-1}}{b}\right)^2\right]$$

Here, $p_n = 0$ since the outer boundary is free from stress. Adding Equations (f) gives the pressure as function of the radii of the various layers.

$$p = \frac{\sigma_0}{2}\left[n - \left\{\left(\frac{a}{r_1}\right)^2 + \left(\frac{r_1}{r_2}\right)^2 + \left(\frac{r_2}{r_3}\right)^2 + \cdots + \left(\frac{r_{n-1}}{b}\right)^2\right\}\right] \quad (7.29)$$

In order to arrive at the maximum pressure, the following derivatives are needed:

$$\frac{\partial p}{\partial r_1} = \frac{\partial p}{\partial r_2} = \cdots = \frac{\partial p}{\partial r_{n-1}} = 0 \quad \text{(f)}$$

Making use of Equation 7.29, the first derivative in Equation (g) yields $(2a^2/r_1^3) - (2r_1/r_2^2) = 0$ or, $r_1^2 = ar_2$, $(r_1/a) = (r_2/r_1)$. Similarly, the second derivative furnishes $r_2^2 = r_1 r_3$ or, $(r_2/r_1) = (r_3/r_2)$, and so on. Thus, the maximum pressure that initiates simultaneous yielding corresponds to equal wall thicknesses of the various layers.

$$\frac{r_1}{a} = \frac{r_2}{r_1} = \frac{r_3}{r_2} = \cdots = \frac{b}{r_{n-1}} \quad \text{(g)}$$

Observe that the radii of the cylinder: a, r_1, r_2, ..., b form a geometric sequence. Also, upon cross multiplying the first by the second by the third ratios in Equation (h), and so on, it is evident that each ratio is equivalent to $(b/a)^{1/n}$, and the magnitude of the maximum pressure causing simultaneous yield in the layers follows from Equation 7.29

$$p_e = n\frac{\sigma_0}{2}\left[1 - \left(\frac{a}{b}\right)^{\frac{2}{n}}\right] \quad (7.30)$$

The pressure that causes the cylinder to be fully plastic is obtained when all the layers become fully plastic by satisfying Tresca's yield criterion, $\sigma_\theta - \sigma_r = \sigma_0$, and equilibrium $(\partial\sigma_r/\partial r) = (\sigma_\theta - \sigma_r)/r = (\sigma_0/r)$, that is, $\sigma_r = \sigma_0 \ln r + C$. The constant is found from the boundary condition. For the first layer, the radial stress at $r = a$ is $-p$, $C = -p - \sigma_0 \ln a$. Hence,

$$\sigma_r = \sigma_0 \ln\frac{r}{a} - p$$

Since at $r = r_1$, $\sigma_r = -p_1$, it follows that for the first layer

$$-p + p_1 = \sigma_0\, \mathrm{Ln}\frac{a}{r_1}$$

TABLE 7.1

Variations of p_e and p_u with n and (b/a)

$\frac{b}{a}$	1.5	2	3	4	7	10
			$\frac{p_e}{(\sigma_0/2)} = n\left[1 - \left(\frac{a}{b}\right)^{\frac{2}{n}}\right]$			
n = 1	0.556	0.750	0.890	0.938	0.980	0.990
2	0.667	1.000	1.333	1.500	1.714	1.800
3	0.711	1.110	1.558	1.809	2.180	2.354
4	0.734	1.122	1.691	2.000	2.488	2.735
7	0.766	1.258	1.886	2.289	2.985	3.374
10	0.779	1.295	1.973	2.421	3.224	3.690
$\frac{p_u}{(\sigma_0/2)} = \mathrm{Ln}\left(\frac{b}{a}\right)^2$	**0.811**	**1.386**	**2.197**	**2.773**	**3.892**	**4.610**

Similarly, for the other layer a

$$-p_1 + p_2 = \sigma_0 \ln \frac{r_1}{r_2}$$
$$-p_2 + p_3 = \sigma_0 \ln \frac{r_2}{r_3}$$
$$\cdots$$
$$\cdots$$
$$-p_{n-1} + p_n = \sigma_0 \ln \frac{r_{n-1}}{b}$$

Adding the above equations and noting that $p_n = 0$, it follows that the fully plastic pressure is

$$p_u = \sigma_0 \ln \frac{b}{a} \tag{7.31}$$

Numerical values of p_e and p_u for various layers and wall thicknesses are given in Table 7.1.

It should be mentioned that in deriving the results of this section significant changes in geometry of the cylinder are neglected (see Reference 2.2).

7.1.6 Influence of Strain Hardening of the Material

The previous discussion is concerned with thick-walled cylinders made of elastic–perfectly plastic material. In this section, we discuss the influence of work-hardening strain laws in finding the stress–strain distributions and the corresponding pressure in the cylinder. The case of thick-walled cylinder subjected to internal pressure discussed in Section 7.1.1 is revisited by including the strain-hardening properties of the material. For general work-hardening strain laws discussed in Chapter 4, numerical methods are needed to complete the analysis. Explicit closed-form solutions are available for materials with elastic-linearly hardening stress–strain relations.

Consider a material with the following stress–strain relation in the standard tension test:

$$\sigma = \sigma_0 + E_p \varepsilon^p \tag{7.32}$$

Here, σ is the current normal stress in the tensile test, σ_0 is the yield stress, and E_p is the plastic modulus, usually a fraction of Young's modulus E (Equation 4.30). In this case, Tresca's yield criterion is

$$\sigma_\theta - \sigma_r = \sigma \tag{a}$$

The constant volume condition, $d\varepsilon_r^p + d\varepsilon_\theta^p + d\varepsilon_z^p = 0$, gives (since $d\varepsilon_z^p = 0$)

$$d\varepsilon_\theta^p = -d\varepsilon_r^p \tag{b}$$

It is shown in Section 7.1.1 that the total hoop strain in the cylinder is given by Equation 7.21. Subtracting the elastic hoop strain, $\varepsilon_\theta = -\upsilon\varepsilon_z + (1/2G)[(1-\upsilon)\sigma_\theta - \upsilon\sigma_r]$, from the hoop strain in Equation 7.21, the plastic hoop strain is found as

$$\varepsilon_\theta^p = \frac{1-\upsilon^2}{E}\left(\sigma_0\frac{c^2}{r^2} - \sigma\right), \quad a \le r \le c \tag{7.33}$$

Substituting expression (7.33) for the plastic hoop strain into Equation 7.32, making use of the yield criterion (a), it is readily confirmed that

$$\sigma_\theta - \sigma_r = \sigma_0\left[\frac{1 + (1-\upsilon^2)\dfrac{E_p c^2}{Er^2}}{1 + (1-\upsilon^2)\dfrac{E_p}{E}}\right] \tag{c}$$

Inserting expression (c) into the equilibrium equation, $\partial\sigma_r/\partial r = (\sigma_\theta - \sigma_r)/r$, performing the straightforward integration, the radial stress is reached

$$\sigma_r = \frac{\sigma_0}{1 + (1-\upsilon^2)\dfrac{E_p}{E}}\left[\ln r - \frac{(1-\upsilon^2)E_p c^2}{2Er^2}\right] + C, \quad a \le r \le c \tag{d}$$

The constant of integration, C, is found from the continuity of the radial stress at $r = c$. Equating the radial stress in Equation (d) to that in the first equation of Equation 7.17 at $r = c$, constant C is found as

$$C = \sigma_0\left[\frac{(1-\upsilon^2)\dfrac{E_p}{2E} - \ln c}{1 + (1-\upsilon^2)\dfrac{E_p}{E}} + \frac{1}{2}\left(\frac{c^2}{b^2} - 1\right)\right] \tag{e}$$

It follows that the radial stress in the plastic region of the cylinder is

$$\left[1 + (1-\upsilon^2)\frac{E_p}{E}\right]\sigma_r = -\frac{\sigma_0}{2}\left[1 - \frac{c^2}{b^2} + \ln\frac{c^2}{r^2} + (1-\upsilon^2)\frac{E_p}{E}\left(\frac{c^2}{r^2} - \frac{c^2}{b^2}\right)\right] \tag{7.34a}$$

The tangential stress follows from Equation (c)

$$\left[1+(1-\upsilon^2)\frac{E_p}{E}\right]\sigma_\theta = \frac{\sigma_0}{2}\left[1+\frac{c^2}{b^2}-\ln\frac{c^2}{r^2}+(1-\upsilon^2)\frac{E_p}{E}\left(\frac{c^2}{r^2}+\frac{c^2}{b^2}\right)\right] \tag{7.34b}$$

And the axial stress for the appropriate end condition is determined from

$$\sigma_z = E\varepsilon_z + \upsilon(\sigma_r + \sigma_\theta) \tag{7.34c}$$

where $E\varepsilon_z$ is given in Equation 7.8. Note that when $E_p = 0$, Equations 7.34 reduce to Equation 7.20 for the elastic–perfectly plastic material.

At $r = a$, $\sigma_r = -p$. Hence, the internal pressure acting on the cylinder when the linear work-hardening property of the material is included in the analysis is given by

$$\left[1+(1-\upsilon^2)\frac{E_p}{E}\right]p = \frac{\sigma_0}{2}\left[1-\frac{c^2}{b^2}+\ln\frac{c^2}{a^2}+(1-\upsilon^2)\frac{E_p}{E}\left(\frac{c^2}{a^2}-\frac{c^2}{b^2}\right)\right] \tag{7.35}$$

As expected, the pressure in Equation 7.35 is greater than that in Equation 7.14 for cylinders with elastic-perfectly materials. If $E_p = 0$, Equation 7.35 reduces to Equation 7.19. Subtracting Equation 7.19 from 7.35, the difference in pressure is

$$\Delta p = (1-\upsilon^2)\frac{\sigma_0}{2}\left[\frac{c^2/a^2 - 1 - \ln(c^2/a^2)}{1-\upsilon^2+E/E_p}\right] \tag{f}$$

In order to assess the influence of the work-hardening property of the material on the response of the cylinder, the pressures in Equations 7.19 and 7.35 are plotted versus a suitable parameter, which describe the deformations in the cylinders. A suitable common parameter is the radial extension of either internal or external boundary of each cylinder, that is, either u_a at $r = a$ or u_b at $r = b$.

Consider a thick-walled cylinder in plane strain condition. Assuming $\upsilon = 0.3$, $E_p = 0.1$ E, and $b/a = 2$, Equation 7.18 gives the radial displacement of the outer boundary, $u_b = 0.91$ $(\sigma_0 c^2/Eb)$. A suitable parameter to describe the expansion of the cylinders is

$$1.1\frac{Eu_b}{\sigma_0 b} = \frac{c^2}{b^2} \tag{g}$$

Denoting c^2/b^2 by x and noting that $\ln(c^2/a^2) = \ln(c^2/b^2) + 2\ln(b/a)$, the internal pressure for the elastic–perfectly plastic material (Equation 7.19) is

$$\frac{p}{\sigma_0} = \frac{1}{2}(2.386 - x + \ln x) \tag{7.36a}$$

And the pressure for the work-hardening material (Equation 7.35)

$$\frac{p}{\sigma_0} = 0.458(2.386 - 0.727x + \ln x) \tag{7.36b}$$

The variations of the pressure in Equations 7.36 are plotted in Figure 7.7 for $E_p = 0$ and $E_p = 0.1$ E. The influence of the assumed work hardening is to increase the pressure by about 0.5 σ_0. The assumption that the axial stress, σ_z, is the intermediate principal stress

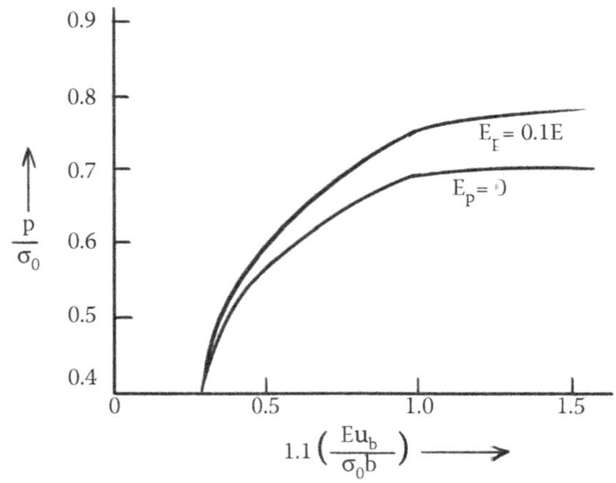

FIGURE 7.7
Pressure expansion curves for a thick-walled cylinder with $b/a = 2$, $\nu = 0\ 3$.

between the tangential and radial stresses has been shown to be valid for larger wall ratios than when work-hardening strains are neglected (References 2.1 and 2.2).

7.2 Thick-Walled Spherical Shell

This section deals with the response of thick-walled spherical vessel subjected to constant internal and external pressures. Figure 7.8 shows a spherical shell of internal and external radii a, b subjected to internal and external pressures p and p_1, respectively. Because of spherical symmetry, it is convenient to use spherical coordinates r, θ, ϕ, with the origin of coordinates at the center of the sphere, θ is the angle made by r with the vertical axis, and ϕ is the angle made by projection of r onto the horizontal plane. The rectangular coordinates x, y, and z are centered at the same origin with the z-axis along the vertical axis and the x and y axes are situated in the horizontal plane. The relations between x, y, and z and the spherical coordinates are as follows:

$$x = r\sin\theta\cos\phi, \quad y = r\sin\theta\sin\phi, \quad z = r\cos\theta$$

Because of symmetry of loads and geometry, the stresses σ_r, $\sigma_\theta = \sigma_\phi$, and the radial displacement, u, are functions of r only. In the absence of body forces, the equation of equilibrium is

$$\frac{d\sigma_r}{dr} = \frac{2}{r}(\sigma_\theta - \sigma_r) \tag{7.37}$$

The stress–strain relations during elastic deformation are given by the usual Hooke's law

$$\varepsilon_r = \frac{\partial u}{\partial r} = \frac{1}{E}(\sigma_r - 2\upsilon\sigma_\theta) \tag{7.38a}$$

$$\varepsilon_\theta = \varepsilon_\phi = \frac{u}{r} = \frac{1}{E}[(1 - \upsilon)\sigma_\theta - \upsilon\sigma_r] \tag{7.38b}$$

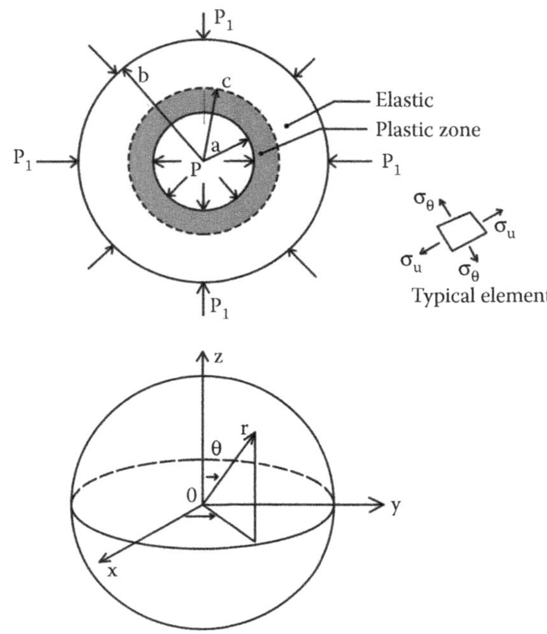

FIGURE 7.8
Thick-walled spherical shell subjected to internal and external pressures. Top image: cross section of spherical shell.

Eliminating u from Equations 7.38, making use of Equation 7.37, and simplifying the algebra, it is readily shown that

$$\frac{d}{dr}(\sigma_r + 2\sigma_\theta) = 0 \tag{7.39}$$

Equation 7.39, which represents compatibility of strains in term of stresses, indicates that

$$\sigma_r + 2\sigma_\theta = 3A \tag{a}$$

where A is constant, substituting Equation (a) into 7.37, the governing equation of equilibrium becomes

$$r\frac{d\sigma_r}{dr} + 3\sigma_r = 3A \tag{b}$$

Solving the first-order differential equation and making use of Equation (a), the elastic stresses are determined in terms of two arbitrary constants

$$\begin{aligned}\sigma_r &= A + \frac{B}{r^3}\\ \sigma_\theta &= A - \frac{B}{2r^3} = \sigma_\phi\end{aligned} \tag{7.40}$$

Here, the constants A and B are determined from the boundary conditions: $\sigma_r = -p$ at $r = a$, and $\sigma_r = -p_1$ at $r = b$, giving $A = (a^3p - b^3p_1)/(b^3 - a^3)$, and $B = (a^3b^3(p_1 - p))/(b^3 - a^3)$. Thus, the elastic stress distribution in the shell is

$$\sigma_r = \frac{1}{b^3 - a^3}\left[(a^3p - b^3p_1) + \frac{a^3b^3}{r^3}(p_1 - p)\right]$$
$$\sigma_\theta = \frac{1}{b^3 - a^3}\left[(a^3p - b^3p_1) - \frac{a^3b^3}{2r^3}(p_1 - p)\right] \tag{7.41}$$

It is interesting to note that the difference in stresses, $\sigma_\theta - \sigma_r = -(3a^3b^3(p_1 - p))/(2(b^3 - a^3)r^3)$, which is typically used in Tresca's yield criterion attains its maximum magnitude at the internal boundary $r = a$, of the sphere where

$$(\sigma_\theta - \sigma_r)_{max} = \frac{3b^3}{2(b^3 - a^3)}(p_1 - p) \tag{c}$$

The radial displacement, $u = r\varepsilon_\theta$, in the shell may be determined using Equation 7.41. The elastic solution of the spherical shell was first obtained by Lame in 1852 [7.1].

7.2.1 Spherical Shell Subjected to Internal Pressure

When the shell is subjected to internal pressure p only, setting $p_1 = 0$ in Equation 7.41, the elastic stresses reduce to

$$\sigma_r = -\frac{a^3p}{b^3 - a^3}\left(\frac{b^3}{r^3} - 1\right)$$
$$\sigma_\theta = \frac{a^3p}{b^3 - a^3}\left(\frac{b^3}{2r^3} + 1\right), \quad a \le r \le c \tag{7.42}$$

The radial stress is compressive throughout the shell while the tangential stresses $\sigma_\theta = \sigma_\phi$ are tensile. The radial displacement is given by

$$u = \frac{a^3pr}{(b^3 - a^3)E}\left[1 - 2\upsilon + (1 + \upsilon)\frac{b^3}{2r^3}\right] \tag{7.43}$$

Increasing the pressure p, yielding will commence at the radius where the yield criterion is first satisfied. In this case, both Tresca's and Mises' criteria reduce to the relation $\sigma_\theta - \sigma_r = \sigma_0$, where σ_0 is the yield stress of the material in tension. It follows from Equation 7.42 that either Tresca's or Mises' criteria are first satisfied at the inner boundary, $r = a$, and the maximum pressure in the elastic stage required for initiating yield is

$$p_e = \frac{2\sigma_0}{3}\left(1 - \frac{a^3}{b^3}\right) \tag{7.44}$$

For a shell with large outer radius, $b \to \infty$, $p_e \approx (2\sigma_0/3)$.

At this stage of expansion, the radial displacements at the inner and outer boundaries, obtained using relation (7.43), are

$$\frac{u_a}{a} = \frac{\sigma_0}{3E}\left[1 + \upsilon + 2(1 - 2\upsilon)\frac{a^3}{b^3}\right], \quad \frac{u_b}{b} = (1 - \upsilon)\left(\frac{a^3}{b^3}\right)\frac{\sigma_0}{E}$$

Note that if the sphere is subjected to both internal and external pressures then relation (c) of the previous section indicates that yielding will first occur at the inner boundary when

$$p_1 - p = p_e \tag{7.45}$$

where p_e is given by Equation 7.44.

Elastic–plastic expansion: As the pressure p increases beyond p_e, the plastic zone around the inner boundary spreads into the shell and an annular plastic zone of radius $r = c$ is formed, as shown in Figure 7.8b. For $r > c$, the shell is elastic. During the initial elastic–plastic expansion, we assume that the radii a and b of the shell have not changed due to yielding to depth c. The influence of yielding on changes in geometry and especially on the maximum internal pressure will be considered later.

In the *elastic region*, $c \leq r \leq b$, the stress field given in relation (7.40) is still valid but the constants A and B are now determined from boundary conditions: $\sigma_r = 0$ at $r = b$ and $\sigma_\theta - \sigma_r = \sigma_0$ at $r = c$. When these conditions are used in conjunction with Equation 7.40, the constants A and B are found to be given by $A = (2\sigma_0 c^3/3b^3)$ and $B = -(2/3)\sigma_0 c^3$, and the stresses in the elastic region are

$$\begin{aligned} \sigma_r &= -\frac{2\sigma_0 c^3}{3b^3}\left(\frac{b^3}{r^3} - 1\right) \\ \sigma_\theta &= \frac{2\sigma_0 c^3}{3b^3}\left(\frac{b^3}{2r^3} + 1\right), \quad c \leq r \leq b \end{aligned} \tag{7.46}$$

The radial displacement in the elastic region is

$$u = \frac{2\sigma_0 c^3 r}{3Eb^3}\left[1 - 2\upsilon + (1 + \upsilon)\frac{b^3}{2r^3}\right], \quad c \leq r \leq b \tag{7.47}$$

In particular, the radial extension of the outer boundary is given by

$$\frac{u_b}{b} = (1 - \upsilon)\frac{\sigma_0 c^3}{Eb^3} \tag{a}$$

In the *plastic region* $a \leq r \leq c$, the stresses are required to satisfy the equilibrium Equation 7.37, and the yield criterion for elastic–perfectly material is $\sigma_\theta - \sigma_r = \sigma_0$. Inserting the yield criterion in Equation 7.37, an ordinary differential equation is obtained, $(d\sigma_r/dr) = (2\sigma_0/r)$, whose integration furnishes $\sigma_r = 2\sigma_0 \ln r + D$, where D is constant. The tangential stress is $\sigma_\theta = \sigma_r + \sigma_0$. In order to determine the constant D, continuity of the radial stress at $r = c$ is used to provide $D = -2\sigma_0 \text{ Ln } c + (2\sigma_0 c^3/3b^3)(1 - (b^3/c^3))$, and the stresses in the plastic region are

$$\sigma_r = -\frac{2\sigma_0}{3}\left(1 - \frac{c^3}{b^3} + \text{Ln}\frac{c^3}{r^3}\right)$$
$$\sigma_\theta = \frac{2\sigma_0}{3}\left(\frac{1}{2} + \frac{c^3}{b^3} - \text{Ln}\frac{c^3}{r^3}\right), \quad a \le r \le c \tag{7.48}$$

The distribution of the stresses in the plastic region of the sphere is shown in Figure 7.9 (see Reference 2.2). The radial stress is compressive and its magnitude decreases as the radius increases from $r = a$ to $r = b$. The tangential stress on the other hand steadily increases with the radius and achieves greatest magnitudes on the elastic/plastic boundary. At $r = a$, the internal pressure necessary to reach this stage of expansion is obtained from Equation 7.48 by setting $p = -\sigma_r$

$$p = \frac{2\sigma_0}{3}\left(1 - \frac{c^3}{b^3} + \text{Ln}\frac{c^3}{a^3}\right) \tag{7.49}$$

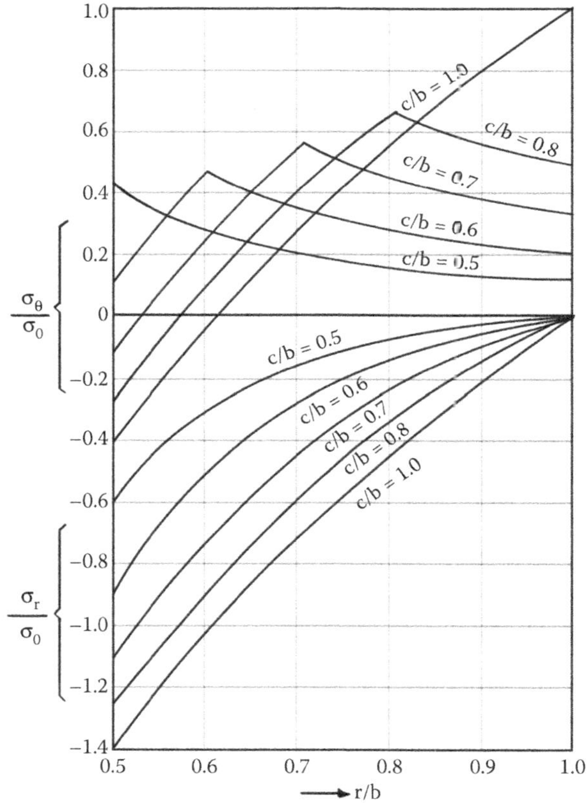

FIGURE 7.9
Plastic stresses in a thick-walled spherical vessel with $b/a = 2$ [2.2].

Equation 7.49 may be used to determine the internal pressure for various values of c and b/a assuming there are no changes in the geometry of the vessel during the elastic–plastic expansion. Maximum pressure occurs when $c = b$

$$p_u = 2\sigma_0 \ln\frac{b}{a} \tag{7.50}$$

Additional expansion causes the material to strain-harden and the maximum pressure in Equation 7.50 will increase somewhat as will be discussed later. The expansion of the sphere following the maximum pressure is unstable and causes eventual bursting.

Next, we find the radial displacement in the plastic region. In spherical coordinates the dilatation, which is the sum of normal strains (Chapter 3, Equation 3.10), is expressed in terms of displacement as $\varepsilon_r + \varepsilon_\theta + \varepsilon_\phi = \varepsilon_r + 2\varepsilon_\theta = (du/dr) + (2u/r)$. Thus, for small strains, using Hooke's law, the elastic compressibility may be expressed in the form, $(du/dr) + (2u/r) = ((1 - 2\upsilon)/E)(\sigma_r + 2\sigma_\theta)$, and upon multiplying both sides by r^2, the equation assumes the form

$$\frac{d}{dr}(r^2 u) = \frac{1 - 2\upsilon}{E} r^2 (\sigma_r + 2\sigma_\theta) \tag{b}$$

At $r = c$, continuity of displacement must be maintained. Thus, inserting the stresses from Equation 7.48 into (b), an ordinary differential equation is obtained

$$\frac{d}{dr}(r^2 u) = 2(1 - 2\upsilon)\frac{\sigma_0}{E} r^2 \left(\frac{c^3}{b^3} + \text{Ln}\frac{r^3}{c^3}\right) \tag{c}$$

Integrating with respect to r results in

$$r^2 u = 2(1 - 2\upsilon)\frac{\sigma_0 r^3}{3E}\left(-1 + \frac{c^3}{b^3} - \text{Ln}\frac{c^3}{r^3}\right) + D \tag{d}$$

The constant of integration, D, is found by satisfying continuity of u at $r = c$. Inserting the displacement u in the elastic region, Equation 7.47, into (d) and setting $r = c$ gives $D = (1 - \upsilon)(\sigma_0 c^3/E)$. Hence, the radial displacement in the plastic region is

$$\frac{u}{r} = \frac{\sigma_0}{E}\left[(1 - \upsilon)\frac{c^3}{r^3} - \frac{2}{3}(1 - 2\upsilon)\left(1 - \frac{c^3}{b^3} + \text{Ln}\frac{c^3}{r^3}\right)\right], \quad a \le r \le c \tag{7.51}$$

When $c = b$, the corresponding radial expansions of the inner and outer boundaries are

$$u_a = \frac{\sigma_0 a}{E}\left[(1 - \upsilon)\frac{b^3}{a^3} - 2(1 - 2\upsilon)\,\text{Ln}\frac{b}{a}\right], \quad u_b = (1 - \upsilon)\frac{\sigma_0 b}{E} \tag{e}$$

In the next section, we consider the influence of changes in the radii of the sphere during the elastic–plastic expansion when the sphere is subjected to the maximum pressure given in Equation 7.50.

7.2.2 Influence of Changes in Geometry of the Sphere

We expect that any changes in the radii of the sphere will affect the maximum pressure predicted by Equation 7.50. At the final stage of full plasticity, leading to the derivation of Equation 7.50 denotes the radii of the shell by a′ and b′.

Since the volume of the shell is conserved, $b'^3 - a'^3 = b^3 - a^3$, it follows that $a'^3 = b'^3 - b^3 + a^3$, and dividing by a we have

$$\left(\frac{a'}{a}\right)^3 = \left(\frac{b'}{a}\right)^3 - \left(\frac{b}{a}\right)^3 + 1 = \left(\frac{b'}{b}\right)^3\left(\frac{b}{a}\right)^3 - \left(\frac{b}{a}\right)^3 + 1$$

Hence,

$$\left(\frac{a'}{a}\right)^3 = \left(\frac{b}{a}\right)^3\left[\left(\frac{b'}{b}\right)^3 - 1\right] + 1$$

Since $(b'/a')^3 = (b'/b)^3(b/a)^3(a/a')^3$, it follows that

$$\left(\frac{b'}{a'}\right)^3 = \frac{\left(\frac{b}{a}\right)^3\left(\frac{b'}{b}\right)^3}{1 + \left(\frac{b}{a}\right)^3\left[\left(\frac{b'}{b}\right)^3 - 1\right]} \tag{a}$$

The radial displacement of the outer fiber of a fully plastic shell is $u_b/b = (1-\upsilon)(\sigma_0/E)$. Also, the hoop strain at the outer fiber is $(b'-b)/b$, it follows that

$$\frac{b'}{b} = 1 + (1-\upsilon)\frac{\sigma_0}{E} \tag{b}$$

For a typical metal, the yield stress is $\sigma_0 = 250$ MPa, $E = 200$ GPa ($\sigma_0/E = 1.25 \times 10^{-3}$), and Poisson's ratio $\upsilon = 0.3$. Using the abbreviation, $x = 0.7\,(\sigma_0/E)$, and substituting Equation (b) into (a), it follows that

$$\left(\frac{b'}{a'}\right)^3 = \frac{(1+x)^3(b/a)^3}{1 + (b/a)^3[(1+x)^3 - 1]} \tag{c}$$

Since x is a small quantity, expanding the terms containing x in series, and neglecting small terms, the magnitude of the final ratio of radii is

$$\left(\frac{b'}{a'}\right)^3 \approx \frac{(b/a)^3}{1 + 3x(b/a)^3} \tag{d}$$

Hence, because of changes in geometry, the maximum internal pressure is given by

$$P'_u = 2\sigma_0 \ln\frac{b'}{a'} = \frac{2\sigma_0}{3}\ln\left[\frac{(b/a)^3}{1 + 3x(b/a)^3}\right] \tag{7.52}$$

TABLE 7.2

Variations of Maximum Internal Pressures with (b/a) and (b′/a′)

$\frac{b}{a}$	2	3	4	5	6	7	8	10
$\frac{P_u}{2\sigma_0} = \text{Ln}\frac{b}{a}$	0.69	1.10	1.39	1.61	1.79	1.95	2.08	2.30
$\frac{b'}{a'}$	1.98	2.93	3.79	4.55	5.17	5.65	6.02	6.51
$\frac{P'_u}{2\sigma_0}$	0.687	1.078	1.34	1.57	1.64	1.76	1.82	1.91

As b/a increases, Equation (a) gives $(b'/a')^3 = (b'/b)^3/[(b'/b)^3 - 1]$, where for a typical metal $(b'/b) = 1 + 0.875(10^{-3})$ and the maximum pressure predicted by Equation 7.52 is $P'_u = 3.97\sigma_0$ while Equation 7.50 predicts infinite pressure. Table 7.2 shows the influence of changes in geometry of the sphere on maximum pressure during the plastic expansion.

7.2.3 Work-Hardening Material

The influence of work-hardening material on the stress field and maximum internal pressure supported by the vessel is considered in this section. Assume that the stress–strain relation of the material of the vessel in the tension test is given by (Equation 7.32)

$$\sigma = \sigma_0 + E_p \varepsilon^p$$

The hoop strain in the plastic region of the vessel, $a \leq r \leq c$, is readily obtained from Equations 7.48

$$\varepsilon_\theta = (1 - 2\upsilon)\frac{\sigma_r}{E} + (1 - \upsilon)\frac{\sigma_0 c^3}{Er^3} \quad \text{(a)}$$

Subtracting the elastic hoop strain, $\varepsilon_\theta = (1/E)[(1 - \upsilon)\sigma_\theta - \upsilon\sigma_r]$, the plastic hoop strain is determined

$$\varepsilon_\theta^p = \frac{1 - \upsilon}{E}\left(\sigma_r - \sigma_\theta + \sigma_0 \frac{c^3}{r^3}\right), \quad a \leq r \leq c \quad (7.53)$$

Using the linear-hardening property of the material, it follows that

$$\sigma - \sigma_0 = (1 - \upsilon)\frac{E_p}{E}\left(\sigma_r - \sigma_\theta + \sigma_0 \frac{c^3}{r^3}\right) \quad \text{(b)}$$

Making use of Tresca's yield criterion, $\sigma_\theta - \sigma_r = \sigma$, it follows that

$$\sigma\left[1 + (1 - \upsilon)\frac{E_p}{E}\right] = \sigma_0\left[1 + (1 - \upsilon)\frac{E_p}{E}\frac{c^3}{r^3}\right] \quad \text{(c)}$$

And the equilibrium expressed by Equation 7.37 results in the simple differential equation, $d\sigma_r/dr = 2\sigma/r$, where σ is given in Equation (c).

Performing the integration

$$\sigma_r = \frac{2\sigma_0}{1 + (1-\upsilon)\dfrac{E_p}{E}}\left[\ln r - (1-\upsilon)\frac{E_p c^3}{3Er^3}\right] + D, \quad a \le r \le c \tag{d}$$

The constant D is found from the condition that the radial stress is continuous at $r = c$. Thus, using Equations 7.46 and (d), we find

$$D = \frac{2\sigma_0}{3}\left(\frac{c^3}{b^3} - 1\right) - \frac{2\sigma_0}{1 + (1-\upsilon)\dfrac{E_p}{E}}\left(\ln c - \frac{1-\upsilon}{3}\frac{E_p}{E}\right)$$

And the stresses in the plastic region are

$$\left.\begin{aligned}\sigma_r &= \frac{-2\sigma_0}{3[1 + (1-\upsilon)E_p/E]}\left[1 - \frac{c^3}{b^3} + \ln\frac{c^3}{r^3} + (1-\upsilon)\frac{E_p}{E}\left(\frac{c^3}{r^3} - \frac{c^3}{b^3}\right)\right]\\ \sigma_\theta &= \frac{2\sigma_0}{3[1 + (1-\upsilon)E_p/E]}\left[\frac{1}{2} + \frac{c^3}{b^3} - \ln\frac{c^3}{r^3} + (1-\upsilon)\frac{E_p}{E}\left(\frac{c^3}{2r^3} + \frac{c^3}{b^3}\right)\right]\end{aligned}\right\}, \quad a \le r \le c \tag{7.54}$$

At $r = a$, the internal pressure associated with this stage of expansion is ($\sigma_r = -p$)

$$p = \frac{2\sigma_0}{3[1 + (1-\upsilon)E_p/E]}\left[1 - \frac{c^3}{b^3} + \ln\frac{c^3}{a^3} + (1-\upsilon)\frac{E_p}{E}\left(\frac{c^3}{a^3} - \frac{c^3}{b^3}\right)\right] \tag{7.55}$$

Subtracting the pressure in the case of elastic-perfectly material, Equation 7.49, and neglecting change in geometry of the vessel, the increment in pressure due to the hardening of the material is

$$\Delta p = \frac{2(1-\upsilon)E_p\sigma_0}{3[E + (1-\upsilon)E_p]}\left[\frac{c^3}{a^3} - 1 - 3\ln\frac{c}{a}\right] \tag{e}$$

Assuming $b/a = 3$, $E_p = 0.1\,E$, and $\upsilon = 0.3$, the increase in pressure for few ratios of c/b is as follows:

$$\frac{c}{b} = 0.5, 0.6, 0.7, 0.8$$

$$\frac{\Delta p}{\sigma_0} = 0.05, 0.14, 0.26, 0.45$$

General work hardening: For general case of work hardening, the stress–strain law of the material is [2.2]

$$\tilde{\sigma} = \sigma_0 + H(\tilde{\varepsilon}) \tag{7.56}$$

where $\tilde{\sigma}$, $\tilde{\varepsilon}$ denote representative stress and strain points in the work-hardening region of the tensile test and $H(\tilde{\varepsilon})$ describes the amount of work hardening.

In this case, the yield criterion, $\sigma_\theta - \sigma_r = \tilde{\sigma}$, together with the equilibrium equation and continuity of stresses at $r = c$ yields the result

$$\sigma_r = 2\sigma_0 \,\mathrm{Ln}\frac{r}{c} - \frac{2}{3}\sigma_0\left(1 - \frac{c^3}{b^3}\right) - 2\int_r^c H(\tilde{\varepsilon})\frac{dr}{r}$$

In general, numerical procedures are required to evaluate the integral and determine the pressure $p = -\sigma_r$. Explicit results can be obtained if $H(\tilde{\varepsilon}) = H'\tilde{\varepsilon}$, where H' is constant. The pressure needed to expand a small shell in an infinitely large sphere is

$$\frac{p}{2\sigma_0} = \frac{1}{3}\left[1 + \ln\left(\frac{2E}{3\sigma_0}\right) + \frac{\pi H'}{9\sigma_0}\right] \tag{7.57}$$

For $E/\sigma_0 = 1/1000$ and $H'/\sigma_0 = 1/3$, $p = 5.2\sigma_0$.

EXAMPLE 7.5

A spherical vessel of internal radius a and external radius b is subjected to internal pressure p. Find the residual stresses after complete unloading from a stage of plastic deformation where $p_e < p < p_u$.

Solution

The unloading is done elastically using a pressure of magnitude $-p$.

Thus, in the elastic zone of the vessel, $c < r < b$, Equations 7.46 and 7.42 give

$$\sigma_r = -\frac{2\sigma_0 c^3}{3b^3}\left(\frac{b^3}{r^3} - 1\right) + \frac{a^3 p}{b^3 - a^3}\left(\frac{b^3}{r^3} - 1\right)$$

Upon using expression (7.44) of the pressure that initiates yield, the above expression of the residual stress takes the form

$$\sigma_r = -\frac{2\sigma_0}{3}\left(\frac{a^3}{r^3} - \frac{a^3}{b^3}\right)\left(\frac{c^3}{a^3} - \frac{p}{p_e}\right) \tag{a}$$

Similarly, the circumferential residual stress is

$$\sigma_\theta = \frac{2\sigma_0}{3}\left(\frac{a^3}{2r^3} + \frac{a^3}{b^3}\right)\left(\frac{c^3}{a^3} - \frac{p}{p_e}\right) \tag{b}$$

In the plastic zone of the vessel, $\mathbf{a} < \mathbf{r} < \mathbf{c}$, Equations 7.48 and 7.42 give the following residual stresses:

$$\begin{aligned} \sigma_r &= -\frac{2\sigma_0}{3}\left[\frac{p}{p_e}\left(1 - \frac{a^3}{r^3}\right) - 3\ln\frac{r}{a}\right] \\ \sigma_\theta &= \frac{2\sigma_0}{3}\left[\frac{3}{2} - \frac{p}{p_e}\left(1 + \frac{a^3}{2r^3}\right) + 3\ln\frac{r}{a}\right] \end{aligned} \tag{c}$$

At $r = a$, $\sigma_\theta = -\sigma_0((p/p_e) - 1)$, indicating that due to the elastic unloading, a compressive tangential stress is induced at the inner boundary of the vessel. Thus, upon reloading the vessel can support greater pressure before experiencing plastic deformation [7.4].

The above solution is valid as long as the maximum value of $(\sigma_r - \sigma_\theta)$, which occurs at $r = a$, is less than the yield stress σ_0. Since at $r = a$, $(\sigma_r - \sigma_\theta)_{r=a} = \sigma_0((p/p_e) - 1) < \sigma_0$, implying that $p/p_e < 2$. Under this condition of P, a reapplication of the pressure will produce elastic deformation only. For $p_u = 2\ p_e$, we get $(b/a) = (2/3)\ [1-(a/b)^3]$, giving $b/a = 1.7$. If $b/a > 1.7$, upon reloading a secondary yielding will occur around the inner surface.

7.3 Stress Concentration Factors

The presence of holes and other geometric irregularities in structural members causes additional stresses in the immediate vicinity of the hole. The ratio of the elevated stress at the edge of the hole to the average (nominal) stress in the element is known as the stress concentration factor (SCF), K. Thus,

$$K = \text{maximum stress at hole/nominal stress} \tag{7.58}$$

Stress elevation occurs frequently around holes of bolts, rivets, notches, and other geometric discontinuities. Also, the presence of such discontinuities in structural members may lead to initiation of cracks followed by crack growth and in some cases even eventual failure of the member. The strength of materials approach cannot predict such stresses and we need to employ the theories of elasticity and plasticity to find the stress and strain distributions around cracks [7.5–7.7].

In this section, we consider the stress distribution around a circular hole induced by loads applied at distances away from the hole. In Section 6.7.1, the expansion of a circular hole in an infinite plate induced by radial pressure applied at the surface of the hole is discussed.

7.3.1 Infinite Plate Containing a Circular Hole

Consider the plane stress problem of a large plate containing a circular hole of radius a subjected to equal biaxial tensile stresses of magnitude t at infinity, as shown in Figure 7.10. Because of axial symmetry in loads and geometry, the applicable stresses and strains are σ_r σ_θ, ε_r, and ε_θ. The elastic stresses are obtained using Lame's solution, $\sigma_r = A/r^2 + C$, $\sigma_\theta = -A/r^2 + C$, where A and C are constants to be determined from the boundary conditions, $\sigma_r = 0$ at $r = a$ and both stresses reduce to t as $r \to \infty$. This gives $C = t$ and $A = -a^2\ t$. Hence, the elastic stresses are

$$\begin{aligned} \sigma_r &= t\left(1 - \frac{a^2}{r^2}\right) \\ \sigma_\theta &= t\left(1 + \frac{a^2}{r^2}\right) \end{aligned} \tag{7.59}$$

The maximum stress is the hoop stress at $r = a$, $(\sigma_\theta)_{max} = 2t$ and the SCF $K = 2$.

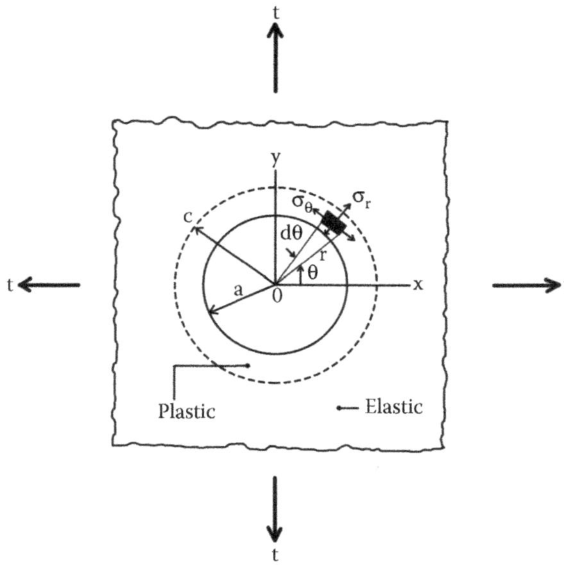

FIGURE 7.10
Circular hole in a large plate subjected to equal biaxial stresses.

If the plate is subjected to a tensile stress in the horizontal direction only, Figure 7.11a, the problem does not possess axial symmetry like the previous one, and the boundary conditions in such a case are $\sigma_r = \tau_{r\theta} = 0$ at $r = a$, $\sigma_x = t$, $\tau_{xy} = 0$ at $x =$ infinity, and $\sigma_y = \tau_{yx} = 0$ at $y =$ infinity. In terms of polar coordinates, the boundary conditions at large distances from the hole are obtained using Equations 2.23 as $\sigma_r = t\cos^2\theta = (t/2)(1 + \cos 2\theta)$ and, $\tau_{r\theta} = -(t/2)\sin 2\theta$. The elastic solution of this problem is obtained by using a stress function consisting of combining two solutions of the compatibility equation, $\nabla^4\phi(r, \theta) = 0$, given in Equations 6.24 and 6.25b with $n = 2$, that is, $\phi(r, \theta) = \phi_0(r) + \phi_2(r, \theta)$ where

$$\begin{aligned} \phi_0(r) &= A\,\mathrm{Ln}\,r + Br^2 \ln r + Cr^2 + D \\ \phi_2(r, \theta) &= \left(A_2 r^2 + \frac{B_2}{r^2} + C_2 r^4 + D_2\right)\cos 2\theta \end{aligned} \tag{a}$$

The corresponding stresses are

$$\begin{aligned} \sigma_r &= \frac{A}{r^2} + B(2\ln r + 1) + 2C - \left(2A_2 + \frac{6B_2}{r^4} + \frac{4D_2}{r^2}\right)\cos 2\theta \\ \sigma_\theta &= -\frac{A}{r^2} + B(2\ln r + 3) + 2C + \left(2A_2 + \frac{6}{r^4}B_2 + 12C_2 r^2\right)\cos 2\theta \\ \tau_{r\theta} &= \left(2A_2 - \frac{6B_2}{r^4} + 6C_2 r^2 - \frac{2D_2}{r^2}\right)\sin 2\theta \end{aligned} \tag{b}$$

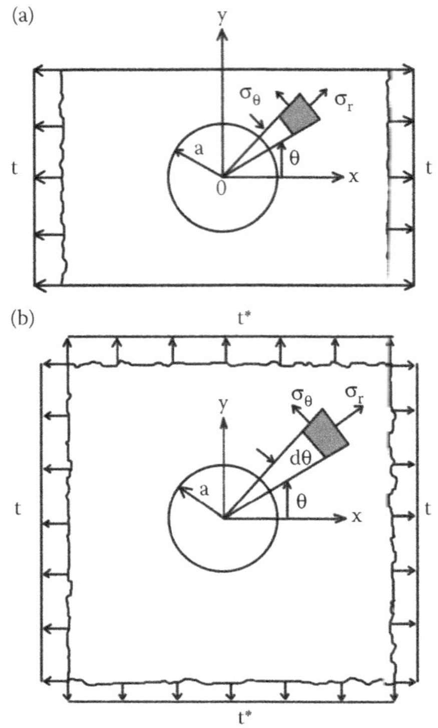

FIGURE 7.11
Circular hole in large plate (a) uniaxial tension; (b) biaxial tension.

Applying the boundary conditions in terms of polar coordinates as mentioned previously, the constants in Equations (b) are found as follows: $A = -D_2 = -(t/2)\ a^2$, $B = C_2 = D = 0$, $C = -A_2 = t/4$, and $B_2 = -t\ (a^4/4)$. Hence, the elastic stresses are

$$\begin{aligned}\sigma_r &= \frac{t}{2}\left[1 - \frac{a^2}{r^2} + \left(1 + \frac{3a^4}{r^4} - \frac{4a^2}{r^2}\right)\cos 2\theta\right]\\ \sigma_\theta &= \frac{t}{2}\left[1 + \frac{a^2}{r^2} - \left(1 + \frac{3a^4}{r^4}\right)\cos 2\theta\right]\\ \tau_{r\theta} &= -\frac{t}{2}\left(1 - \frac{3a^4}{r^4} + \frac{2a^2}{r^2}\right)\sin 2\theta\end{aligned} \tag{7.60}$$

The maximum stress is the tangential stress at $r = a$, that is, $\sigma_\theta = t(1 - 2\cos 2\theta)$. For $\theta = \pi/2$ and $\theta = 3\pi/2$ (ends of the diameter normal to the direction of the load), $\sigma_\theta = 3t$, and the SCF is three times the applied uniform stress. Note that at $\theta = 0$ and π, $\sigma_\theta = -t$, indicating compression at these points of the hole.

If the plate is subjected to constant tensile stresses t* along the y-axis in addition to the horizontal stresses discussed previously (see Figure 7.11b), the new stress distribution can be obtained by superposition. Replacing t by t* and θ by $(\pi/2) + \theta$ in Equations 7.60 and adding the new stresses to those in Equation 7.60, the elastic stresses for the problem shown in

Figure 7.11b are obtained as follows:

$$\sigma_r = \frac{1}{2}(t + t^*)\left(1 - \frac{a^2}{r^2}\right) + \frac{1}{2}(t - t^*)\left(1 - \frac{4a^2}{r^2} + \frac{3a^4}{r^4}\right)\cos 2\theta$$

$$\sigma_\theta = \frac{1}{2}(t + t^*)\left(1 + \frac{a^2}{r^2}\right) - \frac{1}{2}(t - t^*)\left(1 + \frac{3a^4}{r^4}\right)\cos 2\theta \qquad \text{(c)}$$

$$\tau_{r\theta} = -\frac{1}{2}(t - t^*)\left(1 + \frac{2a^2}{r^2} - \frac{3a^4}{r^4}\right)\sin 2\theta$$

When $t^* = -t$, the maximum stress is the tangential stress at $r = a$, $\sigma_\theta = -4t\cos 2\theta$. For $\theta = \pi/2$ or $3\pi/2$, $(\sigma_\theta)_{max} = 4t$ while for $\theta = 0$ or π the maximum stress is -4 t. Thus, for a pure shear loading, the maximum tangential stress at the hole is four times the applied stress and the SCF is $K = 4$.

Next, we consider the elastic/plastic expansion of the circular hole subjected to biaxial tensile stresses at infinity, Figure 7.10, the elastic stress distribution of such problem is given in Equations 7.59.

Elastic–plastic analysis: Consider the stress distribution in Equations 7.59 pertaining to the problem of a large plate containing a circular hole subjected to uniform radial stress of magnitude t at large distances from the hole (Figure 7.10). As the applied load is increased, yield is initiated at the edge of the hole, $r = a$, according to either Tresca's or Mises' criteria. The load initiating yield is $t_e = \sigma_0/2$, σ_0 is the yield stress of the material. When t is increased beyond t_e an annular plastic region is formed around the hole as shown shaded in Figure 7.10, while the deformation in the region $r > c$ is still elastic. It should be noted that we cannot make such conclusion about the geometry of the plastic region of any of the problems shown in Figure 7.11 because there is no radial symmetry, and a step-by-step numerical analysis is required for such problems.

We consider both elastic–perfectly plastic and elastic–strain-hardened materials and neglect significant change in the geometry of the element.

1. *Elastic–perfectly plastic material:* The elastic stresses in the element are governed by Lame's solution mentioned previously but now the constants A and C are determined from the following boundary conditions: at the edge of the plastic region, $r = c$, $\sigma_\theta = \sigma_0$ while at $r \to \infty$, $\sigma_\theta = \sigma_r = t$. Thus,

$$\sigma_r = t - (\sigma_0 - t)\frac{c^2}{r^2}$$
$$\sigma_\theta = t + (\sigma_0 - t)\frac{c^2}{r^2}, \quad r \geq c \qquad \text{(7.61a)}$$

   In the plastic region, $a \leq c \leq r$, the stresses must satisfy the equilibrium equation, $(d/dr)(r\sigma_r) = \sigma_\theta$ and the yield criterion, $\sigma_\theta - \sigma_r = \sigma_0$. This furnishes the simple differential equation, $(d\sigma_r/dr) = (\sigma_0/r)$ subject to the boundary condition $\sigma_r = 0$ at $r = a$. Solving the differential equation gives

$$\sigma_r = \sigma_0 \ln\frac{r}{a}$$
$$\sigma_\theta = \sigma_0\left(1 + \ln\frac{r}{a}\right), \quad a \leq r \leq c \qquad \text{(7.61b)}$$

   Maximum stress occurs at the hole, $\sigma_\theta = \sigma_0$ and the SCF is $K = (\sigma_0/t)$.

2. *Strain-hardened material:* We consider an incompressible material in which the strain hardens according to the relation, $\tilde{\sigma} = k\tilde{\varepsilon}^n$, $\tilde{\sigma} > \sigma_0$, where $\tilde{\sigma}$ and $\tilde{\varepsilon}$ are representative points on the uniaxial stress–strain diagram and k and n are constants.

   In the elastic region, $r \geq c$, the stresses are given by Equations 7.61a while in the plastic region, $a \leq r \leq c$, we know that Tresca's criterion requires $\sigma_\theta = \tilde{\sigma}$ at the elastic/plastic boundary $r = c$. Making use of the deformation theory of plasticity, Henky's stress–strain relations (Equations 4.52) provide

$$\varepsilon_r = \frac{1}{E_s}\left(\sigma_r - \frac{1}{2}\sigma_\theta\right) = \frac{\tilde{\sigma}}{E_s}\left(\frac{\sigma_r}{\sigma_\theta} - \frac{1}{2}\right)$$

$$\varepsilon_\theta = \frac{1}{E_s}\left(\sigma_\theta - \frac{1}{2}\sigma_r\right) = \frac{\tilde{\sigma}}{E_s}\left(1 - \frac{\sigma_r}{2\sigma_\theta}\right)$$

   where E_s is the secant modulus of the material. Noting that $\tilde{\varepsilon} = (\tilde{\sigma}/E_s)$, and introducing the abbreviation, $s = (\sigma_r/\sigma_\theta)$, the strains in the plastic region are expressed as

$$\begin{aligned} \varepsilon_r &= \tilde{\varepsilon}\left(s - \frac{1}{2}\right) \\ \varepsilon_\theta &= \tilde{\varepsilon}\left(1 - \frac{s}{2}\right), \quad a \leq r \leq c \end{aligned} \tag{7.62}$$

For convenience, the strain of the hardened material is written in the following non-dimensional form:

$$\tilde{\varepsilon} = \frac{\sigma_0}{E}\left(\frac{\tilde{\sigma}}{\sigma_0}\right)^{\frac{1}{n}} \tag{d}$$

Here, E is Young's modulus of elasticity.

Differentiating $\tilde{\varepsilon}$ with respect to r leads to the expression, $(d\tilde{\varepsilon}/dr) = (\tilde{\varepsilon}/n\tilde{\sigma})(d\tilde{\sigma}/dr)$, making use of this expression in conjunction with the strain compatibility relation, $(d/dr)(r\varepsilon_\theta) = \varepsilon_r$ and Equations 7.62, the following differential equation is obtained:

$$\frac{1}{\tilde{\sigma}}\frac{d\tilde{\sigma}}{dr} = \left(\frac{n}{2-s}\right)\frac{ds}{dr} - \frac{3n}{r}\left(\frac{1-s}{2-s}\right) \tag{e}$$

Also, inserting the expressions $\sigma_\theta = \tilde{\sigma}$ and $\sigma_r = s\tilde{\sigma}$ into the equilibrium equation $(d/dr)(r\sigma_r) = \sigma_\theta$ leads to the equation

$$\frac{1}{\tilde{\sigma}}\frac{d\tilde{\sigma}}{dr} = -\frac{1}{s}\frac{ds}{dr} + \frac{1-s}{rs} \tag{f}$$

Eliminating r and ds/dr between Equations (e) and (f) furnishes the differential equation

$$\frac{1}{\tilde{\sigma}}\frac{d\tilde{\sigma}}{ds} = -\frac{2n}{2-(1-3n)s} \tag{g}$$

The solution of Equation (g) involves one constant of integration to be found from the condition that at $r = c$, $\tilde{\sigma} = \sigma_0$, and $s = (2t/\sigma_0) - 1$ (derived using the first of Equation 7.61a). Hence,

$$\frac{\tilde{\sigma}}{\sigma_0} = \left[\frac{1 - (1 - 3n)(s/2)}{(3/2)(1 - n) - (1 - 3n)(t/\sigma_0)}\right]^{\frac{2n}{1-3n}}, \quad n \neq (1/3) \tag{7.63a}$$

Equation 7.63a can be written as follows:

$$\frac{\tilde{\sigma}}{t} = K[1 - (1 - 3n)(s/2)]^{\frac{2n}{1-3n}}, \quad n \neq (1/3) \tag{7.63b}$$

where K, the elastic/plastic SCF, is given by

$$K = \frac{\sigma_0}{t}[(3/2)(1 - n) - (1 - 3n)(t/\sigma_0)]^{-\frac{2n}{1-3n}}, \quad n \neq (1/3) \tag{7.63c}$$

When $n = (1/3)$, Equation (g) is reduced to $(1/\tilde{\sigma})(d\tilde{\sigma}/ds) = -(1/3)$ with the solution

$$\frac{\tilde{\sigma}}{\sigma_0} = e^{(1/3)(2t/\sigma_0 - 1)} e^{-(s/3)} \tag{7.64a}$$

Equation 7.64b can be written as

$$\frac{\tilde{\sigma}}{t} = Ke^{-s/3}, \quad n = (1/3) \tag{7.64b}$$

In which, the SCF is given by

$$K = \frac{\sigma_0}{t} e^{\frac{1}{3}\left(\frac{2t}{\sigma_0} - 1\right)}, \quad n = (1/3) \tag{7.64c}$$

Equations 7.63c and 7.64c indicate that as the load t is increased from $\sigma_0/2$ to σ_0, the SCF is decreased from 2 to the following limiting values:

$K = e^{1/3} = 1.4$, $n = 1/3$,
$K = (2/(1 + n))^{2n/(1-3n)}$, $n =$ all other values.

Table 7.3 contains numerical values of K for $(t/\sigma_0) = (0.5 - 1.0)$ and $n = (0 - 0)$.

Next, the relation between the radius of the plastic zone c and the applied load t is determined. Eliminating $\tilde{\sigma}$ from Equations (e) and (f) gives

$$r\frac{ds}{dr} = (1 - s)\left[\frac{2 - (1 - 3n)s}{2 - (1 - n)s}\right]$$

TABLE 7.3

Variations of K with Load and n

$t/\sigma_0 n$	0	0.1	0.2	0.3	1/3	0.4	0.5	0.6
0.5	2	2	2	2	2	2	2	2
0.6	1.67	1.70	1.73	1.75	1.78	1.80	1.84	1.87
0.7	1.42	1.49	1.55	1.61	1.63	1.67	1.73	1.78
0.8	1.25	1.34	1.42	1.50	1.53	1.58	1.65	1.73
0.9	1.11	1.22	1.32	1.42	1.45	1.51	1.60	1.69
1.0	1.00	1.13	1.25	1.36	1.4	1.46	1.56	1.66

or

$$\frac{dr}{r} = \left\{ \frac{(1+n)/(1+3n)}{1-s} + \frac{4n/(1+3n)}{[2-(1-3n)s]} \right\} ds \tag{h}$$

Performing the integration and evaluating the constant of integration from the condition that $s = 0$ at $r = a$, the following relation between r and s is obtained:

$$\frac{a}{r} = (1-s)^{\frac{1+n}{1+3n}} \left[1 - (1-3n)\frac{s}{2} \right]^{\frac{4n}{1-9n^2}}, \quad n \neq 1/3 \tag{7.65a}$$

When $n = 1/3$, Equation (h) is reduced to $3(dr/r) = (1 + (2/(1-s)))ds$ with the solution

$$\frac{a}{r} = (1-s)^{\frac{2}{3}} e^{-\frac{s}{3}}, \quad n = 1/3 \tag{7.65b}$$

At the elastic/plastic boundary, $r = c$, $s = (2t/\sigma_0) - 1$, and Equations 7.65 give

$$\frac{a}{c} = \left[2\left(1 - \frac{t}{\sigma_0}\right) \right]^{\frac{1+n}{1+3n}} \left[\frac{3}{2}(1-n) - (1-3n)\frac{t}{\sigma_0} \right]^{\frac{4n}{1-9n^2}}, \quad n \neq 1/3 \tag{7.66a}$$

$$\frac{a}{c} = \left[2\left(1 - \frac{t}{\sigma_0}\right) \right]^{1/3} e^{(1/3)\left(1 - 2\frac{t}{\sigma_0}\right)}, \quad n = 1/3 \tag{7.66b}$$

Equations 7.66a and b give the relation between the radius of the plastic zone and the magnitude of the applied load. When $t = \sigma_0$, Equation 7.66 gives $c \to \infty$ and the entire plate is rendered plastic. The boundary condition at the outer boundary becomes $s = 1$ at $r = \infty$ and the stresses are proportional to the applied loading as long as the strains remain small. (The magnitude of the tangential strain at the hole is comparable to its value at initiation of yield.)

Values of the stress concentration factors using von Mises' criterion have been obtained by Chakrabarty [2.2]. The results are very close to those obtained using Tresca's criterion.

Budiandky and Mangasrian [4.8] presented a more general solution for the SCFs around a circular hole by using the Ramberg–Osgood equation to model the stress–strain law of the material (see Equation 4.36).

$$\varepsilon = \frac{\sigma}{E}\left[1 + \frac{3}{7}\left(\frac{\sigma}{\sigma_0}\right)^{m-1}\right] \tag{7.67}$$

Here, m is a constant. The solution required solving a nonlinear differential equation. It was shown in Reference 7.8 that the SCF varies from 2 in the elastic behavior to limiting values associated with a material having a pure power law of the form $\varepsilon = K\sigma^m$. The formula of the SCF associated with this limiting case of the hardened material provided in Reference 4.8 is

$$K = \left[\left(\frac{m+3}{2m}\right)^{\frac{m+3}{m^2+3}}\right] e^{\frac{(m-1)\pi}{(m^2+3)\sqrt{3}}} \tag{7.68}$$

Numerical values of K for few values of m are obtained from Equation 7.68 as follows:

$$\begin{aligned} m &= 10; 5; 3.33; 2.5; 2 \\ K &= 1.11; 1.22; 1.34; 1.45; 1.53 \end{aligned} \tag{i}$$

These values of the SCF can be compared to those derived from the previous model (see Equations 7.63c and 7.64c). Note that $n = 1/m$ and the values of K in Equation (i) correspond to those in Table 7.3 for $n = 0.1$ through 0.5 and $t = \sigma_0$.

7.3.2 Infinite Plate Containing an Elliptical Hole

In this section, we consider the problem of an elliptical hole in a large thin plate subjected to uniform tensile stress of magnitude, t, along the normal to the major axis of the hole (Figure 7.12). In terms of rectangular coordinates with the origin located at the center of

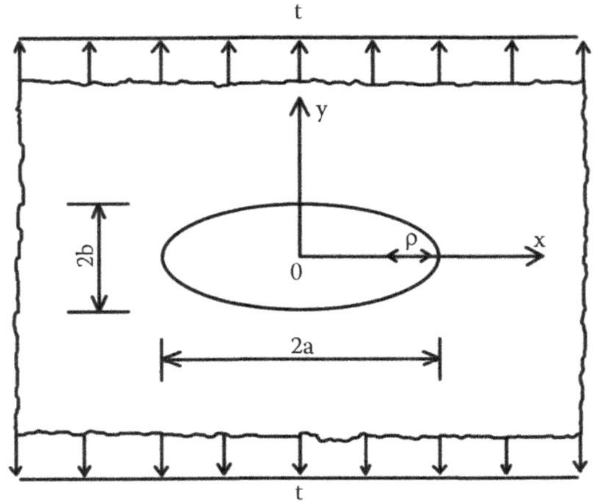

FIGURE 7.12
Elliptical hole in a large plate.

the hole, the boundary of the hole is

$$\frac{x^2}{a^2} + \frac{y^2}{b^2} = 1 \tag{a}$$

where a, b denote the length of the major and minor semiaxes of the ellipse, respectively. The elastic solution of this problem was first presented by Inglis in 1913. Inglis showed that the maximum stress is located at the ends of the major axis, points A and B in Figure 7.12, [7.8].

$$\sigma_{max} = \sigma_y(x = a, y = 0) = t\left(1 + 2\frac{a}{b}\right) \tag{7.69}$$

It follows that the SCF is

$$K = 1 + 2\frac{a}{b} \tag{7.70}$$

When $a = b$, the hole becomes circular and $K = 3$ as was derived previously in Section 7.3.1. If semiaxis "a" is increased relative to b, the magnitude of the SCF increases. In the limit as $b \to 0$, the elliptical hole degenerates into a line crack and the maximum stress becomes unbounded, a purely mathematical abstraction because all structural materials are either ductile, which yield when subjected to increasing load and elastic behavior is no longer valid, or brittle, inducing the crack to blunt or to propagate.

Equation 7.70 for the SCF can be expressed in terms of the radius of curvature of the elliptical curve, distance ρ at point A of Figure 7.12. We know from analytical geometry that the radius of curvature of a curve is defined as

$$\frac{1}{\rho} = \frac{\dfrac{d^2y}{dx^2}}{\left[1 + \left(\dfrac{dy}{dx}\right)^2\right]^{\frac{3}{2}}} \tag{b}$$

It follows by differentiating (a) with respect to x and setting $x = a, y = 0$, that $b = \sqrt{a\rho}$, and

$$K = 1 + 2\sqrt{\frac{a}{\rho}} \tag{c}$$

When $\rho \to 0$, $K \to \infty$. Additional information about SCFs are available in References 7.9 and 7.10.

7.4 Structural Members Containing Cracks

It is clear from the previous section that for members containing cracks, the elevation of stress at the crack tip is very large and the use of SCF in the design process is no longer appropriate. A different approach is needed to relate the magnitudes and distributions of elastic and plastic stress fields at the crack tip to the applied stress, crack size, and material properties. Such an approach requires tools of fracture mechanics, the original references and summaries of analytical and experimental works are available in many standard books, for example, References 7.9 through 7.11.

In this section, a brief introduction to linear elastic fracture mechanics is given. The basic principle of the approach is the introduction of a single parameter, known as the stress-intensity factor, k, which describes the stress and displacement fields ahead of the crack tip in terms of the applied stress and crack size. The units of k are $MPa \cdot m^{1/2}$ or $Ksi \cdot in.^{1/2}$ This parameter is obtained using mathematical methods of the theory of elasticity and determined experimentally in the laboratory using cracked specimens to arrive at its critical values for various structural materials. Thus, the roles of the stress intensity factor, k, and its critical value, k_c, known as fracture toughness, are similar to that of applied and allowable stresses in the traditional design process.

Consider an infinite plate containing a through-thickness crack of length 2a subjected to uniform tensile stress of magnitude t at infinity, as shown in Figure 7.13. There are several analytical techniques to determine the elastic stresses and displacements induced in the plate. In this case, the applied load tends to open out the surfaces of the crack and the stress intensity factor is known as the opening-mode factor and commonly denoted by k_1 (mode I). For shear loading, the surfaces of the crack tend to slide along the surfaces, for in-plane shear the stress intensity factor is denoted by k_2 (mode II), and for out-of-plane shear loading, the stress intensity factor by k_3 (mode III).

Using polar coordinates r, θ, measured from the crack tip, the stress and displacement fields for plane strain condition of the problem in Figure 7.13 can be expanded in infinite series in terms of r and θ, the first term of the series for each component

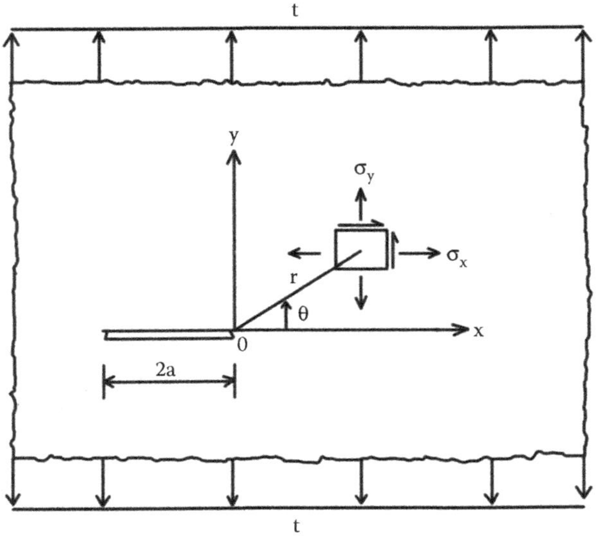

FIGURE 7.13
Infinite sheet containing through-thickness crack.

is given by

$$
\begin{aligned}
\sigma_x &= \frac{k_1}{\sqrt{2\pi r}}\cos\frac{\theta}{2}\left(1 - \sin\frac{\theta}{2}\sin\frac{3\theta}{2}\right) \\
\sigma_y &= \frac{k_1}{\sqrt{2\pi r}}\cos\frac{\theta}{2}\left(1 + \sin\frac{\theta}{2}\sin\frac{3\theta}{2}\right) \\
\tau_{xy} &= \frac{k_1}{\sqrt{2\pi r}}\sin\frac{\theta}{2}\cos\frac{\theta}{2}\cos\frac{3\theta}{2} \\
\sigma_z &= \upsilon(\sigma_x + \sigma_y) \\
\tau_{xz} &= \tau_{yz} = 0
\end{aligned}
\tag{7.71a}
$$

and

$$
\begin{aligned}
u &= \frac{k_1}{G}\sqrt{\frac{r}{2\pi}}\cos\frac{\theta}{2}\left(1 - 2\upsilon + \sin^2\frac{\theta}{2}\right) \\
v &= \frac{k_1}{G}\sqrt{\frac{r}{2\pi}}\sin\frac{\theta}{2}\left(2 - 2\upsilon - \cos^2\frac{\theta}{2}\right) \\
w &= 0
\end{aligned}
\tag{7.71b}
$$

where G is the modulus of rigidity and k_1 is the opening mode stress-intensity factor given by

$$k_1 = t\sqrt{\pi a} \tag{7.72}$$

Because higher-order terms in distance "r" are neglected, the solution (Equation 7.71) is valid in small region near the crack tip. There are similar formulas for k_2 and k_3, which describe the deformations of modes II and III crack surface displacements.

It turns out that the stresses near the crack tip always have square root singularity in r, and the stress fields ahead of the crack tip are similar regardless of configuration of the crack and geometry of the component. The stress intensity factor, which is related to both the applied stress and crack size, describes the amplitude of the stress field ahead of the crack tip. The general form of the stress intensity factor is as follows:

$$K = t \cdot (a)^{1/2} \cdot f(g) \tag{7.73}$$

where f(g) is a functional parameter that describes the structural component and geometry of the crack. As mentioned previously, standard experimental techniques exist to determine critical values of the opening mode stress intensity factor, commonly denoted by k_{1c}, and is commonly referred to as the "fracture toughness" of the material.

Linear elastic analysis of cracked elements predicts infinite stresses near the crack tip, however, in real materials, plastic zone is formed in that region. An estimate of the extent of the plastic zone at the crack can be obtained by using one of the yield criteria. This is discussed in Section 7.4.1.

EXAMPLE 7.6

A finite thickness plate of width $2b = 4.8$ in. containing a through-thickness crack of length $2a = 1$ in. is required to support a tensile stress of magnitude t ksi at the ends, as shown in Figure 7.14.

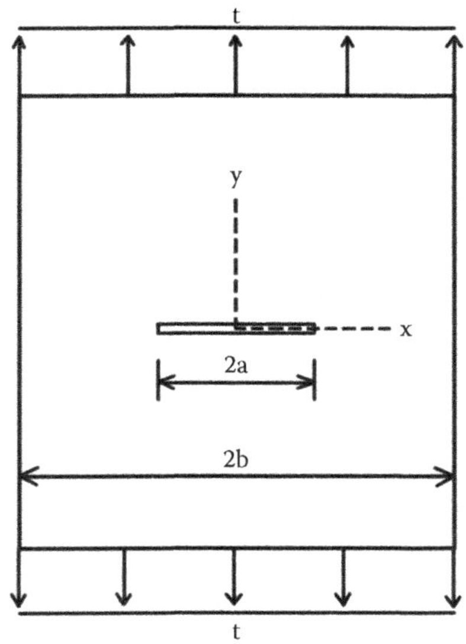

FIGURE 7.14
Example 7.6.

If the fracture toughness of the material is $k_{1c} = 70$ ksi $(\text{in.})^{1/2}$, what is the allowable value of t using a factor of safety = 2?

Solution

The stress intensity factor for a finite width plate containing a through-thickness central crack is given by the following formula:

$$K_1 = F(t\sqrt{\pi a}) \tag{a}$$

Note that F is a correction factor on the stress intensity factor given in Equation 7.72 for a similar crack geometry in a large plate. F is given in terms of the ratio a/b

$$F = \left[\frac{2b}{\pi a}\tan\left(\frac{\pi a}{2b}\right)\right]^{\frac{1}{2}}$$

In this case, $a/b = 0.5/2.4 = 0.208$, so $F = 1.022$ and $K_1 = 1.281$ t.

Using $K_1 = K_{1c}$ gives 1.281 t = 70, or t = 54.65 ksi, and hence the allowable stress supported by the plate is

$$t_{all} = t/2 = 27.32 \text{ ksi}$$

7.4.1 Plastic Zone near Crack Tip

The principal stresses are determined using the relations

$$\begin{aligned} \frac{\sigma_1}{\sigma_2} &= \frac{\sigma_x + \sigma_y}{2} \pm \left[\left(\frac{\sigma_x - \sigma_y}{2}\right)^2 + \tau_{xy}^2\right] \\ \sigma_3 &= 0 \quad \text{plane stress} \\ \sigma_3 &= \upsilon(\sigma_x + \sigma_y) \quad \text{plane strain} \end{aligned} \tag{a}$$

Using Equations 7.71a, the principal stresses are found to be given by

$$\sigma_1 = \frac{k_1}{\sqrt{2\pi r}}\cos\left(\frac{\theta}{2}\right)\left[1+\sin\left(\frac{\theta}{2}\right)\right]$$

$$\sigma_2 = \frac{k_1}{\sqrt{2\pi r}}\cos\left(\frac{\theta}{2}\right)\left[1-\sin\left(\frac{\theta}{2}\right)\right] \tag{b}$$

$$\sigma_3 = 0 \quad \text{plane stress}$$

$$\sigma_3 = \frac{2\upsilon k_1}{\sqrt{2\pi r}}\cos\left(\frac{\theta}{2}\right) \quad \text{plane strain}$$

Substituting expression (b) in Mises' yield criterion, Equation 4.42b, and solving for r gives

$$1+\cos\theta+\frac{3}{2}\sin^2\theta \quad \text{plane stress}$$

$$r = \frac{1}{4\pi}\left(\frac{k_1}{\sigma_0}\right)^2 \tag{7.74}$$

$$(1-2\upsilon)^2(1+\cos\theta)+\frac{3}{2}\sin^2\left(\frac{\theta}{2}\right) \quad \text{plane strain}$$

Equations 7.74 are plotted in Figure 7.15 for Poisson's ratio = 0.3, which shows the boundaries between elastic and plastic behaviors. The plane strain condition results in a smaller plastic zone [7.10].

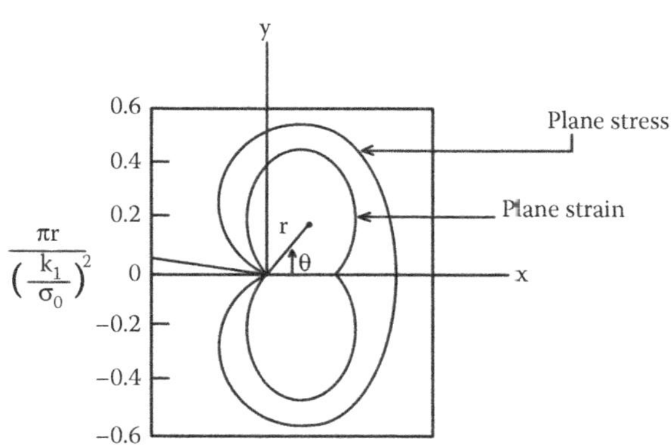

FIGURE 7.15
Estimates of plastic zones around mode I crack tip.

Problems

7.1 a. A thick-walled cylindrical vessel with closed ends is subjected to an internal pressure p. The internal radius is a and the external radius is b. Show that if the thickness $t = b - a$ is small compared to the radii, the elastic stresses in the vessel reduce to those of a thin-walled pressure vessel.

b. Do the same for a spherical pressure vessel.

Ans: (a) $\sigma_\theta = pr/t$, $\sigma_z = pr/2t$ (b) $\sigma_\theta = \sigma_\phi = pr/2t$.

7.2 A long thin-walled cylinder of internal radius a, external radius b, and wall thickness T is subjected to an internal pressure p. If the circumferential stress is independent of r, prove that the elastic stresses at any thickness t are given by the following expressions:

$$\sigma_r = -p\frac{a}{T}\left(\frac{T-t}{T+t}\right)$$

$$\sigma_\theta = p\frac{a}{T}$$

(Hint: Assume the tangential stress is constant and use equilibrium equation.)

7.3 A closed ends thick-walled cylinder is subjected to an internal pressure of 90 MPa. The internal and external radii are 10 and 20 cm, respectively. Given that E = 200 GPa, yield stress = 250 MPa, and Poisson's ratio = 0.3, determine the change in length of the outside circumference of the cylinder.

Ans: 0.321 mm.

7.4 If, in problem 7.3, the internal pressure is p = 150 Mpa, find

a. Extent of the plastic zone c

b. Change in length of the outside circumference of the cylinder

Ans: 142.5 mm, 0.884 mm.

7.5 A thick-walled cylindrical vessel, inner and outer radii, a and b, respectively, is made of an elastic–perfectly plastic material with yield stress σ_0, and modulus of elasticity E. The vessel is subjected to an internal pressure p such that according to Tresca's criterion, extent of the plastic zone is c. If, at r = c, the radial stress is denoted by $\sigma_r = -q$, and u_b is the radial displacement at r = b, prove that

$$p = q + \sigma_0 \,\mathrm{Ln}\frac{c}{a}$$

$$\frac{u_b}{b} = \frac{3\sigma_0}{4E}\left(\frac{c}{b}\right)^2$$

$$\frac{a^2}{b^2} \le \frac{4Eu_b}{3b\sigma_0} \le 1$$

7.6 A thick-walled cylinder in plane strain condition has radii, a and b, b > a. Use the following stress function:

$$\phi(r) = -A\,\mathrm{Ln}\,r + \frac{pb^2r^2}{2(b^2 - a^2)}$$

where A, p are constants to determine the elastic stresses in the cylinder assuming $\sigma_r = 0$.
Find the maximum elastic stress and consider its behavior as b approaches infinity.

Ans: S.C.F. = 2 at r = a.

7.7 Repeat the previous problem with –A replaced by A and p replaced by –p.

7.8 A thick-walled cylinder, inner radius = a and outer radius = b = 2a, is in plane strain condition. If the inner surface is stress-free and the outer surface is stretched a constant amount u_0, find the maximum elastic stress in the cylinder. Assume $\nu = 1/3$, $E = 30 \times 10^3$ ksi, $u_0 = 0.003$ in.

Ans: 72 ksi.

7.9 A thick-walled cylinder with closed ends is subjected to an internal pressure p and simultaneously a torque T about its longitudinal axis. The internal and external radii of the cylinder are, a and b, b > a. Use Mises' yield criterion to find the ratio T/p, which causes yielding to occur simultaneously at the internal and external surfaces of the cylinder.

Ans: $\dfrac{T}{p} = \dfrac{\pi(a^2 + b^2)^{\frac{3}{2}} a^2}{32b^2}$.

7.10 A closed ends thick-walled cylinder with internal and external radii a and b, respectively, is subjected to internal pressure p. If the cylinder is experiencing elastic–plastic expansion to radius $c = (a + b)/2$, find the ratio p/p_e assuming Mises' yield criterion.

Ans: 1.9.

7.11 A compound member is made of a solid circular cylinder of cross-sectional area A_1, length L surrounded by a thick cylinder with cross-sectional area A_2 and same length. The two cylinders are firmly glued together at their common interface. The true stress–true strain curves of the materials are given as follows:

$$\sigma_1 = k_1 \varepsilon^{n_1}$$
$$\sigma_2 = k_2 \varepsilon^{n_2}$$

Here, k_j, n_j, $j = 1, 2$ are constants and ε is the true strain. Find an equation that relates the true strain to the areas A_1/A_2 at which tensile instability will occur when the compound member is stretched by an axial load P.

If $k_2/k_1 = 5$, $n_1 = 0.25$, $n_2 = 0.5$ and the true strain at instability = 0.375, find the ratio A_1/A_2.

Ans: 3.91.

7.12 For the compound member in Problem 7.11 suppose that the material curves represent true stress versus engineering strain, find the relations that give the ratio A_1/A_2 in terms of the corresponding strain.

Ans:

$$n_1 \left[\frac{1 + \dfrac{n_2 k_2 A_2}{n_1 k_1 A_1} e^{n_2 - n_1}}{1 + \dfrac{k_2 A_2}{k_1 A_1}} \right] = \frac{e}{1 + e}$$

e stands for the engineering strain.

7.13 A closed ends thick-walled cylinder with internal radius $a = 0.15$ m and an external radius $b = 0.3$ m is subjected to internal pressure p. The material of the cylinder is elastic–perfectly plastic with yield stress $\sigma_0 = 250$ MPa. If $p = 150$ MPa, find extent of the plastic zone radius, c, assuming Tresca's yield criterion.

Ans: $c = 0.215$ m.

7.14 A thick-walled cylinder with internal radius $a = 4$ in. and external radius $b = 8$ in. is subjected to an internal pressure p. Knowing that for elastic behavior the allowable normal stress is 20 ksi, the allowable shearing stress is 12 ksi, and Poisson's ratio = 0.3, determine the allowable value of p using a factor of safety = 2 assuming that the ends of the cylinder are (i) closed, (ii) open, and (iii) plane strain conditions.

Ans: 4.5, 1.3, 1.3 ksi.

7.15 A thick-walled spherical shell, internal and external radii a and b, is experiencing plastic deformation to radius c under pressure p. Show that

$$a\left(\frac{p}{p_e}\right)^{\frac{1}{3}} < c$$

7.16 A thin-walled spherical shell of mean radius r_0 and thickness t_0 is subjected to internal pressure p. The true stress–strain curve of the material is $\sigma = k\,\varepsilon^n$ where k and n are constants.

At the onset of instability, show that $(d\sigma/d\varepsilon) = (3/2)\sigma$ where $\sigma = \sigma_\theta = \sigma_\phi = p\,r/2t_0$, and find an expression for the bursting pressure of the shell.

Ans: $(p/k) = 2(t_0/r_0)(2n/3)^n e^{-n}$, e is the base of natural logarithm.

7.17 A thick-walled spherical shell with internal radius = 0.2 m and external radius = 0.4 m is subjected to internal pressure p. The material is elastic–perfectly plastic with yield stress = 400 MPa. If $p = 350$ MPa, find extent of the plastic zone radius c.

Ans: $c = 0.238$ m.

7.18 A thin large plate having the upper and lower sides fixed contains a central hole of radius a. The plate is subjected to uniform horizontal pressure p, as shown in Figure P7.18. Assuming the elastic behavior, Poisson's ratio = 0.3, show the SCF at $\theta = \pi/2$ is equal to 2.7.

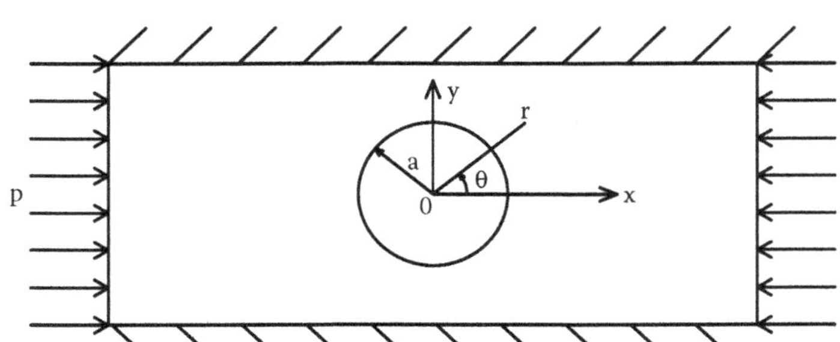

FIGURE P7.18

7.19 A finite-width rectangular plate containing an edge crack of length = 1.2 in. is subjected to a tensile stress σ as shown in Figure P7.19. The fracture toughness of the material is $k_{1c} = 80$ ksi $(\text{in.})^{1/2}$.

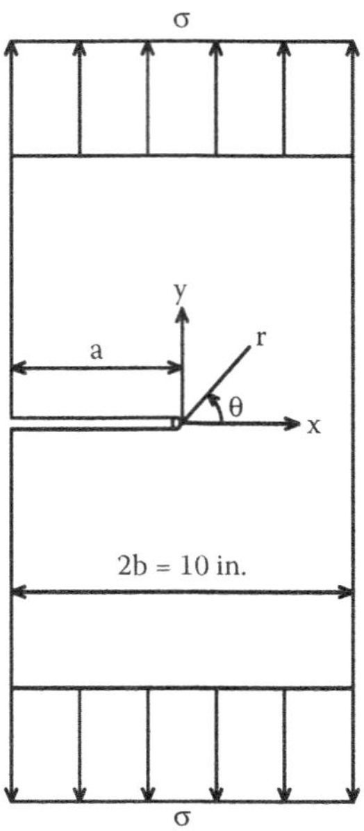

FIGURE P7.19

Use the following stress function to determine the stress field ahead of the crack tip:

$$\phi(r, \theta) = \frac{2}{3}\left(\frac{2}{\pi}\right)^{\frac{1}{2}} K_1 r^{3/2} \cos^3\left(\frac{\theta}{2}\right)$$

Here, K_1 denotes the mode 1 stress-intensity factor given by the formula

$$K_1 = F\left(\frac{a}{b}\right)\sigma\sqrt{\pi a}$$

and F(a/b) is a correction factor whose values are tabulated below:

a/b	F(a/b)
0.1	1.154
0.2	1.198
0.3	1.288
0.4	1.386
0.5	1.512

a. Show that the crack surfaces are stress-free.

b. Determine the allowable value of σ using a safety factor of 2.

Ans: 16.74.

References

7.1. G. Lame, *Lecons sur La theorie de L'elastitecite* [Lessons in the theory of elasticity], Gauthier-Villars, Paris, 1852.
7.2. W. T. Koiter, *Biezeno Anniversary Volume*, Starn, Haarlem, The Netherland, Gauthier-Villars, 1953.
7.3. T. Ranov and F. R. Park, On the numerical value of the tangential stress in thick-walled cylinders, *J. Appl. Mech. ASME*, March, 1953.
7.4. R. Hill, *The Mathematical Theory of Plasticity*, Oxford University Press, Oxford, 1950.
7.5. R. E. Peterson, *Stress Concentration Design Factors*, John Wiley, New York, 1958.
7.6. H. P. Neuber, *Kerbspannungsleter*, 2nd edition, Springer-Verlag, New York, 1958.
7.7. G. N. Savin, *Stress Concentration around Holes*, Pergamon Press, New York, 1961.
7.8. C. E. Inglis, *Transactions of the Institute of Naval Architecture*, London, 1913.
7.9. J. M. Barsom and S. T. Rolfe, *Fracture and Fatigue Control in Structures*, 2nd edition, Prentice-Hall, Englewood, NJ, 1987.
7.10. T. L. Anderson, *Fracture Mechanics*, CRC Press Inc., Boca Raton, FL, 1995.
7.11. C. P. Paris and G. C. Sih, Stress analysis of cracks, *Symposium on Fracture Toughness Testing and Its Application*, ASTM STP special publication no. 381, American Society for Testing of Materials, Philadelphia, PA, 1965.

8

Thermal Stresses

This chapter deals with determining thermal stresses induced in structural components due to nonuniform change in temperature. A uniform change in temperature causes a component to expand or contract, the shape of each element in the component is preserved, normal strains are developed without shearing strains, and if the component is free to expand and contract, no stress is developed. However, thermal stresses will develop if the component is constrained when subjected to uniform heating or if there is a nonuniform change in temperature. Also, thermal stresses are developed in materials that exhibit anisotropy in uniform heating environment. In many cases, the effect of such stresses is severe. It is essential to consider thermal stresses in the design process of components subject to adverse thermal environment like those encountered in the aerospace and chemical industries.

When a linear element of length L is heated, it elongates by an amount $\delta L = \alpha TL$ where α is the coefficient of thermal expansion, a dimensionless positive number, expressed in units of length per length per degree of temperature change. Assuming an isotropic solid, the corresponding thermal strain is

$$\varepsilon_{th} = \alpha T \tag{8.1}$$

In rectangular coordinates, the strains associated with a change in temperature given by T (x, y, z) at any point of the body are $\varepsilon_x = \varepsilon_y = \varepsilon_z = \alpha T$ while the shearing strains are unaffected by the temperature. These strains are superimposed on those due to external loads when considering thermal stress problems. Accordingly, the modified 3D Hooke's law becomes

$$\begin{aligned} \varepsilon_x &= \frac{1}{E}[\sigma_x - \upsilon(\sigma_y + \sigma_z)] + \alpha T \\ \varepsilon_y &= \frac{1}{E}[\sigma_y - \upsilon(\sigma_z + \sigma_x)] + \alpha T \\ \varepsilon_z &= \frac{1}{E}[\sigma_z - \upsilon(\sigma_y + \sigma_y)] + \alpha T \end{aligned} \tag{a}$$

The equations of equilibrium and the strain–displacement relations are not affected by changes in temperature. This is because equilibrium equations are based on principles of statics while strain–displacement relations are purely geometrical.

For a plane element, we have

$$\begin{aligned} E\varepsilon_x &= \sigma_x - \upsilon\sigma_y + E\alpha T \\ E\varepsilon_y &= \sigma_y - \upsilon\sigma_x + E\alpha T \end{aligned} \tag{8.2a}$$

Solving for the stresses yields

$$\begin{aligned} \sigma_x &= \frac{E}{1-\upsilon^2}(\varepsilon_x + \upsilon\varepsilon_y) - \frac{E\alpha T}{1-\upsilon} \\ \sigma_y &= \frac{E}{1-\upsilon^2}(\varepsilon_y + \upsilon\varepsilon_x) - \frac{E\alpha T}{1-\upsilon} \\ \tau_{xy} &= G\gamma_{xy} \end{aligned} \tag{8.2b}$$

Inserting Equations 8.2a into the compatibility Equation 3.21, and making use of equilibrium Equations 2.6, it is readily confirmed that the modified compatibility equation in terms of stresses is

$$\nabla^2(\sigma_x + \sigma_y + E\alpha T) = 0 \tag{8.3}$$

where $\nabla^2 = (\partial^2/\partial x^2) + (\partial^2/\partial y^2)$ denotes Laplace operator in two dimensions. In terms of the stress function, $\varphi(x, y, z)$, introduced in Equation 6.6, the above compatibility equation, which is valid for both plane stress and plane strain problems, assumes the form

$$\nabla^4\phi + E\alpha\nabla^2 T = 0 \tag{8.4}$$

Equations 8.3 and 8.4 provide a simple analytical formulation to determine the temperature field in structural components. Using the physical properties of heat transfer, the temperature field in isotropic elastic components is governed by the following transient heat conduction equation:

$$\frac{\partial^2 T}{\partial x^2} + \frac{\partial^2 T}{\partial y^2} + \frac{\partial^2 T}{\partial z^2} = \frac{\rho c}{k}\frac{\partial T}{\partial t} + \frac{\eta T^*}{k}\frac{\partial e}{\partial t} \tag{8.5}$$

Here, ρ denotes density of the solid; c, heat capacity; k, coefficient of heat conduction ($k/\rho c$ is diffusivity); t, time; T^*, absolute temperature; η, material constant; and e, dilatation. When solving stress problems involving temperature, it is usual to assume that the coefficients of thermal expansion and heat conduction as well as the material constants, Young's modulus, Poisson's ratio, and yield stress, are not affected by changes in temperature of the body.

A comprehensive treatment of thermoelastic problems including the influence of inertia and coupling between strain and temperature, governed by Equation 8.5, is available in References 8.1, 8.2. It turns out that for most practical engineering problems the right-hand side of Equation 8.5 is not large enough and may be neglected for many applications.

When the influence of both inertia and coupling between temperature and strain is neglected, determining the stresses in structural components due to change in temperature is reduced to two distinct problems. The first is a problem in the theory of steady-state heat flow, which requires determining the temperature distribution, T(x, y, z), at every point of the body using Laplace's equation

$$\nabla^2 T(x, y, z) = 0 \tag{8.6}$$

And the second involves determining the resulting stresses, strains, and displacements in the usual manner discussed in previous chapters.

In order to illustrate the procedure, consider the plane stress problem of a thin rectangular plate (or beam) of length 2L, depth 2d, and thickness t subjected to an arbitrary variation of temperature throughout the depth, T(y). There are no external forces acting on the plate (Figure 8.1).

The applied temperature, which is independent of x and z, gives rise to longitudinal thermal strain $\alpha\,T(y)$, which is accompanied by longitudinal compressive stress at each element of the plate.

$$\sigma_x(y) = -E(\alpha T) \tag{a}$$

Here, E is Young's modulus of the material. Since the plate is free to expand there will be no stress in the plate provided that at the ends there are compressive stresses of magnitude (a) to suppress the longitudinal strain induced by the temperature.

As in beam theory, assume that $\sigma_y = \tau_{xy} = 0$, which satisfies the boundary conditions across $y = \pm d/2$. Inserting the stresses into Equation 8.3 gives $(d^2/dy^2)(\sigma_x + \alpha ET) = 0$, and after integration

$$\sigma_x = -\alpha ET + c_1 y + c_2 \tag{b}$$

where c_1 and c_2 are constants of integration to be determined from the boundary conditions, namely, zero resultant force and zero moment at the ends $x = \pm L$, or

$$\int_{-d}^{d} \sigma_x t dy = 0, \qquad \int_{-d}^{d} \sigma_x y(t dy) = 0 \tag{c}$$

Inserting Equation (b) into (c), performing the integration, it follows that

$$c_1 = \frac{3E\alpha}{2d^3}\int_{-d}^{d} yT(y)\,dy, \quad c_2 = \frac{E\alpha}{2d}\int_{-d}^{d} T(y)\,dy$$

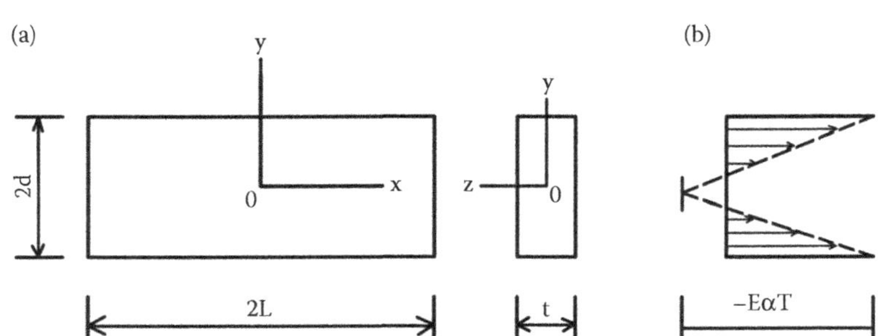

FIGURE 8.1
Thin rectangular plate subjected to T(y), (a) thin plate subjected to T(y) and (b) compressive stress at each element of plate.

Hence, the longitudinal stress is

$$\sigma_x = E\alpha\left[-T(y) + \frac{1}{2d}\int_{-d}^{d} T(y)\,dy + \frac{3y}{2d^3}\int_{-d}^{d} yT(y)\,dy\right] \tag{8.7a}$$

The corresponding strains are

$$\varepsilon_x = \frac{\sigma_x}{E} + \alpha T, \quad \varepsilon_y = -\frac{\upsilon\sigma_x}{E} + \alpha T, \quad \gamma_{xy} = 0 \tag{8.7b}$$

and the associated displacements follow from Equations 3.7.

Note that if $T(y) = T_0$, constant, then Equation 8.7a gives $\sigma_x = 0$ as expected.

Next, we determine the longitudinal stress when simple variations in temperature are applied to the plate.

Linear variation: If the surfaces $y = \pm d$ of the plate are maintained at two different constant temperatures, say T_1 at the bottom surface and T_2 at the top, then a steady-state heat flow is established after a certain time and the temperature is given by the function

$$T(y) = \frac{1}{2}(T_1 + T_2) + \frac{1}{2}(T_1 - T_2)\frac{y}{d} \tag{d}$$

Substituting Equation (d) into 8.7a and performing the integration, the longitudinal stress is found to be zero and the strains are αT because the beam is free to expand.

Parabolic variation: If the beam is heated by a parabolic distribution of temperature,

$$T(y) = T_0\left(1 - \frac{y^2}{d^2}\right), \quad T_0 = \text{constant} \tag{e}$$

Inserting the temperature function in Equation 8.7a and performing the integration, it is found that the longitudinal stress is given by

$$\sigma_x = E\alpha T_0\left[-\frac{1}{3} + \frac{y^2}{d^2}\right] \tag{f}$$

and the strains follow from Equation 8.7b.

If the ends $x = \pm L$ of the plate are prevented from expansion, Figure 8.1b, then $\varepsilon_x = 0$ throughout for any temperature $T(y)$ and Equation 8.7b gives

$$\sigma_x = -E\alpha T \tag{8.8}$$

The corresponding strains are $\varepsilon_x = \gamma_{xy} = 0$, $\varepsilon_y = (1 + \upsilon)\alpha T$.

EXAMPLE 8.1

A composite bar ABC made up of steel and aluminum is fixed at ends A and C as shown in Figure 8.2. The bar is unstressed at 20°C.

a. If the temperature of the bar is uniformly raised to 120°C, find the normal stress.
b. If the unstressed bar is heated nonuniformly so that the temperature varies linearly from 20°C at end A to 100°C at end C, find the normal stress.

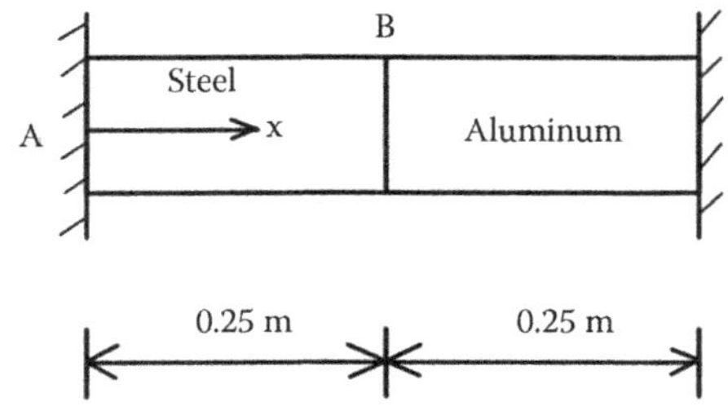

FIGURE 8.2
Example 8.1.

c. If the connection between the segments fails in tension at a stress of 70 MPa, find the temperature required to cause failure.

Assume

$$E_S = 200\,\text{GPa}, E_{AL} = 70\,\text{GPa}$$
$$\alpha_S = 11.7 \times 10^{-6}/°\text{C}, \alpha_{AL} = 23 \times 10^{-6}/°\text{C}$$

Solution

a. Since the ends are fixed, the basic equations are

$$\delta_S + \delta_{AL} = 0 \text{ or } \varepsilon_s + \varepsilon_{AL} = 0, \quad \text{and} \quad F_S = F_{AL} \text{ or } \sigma_S = \sigma_{AL} = \sigma$$

The extension of each segment consists of a thermal component due to change in temperature and a mechanical component due to the induced normal stress. The two segments have equal length. It follows that

$$\alpha_S \Delta T + \frac{\sigma}{E_S} + \alpha_{AL} \Delta T + \frac{\sigma}{E_{AL}} = 0$$

Inserting the numerical values,

$$11.7 \times 10^{-6}(100) + (\sigma/200 \times 10^9) + 23 \times 10^{-6}(100) + (\sigma/70 \times 10^9) = 0$$

Solving for the normal stress, we have $\sigma = -180\,\text{MPa}$.

b. In this case, we have the variation in temperature $= (80°\text{C}/0.5\,\text{m})\ x = 160x$, where x is measured from end A. The basic equation becomes

$$\int_0^{0.25} \alpha_S(160x)dx + \frac{\sigma_S L_s}{E_S} + \int_{0.25}^{0.5} \alpha_{AL}(160x)dx + \frac{\sigma_{AL} L_{Al}}{E_{Al}} = 0$$

That is,

$$\int_0^{0.25} 11.7 \times 10^{-6}(160x)\,dx + \frac{0.25\sigma}{200 \times 10^9} + \int_{0.25}^{0.5} 23 \times 10^{-6}(160x)\,dx + \frac{0.25\sigma}{70 \times 10^9} = 0$$

Performing the integration and solving for the normal stress, we obtain

$$\sigma = -83.7\,\text{MPa}$$

c. In this case, the basic equation is

$$\alpha_S \Delta T(0.25) + \frac{\sigma(0.25)}{E_s} + \alpha_{AL}\Delta T(0.25) + \frac{\sigma(0.25)}{E_{AL}} = 0$$

Inserting numerical values of the material, observing that $\sigma = 70$ Mpa, and solving for the change in temperature, we find

$$\Delta T = -38.9°C$$

8.1 Thermal Stresses in Uniform Circular Disks

Consider the plane stress problem of a circular disk subjected to change in temperature T(r). The modified elastic strains are given by

$$\begin{aligned} \frac{du}{dr} &= \varepsilon_r = \frac{1}{E}(\sigma_r - \upsilon\sigma_\theta) + \alpha T \\ \frac{u}{r} &= \varepsilon_\theta = \frac{1}{E}(\sigma_\theta - \upsilon\sigma_r) + \alpha T \end{aligned} \tag{8.9}$$

Differentiating the second equation of Equations 8.9 with respect to r and equating the result to the first equation, it is readily confirmed that

$$(1 + \upsilon)(\sigma_\theta - \sigma_r) + r\frac{d}{dr}(\sigma_\theta - \upsilon\sigma_r + E\alpha T) = 0 \tag{a}$$

Since the disk is in a state of equilibrium, that is the hoop and radial stresses satisfy the equation, $\sigma_\theta - \sigma_r = r(d\sigma_r/dr)$ (see Equation 7.1), it follows that Equation (a) is equivalent to the compatibility equation in terms of stresses

$$\frac{d}{dr}[\sigma_r + \sigma_\theta + E\alpha T] = 0 \tag{8.10}$$

The integral of Equation 8.10 gives

$$\sigma_r + \sigma_\theta = 2A - E\alpha T \tag{8.11}$$

Here, 2A is a convenient constant of integration. Equation 8.11 indicates that the sum of the radial and tangential stresses is proportional to the applied temperature. Eliminating the hoop stress by using the equilibrium equation given above results in

$$2\sigma_r + r\frac{d\sigma_r}{dr} = 2A - E\alpha T \tag{b}$$

Multiplying both sides of the last equation by r, the result is equivalent to

$$\frac{d}{dr}(r^2\sigma_r) = 2Ar - E\alpha rT(r) \tag{c}$$

Integrating Equation (c), using B as constant of integration, the radial stress is immediately obtained, and the hoop stress follows from Equation 8.11. Hence, the stresses in the disk are given by

$$\begin{aligned}\sigma_r &= A + \frac{B}{r^2} - \frac{E\alpha}{r^2}\int rT(r)dr \\ \sigma_\theta &= A - \frac{B}{r^2} - E\alpha T + \frac{E\alpha}{r^2}\int rT(r)dr\end{aligned} \tag{8.12}$$

Equations 8.12 apply to both solid and annular disks. For the solid disk, B must vanish and A is determined from the condition that $\sigma_r = 0$ at the outer boundary. For the annular disk, A and B are determined from the same boundary condition at the inner and outer edges.

8.1.1 Solid Disk

Consider a solid disk of radius $r = a$, subjected to a temperature that varies linearly with the radial distance

$$T(r) = T_0 + rT_1 \tag{a}$$

where T_0 and T_1 are constants. It is required to find the stresses and strains produced by the temperature.

In this case, constant B must vanish and A is determined from the condition that $\sigma_r = 0$ at $r = a$. Inserting expression (a) into Equation 8.12 evaluating the integral, the boundary condition gives, $A = E\alpha((T_0/2) + (T_1/3)a)$, and the stresses are

$$\begin{aligned}\sigma_r &= \frac{E\alpha}{3}T_1(a - r) \\ \sigma_\theta &= \frac{E\alpha}{3}T_1(a - 2r)\end{aligned} \tag{8.13}$$

The strains and radial displacement u follow immediately from Equations 8.9. Observe that the constant temperature term T_0 produces free expansion with no stresses induced in the disk. Also, both radial and hoop stresses vary linearly with r and T_1. The radial stress is tensile throughout the disk while the hoop stress is tensile for $r < a/2$ and compressive for $r > a/2$. At the boundary, $r = a(\sigma_\theta)_{r=a} = -(E\alpha T_1/3)a$.

The maximum stress occurs at the center $r = 0$ and has the magnitude

$$(\sigma_r)_{max} = (\sigma_\theta)_{max} = \frac{E\alpha}{3}T_1 a \tag{8.14}$$

8.1.2 Annular Disk

If the temperature gradient, $T(r) = T_0 + rT_1$, is applied to an annular disk with inner and outer radii denoted by a, b, respectively, the constants A and B in Equation 8.12 are obtained from the conditions $\sigma_r = 0$ at $r = a$, and at $r = b$. It is readily confirmed that

$$A = \frac{E\alpha}{3}\left[\frac{3}{2}T_0 + \left(\frac{a^2 + ab + b^2}{b + a}\right)T_1\right], \quad B = -\frac{E\alpha}{3}\left(\frac{a^2b^2}{b + a}\right)T_1$$

and the stresses are

$$\sigma_r = \frac{E\alpha T_1}{3}\left[b - r + \frac{a^2}{b+a}\left(1 - \frac{b^2}{r^2}\right)\right]$$
$$\sigma_\theta = \frac{E\alpha T_1}{3}\left[b - 2r + \frac{a^2}{b+a}\left(1 + \frac{b^2}{r^2}\right)\right] \tag{8.15}$$

The strains are obtained from relations (8.9). As in the case of a solid disk, applying constant temperature, T_0, produces free expansion of the disk without stress.

The values of the hoop stress at the boundaries are

$$(\sigma_\theta)_{r=a} = \frac{E\alpha T_1}{3}\left(-a + \frac{2b^2}{b+a}\right)$$
$$(\sigma_\theta)_{r=b} = \frac{E\alpha T_1}{3}\left(-b + \frac{2a^2}{b+a}\right) \tag{8.16}$$

The maximum hoop stress occurs at the inner surface $r = a$. Note that by setting $a = 0$ in the first equation of Equation 8.16, the magnitude of the hoop stress, $(\sigma_\theta)_{max} = 2E\alpha bT_1/3$, is twice the magnitude of the maximum hoop stress found in solid disk of radius b (Equation 8.14). This is due to the phenomenon of stress concentration around holes discussed previously in Chapters 6 and 7.

The maximum radial stress, obtained by setting $d\sigma_r/dr = 0$, is found to be at a radius given by

$$r = \left(\frac{2a^2b^2}{b+a}\right)^{1/3} \tag{8.17}$$

8.1.3 Combined Thermal and Rotational Stresses

This section deals with the stresses induced in a circular disk, which is rotating with an angular velocity ω rad/s while experiencing steady-state heat flow.

When a circular disk is subjected to a temperature rise of T(r), it was shown in Section 8.1 that the stresses satisfy the equation

$$(1 + \upsilon)(\sigma_\theta - \sigma_r) + r\frac{d}{dr}(\sigma_\theta - \sigma_r + E\alpha T) = 0 \tag{a}$$

These stresses must also satisfy the equilibrium equation of a rotating disk, namely

$$\sigma_\theta - \sigma_r - r\frac{d\sigma_r}{dr} = \rho r^2\omega^2 \tag{b}$$

Combining Equations (a) and (b) yields the following compatibility equation in terms of stresses:

$$\frac{d}{dr}\left[\sigma_r + \sigma_\theta + E\alpha T + \frac{1+\upsilon}{2}\rho r^2\omega^2\right] = 0 \tag{8.18}$$

Integrating the compatibility equation gives

$$\sigma_r + \sigma_\theta = 2A - E\alpha T - \frac{1+\upsilon}{2}\rho r^2\omega^2, \quad A = \text{constant} \tag{8.19}$$

Eliminating the hoop stress by subtracting Equation (b) from Equation 8.19, multiplying both sides of the result by r, the following equation is obtained:

$$\frac{d}{dr}(r^2\sigma_r) = r\left[2A - \frac{3+\upsilon}{2}\rho\omega^2 r^2 - E\alpha T\right] \tag{c}$$

Performing the integration, using B as constant, the radial stress is immediately obtained and the hoop stress follows from Equation 8.19. The stresses are

$$\begin{aligned} \sigma_r &= A + \frac{B}{r^2} - \frac{3+\upsilon}{8}\rho\omega^2 r^2 - \frac{E\alpha}{r^2}\int rT(r)dr \\ \sigma_\theta &= A - \frac{B}{r^2} - \frac{1+3\upsilon}{8}\rho\omega^2 r^2 - E\alpha T + \frac{E\alpha}{r^2}\int rT(r)dr \end{aligned} \tag{8.20}$$

Note that Equations 8.20 could have been obtained by simply adding Equations 6.51 and 8.12 to obtain the combined thermal and rotational stresses induced in uniform circular disks.

As an example, suppose that a solid disk of radius a is rotating with an angular velocity ω rad/s and at the same time experiencing a temperature rise of $T(r) = T_0 + T_1 r$, the radial and hoop stresses are obtained simply by adding Equations 6.52 and 8.13 to give the following result without the need to evaluate the constants A and B in Equation 8.20:

$$\begin{aligned} \sigma_r &= \frac{3+\upsilon}{8}\rho\omega^2(a^2 - r^2) + \frac{E\alpha T_1}{3}(a - r) \\ \sigma_\theta &= \frac{\rho\omega^2}{8}[(3+\upsilon)a^2 - (1+3\upsilon)r^2] + \frac{E\alpha T_1}{3}(a - 2r) \end{aligned} \tag{8.21}$$

Numerical examples are given below to illustrate the procedure.

EXAMPLE 8.2

A solid steel disk, 10 in. diameter, is subjected to a linear temperature rise that varies from 70°F at the center to T_a at the edge. If the maximum radial expansion is not to exceed 5×10^{-3} in., find T_a and the maximum stress in the disk. For steel, $E = 30 \times 16^6$ psi and $\alpha = 6.5 \times 10^{-6}$/°F.

Solution

The temperature rise is given by $T(r) = 70 + r\,T_1$.

At $r = 0$, $T = 70$°F, while at $r = 5$ in., $T = T_a$, it follows that

$$T_1 = (T_a - 70)/5 \tag{a}$$

The radial displacement is obtained from Equations 8.9, $u = (r/E)(\sigma_\theta - \upsilon\sigma_r) + \alpha r T(r)$.

At $r = a$, $\sigma_r = 0$ and from Equation 8.12, $\sigma_\theta = -(E\alpha T_1/3)a$, it follows that

$$u_a = \frac{a}{E}\left(-\frac{E\alpha T_1 a}{3}\right) + \alpha a(70 + T_1 a) = \alpha a\left(\frac{2T_1 a}{3} + 70\right)$$

and

$$5 \times 10^{-3} = 6.5 \times 10^{-6}\left(350 + \frac{50T_1}{3}\right)$$

Hence, $T_1 = 25.2°F/in.$ and $T_a = 195.8°F$.

The maximum stress from Equation 8.13 is $\frac{E\alpha}{3}T_1 a = \frac{1}{3}(30 \times 10^6)(6.5 \times 10^{-6})(25.2)(5) =$ 8.19 ksi.

EXAMPLE 8.3

A uniform disk, 0.4 m outside diameter with a central hole, 0.2 m diameter, is subjected to a linear temperature gradient such that the temperature at the outer edge is 60°C higher than that at the inner edge. If the disk rotates at 1200 RPM, find the hoop stresses induced at both edges. For steel, $E = 200$ GPa, $\upsilon = 0.3$, $\alpha = 11.7 \times 10^{-6}/°C$, and $\rho = 7800\ kg/m^3$.

Solution

Here, $a = 0.1$, $b = 0.2$ m, and $T(r) = T_0 + r\,T_1$.

At $r = b$, $T_b = T_0 + 0.2\,T_1$ and at $r = a$, $T_a = T_0 + 0.1\,T_1$; $T_b - T_a = 60°C = 0.1\,T_1$. Hence, $T_1 = 600°C/m$

$$\omega = 1200\left(\frac{2\pi}{60}\right) = 125.66\ \text{rad/s}$$

Using Equations 8.16 and 6.60, we have

$$\begin{aligned}
(\sigma_\theta)_{r=a} &= \frac{E\alpha T_1}{3}\left(-a + \frac{2b^2}{b+a}\right) + \frac{\rho\omega^2}{4}[(3+\upsilon)b^2 + (1-\upsilon)a^2] \\
&= (1/3)(200 \times 10^9)(11.7 \times 10^{-6})(600)[-0.1 + 2(0.2)^2/(0.3)] \\
&\quad + (1/4)(7800)(125.66)^2\left[3.3(0.2)^2 + 0.7(0.1)^2\right] \\
&= 78\ \text{MPa} + 4.29\ \text{MPa} \\
&= 82.3\ \text{MPa}
\end{aligned}$$

$$\begin{aligned}
(\sigma_\theta)_{r=b} &= \frac{E\alpha T_1}{3}\left(-b + \frac{2a^2}{b+a}\right) + \frac{\rho\omega^2}{4}[(3+\upsilon)a^2 + (1-\upsilon)b^2] \\
&= (1/3)(200 \times 10^9)(11.7 \times 10^{-6})(600)[-0.2 + 2(0.1)^2/(0.3)] \\
&\quad + (1/4)(7800)(125.66)^2\left[3.3(0.1)^2 + 0.7(0.2)^2\right] \\
&= -62.38\ \text{MPa} - 0.54\ \text{MPa} \\
&= -63\ \text{MPa}
\end{aligned}$$

8.2 Thermal Stresses in Cylindrical Vessels

In this section, we consider a thick-walled cylindrical vessel of internal and external radii a and b, respectively, subjected to heat flow T(r). The z-axis is taken along the axis of the cylinder. The modified stress–strain relations are

$$\begin{aligned}\varepsilon_r &= \frac{1}{E}[\sigma_r - \upsilon(\sigma_\theta + \sigma_z)] + \alpha T = \frac{du}{dr}\\ \varepsilon_\theta &= \frac{1}{E}[\sigma_\theta - \upsilon(\sigma_z + \sigma_r)] + \alpha T = \frac{u}{r}\\ \varepsilon_z &= \frac{1}{E}[\sigma_z - \upsilon(\sigma_r + \sigma_\theta)] + \alpha T\end{aligned} \tag{8.22}$$

The axial stress is obtained from the third equation in Equation 8.22

$$\sigma_z = E\varepsilon_z + \upsilon(\sigma_r + \sigma_\theta) - E\alpha T \tag{a}$$

Inserting Equation (a) into the first two equations of Equation 8.22, we get

$$\begin{aligned}\frac{du}{dr} &= -\upsilon\varepsilon_z + \frac{1+\upsilon}{E}[(1-\upsilon)\sigma_r - \upsilon\sigma_\theta] + (1+\upsilon)\alpha T\\ \frac{u}{r} &= -\upsilon\varepsilon_z + \frac{1+\upsilon}{E}[(1-\upsilon)\sigma_\theta - \upsilon\sigma_r] + (1+\upsilon)\alpha T\end{aligned} \tag{b}$$

Similar to the procedure used previously in Section 8.1, eliminating u from Equations (b) by assuming that the axial strain does not vary with r and making use of equilibrium, Equation 7.1, the following compatibility equation is obtained:

$$\frac{d}{dr}\left[\sigma_r + \sigma_\theta + \frac{E\alpha T}{1-\upsilon}\right] = 0 \tag{8.23}$$

which is identical to the corresponding disk result, Equation 8.10, provided that $E\alpha$ is replaced by $E\alpha/(1-\upsilon)$. The integral of Equation 8.23 is

$$\sigma_r + \sigma_\theta = 2A - \frac{E\alpha T}{1-\upsilon} \tag{8.24}$$

Equation 8.24 indicates that the sum of the radial and hoop stresses is proportional to the applied temperature. Following the procedure used in Section 8.1, the radial and hoop stresses induced in a cylindrical vessel are

$$\begin{aligned}\sigma_r &= A + \frac{B}{r^2} - \frac{E\alpha}{(1-\upsilon)r^2}\int rT(r)dr\\ \sigma_\theta &= A - \frac{B}{r^2} - \frac{E\alpha T(r)}{1-\upsilon} + \frac{E\alpha}{(1-\upsilon)r^2}\int rT(r)dr\end{aligned} \tag{8.25a}$$

And the axial stress in the cylinder is obtained by inserting Equation 8.24 into (a) to give

$$\sigma_z = E\varepsilon_z + 2\upsilon A - \frac{E\alpha T}{1-\upsilon} \tag{8.25b}$$

The constants A and B are determined from the boundary condition, $\sigma_r = 0$ at $r = a$ and $r = b$. Since the integrals in Equation 8.25a are valid in the region $r = a$ to $r = b$, it follows that

$$\begin{aligned} A + \frac{B}{a^2} &= \frac{E\alpha}{(1-\upsilon)a^2}\int_a^{r=a} rTdr = 0 \\ A + \frac{B}{b^2} &= \frac{E\alpha}{(1-\upsilon)b^2}\int_a^{r=b} rTdr \end{aligned} \tag{c}$$

Solving the simultaneous Equations (c) gives

$$A = \frac{E\alpha}{(1-\upsilon)(b^2-a^2)}\int_a^b rTdr$$

and

$$B = -\frac{a^2E\alpha}{(1-\upsilon)(b^2-a^2)}\int_a^b rTdr$$

Hence, the stresses induced in a thick cylindrical vessel subjected to a steady-state temperature field T(r) are

$$\begin{aligned} \sigma_r &= \frac{E\alpha}{(1-\upsilon)r^2}\left[\left(\frac{r^2-a^2}{b^2-a^2}\right)\int_a^b rTdr - \int_a^r rTdr\right] \\ \sigma_\theta &= \frac{E\alpha}{(1-\upsilon)r^2}\left[\left(\frac{r^2+a^2}{b^2-a^2}\right)\int_a^b rTdr + \int_a^r rTdr - r^2T\right] \\ \sigma_z &= E\varepsilon_z + \frac{E\alpha}{(1-\upsilon)}\left[\left(\frac{2\upsilon}{b^2-a^2}\right)\int_a^b rTdr - T\right] \end{aligned} \tag{8.26}$$

8.2.1 Steady-State Heat Flow

A steady-state temperature distribution is governed by Laplace's Equation 8.6. In cylindrical coordinates, the general equation for the temperature distribution assumes the form

$$\frac{\partial^2 T}{\partial r^2} + \frac{1}{r}\frac{\partial T}{\partial r} + \frac{1}{r^2}\frac{\partial^2 T}{\partial \theta^2} + \frac{\partial^2 T}{\partial z^2} = 0 \tag{a}$$

Because of symmetry of the cylinder, we have in this case, $\partial T/\partial\theta = \partial T/\partial z = 0$, and Equation (a) is reduced to

$$\frac{d^2 T}{dr^2} + \frac{1}{r}\frac{dT}{dr} = 0 \tag{8.27}$$

The solution of Equation 8.27 is

$$T(r) = C_1 \mathrm{Ln}\, r + C_2 \tag{b}$$

where C_1 and C_2 are constants. Denoting the temperatures at the inner and outer surfaces of the cylinder by T_i and T_O, respectively, the boundary conditions for the heat flow are $T(r) = T_i$ at $r = a$ and $T(r) = T_O$ at $r = b$. Hence, the temperature distribution for constant temperature difference between vessel walls is

$$T(r) = T_O + (T_i - T_O)\frac{\mathrm{Ln}\frac{b}{r}}{\mathrm{Ln}\frac{b}{a}} \tag{8.28}$$

Inserting Equation 8.28 into Equations 8.26, introducing the dimensionless parameter,

$$\beta = \frac{\alpha(T_i - T_O)}{2(1 - \upsilon)} \tag{8.29}$$

And making use of the result,

$$\int_a^r r\mathrm{Ln}\frac{b}{r}dr = \frac{1}{4}(r^2 - a^2) + \frac{r^2}{2}\mathrm{Ln}\frac{b}{r} - \frac{a^2}{2}\mathrm{Ln}\frac{b}{a}$$

the stresses in the cylinder are found to be given by

$$\begin{aligned}
\sigma_r &= \beta E\left[\frac{(b^2/r^2) - 1}{(b^2/a^2) - 1} - \frac{\mathrm{Ln}(b/r)}{\mathrm{Ln}(b/a)}\right] \\
\sigma_\theta &= \beta E\left[-\frac{(b^2/r^2) + 1}{(b^2/a^2) - 1} + \frac{1 - \mathrm{Ln}(b/r)}{\mathrm{Ln}(b/a)}\right] \\
\sigma_z - E\varepsilon_z &= \beta E\left[\frac{\upsilon - 2\mathrm{Ln}(b/r)}{\mathrm{Ln}(b/a)} - \frac{2\upsilon}{(b^2/a^2 - 1)}\right] - E\alpha T_O
\end{aligned} \tag{8.30}$$

The axial strain in the expression for σ_z is obtained from the end condition of the tube.

The hoop stresses at the cylindrical boundaries of the vessel are given by

$$(\sigma_\theta)_{r=a} = -\beta E\left[\frac{2}{1-(a^2/b^2)} - \frac{1}{\mathrm{Ln}(b/a)}\right]$$
$$(\sigma_\theta)_{r=b} = \beta E\left[\frac{1}{\mathrm{Ln}(b/a)} - \frac{2}{(b^2/a^2)-1}\right] \tag{8.31}$$

For $\beta > 0$, the hoop stress is compressive at $r = a$, its magnitude varies between (βE) for wall thickness ratio b/a approximately equal to unity and $2(\beta E)$ for $(b/a) \approx \infty$. At $r = b$, the hoop stress is tensile and its magnitude varies between (βE) for small values of (b/a) and zero for large $(b/a) \approx \infty$.

The location of the maximum radial stress is obtained by requiring, $d\sigma_r/dr = 0$, that is,

$$\frac{(-2b^2/r^3) - 0}{\left(\frac{b^2}{a^2} - 1\right)} - \frac{0 - (1/r)}{\mathrm{Ln}(b/a)} = 0$$

which gives

$$r = \left[\frac{2\mathrm{Ln}(b/a)}{1-(a^2/b^2)}\right]^{1/2} a \tag{8.32}$$

Initiation of yield: As the magnitude of the steady-state heat flow increases, yield is initiated at critical elements of the vessel. Using Tresca's criterion and considering cases where the principal axial stress is between the principal radial and hoop stresses, the maximum value of the stress difference from Equation 8.30

$$\sigma_\theta - \sigma_r = \beta E\left[\frac{1}{\mathrm{Ln}(b/a)} - \frac{2(b^2/r^2)}{(b^2/a^2) - 1}\right] \tag{c}$$

occurs at $r = a$. Hence, for $\beta > 0$, yield is initiated when

$$\sigma_0 = E\beta_e\left[\frac{2}{1-(a^2/b^2)} - \frac{1}{\mathrm{Ln}(b/a)}\right] \tag{d}$$

It follows that the temperature difference $T_i - T_0$ between the inner and outer surfaces required to initiate yield in the vessel is given by

$$(T_i - T_0)_e = \left\{\frac{2(1-\upsilon)}{[2/(1-a^2/b^2) - 1/\mathrm{Ln}(b/a)]}\right\}\frac{\sigma_0}{\alpha E} \tag{8.33}$$

Numerical values of the temperature difference required to initiate yield in a vessel constructed of mild steel with $E = 200$ GPa, $\nu = 0.3$, $\sigma_0 = 250$ MPa, and $\alpha = 11.7 \times 10^{-6}/°C$ are given in Table 8.1.

It is worth noting that we may use different procedures than the one used above to derive formulas for the elastic stresses in cylindrical vessels subjected to steady-state heat flow. Two such procedures are given below:

TABLE 8.1

Values of Temperature Required to Initiate Yield

Wall Thickness b/a	$(T_i - T_0)_e$°C
2	122
3	112
4	106
5	102
Large	84

1. Upon inserting Equations 8.28 and 8.29 into Equation 8.24 and absorbing the resulting constant terms in A, results in

$$\sigma_r + \sigma_\theta = 2A - 2\beta E \frac{\mathrm{Ln}(b/r)}{\mathrm{Ln}(b/a)} \tag{e}$$

Invoking equilibrium to eliminate σ_θ, and multiplying by r, Equation (e) gives

$$\frac{d}{dr}(r^2\sigma_r) = 2A - 2\beta E \frac{r\mathrm{Ln}(b/r)}{\mathrm{Ln}(b/a)} \tag{f}$$

Integrating Equation (f), it is readily confirmed that

$$\begin{aligned} \sigma_r &= A + \frac{B}{r^2} - \beta E\left[\frac{(1/2) + \mathrm{Ln}(b/r)}{\mathrm{Ln}(b/a)}\right] \\ \sigma_\theta &= A - \frac{B}{r^2} + \beta E\left[\frac{(1/2) - \mathrm{Ln}(b/r)}{\mathrm{Ln}(b/a)}\right] \end{aligned} \tag{8.34}$$

The constants A and B are determined from the boundary conditions $\sigma_r = 0$ at $r = a$ and $r = b$, which gives the stress expressions stated in Equation 8.30.

2. Inserting Equation (b) into 8.25a and using the result

$$\int r(C_1\mathrm{Ln}r + C_2)dr = C_1\left[\frac{r^2}{2}\mathrm{Ln}r - \frac{r^2}{4}\right] + C_2\frac{r^2}{2} \tag{g}$$

it is readily confirmed that Equations 8.25a become

$$\begin{aligned} \sigma_r &= A + \frac{B}{r^2} - \frac{E\alpha T(r)}{2(1-\upsilon)} \\ \sigma_\theta &= A - \frac{B}{r^2} - \frac{E\alpha T(r)}{2(1-\upsilon)} - \frac{E\alpha C_1}{2(1-\upsilon)} \end{aligned} \tag{8.35}$$

Equations 8.34 and 8.35 apply to uniform disks subjected to steady-state heat flow provided that $E\alpha/(1-\nu)$ is replaced by $E\alpha$.

EXAMPLE 8.4

An unloaded steel cylinder, internal diameter = 0.3 m and external diameter = 0.5 m, is subjected to steady-state temperatures of 150°F and 50°F at the inner and outer surfaces, respectively. Assume plain strain condition, determine

a. The circumferential stresses induced at the inner and outer surfaces
b. The radial expansion of the inner and outer boundaries

For steel, $E = 207 \times 10^9\ N/m^2$, $\upsilon = 0.3$, and $\alpha = 11 \times 10^{-6}/°C$.

Solution

a. $T = C_1 \ Ln\ r + C_2$
$150 = C_1 \ Ln\ 0.15 + C_2$
$50 = C_1 \ Ln\ 0.25 + C_2$

It follows that $C_1 = -196$, and $\dfrac{E\alpha}{2(1-\upsilon)} = \dfrac{207 \times 10^9 \times 11 \times 10^{-6}}{1.4} = 1.626 \times 10^6\ N/m^2$.

Next, we find the constants A and B in Equations 8.33 from the thermal and mechanical boundary conditions at the inner and outer boundaries of the vessel where the radial stress is zero and at r = 0.15 m, T = 150°C while at r = 0.3 m, T = 50°C, that is,

$A + B/0.0225 - 1.626 \times 10^6 \times 150 = 0$
$A - B/0.0625 - 1.626 \times 10^6 \times 50 = 0$

which gives $A = -10.2 \times 10^6$ and $B = 5.72 \times 10^6$.

The circumferential stress in Equation 8.34 becomes

$$\sigma_\theta = -10.2 \times 10^6 - \frac{5.72 \times 10^6}{r^2} - 1.63 \times 10^6 T - 1.62 \times 10^6 \times (-196)$$

Hence, at r = 0.15 m, T = 150°C, $\sigma_\theta = -\ 90\ MN/m^2$ and at r = 0.25 m, T = 50°C, $\sigma_\theta = 135\ MN/m^2$.

b. The radial displacements of the inner and outer surfaces are computed using Equation 8.22.
Thus, at the inner boundary, a = r = 0.15 m

$$u_i = \frac{a}{E}[\sigma_\theta - \upsilon(\sigma_z + \sigma_r)] + a(\alpha T)$$

where $\sigma_\theta = -\ 90 \times 10^6\ N/m^2$, T = 150°C, $\sigma_r = 0$, and the axial stress is

$$\begin{aligned}\sigma_z &= 0.3\sigma_\theta - E\alpha T = -0.3(-90 \times 10^6) - 207 \times 10^9 \times 11 \times 10^{-6} \times 150 \\ &= -371.5 \times 10^6\ N/m^2\end{aligned}$$

Hence,

$$\begin{aligned}u_i &= \frac{0.15}{207 \times 10^9}[-90 \times 10^6 - 0.3(-371.5 \times 10^6) + 0] + 0.15(11 \times 10^{-6})150 \\ &= 0.27\ mm\end{aligned}$$

Similarly, at the outer boundary, b = r = 0.25 m

$$u_0 = \frac{b}{E}[\sigma_\theta - \upsilon(\sigma_z - \sigma_r) + b(\alpha T)$$

where, $\sigma_\theta = 135 \times 10^6\,\mathrm{N/m^2}$, $T = 50°C$, $\sigma_r = 0$, and the axial stress

$$\sigma_z = 0.3 \times 135 \times 10^6 - 207 \times 10^9 \times 11 \times 10^{-6} \times 50 = -73.35\,\mathrm{N/m^2}$$

Hence,

$$u_0 = \frac{0.25}{207 \times 10^9}[135 \times 10^6 - 0.3(-73.35 \times 10^6) + 0] + 0.25(11 \times 10^{-6}) \times 50$$
$$= 0.33\,\mathrm{mm}$$

8.2.2 Thermal Stresses in Pressurized Vessels Subject to Steady-State Heat Flow

When the vessel is subject to simultaneous steady-state heat flow and internal pressure p, the induced elastic stresses may be obtained by finding constants A and B in Equation 8.34 using boundary conditions $\sigma_r = 0$ at $r = b$ and $\sigma_r = -p$ at $r = 0$, which give

$$A = \frac{p - \beta E}{(b^2/a^2) - 1} + \frac{\beta E}{2\mathrm{Ln}(b/a)}$$

$$B = -b^2 \frac{p - \beta E}{(b^2/a^2) - 1}$$

Or simply by adding Equations 7.7 and 8.30 to yield

$$\sigma_r = (\beta E - p)\frac{(b^2/r^2) - 1}{(b^2/a^2) - 1} - \beta E\frac{\mathrm{Ln}(b/r)}{\mathrm{Ln}(b/a)}$$

$$\sigma_\theta = (p - \beta E)\frac{(b^2/r^2) + 1}{(b^2/a^2) - 1} + \beta E\frac{1 - \mathrm{Ln}(b/r)}{\mathrm{Ln}(b/a)} \tag{8.36}$$

$$\sigma_z = E\varepsilon_z + \frac{2\upsilon(p - \beta E)}{(b^2/a^2) - 1} + \beta E\frac{\upsilon - 2\mathrm{Ln}(b/r)}{\mathrm{Ln}(b/a)} - \alpha E T_0$$

The magnitude of ε_z, which is independent of r, depends on the end conditions of the vessel. For plane strain condition $\varepsilon_z = 0$ and for open ends $\sigma_z = 0$.

For vessels with closed ends, the resultant axial force acting on the vessel is $F = \pi a^2 p$. Since

$$F = 2\pi\int_a^b r\sigma_z dr = 2\pi\int_a^b r[E\varepsilon_z + \upsilon(\sigma_r + \sigma_\theta) - E\alpha T]dr \tag{a}$$

and $\sigma_\theta - \sigma_r = r\frac{d\sigma_r}{dr}$, it follows that

$$(\sigma_r + \sigma_\theta) = \frac{1}{r}\frac{d}{dr}(r^2\sigma_r) \tag{b}$$

Inserting Equation (b) into (a) and evaluating the integral, we find

$$(1 - 2\upsilon)a^2 p = (b^2 - a^2)E\varepsilon_z - 2E\alpha\int_a^b rT(r)dr \tag{c}$$

Inserting Equation 8.28 in (c) and evaluating the resulting integral, the axial strain for pressurized closed tube is found to be given by

$$E\varepsilon_z = \frac{(1-2\upsilon)p - 2(1-\upsilon)E\beta}{(b^2/a^2)-1} + \frac{(1-\upsilon)E\beta}{Ln(b/a)} + E\alpha T_0 \tag{d}$$

And the corresponding axial stress follows from the third equation in Equation 8.36 as

$$\sigma_z = \frac{p - 2\beta E}{(b^2/a^2)-1} + \beta E\frac{1-2Ln(b/r)}{Ln(b/a)} \tag{8.37}$$

8.2.3 Initiation of Ductile Yield

As the magnitudes of the internal pressure and heat flow increase, yield may initiate anywhere in the tube depending on the ratio p/β [8.4,8.5]. Based on Tresca's criterion and the fact that the magnitude of the principal axial stress is always intermediate between the other two, the onset of yield depends on the magnitude of the stress difference

$$\sigma_\theta - \sigma_r = (p-\beta E)\frac{2(b^2/r^2)}{(b^2/a^2)-1} + \frac{\beta E}{Ln(b/a)} \tag{a}$$

In the case, when p is greater than βE, then the stress difference in Equation (a) becomes maximum at $r = a$. It follows that the pressure required to initiate yield is

$$p_e = \frac{\sigma_0}{2}\left(1 - \frac{a^2}{b^2}\right) + \beta E\left[1 - \frac{1-(a^2/b^2)}{2Ln(b/a)}\right] \tag{8.38}$$

Equation 8.38 indicates that p_e is increased when $\beta > 0$ and is decreased if $\beta < 0$. It can be shown that formula (8.38) gives the correct yield initiation pressure provided that the wall ratio b/a, denoted by m, satisfies the following inequality:

$$\frac{-Ln(m)}{m^2Ln(m^2)/(m^2-1)-1} < \frac{\beta E}{\sigma_0} < Ln(m) \tag{b}$$

The onset of yield in cylindrical tubes subject to other combinations of internal pressure, steady-state temperature gradient, and various end conditions is available in Reference 8.3.

When $p = 0$, the formula in Equation 8.38 is reduced to that in Equation 8.33.

8.2.4 Elastic–Plastic Expansion

When p is increased beyond p_e in Equation 8.38, an annular plastic zone of radius c is formed around the inner surface, $a \le r \le c$, while the remaining cross section remains elastic.

In the elastic zone, $c \le r \le b$, the stresses derived in Equations 8.25a are still valid but now the constants A and B are determined from the boundary conditions:

i. $\sigma_r = 0$ at $r = b$, and (ii) $\sigma_\theta - \sigma_r = \sigma_0$ at $r = c$. Inserting the steady-state temperature distribution found in Equation 8.28 into Equation 8.25a, performing the integration and absorbing any constant term into A and B, the following stress expressions are obtained

$$\left.\begin{aligned}\sigma_r &= A + \frac{B}{r^2} - \beta E\left[\frac{(1/2) + \mathrm{Ln}(b/r)}{\mathrm{Ln}(b/a)}\right] \\ \sigma_\theta &= A - \frac{B}{r^2} + \beta E\left[\frac{(1/2) - \mathrm{Ln}(b/r)}{\mathrm{Ln}(b/a)}\right]\end{aligned}\right\}, \quad c \le r \le b \tag{8.39}$$

The above-mentioned boundary conditions (i) and (ii) yield

$$\begin{aligned} A + \frac{B}{b^2} - \frac{\beta E}{2\mathrm{Ln}(b/a)} &= 0 \\ -\frac{2B}{c^2} + \frac{\beta E}{2\mathrm{Ln}(b/a)} &= \sigma_0 \end{aligned} \tag{a}$$

which gives

$$A = \frac{\sigma_0 c^2}{2b^2} + \frac{\beta E}{2\mathrm{Ln}(b/a)}\left(1 - \frac{c^2}{b^2}\right), \quad \text{and} \quad B = -\frac{c^2\sigma_0}{2} + \frac{c^2\beta E}{2\mathrm{Ln}(b/a)}$$

Hence, the expressions of the radial and hoop stresses in the elastic region are

$$\left.\begin{aligned}\sigma_r &= \frac{1}{2}\left[\sigma_0 - \frac{\beta E}{\mathrm{Ln}(b/a)}\right]\left[\frac{c^2}{b^2} - \frac{c^2}{r^2}\right] - \beta E\frac{\mathrm{Ln}(b/r)}{\mathrm{Ln}(b/a)} \\ \sigma_\theta &= \frac{1}{2}\left[\sigma_0 - \frac{\beta E}{\mathrm{Ln}(b/a)}\right]\left[\frac{c^2}{b^2} + \frac{c^2}{r^2}\right] + \beta E\frac{1 - \mathrm{Ln}(b/r)}{\mathrm{Ln}(b/a)}\end{aligned}\right\}, \quad c \le r \le b \tag{8.40}$$

Note that the stress difference, $\sigma_\theta - \sigma_r = \left[\sigma_0 - \frac{\beta E}{\mathrm{Ln}(b/a)}\right]\frac{c^2}{r^2} + \frac{\beta E}{\mathrm{Ln}(b/a)}$, attains its maximum value of σ_0 provided that $(\beta E/\sigma_0) < \mathrm{Ln}(b/a)$.

In the plastic zone, $a \le r \le c$, using equilibrium equation and yield criterion, the radial stress is found to be governed by the equation $d\sigma_r/dr = \frac{\sigma_0}{r}$, which gives

$$\sigma_r = \sigma_0 \mathrm{Ln}r + C_1, \quad a \le r \le c \tag{b}$$

and $\sigma_\theta = \sigma_r + \sigma_0$. The constant C_1 is found from the condition that the radial stress has to be continuous at $r = c$. Equating expression (b) and the first equation of Equation 8.40 at $r = c$ yields

$$C_1 = -\sigma_0 \mathrm{Lnc} + \frac{1}{2}\left(\sigma_0 - \frac{\beta E}{\mathrm{Ln}(b/a)}\right)\left(\frac{c^2}{b^2} - 1\right) - \beta E\frac{\mathrm{Ln}(b/c)}{\mathrm{Ln}(b/a)} \tag{c}$$

Inserting expression (c) into (b) and simplifying the algebra, the radial and hoop stresses in the plastic region are determined as

$$\begin{aligned}\sigma_r &= -\frac{1}{2}\left(\sigma_0 - \frac{\beta E}{\mathrm{Ln}(b/a)}\right)\left(1 - \frac{c^2}{b^2} + \mathrm{Ln}\frac{c^2}{r^2}\right) - \beta E\frac{\mathrm{Ln}(b/r)}{\mathrm{Ln}(b/a)} \\ \sigma_\theta &= \frac{1}{2}\left(\sigma_0 - \frac{\beta E}{\mathrm{Ln}(b/a)}\right)\left(1 + \frac{c^2}{b^2} - \mathrm{Ln}\frac{c^2}{r^2}\right) + \beta E\frac{1 - \mathrm{Ln}(b/r)}{\mathrm{Ln}(b/a)}\end{aligned}, \quad a \le r \le c \tag{8.41}$$

Since at $r = a$, $\sigma_r = -p$, it follows that the internal pressure necessary to create a plastic zone within a radius c of the tube is given by

$$p = \frac{1}{2}\left(\sigma_0 - \frac{\beta E}{\mathrm{Ln}(b/a)}\right)\left(1 - \frac{c^2}{b^2} + \mathrm{Ln}\frac{c^2}{a^2}\right) + \beta E \tag{8.42}$$

For typical values of wall ratio b/a, Equation 8.42 indicates that the pressure is increased when $\beta > 0$ and decreased if $\beta < 0$.

The ultimate pressure is reached when $c = b$, that is,

$$p_u = \sigma_0 \mathrm{Ln}\frac{b}{a} \tag{8.43}$$

Formula 8.43 is independent of the heat flow, it is identical to Equation 7.19.

The above analysis is based on the assumption that the principal axial stress is situated between the tangential and radial principal stresses. This fact gives rise to limits on the wall thickness ratio b/a, which depend on the end conditions of the tube as was discussed in Chapter 7 for cylindrical vessels subjected to internal pressure only. Reference 2.2 gives limits on the wall thickness of a pressurized tube subjected to steady-state heat flow.

8.3 Thermal Stresses in Thick-Walled Spherical Shell

This section is concerned with determining the stresses induced in thick-walled spherical shells when subjected to changes in temperature. Consider a spherical shell of internal radius a, external radius b, subjected to temperature distribution $T(r)$. The elastic strains consist of two components, one due to stress and the other due to temperature. In spherical coordinates (r, θ, φ), we have (Chapter 7)

$$\begin{aligned}\varepsilon_r &= \frac{du}{dr} = \frac{1}{E}(\sigma_r - 2\upsilon\sigma_\theta) + \alpha T \\ \varepsilon_\theta &= \frac{u}{r} = \frac{1}{E}[(1-\upsilon)\sigma_\theta - \upsilon\sigma_r] + \alpha T\end{aligned} \tag{8.44}$$

Eliminating the radial displacement $u(r)$ by differentiating the second equation in Equation 8.44 with respect to r and equating the result to the first equation results in

$$(1+\upsilon)(\sigma_r - \sigma_\theta) + r\frac{d}{dr}[\upsilon\sigma_r - (1-\upsilon)\sigma_\theta - E\alpha T] = 0 \tag{a}$$

Using equation of equilibrium of the shell (7.37), $\sigma_r - \sigma_\theta = -(r/2)(d\sigma_r/dr)$, the following compatibility equation is obtained:

$$\frac{d}{dr}\left(\sigma_r + 2\sigma_\theta + \frac{2E\alpha T}{1-\upsilon}\right) = 0 \tag{8.45}$$

The integral of Equation 8.45 is

$$\sigma_r + 2\sigma_\theta = 3A - \frac{2E\alpha T(r)}{1-\upsilon} \tag{8.46}$$

where 3A is a convenient constant of integration. Equation 8.46 indicates that the sum $\sigma_r + 2\sigma_\theta$ is linearly proportional to the temperature.

General expressions of the radial and hoop stresses in terms of T(r) and two arbitrary constants can be derived as follows: using equilibrium to eliminate the hoop stress, Equation 8.46 is reduced to the following differential equation:

$$r\frac{d\sigma_r}{dr} + 3\sigma_r = 3A - \frac{2E\alpha T(r)}{1-\upsilon} \tag{b}$$

Multiplying both sides by r^2, Equation (b) becomes

$$\frac{d}{dr}(r^3\sigma_r) = 3Ar^2 - \frac{2E\alpha r^2 T(r)}{1-\upsilon} \tag{c}$$

Carrying out the integration indicated in Equation (c), using B as constant of integration, σ_r is obtained and the hoop stress follows from Equation 8.46. In this manner, we find the following general expressions of the stresses for any temperature T(r):

$$\begin{aligned} \sigma_r &= A + \frac{B}{r^3} - \frac{2E\alpha}{(1-\upsilon)r^3}\int_a^r r^2 T(r)dr \\ \sigma_\theta &= A - \frac{B}{2r^3} - \frac{E\alpha T(r)}{1-\upsilon} + \frac{E\alpha}{(1-\upsilon)r^3}\int_a^r r^2 T(r)dr \end{aligned} \tag{8.47}$$

where constants A and B are determined from boundary conditions that the radial stress vanishes at $r = a$, and $r = b$, giving $A = [2E\alpha/(1-\upsilon)(b^3 - a^3)]\int_a^b r^2 T dr$ and $B = -a^3 A$. Hence, the general expressions of the stresses in the elastic range are

$$\begin{aligned} \sigma_r &= \frac{2E\alpha}{1-\upsilon}\left[\frac{r^3 - a^3}{(b^3 - a^3)r^3}\int_a^b r^2 T dr - \frac{1}{r^3}\int_a^r r^2 T dr\right] \\ \sigma_\theta &= \frac{2E\alpha}{1-\upsilon}\left[\frac{2r^3 + a^3}{2(b^3 + a^3)r^3}\int_a^b r^2 T dr + \frac{1}{2r^3}\int_a^r r^2 T dr - \frac{T}{2}\right] \end{aligned} \tag{8.48}$$

Next, we determine the stresses when the vessel is subject to steady-state temperature distribution governed by Laplace equation in spherical coordinates.

8.3.1 Steady-State Temperature Distribution in Spherical Vessels

Consider a spherical vessel of internal radius a, external radius b, which is subjected to steady-state temperature distribution such that the temperature at the inner and outer surfaces is given by T_i and T_0, respectively. The steady-state temperature is governed by Laplace's equation. In spherical coordinates (r, φ, θ), Laplace equation assumes the form

$$\frac{1}{r}\frac{\partial}{\partial r}\left(r^2\frac{\partial T}{\partial r}\right)+\frac{1}{r^2\sin\phi}\frac{\partial}{\partial\phi}\left(\sin\phi\frac{\partial T}{\partial\phi}\right)+\frac{1}{r^2\sin\phi}\left(\frac{\partial T}{\partial\theta^2}\right)=0 \tag{a}$$

Because of radial symmetry, we have $\partial T/\partial\varphi=\partial T/\partial\theta=0$ and T is function of r only. Hence, Equation (a) is reduced to an ordinary differential equation

$$r\frac{d^2T}{dr^2}+2\frac{dT}{dr}=0 \tag{8.49}$$

which gives $T(r)=C_1+(C_2/r^2)$, where constants C_1 and C_2 are found from the conditions that $T=T_i$ at $r=a$ and $T=T_0$ at $r=b$. Thus, the steady-state temperature distribution in the vessel is given by

$$T(r)=T_0+(T_i-T_0)\frac{(b/r)-1}{(b/a)-1} \tag{8.50}$$

Equation 8.50 can now be inserted into Equations 8.48, evaluating the resulting integrals, the stresses in the shell are obtained. Although the procedure is straightforward, a different approach is used in the sequel to derive the stresses.

Making use of the dimensionless parameter

$$\beta=\frac{\alpha(T_i-T_0)}{1-\upsilon} \tag{8.51}$$

the steady-state temperature in Equation 8.50 when inserted into Equation (c) of the previous section, the radial stress is found to be governed by a first-order differential equation. (Note that constant terms associated with the temperature have been absorbed in A.)

$$r\frac{d\sigma_r}{dr}+3\sigma_r=3A-2\beta E\frac{b/r}{(b/a)-1} \tag{8.52}$$

Equation 8.52 is a nonhomogeneous differential equation, whose solution consists of a solution to the homogeneous part, $r(d\sigma_r/dr)+3\sigma_r=0$, which is $\sigma_r=B/r^3$, and a particular solution, $\sigma_r=A-\beta E((b/r)/(b/a)-1)$. The circumferential stress is then obtained from Equation 8.46.

Hence, using the abbreviation $m=b/a$, we have

$$\begin{aligned}\sigma_r&=A+\frac{B}{r^3}-\frac{\beta E}{m-1}\left(\frac{b}{r}\right)\\ \sigma_\theta&=A-\frac{B}{2r^3}-\frac{\beta E}{m-1}\left(\frac{b}{2r}\right)\end{aligned} \tag{8.53}$$

The constants A and B are obtained from the conditions that the radial stress vanishes at $r=a$, and $r=b$, giving $A=(m(m+1)/(m^3-1))\beta E$, $B=(b^3/(m^3-1))\beta E$. Hence, the elastic stresses are given by

$$\left.\begin{aligned}\sigma_r &= -\beta E\left[\frac{(b/r)-1}{m-1}-\frac{(b^3/r^3)-1}{m^3-1}\right]\\ \sigma_\theta &= -\beta E\left[\frac{(b/2r)-1}{m-1}+\frac{(b^3/2r^3)+1}{m^3-1}\right]\end{aligned}\right\} \quad a \le r \le b \tag{8.54}$$

The magnitudes of the hoop stress at the boundaries of the sphere are

$$\begin{aligned}(\sigma_\theta)_{r=a} &= -\beta E\left[\frac{2m^3-m^2-m}{2(m^3-1)}\right]\\ (\sigma_\theta)_{r=b} &= -\beta E\left[\frac{2-m^2-m}{2(m^3-1)}\right]\end{aligned} \tag{8.55}$$

Equations 8.55 indicates that the hoop stress is compressive in the inner region of the sphere, and tensile in the outer region.

The radial stress attains its maximum magnitude at a radius obtained by requiring

$$\frac{d\sigma_r}{dr} = 0 = \frac{1}{m-1}\left(-\frac{b}{r^2}\right) - \frac{1}{m^3-1}\left(-\frac{3b}{r^4}\right)$$

which gives

$$r = \left(\frac{3m^2}{m^2+m+1}\right)^{\frac{1}{2}} a \tag{8.56}$$

For $\beta > 0$, the magnitude of the radial stress in the inner part of the vessel is larger than that in the outer part. For a vessel with $m = 2$, the radial stress at $r = (1.2\ a)$ is $-0.15\ \beta E$, while at $r = (1.8\ a)$, the radial stress is $-0.06\ \beta E$. This implies that the inner boundary will contract more than the outer one because of the larger compressive stress.

The radial displacement is obtained from Equation 8.44 as

$$u = r\left\{(1-2\upsilon)\frac{A}{E} - (1+\upsilon)\frac{Bb^3}{2Er^3} + \beta\left[\frac{(1+\upsilon)(b/2r)-(1-\upsilon)}{m-1}\right] + \alpha T_0\right\} \tag{8.57}$$

where the constants A and B are given above in terms of m and β.

8.3.2 Initiation of Yield

As the steady-state temperature difference increases, yielding according to both Tresca's and Mises' criterion will occur at a radius such that the stress difference satisfies the following criterion:

$$|\sigma_r - \sigma_\theta| = \sigma_0 \tag{a}$$

TABLE 8.2

Values of Temperature Difference Required to Initiate Yield at the Inner Surface of the Sphere

Wall Thickness $m = b/a$	Temperature Difference $(T_i - T_0)_e$ °C
2	105
3	93
4	87
5	84
Large	75

Equations 8.54 gives

$$\sigma_r - \sigma_\theta = \beta E\left[\frac{3(b^3/2r^3)}{m^3 - 1} - \frac{(b/r)}{m - 1}\right]$$

which has the greatest numerical value at the inner surface $r = a$. Hence, the hoop compression at the inner surface of the vessel will cause initial yielding when β_e reaches the magnitude

$$\beta_e = \left[\frac{m^2 + m + 1}{m^2 + (m/2)}\right]\frac{\sigma_0}{E} = \left[\frac{1 + (a/b) + (a^2/b^2)}{1 + (a/2b)}\right]\frac{\sigma_0}{E} \tag{8.58}$$

and the temperature difference between the walls required to initiate yield is

$$(\Delta T)_e = (T_i - T_0) = (1 - \upsilon)\frac{\sigma_0}{\alpha E}\left[\frac{1 + a/b + (a/b)^2}{1 + (a/2b)}\right] \tag{8.59}$$

For mild steel vessels with material properties, $E = 200$ GPa, $\upsilon = 0.3$, $\sigma_0 = 250$ MPa, and $\alpha = 11.7 \times 10^{-6}$/°C, the magnitudes of the temperature difference required for initiating the yield are given in Table 8.2.

The values of the temperature difference between the walls required to initiate the yield decrease as the wall ratio increase, the same as in cylindrical vessels.

8.3.3 Elastic–Plastic Expansion

When the temperature difference between the walls increases beyond the value that initiates yield in the vessel, the sphere becomes plastic within a radius, say $r = c$. Assuming the elastic–perfectly plastic material, the stresses in the plastic region, $a \le r \le c$, must satisfy the yield criterion $|\sigma_r - \sigma_\theta| = 0$ and equilibrium equation, $d\sigma_r/dr = -(2/r)\sigma_0$, giving $\sigma_r = -2\sigma_0 \text{Ln} r +$ constant. Since the radial stress must vanish at $r = a$, the stresses in the plastic region are

$$\left.\begin{aligned}\sigma_r &= -2\sigma_0 \mathrm{Ln}\frac{r}{a}\\ \sigma_\theta &= -\sigma_0\left(1+2\mathrm{Ln}\frac{r}{a}\right)\end{aligned}\right\}, \quad a \le r \le c \tag{8.60}$$

which are independent of the temperature.

In the elastic region, $c \le r \le b$, the stresses in Equation 8.53 are still valid but now constants A and B are found from the boundary conditions, radial stress vanishes at $r = b$ and elastic and plastic stresses are continuous at the interface $r = c$. Let us use Equations 8.53 to find A in terms of B by applying the condition that $\sigma_r = 0$ at $r = b$. Hence, $A = -(B/b^3) + (\beta E/m - 1)$, and the stresses in Equation 8.53 become

$$\left.\begin{aligned}\sigma_r &= \frac{B}{b^3}\left(\frac{b^3}{r^3}-1\right) - \frac{\beta E}{m-1}\left(\frac{b}{r}-1\right)\\ \sigma_\theta &= -\frac{B}{b^3}\left(\frac{b^3}{2r^3}+1\right) - \frac{\beta E}{m-1}\left(\frac{b}{2r}-1\right)\end{aligned}\right\}, \quad c \le r \le b \tag{8.61}$$

The continuity of stresses in Equations 8.60 and 8.61 imply that

$$\begin{aligned}-2\sigma_0 \mathrm{Ln}\frac{c}{a} &= \frac{B}{b^3}\left(\frac{b^3}{c}-1\right) - \frac{\beta E}{m-1}\left(\frac{b}{c}-1\right)\\ -\sigma_0\left(1+2\mathrm{Ln}\frac{c}{a}\right) &= -\frac{B}{b^3}\left(\frac{b^3}{2c^3}+1\right) - \frac{\beta E}{m-1}\left(\frac{b}{2c}-1\right)\end{aligned} \tag{a}$$

Solving Equations (a) give the unknown B and a relation between β and c

$$\left.\begin{aligned}B &= 2c^3\left[\frac{1-(c/b)+\mathrm{Ln}(c/a)}{(2+c/b)(1-c/b)^2}\right]\sigma_0\\ \beta E &= 2[(c/a)-(c/b)]\left[\frac{1-(c/b)^2+\mathrm{Ln}(c/a)^2}{(2+c/b)(1-c/b)^2}\right]\sigma_0\end{aligned}\right\} \quad c \le r \le b \tag{8.62}$$

Inserting the above values of B and βE into Equations 8.61 provides expressions of the stresses in the elastic region in terms of the yield stress and c.

The radial displacement at any radius r is obtained from the compressibility equation, $\varepsilon_r + \varepsilon_\theta + \varepsilon_\varphi = (du/dr) + 2(u/r) = 0$, that is,

$$\frac{du}{dr} + 2\frac{u}{r} = \frac{1}{E}(1-2\upsilon)(\sigma_r + 2\sigma_\theta) + 3\alpha T \tag{b}$$

Using equilibrium equation to replace the hoop stress by $2\sigma_\theta = 2\sigma_r + r(d\sigma_r/dr)$, and multiplying both sides of Equation (b) by r^2, there results the simple differential equation

$$\frac{d}{dr}(r^2 u) = \frac{1-2\upsilon}{E}\frac{d}{dr}(r^3\sigma_r) + 3\alpha r^2 T(r) \tag{c}$$

where T(r) is given by Equation 8.50. The integral of the derivatives in Equation (b) is

$$\frac{u}{r} = \frac{1-2\upsilon}{E}\sigma_r + (1-\upsilon)\left[\beta\left(\frac{(3b/2r)-1}{m-1}\right) - K\frac{b^3}{r^3}\right] + \alpha T_0, \quad K = \text{constant} \tag{d}$$

Equation (d) is valid in both elastic and plastic regions of the vessel. The constant of integration, K, is determined by equating the radial displacement in the last equation to that found for the elastic region in Equation 8.57 at $r = b$. This gives $K = 3B/2E$. Hence,

$$\frac{u}{r} = \frac{1-2\upsilon}{E}\sigma_r + (1-\upsilon)\left[\beta\left(\frac{3b/2r-1}{m-1}\right) - \frac{3B}{2E}(b^3/r^3)\right] + \alpha T_0 \tag{8.63}$$

where β and B are given in Equation 8.62. In particular,

$$\frac{u_a}{a} = \frac{1-\upsilon}{2}\left[\beta\left(\frac{3m-2}{m-1}\right) - \frac{3m^3 B}{E}\right] + \alpha T_0$$

and

$$\frac{u_b}{b} = \frac{1-\upsilon}{2}\left[\frac{\beta}{m-1} - 3\frac{B}{E}\right] + \alpha T_0$$

Further increase in the heat flow causes the elastic–plastic boundary to advance outward, say to a radius c_1, and a second plastic zone is initiated either at the outer boundary or somewhere in the interior of the elastic region between radius c and the outer boundary. With further increase in the heat flow, the second plastic zone spreads inward if formed at the outer boundary, or in both directions if formed in the interior, while the original plastic zone continues to spread outward. Further details are available in References 2.2, 8.4, 8.5.

8.3.4 Combined Steady-State Heat Flow and Internal Pressure

When the sphere is subjected to internal pressure p, steady-state heat flow expressions of σ_r and σ_θ are simply obtained by adding expressions (7.42) and (8.54). Thus,

$$\begin{aligned}\sigma_r &= -(p-\beta E)\frac{(b^3/r^3)-1}{m^3-1} - \beta E\frac{(b/r)-1}{m-1}\\ \sigma_\theta &= (p-\beta E)\frac{(b^3/2r^3)+1}{m^3-1} - \beta E\frac{(b/2r)-1}{m-1}\end{aligned} \tag{8.64}$$

Here, $m = b/a$. It follows that

$$\sigma_\theta - \sigma_r = (p-\beta E)\frac{(3b^3/2r^3)}{m^3-1} + \beta E\frac{(b/2r)}{m-1} \tag{8.65}$$

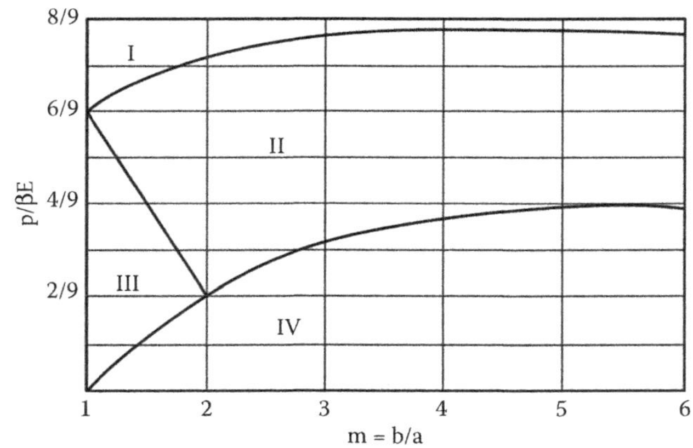

FIGURE 8.3
Regions of onset of initial yield.

The maximum value of the stress difference in Equation 8.65 is obtained by differentiation with respect to r. Thus,

$$(p - \beta E)\left(\frac{3b^3/2}{m^3 - 1}\right)\left(-\frac{3}{r^4}\right) + \frac{\beta E}{m - 1}\left(-\frac{b}{2r^2}\right) = 0$$

which furnishes

$$r_{max} = 3a\sqrt{\frac{1 - (p/\beta E)}{m^2 + m + 1}} \tag{8.66}$$

Yielding will occur according to Tresca's or Mises' criterion at any radius in the shell depending on the values of the pressure, steady-state heat flow, and wall ratio m. Figure 8.3 identifies four regions where yield first occurs in terms of $p/\beta E$ versus m as follows:

Region I—Yield occurs on inner surface $r = a$.
Region II—Yield occurs when the radius is $r = r_{max}$, given in Equation 8.66.
Region III—Yield occurs at outer surface $r = b$.
Region IV—Yield occurs at inner surface $r = a$.

Reference 8.5 provide more detail.

EXAMPLE 8.5

A spherical vessel, radii a and b, $b > a$, is made of an elastic–perfectly plastic material with yield stress σ_0. The vessel is subjected to steady-state heat flow greater than that which initiates yield, $\Delta T > (\Delta T)_e$. Find the induced pressure in the plastic region, $r = c$, $a < c < b$.

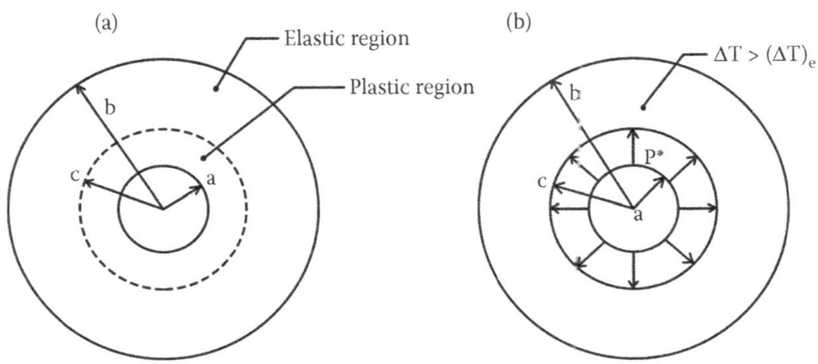

FIGURE 8.4
Spherical shell subjected to steady-state heat flow Example 8.5; (a) $(\Delta T)_e$ at $r = c$ and (b) pressure p^* at $r = c$ and $\Delta T > (\Delta T)e$.

Solution

Consider the outer shell of the vessel bounded by $r = c$ and $r = b$ (Figure 8.4). At $r = c$, the material is yielding $\sigma_\theta - \sigma_r = -\sigma_0$, and the temperature T_c is determined from Equation 8.50 with $T_0 = 0$ as follows:

$$T_c = T_i \frac{b/c - 1}{b/a - 1} = T_i \frac{m_1 - 1}{m - 1}$$

Here, $m = b/a$ and $m_1 = b/c$. Using Equation 8.60, the radial pressure p^* at $r = c$ is

$$p* = -2\sigma_0 \text{Ln}(c/a) = -2\sigma_0 \text{Ln}(m/m_1)$$

The inner surface of the outer shell is subjected to pressure and steady-state heat flow. Using the result given in Equation 8.65 with p replaced by p^*, r by c, m by m_1, and β by $\beta^* = \alpha T_c/1 - \upsilon$, we obtain

$$-\sigma_0 = (p^* - \beta^* E)\left[\frac{(3b^3)/(2c^3)}{m_1^3 - 1}\right] + \beta^* E\left[\frac{b/(2c)}{m_1 - 1}\right] \quad \text{(a)}$$

or

$$-\frac{\sigma_0}{\beta^* E} = \frac{3m_1^3[p^*/(\beta^* E) - 1] + m_1(m_1^2 + m_1 + 1)}{2(m_1^3 - 1)}$$

Equation (a) can be simplified to give the induced pressure

$$\frac{p^*}{\beta^* E} = \frac{m_1(2m_1^2 - m_1 - 1)\text{Ln}(m/m_1)}{m_1^3[3\text{Ln}(m/m_1) + 1] - 1}$$

Problems

8.1 A prismatic bar is restrained in the x (axial) and y (vertical) directions but free to expand in the z (lateral) direction, as shown in Figure P8.1. The bar is subjected to a uniform temperature rise of T_0 = constant. Assuming the elastic behavior, find the stresses and strains induced in the bar in terms of T_0, α, and ν.

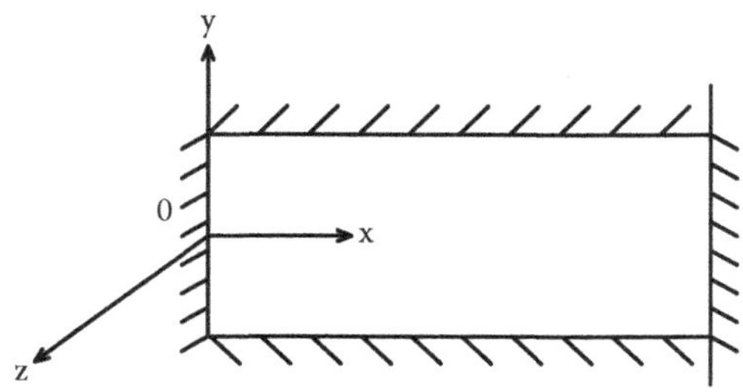

FIGURE P8.1

Ans: $-ET_0\,\alpha/1-\nu$.

8.2 A uniform aluminum disk of radius 10 cm is subjected to linear temperature rise from 20°C at the center to 100°C at the edge. If the coefficient of thermal expansion = 23.6×10^{-6}/°C, find (a) maximum radial stress, (b) radial expansion of the edge.

Ans: 44 M N/m^2, 0.12 mm.

8.3 A thin steel disk 4 in. diameter is subjected to linear temperature rise, which varies from 70°F at the center to Ta at the edge. If the maximum radial expansion is not to exceed 2×10^{-3} in., find the maximum stress in the disk. For steel, $E = 30\times10^6$ psi and $\alpha = 6.5\times10^{-6}$/°F.

Ans: 8.2 ksi.

8.4 Determine the maximum stress and radial displacement in a uniform disk, 10 in. diameter, when the temperature rise is linear from 70°F at the center to 200°F at the edge. Assume $E = 30\times10^6$ psi and $\alpha = 6.5\times10^{-6}$°F.

Ans: 8.5 ksi, 5.1×10^{-3} in.

8.5 Find maximum radial stress and displacement of the disk in the previous problem when it is subjected to a linear thermal gradient with zero temperature at the origin and 300°F at the edge.

Ans: 19.5 ksi, 0.65×10^{-3} in.

8.6 A uniform disk of radius R is subjected to a temperature that varies linearly from constant value at the center to zero at the outer surface. Find the expression for the stress ratio σ_θ/σ_r.

Ans: $(2r/R-1)/(r/R-1)$.

8.7 A thin steel disk with 10 cm radius is rotating with an angular velocity ω rad/s. If the disk is subjected to a linear temperature that varies from 20°C at the center to 90°C at the edge, what is the maximum value of ω so that the disk will remain elastic? For steel, $E = 200$ GPa, $\alpha = 11.7\times10^{-6}$/°C, $\rho = 7860$ kg/m^3, and the yield stress in tension = 250 MPa.

Ans: 15,664 RPM.

8.8 A uniform disk of 150 mm in diameter is rotating at 6000 RPM. If the disk is experiencing a linear heat flow that varies from 10°C at the center to 70°C at the edge, find the radial and circumferential stresses at $r = 50$ mm. Assume $E = 129$ GPa, $\alpha = 16.8 \times 10^{-6}/°C$, $\nu = 0.3$, and $\rho = 8860$ kg/m^3.

Ans: 18, −13.4 MN/m^2.

8.9 A hollow disk with internal and external diameters of 200 mm and 500 mm, respectively, is subjected to a linear temperature that varies from 40°C at the inner surface to 0°C at the outer surface. Find the maximum stress in the disk. Assume $E = 120$ GPa, $\alpha = 16.8 \times 10^{-6}/°C$, and $\nu = 0.42$.

Ans: −46 MPa.

8.10 The internal and external radii of a hollow disk are 40 mm and 70 mm, respectively. Determine the maximum stress induced in the disk when it is subjected to a temperature gradient of the form $T(r) = kr$, $k =$ constant, such that the temperature at the outside edge is −5 0°C. The material is brass with $E = 105$ GPa and $\alpha = 20 \times 10^{-6}/°C$.

Ans: −24.55 M N/m^2.

8.11 Repeat the previous problem assuming that the disk is rotating at angular speed of 9000 rev/min. The density of the material is $\rho = 8460$ kg/m^3 and Poisson's ratio = 0.35.

Ans: 36.51 M N/m^2.

8.12 A thin steel disk, 250 mm outside diameter with a central hole 100 mm diameter, rotates at 9000 RPM. The disk is subjected to a temperature gradient that varies linearly such that the difference in temperature between the outer and inner (cooler) edges of the disk is 50°C. Calculate the circumferential stresses induced at the edges. Assume $E = 200$ GPa, $\nu = 0.3$, $\alpha = 12 \times 10^{-6}/°C$, and $\rho = 7700$ kg/m^3.

Ans: 160 MN/m^2, −19 MN/m^2.

8.13 The internal and external diameters of a cylinder are 400 mm and 600 mm, respectively. The cylinder is subjected to steady-state heat flow such that the temperatures at the inside and outside surfaces are maintained at 200°C and 100°C, respectively. Find the maximum hoop stress induced in the cylinder using Equation 8.35. Assume $E = 205$ GPa, $\alpha = 11 \times 10^{-6}/°C$, and Poisson's ratio = 0.3.

Ans: −182, 140 MN/m^2.

8.14 A cast iron cylinder with internal radius $a = 10$ mm and external radius $b = 20$ mm is subjected to steady-state heat flow such that the temperature at the inner surface is −25°C and zero at the outer surface. $E = 90$ GPa, Poisson's ratio = 0.3, and the coefficient of linear thermal expansion = $10.4 \times 10^{-6}/°C$. Find (a) the hoop stress at the inner surface and (b) the maximum radial stress.

Ans: 20.5, 2.4 MN/m^2.

8.15 A closed ends pressurized cylindrical vessel, radii a and b, $b > a$, is subjected to steady-state temperature field $T(r)$. The yield stress of the material is σ_0. For small plastic strains, the expressions of ϵ_r, ϵ_θ, and ϵ_z are identical. Show that the expressions for the radial displacement is given by

$$\frac{u}{r} = -\upsilon\varepsilon_z + (1+\upsilon)\alpha T(r) + (1+\upsilon)(1-2\upsilon)\frac{\sigma_r}{E} + (1-\upsilon^2)\left[\frac{\sigma_0 c^2}{Er^2} - \frac{\beta(c^2/r^2 - 1)}{Ln(b/a)}\right]$$

where c is the radius of the plastic zone and σ_r is given by Equation 8.40 in the elastic zone and by Equation 8.41 in the plastic zone.

8.16 The internal and external diameters of a spherical vessel are 0.3 m and 0.6 m, respectively. The vessel is subjected to steady-state heat flow such that the temperature difference between the outer (hotter) and inner surfaces is 100°C. Find the maximum radial and circumferential stresses in the vessel. Assume $E = 200$ GPa, $\nu = 0.3$, and $\alpha = 11.7 \times 10^{-6}/°C$.

Ans: $-54\ MN/m^2$, $-239\ MN/m^2$.

8.17 The material of a spherical shell is elastic–perfectly plastic with yield stress = 260 MPa. The internal and external radii are 200 mm and 300 mm, respectively. The shell is subjected to steady-state heat flow. Assuming that $E = 70$ GPa, $\alpha = 23 \times 10^{-6}/°C$. Find (a) the temperature difference required to initiate the yield at the inner surface of the shell; (b) the stresses in the plastic region.

Ans: 179°C, $-2\sigma_0\ Ln(c/a)$, $-\sigma_0\ [1 + 2\ Ln\ (c/a)]$.

References

8.1. B. A. Boley and J. H. Weiner, *Theory of Thermal Stresses*, John Wiley, New York, 1960.
8.2. N. Nowacki, *Thermoelasticity*, Addison-Wiley Publishing Co., Reading, MA, 1962.
8.3. M. G. Derrington, The onset of yield in a thick-walled cylinder subjected to uniform internal or external pressure and steady heat flow, *Int. J. Mech. Sci.*, vol. 4, No. 43, 1962.
8.4. F. Drabble and W. Johnson, The development of the zones of yielding in thick-walled spherical shells of non-work hardening material subjected to a steady state radial temperature gradient and internal or external pressure, *Conference on Thermal Loading and Creep*, paper 19, Inst. Mech. Eng., London, 1964.
8.5. G. R. Cowper, The elastoplastic thick-walled spherical shell subjected to a radial temperature gradient, *J. Appl. Mech.*, vol. 47, 1960.

9

Three-Dimensional Problems

In this chapter, we discuss elastic and plastic methods of obtaining analytical solutions of few basic structural components that satisfy 3D equations of equilibrium given in Chapter 2, Equations 2.5 and 2.7, the associated boundary conditions in Equations 2.13, and the compatibility equations in Equations 3.14.

9.1 Elongation of a Prismatic Bar due to Its Own Weight

Figure 9.1 shows a prismatic bar of length L and cross-sectional area A supported at the top and hanging freely in the vertical plane. Point C is the midpoint of the support. The resultant force at the support is the weight of the bar W = volume × weight per unit volume of the bar. If ρg denotes weight per unit volume of the bar then $W = (AL) \times \rho g$, and the body forces acting on the bar in rectangular coordinates are

$$B_x = B_y = 0 \quad \text{and} \quad B_z = -\rho g \tag{a}$$

Equations 2.7 are satisfied by writing the following stresses:

$$\sigma_z = \rho g z, \sigma_x = \sigma_y = \tau_{xy} = \tau_{yz} = 0 \tag{9.1a}$$

Equations 9.1a imply that on each horizontal section of the bar, there is a uniform tension in the z-direction produced by weight of the bar below that section. The lateral and lower surfaces of the bar are free from stress. At the upper section, there is a uniformly distributed tensile stress of magnitude

$$\sigma_z = \rho g L$$

Compatibility Equations 3.14 are also satisfied. It follows that the stresses in Equation 9.1a are the exact elastic stresses in the bar. The strains and displacements u (x, y, z), v (x, y), and w (x, y, z) are obtained from Hooke's law in three dimensions as follows:

$$\begin{aligned}
\varepsilon_z &= \frac{\partial w}{\partial z} = \frac{\sigma_z}{E} = \frac{\rho g z}{E} \\
\varepsilon_x &= \frac{\partial u}{\partial x} = -\upsilon \frac{\rho g z}{E} = \varepsilon_y = \frac{\partial v}{\partial y} \\
\gamma_{xy} &= \frac{\partial u}{\partial y} + \frac{\partial v}{\partial x} = 0 \\
\gamma_{yz} &= \frac{\partial v}{\partial z} + \frac{\partial w}{\partial y} = 0 \\
\gamma_{zx} &= \frac{\partial w}{\partial x} + \frac{\partial u}{\partial z} = 0
\end{aligned} \tag{9.1b}$$

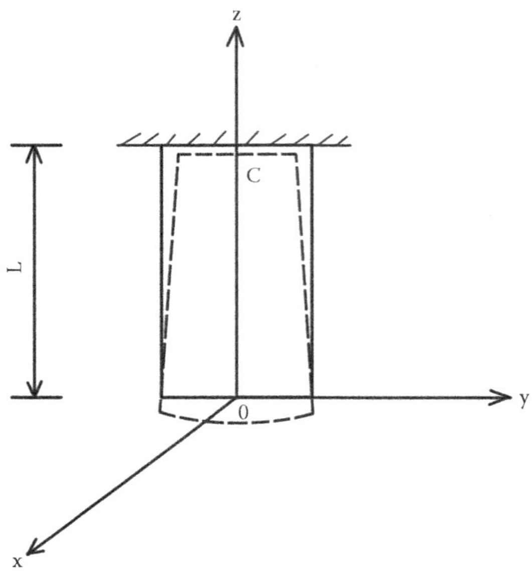

FIGURE 9.1
Prismatic bar hanging vertically.

Upon integration of the normal strains relations in the above equations, we find

$$\begin{aligned} w &= (\rho g z^2/2E) + w_0(x, y) \\ u &= -z(\delta w_0/\delta x) + u_0(x, y) \\ v &= -z(\delta w_0/\delta y) + v_0(x, y) \end{aligned} \tag{b}$$

where u_0, v_0, and w_0 are constants of integration. Making use of Equations 9.1b, it is readily confirmed that Equations (b) yield

$$\begin{aligned} -z\frac{\partial^2 w_0}{\partial x^2} + \frac{\partial u_0}{\partial x} &= -\upsilon\frac{\rho g z}{E} \\ -z\frac{\partial^2 w_0}{\partial y^2} + \frac{\partial v_0}{\partial y} &= -\upsilon\frac{\rho g z}{E} \end{aligned} \tag{c}$$

Since u_0 and v_0 are functions of x and y only, Equations (c) can only be satisfied if

$$\begin{aligned} \frac{\partial u_0}{\partial x} &= \frac{\partial v_0}{\partial y} = 0 \\ \frac{\partial^2 w_0}{\partial x^2} &= \frac{\partial^2 w_0}{\partial y^2} = \upsilon\frac{\rho g}{E} \end{aligned} \tag{d}$$

Also, by inserting relation (b) into $\gamma_{xy} = 0$ we get

$$-2z\frac{\partial^2 w_0}{\partial x \partial y} + \frac{\partial u_0}{\partial y} + \frac{\partial v_0}{\partial x} = 0 \tag{e}$$

It follows that

$$\frac{\partial^2 w_0}{\partial x \partial y} = 0, \quad \frac{\partial u_0}{\partial y} + \frac{\partial v_0}{\partial x} = 0 \tag{f}$$

Equations (e) and (f) yield

$$u_0 = c_1 y + c_2$$
$$v_0 = c_1 x + c_3$$
$$w_0 = (\rho g z^2/2E)(x^2 + y^2) + c_4 x + c_5 + c_6$$

where c_j, $j = 1, 2, \ldots, 6$ are constants.

The general expressions for the displacements follow from Equations (b):

$$\begin{aligned} u &= -\frac{\upsilon \rho g x z}{E} + c_1 y - c_4 z + c_2 \\ v &= -\frac{\upsilon \rho g y z}{E} - c_1 x - c_5 z + c_3 \\ w &= \frac{\rho g z^2}{2E} + \frac{\rho g z}{2E}(x^2 + y^2) + c_4 x + c_5 y + c_6 \end{aligned} \tag{g}$$

The arbitrary constants c_j, $j = 1, 2, \ldots, 6$ are determined from the conditions required to prevent the movement of the bar at the support. (i) To prevent a translatory motion of the bar, point C in Figure 9.1 is fixed, which requires $u = v = w = 0$ at $x = y = 0$, $z = L$. (ii) To prevent the rotation of the bar about axes through point C parallel to the x and y axes requires $\delta u/\delta z = \delta v/\delta z = 0$ at point C. (iii) To prevent the rotation of the bar about the z-axis, an elemental area through point C parallel to the xz-plane needs to be fixed, which requires $\delta v/\delta x = 0$ at point C.

Using Equations (g), we find $c_1 = c_2 = c_3 = c_4 = c_5 = 0$, and $c_6 = -(\rho g L^2/2E)$.

The final expressions of the displacements are

$$\begin{aligned} u &= -\frac{\upsilon \rho g x z}{E} \\ v &= -\frac{\upsilon \rho g y z}{E} \\ w &= \frac{\rho g z^2}{2E} + \frac{\upsilon \rho g}{2E}(x^2 + y^2) - \frac{\rho g L^2}{2E} \end{aligned} \tag{9.2}$$

The deformed shape of the bar is shown in the dotted red lines in Figure 9.1. Points on the z-$w = -\frac{\rho g}{2E}(L^2 - z^2)$ axis, that is, $x = y = 0$, have only vertical displacement, other points of the bar have vertical and horizontal displacements.

After deformation, cross sections of the bar normal to the z-axis become curved and form the surfaces of a paraboloid. For example, points on the cross section at $z = c$, after deformation will be on the surface

$$z = c + \frac{\rho g c^2}{2E} + \frac{\upsilon \rho g}{2E}(x^2 + y^2) - \frac{\rho g L^2}{2E}$$

This surface is normal to all longitudinal fibers of the bar, the longitudinal bars become inclined to the z-axis as shown in the dotted red lines in Figure 9.1, so that there are no shearing strains γ_{xy} or γ_{xz}.

EXAMPLE 9.1

Derive expressions for the displacements in a 3D beam when it is bent in one of its principal planes by equal and opposite moments M applied at the ends, as shown in Figure 9.2. The origin of rectangular coordinates is taken at the centroid of the cross section in the xz-plane, the principal plane of bending.

The stress components in the elementary theory of bending are

$$\begin{aligned} \sigma_0 &= Ex/R \\ \sigma_y &= \sigma_x = \tau_{xy} = \tau_{yz} = \tau_{xz} = 0 \end{aligned} \qquad \text{(a)}$$

Here, R stands for the radius of curvature of the bar after bending. Assuming there are no body forces, the stresses in Equation (a) satisfy the Equations of equilibrium (2.5) and (2.7) and the boundary conditions (2.13) for the lateral surfaces, which are free from stress. In order for the stresses (a) to represent an exact solution, the surface forces at the ends must be distributed over the end sections in the same manner as the stresses σ_z.

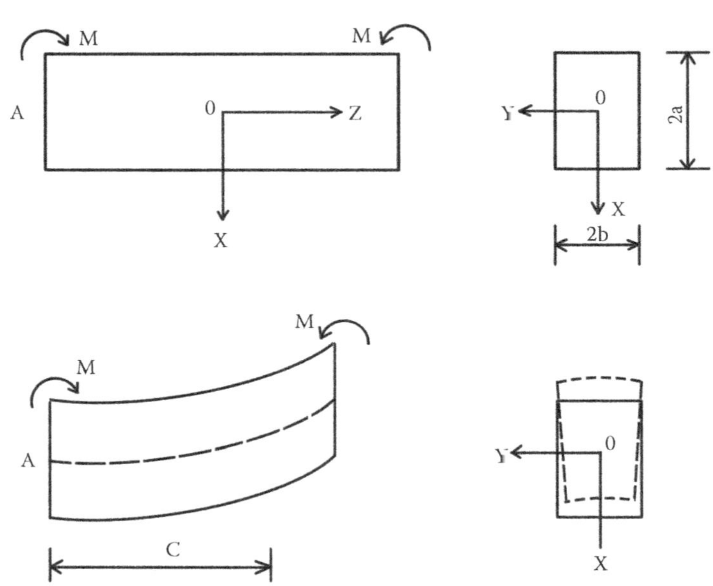

FIGURE 9.2
Example 9.1.

The bending moment M is given by the following equation:

$$M = \iint_A \sigma_z x dA = \iint_A \frac{Ex^2 dA}{R} = \frac{EI_y}{R}$$

where I_y stands for the moment of inertia of the cross section of the beam with respect to the neutral axis parallel to the y axis. Hence, the strength of materials formula of pure bending is obtained.

$$\frac{1}{R} = \frac{M}{EI_y}$$

The displacements are found using Hooke's law and Equations (a) as follows:

$$\begin{aligned}
&\varepsilon_z = \frac{\sigma_z}{E} = \frac{\partial w}{\partial z} = \frac{x}{R} \\
&\varepsilon_x = \frac{\partial u}{\partial x} = -\upsilon \frac{x}{R} = \varepsilon_y = \frac{\partial v}{\partial y} \\
&\gamma_{xy} = \frac{\partial u}{\partial y} + \frac{\partial v}{\partial x} = \gamma_{yz} = \frac{\partial v}{\partial z} + \frac{\partial w}{\partial y} = \gamma_{zx} = \frac{\partial w}{\partial x} + \frac{\partial u}{\partial z} = 0
\end{aligned} \tag{b}$$

Integrating the differential Equations (b), we find

$$\begin{aligned}
w &= \frac{xz}{R} + w_0 \\
u &= -\frac{z^2}{2R} - z\frac{\partial w_0}{\partial x} + u_0 \\
v &= -z\frac{\partial w_0}{\partial y} + v_0
\end{aligned} \tag{c}$$

where u_0, v_0, and w_0 are arbitrary functions of x and y only. Using Equations (c) and (b), it is readily shown that

$$\begin{aligned}
-z\frac{\partial^2 w_0}{\partial x^2} + \frac{\partial u_0}{\partial x} &= -\frac{\upsilon x}{R} \\
-z\frac{\partial^2 w_0}{\partial y^2} + \frac{\partial v_0}{\partial y} &= -\frac{\upsilon x}{R}
\end{aligned} \tag{d}$$

In order for Equations (d) to hold for all values of z, we must have

$$\frac{\partial^2 w_0}{\partial x^2} = \frac{\partial^2 w_0}{\partial y^2} = 0 \tag{e}$$

Upon integration, we have

$$\begin{aligned}
u_0 &= -\frac{\upsilon x^2}{2R} + f_1(y) \\
v_0 &= -\frac{\upsilon xy}{R} + f_2(x)
\end{aligned} \tag{f}$$

Using Equations (c), (f), and (b), we find

$$2z\frac{\partial^2 w_0}{\partial x \partial y} - \frac{\partial f_1(y)}{\partial y} - \frac{\partial f_2(x)}{\partial x} + \frac{\upsilon y}{R} = 0$$

which requires

$$\frac{\partial^2 w_0}{\partial x \partial y} = 0$$

$$\frac{\partial f_1(y)}{\partial y} + \frac{\partial f_2(x)}{\partial x} - \frac{\upsilon y}{R} = 0$$

Hence, we must have

$$w_0 = c_1 x + c_2 y + c_3$$

$$f_1(y) = (\nu y^2)/2R + c_4 y + c_5$$

$$f_2(x) = -c_4 x + c_6$$

where c_j, $j = 1, 2, \ldots, 6$ are constants. The expressions for the displacements become

$$\begin{aligned} u &= -\frac{z^2}{2R} - \frac{\upsilon x^2}{2R} + \frac{\upsilon y^2}{2R} - c_1 z + c_4 y + c_5 \\ v &= -\frac{\upsilon xy}{R} - c_2 z - c_4 x + c_6 \\ w &= \frac{xz}{R} + c_1 x + c_2 y + c_3 \end{aligned} \tag{g}$$

where the constants c_j, $j = 1, 2, \ldots, 6$ are determined from the conditions of fastening of the beam. Assuming that point A at the centroid of the left end of the beam, an element of the z-axis, and an element of the xz-plane are fixed, then at $x = y = z = 0$, we have the conditions $u = v = w = 0$ and $du/dz = dv/dz = dv/dx = 0$.

It follows that all constants c_j, $j = 1, 2, \ldots, 6$ vanish and the displacements in the beam are

$$\begin{aligned} u &= -\frac{1}{2R}[z^2 + \upsilon(x^2 - y^2)] \\ v &= -\frac{\upsilon xy}{R} \\ w &= \frac{xz}{R} \end{aligned} \tag{h}$$

Upon setting $x = y = 0$ in Equations (h), we get the deflection of the beam

$$u = -\frac{z^2}{2R} = -\frac{Mz^2}{2EI_y}$$

$$v = w = 0$$

which is identical to the strength of materials solution.

The exact deflection curve has been studied in detail, its geometrical properties provided a method to determine the values of Poisson's ratio [5.3].

9.2 3D Axisymmetric Deformations of Solids

This section presents basic mathematical solutions to determine the axisymmetric stress and deformation in solids. The displacements in cylindrical coordinates r, θ, z are u, v, and w, respectively. For axisymmetric deformation, v vanishes and u and w are independent of θ. It follows that the stress components are independent of θ, and two of the shearing stresses, $\tau_{r\theta}$ and $\tau_{\theta z}$, vanish.

The strain–displacement relations are

$$\begin{aligned} \varepsilon_r &= \frac{\partial u}{\partial r}, \quad \varepsilon_\theta = \frac{u}{r}, \quad \varepsilon_z = \frac{\partial w}{\partial z} \\ \gamma_{rz} &= \frac{\partial u}{\partial z} + \frac{\partial w}{\partial r} \end{aligned} \tag{9.3}$$

The equations of equilibrium are

$$\begin{aligned} &\frac{\partial \sigma_r}{\partial r} + \frac{\partial \tau_{rz}}{\partial z} + \frac{\sigma_r - \sigma_\theta}{r} = 0 \\ &\frac{\partial \tau_{rz}}{\partial r} + \frac{\partial \sigma_z}{\partial z} + \frac{\tau_{rz}}{r} = 0 \end{aligned} \tag{9.4}$$

Equations 9.4 is satisfied by introducing a stress function $\phi(r, z)$, known as Love's stress function [9.1], such that

$$\begin{aligned} \sigma_r &= \frac{\partial}{\partial z}\left(\upsilon \nabla^2 \phi - \frac{\partial^2 \phi}{\partial r^2}\right) \\ \sigma_\theta &= \frac{\partial}{\partial z}\left(\upsilon \nabla^2 \phi - \frac{1}{r}\frac{\partial \phi}{\partial r}\right) \\ \sigma_z &= \frac{\partial}{\partial z}\left[(2-\upsilon)\nabla^2 \phi - \frac{\partial^2 \phi}{\partial z^2}\right] \\ \tau_{rz} &= \frac{\partial}{\partial r}\left[(1-\upsilon)\nabla^2 \phi - \frac{\partial^2 \phi}{\partial z^2}\right] \end{aligned} \tag{9.5a}$$

The stress function in Equations 9.5 satisfies the bi-harmonic equation

$$\begin{aligned} &\nabla^2 \nabla^2 \phi = 0 \\ &\quad \nabla^2 = \frac{\partial^2 \phi}{\partial r^2} + \frac{1}{r}\frac{\partial \phi}{\partial r} + \frac{\partial^2 \phi}{\partial z^2} \end{aligned} \tag{9.5b}$$

The corresponding displacements in terms of the stress function are readily found as follows:

$$u = r\varepsilon_\theta = \frac{r}{E}[\sigma_\theta - \upsilon(\sigma_r + \sigma_z)]$$

Making use of Equations 9.5a, we find that

$$2Gu = -\frac{\partial^2 \phi}{\partial r \partial z} \tag{9.6a}$$

For component w, making use of Hooke's law and Equations 9.5a, we arrive at

$$E\frac{\partial w}{\partial z} = \sigma_z - (\sigma_r + \sigma_\theta) = \frac{\partial}{\partial z}\left[2(1-\upsilon^2)\nabla^2\phi - (1+\upsilon)\frac{\partial^2\phi}{\partial z^2}\right]$$

Performing the integration, we obtain

$$Ew = (1+\upsilon)\left[2(1-\upsilon)\nabla^2\phi - \frac{\partial^2\phi}{\partial z^2}\right] + f(r) \tag{a}$$

Here, f(r) is arbitrary constant. Making use of Hooke's law and the second equation in Equation 9.3 in conjunction with Equations 9.5a and 9.6a, we have the following relation:

$$G\frac{\partial w}{\partial r} = \tau_{rz} - G\frac{\partial u}{\partial z} = \frac{\partial}{\partial r}\left[(1-\upsilon)\nabla^2\phi - \frac{1}{2}\frac{\partial^2\phi}{\partial z^2}\right]$$

Performing the integration, we obtain

$$Ew = (1+\upsilon)\left[2(1-\upsilon)\nabla^2\phi - \frac{\partial^2\phi}{\partial z^2}\right] + g(z) \tag{b}$$

where g(z) is arbitrary constant. Since Equations (a) and (b) are identical, it can be concluded that $f(r) = g(z) = \text{constant}$, which represents rigid body displacement, which can be neglected. Hence, we have for displacement w the following expression:

$$2Gw = 2(1-\upsilon)\nabla^2\phi - \frac{\partial^2\phi}{\partial z^2} \tag{9.6b}$$

Once a stress function, which satisfies the boundary conditions, is found, the displacements follow from Equations 9.6a, 9.6b, and the strains and stresses in the solid are obtained.

For some problems, it is easier to use polar coordinates than cylindrical coordinates. This is discussed in the next section.

9.2.1 Polar Coordinates and Legendre Functions

Instead of cylindrical coordinates r and z, we consider polar coordinate R,η where $z = R\cos\eta$ and $r = R\sin\eta$, as shown in Figure 9.3. It follows that

$$\begin{aligned}
&\frac{\partial^2}{\partial r^2} + \frac{\partial^2}{\partial z^2} = \frac{\partial^2}{\partial R^2} + \frac{1}{R}\frac{\partial}{\partial R} + \frac{1}{R^2}\frac{\partial^2}{\partial \eta^2} \\
&\frac{1}{r}\frac{\partial}{\partial r} = \frac{1}{R\sin\eta}\left(\frac{\partial}{\partial R}\sin\eta + \frac{\cos\eta}{R}\frac{\partial}{\partial\eta}\right) = \frac{1}{R}\frac{\partial}{\partial R} + \frac{\cot\eta}{R^2}\frac{\partial}{\partial\eta}
\end{aligned} \tag{a}$$

Inserting expression (a) in Laplace's equation, Equation 9.5b, results in

$$\frac{\partial^2\phi}{\partial R^2} + \frac{2}{R}\frac{\partial\phi}{\partial R} + \frac{1}{R^2}\text{ctn}\,\phi\frac{\partial\phi}{\partial\eta} + \frac{1}{R^2}\frac{\partial^2\phi}{\partial\eta^2} = 0 \tag{9.7}$$

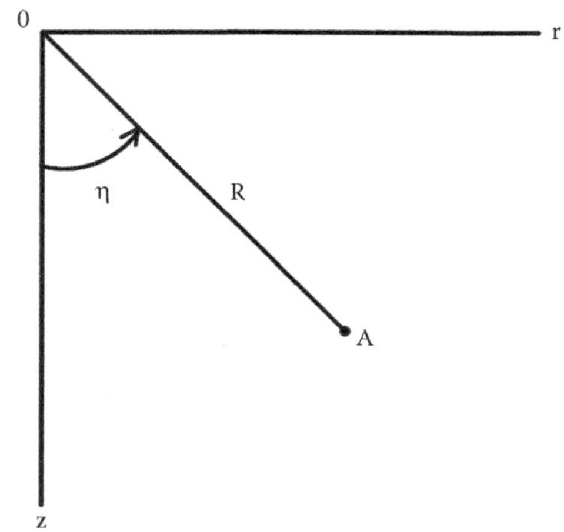

FIGURE 9.3
Polar coordinates (R, η).

A particular solution of Equation 9.7 takes the form

$$\phi_n = R^n \eta_n \tag{b}$$

Here, n stands for 0, 1, 2, ... , etc. Inserting Equations (b) into 9.7, the following ordinary differential equation is obtained:

$$\frac{1}{\sin\eta}\frac{d}{d\eta}\left(\sin\eta\frac{d\eta_n}{d\eta}\right) + n(n+1) = 0 \tag{c}$$

Introducing a new independent variable $x = \cos\eta$, Equation (c) becomes

$$(1 - x^2)\frac{d^2\eta_n}{dx^2} - 2x\frac{d\eta_n}{dx} + n(n+1)\eta_n = 0 \tag{9.8}$$

which is *Legendre's Equation*. Its two basic solutions are Legendre's functions of the first and second kinds, usually denoted by $P_n(x)$ and $Q_n(x)$, respectively [9.2].

For $n = 0, 1, 2, 3, 4, \ldots$, the $P_n(x)$ are polynomials as follows:

$$P_0(x) = 1, P_1(x) = x, P_2(x) = \frac{1}{2}(3x^2 - 1)$$

$$P_3(x) = \frac{1}{2}(5x^3 - 3x), P_4(x) = 1/8(35x^4 - 30x^2 + 3)$$

$$P_5(x) = 1/8(63x^5 - 70x^3 + 15x), \text{etc.}$$

Multiplying each of these polynomials by arbitrary constants A_n and using the results for η_n in Equation (b), we have solutions of Laplace's Equation 9.7.

In terms of the original variables, $x = \cos\eta$, $Rx = z$, $R = (r^2 + z^2)^{1/2}$, it follows that the following polynomials are solutions of Equation 9.7:

$$
\begin{aligned}
\phi_0 &= A_0 \\
\phi_1 &= A_1 z \\
\phi_2 &= A_2\left[z^2 - \frac{1}{3}(r^2 + z^2)\right] \\
\phi_3 &= A_3\left[z^3 - \left(\frac{3}{5}\right)z(r^2 + z^2)\right] \\
\phi_4 &= A_4\left[z^4 - \left(\frac{6}{7}\right)z^2(r^2 + z^2) + \left(\frac{3}{35}\right)(r^2 + z^2)^2\right] \\
\phi_5 &= A_5\left[z^5 - \left(\frac{10}{9}\right)z^3(r^2 + z^2) + \left(\frac{5}{21}\right)(r^2 + z^2)^2\right]
\end{aligned}
\tag{9.9}
$$

The polynomial (9.9) is also solutions of the bi-harmonic equation

$$
\left(\frac{\partial^2}{\partial R^2} + \frac{2\partial}{R\partial R} + \frac{1}{R^2}\operatorname{ctn}\eta\frac{\partial}{\partial\eta} + \frac{\partial^2}{R^2\partial\eta^2}\right)\left(\frac{\partial^2\phi}{\partial R^2} + \frac{2}{R}\frac{\partial\phi}{\partial R} + \frac{1}{R^2}\operatorname{ctn}\eta\frac{\partial\phi}{\partial\eta} + \frac{1}{R^2}\frac{\partial^2\phi}{\partial\eta^2}\right) = 0 \tag{9.10}
$$

Additional solutions of Equation 9.10, which are not solutions of Equation 9.7, can be derived from Equation 9.9. Since $R^n\ \eta_n$ is a solution of Equation 9.7, it can be shown that $R^{n+2}\ \eta_n$ is a solution (9.10). Multiplying the polynomials in Equation 9.9 by $R^2 = r^2 + z^2$, we establish the following new solutions of Equation 9.10:

$$
\begin{aligned}
\phi_2 &= B_2(r^2 + z^2) \\
\phi_3 &= B_3 z(r^2 + z^2) \\
\phi_4 &= B_4(2z^2 - r^2)(r^2 + z^2) \\
\phi_5 &= B_5(2z^3 - 3r^2 z)(r^2 + z^2)
\end{aligned}
\tag{9.11}
$$

Legendre's polynomials, Equations 9.9 and 9.10, are used in the next few sections to solve few basic 3D problems in elasticity.

Additional solutions of Love's stress function, Equations 9.5, can be obtained as follows: In Equation (b) above replace n by $-n-1$, the stress function becomes

$$
\phi_{-n-1} = R^{-n-1}\eta_{-n-1} \tag{d}
$$

and the coefficient $n(n+1)$ is modified to $(-n-1)(-n)$. Hence, η_{-n-1} has the same value as η_n. Instead of using Equation (d), we can use

$$
\phi_{-n-1} = R^{-n-1}\eta_n \tag{e}
$$

Using polynomials $P_n(x)$ for η_n as was done in deriving the solutions in Equations 9.10 and 9.11, we have the following set of solutions for the bi-harmonic stress function:

$$\begin{aligned}\phi_1 &= A_1(r^2 + z^2)^{-1/2}\\ \phi_2 &= A_2 z(r^2 + z^2)^{-3/2}\\ \phi_3 &= A_3\left[z^2(r^2 + z^2)^{-5/2} - \frac{1}{3}(r^2 + z^2)^{-3/2}\right]\end{aligned} \tag{9.12}$$

Multiplying expression (9.12) by $(r^2 + z^2)$, we arrive at another set of solutions for the stress function

$$\begin{aligned}\phi_1 &= B_1(r^2 + z^2)^{1/2}\\ \phi_2 &= B_2 z(r^2 + z^2)^{-1/2}\end{aligned} \tag{9.13}$$

Each of the solutions in Equations 9.9 through 9.13, and any combinations of them, can be used as a stress function. By adjusting the constants in the stress functions, solutions of practically important problems may be found. As an example, for the bending of a circular plate of finite thickness, polynomials of the third degree from Equations 9.9 and 9.11 may be combined to give the stress function

$$\phi(r, z) = A_3(2z^2 - 3r^2 z) + B_3(r^2 z + z^3) \tag{f}$$

where A_3 and B_3 are constants and z is the distance from the middle plane of the plate. By adjusting the constants in Equation (f), it can be shown that the stresses in the plate are constants. See Problem P9.1.

Polynomials of the fourth degree from Equations 9.9 and 9.11 are combined to yield the stress function

$$\phi(r, z) = A_4(8z^2 - 24r^2z^2 + 3r^4) + B_4(2z^4 + r^2z^2 + z^4 - r^4) \tag{g}$$

The induced stresses represent pure bending of the circular plate by moments uniformly distributed along the boundary. See Problem P9.2.

Reference 5.3 provides solution of a circular plate uniformly loaded across the top surface by load q and compares the result to that obtained in the elementary theory of bending of plates. The additional terms in the exact solution are small if the thickness of the plate, 2c, is small compared to the radius of the plate.

9.3 Rotation of a Circular Disk

In Chapter 6, Section 6.8, the stresses induced in a rotating circular disk, σ_r and σ_θ, are determined based on the assumption that the thickness of the disk, z, is small so the stresses are independent of z.

In this section, the same problem is reexamined assuming that the stress distribution is symmetrical with respect to the axis of rotation. Using cylindrical coordinates, r, θ, and z, the equations of equilibrium are obtained by adding to Equation 9.4 the centrifugal force due to rotation of the disk.

$$\begin{aligned}&\frac{\partial \sigma_r}{\partial r} + \frac{\partial \tau_{rz}}{\partial z} + \frac{\sigma_r - \sigma_\theta}{r} + \rho\omega^2 r = 0\\&\frac{\partial \tau_{rz}}{\partial r} + \frac{\partial \sigma_z}{\partial z} + \frac{\tau_{rz}}{r} = 0\end{aligned} \tag{9.14}$$

where ρ is the mass density per unit volume of the material and ω is the angular velocity of the disk.

The compatibility equations in terms of stresses are obtained by including body forces $B_x = \rho\omega^2 x$, $B_y = \rho\omega^2 y$, and $B_z = 0$ (see Chapter 4, solved Example 4.2). For the normal forces, we have

$$\begin{aligned}&\nabla^2\sigma_r - \frac{2}{r^2}(\sigma_r - \sigma_\theta) + \frac{1}{1+\upsilon}\frac{\partial^2 I_1}{\partial r^2} = -\frac{2\rho\omega^2}{1-\upsilon}\\&\nabla^2\sigma_\theta + \frac{2}{r^2}(\sigma_r - \sigma_\theta) + \frac{1}{1+\upsilon}\frac{1}{r}\frac{\partial I_1}{\partial r} = -\frac{2\rho\omega^2}{1-\upsilon}\\&\nabla^2\sigma_z + \frac{1}{1+\upsilon}\frac{\partial^2 I_1}{\partial z^2} = -\frac{2\upsilon\rho\omega^2}{1-\upsilon}\end{aligned} \tag{9.15}$$

where ∇^2 is Laplace's operator in cylindrical coordinates and I_1 is the first stress invariant,

$$I_1 = \sigma_r + \sigma_\theta + \sigma_z$$

The compatibility equation containing shearing stresses remains the same as in Chapter 4.

Solution for the stresses: In order to find the stresses in the disk, we start with a particular solution of Equations 9.14 satisfying the compatibility equations (Equations 9.15). On this solution, we superpose polynomial solutions from Equations 9.9 and 9.11 and adjust the constants of the polynomial to ensure that the boundary conditions are satisfied.

For the particular solution, consider the expressions

$$\sigma_z = Ar^2, \sigma_r = Br^2 + Cz^2, \sigma_{\theta_r} = Cz^2 + Dr^2, \tau_{rz} = 0 \tag{a}$$

where A, B, C, and D are arbitrary constants. Inserting Equation (a) into Equations 9.1 and 9.15, we find

$$\begin{aligned}&A = (1+3\upsilon)\frac{\rho\omega^2}{6\upsilon}\\&B = -\frac{\rho\omega^2}{3}\\&C = -(1+\upsilon)(1+2\upsilon)\frac{\rho\omega^2}{6\upsilon(1-\upsilon)}\\&D = 0\end{aligned} \tag{b}$$

and the stresses in Equation (a) become

$$\begin{aligned}
\sigma_r &= -\frac{\rho\omega^2}{3}r^2 - \frac{(1+\upsilon)(1+2\upsilon)}{6\upsilon(1-\upsilon)}\rho\omega^2 z^2 \\
\sigma_\theta &= -\frac{(1+\upsilon)(1+2\upsilon)}{6\upsilon(1-\upsilon)}\rho\omega^2 z^2 \\
\sigma_z &= \frac{(1+3\upsilon)}{6\upsilon}\rho\omega^2 r^2 \\
\tau_{rz} &= 0
\end{aligned} \tag{9.16}$$

Note that the stresses in Equation 9.16 can be used to discuss the stress distribution in any solid body rotating about an axis of generation.

In the case of a circular disk of radius a and thickness 2c, we superpose on Equation 9.16 stresses generated from polynomials of the fifth degree in Equations 9.9 and 9.11. Using the stress function

$$\phi(r,z) = A_5(8z^5 - 40r^2z^3 + 15r^4z) + B_5(2z^5 - r^2z^3 - 3r^4z) \tag{c}$$

where A_5 and B_5 are arbitrary constants. Using Equations 9.5, the following stresses are obtained:

$$\begin{aligned}
\sigma_r &= -A_5(180r^2 - 240z^2) + B_5[(36-54\upsilon)r^2 + 6(1+18\upsilon)z^2] \\
\sigma_z &= -A_5(-240r^2 + 480z^2) + B_5[(-102+54\upsilon)r^2 + (96-108\upsilon)z^2] \\
\sigma_\theta &= A_5(-60r^2 + 240z^2) + B_5[(12-54\upsilon)r^2 + (6+108\upsilon)z^2] \\
\tau_{rz} &= 480A_5rz - B_5(96-108\upsilon)rz
\end{aligned} \tag{9.17}$$

Adding Equations 9.16 and 9.17 and determining the constants A_5 and B_5 from the requirements that the resultant σ_z and τ_{rz} vanishes furnish the following values for the constants:

$$\begin{aligned}
A_5 &= \frac{(1+3\upsilon)(8-9\upsilon)}{480\rho(51-27\upsilon)}\rho\omega^2 \\
B_5 &= \frac{(1+3\upsilon)}{12\upsilon(51-27\upsilon)}\rho\omega^2
\end{aligned} \tag{d}$$

and stresses

$$\begin{aligned}
\sigma_r &= -\left[\frac{3+\upsilon}{8}r^2 + \frac{\upsilon(1+\upsilon)}{2(1-\upsilon)}z^2\right]\rho\omega^2 \\
\sigma_\theta &= -\left[\frac{(1+3\upsilon)}{8}r^2 + \frac{\upsilon(1+\upsilon)}{2(1-\upsilon)}z^2\right]\rho\omega^2
\end{aligned} \tag{e}$$

In order to eliminate the resultant radial compression along the boundary, that is to require

$$\left\{\int_{-c}^{c} \sigma_r dz\right\}_{r=a} = 0$$

we need to superpose on stresses (e) a uniform radial tension of magnitude

$$\left(\frac{3+\upsilon}{8}\right)a^2\rho\omega^2 + \frac{\upsilon(1+\upsilon)}{6(1-\upsilon)}c^2\rho\omega^2$$

and the final stresses in the disk become

$$\begin{aligned}
\sigma_r &= \left[\frac{3+\upsilon}{8}(a^2-r^2) + \frac{\upsilon(1+\upsilon)}{6(1-\upsilon)}(c^2-3z^2)\right]\rho\omega^2 \\
\sigma_\theta &= \left[\frac{3+\upsilon}{8}a^2 - \frac{1+3\upsilon}{8}r^2 + \frac{\upsilon(1+\upsilon)}{6(1-\upsilon)}(c^2-3z^2)\right]\rho\omega^2 \\
\sigma_z &= 0 \\
\tau_{rz} &= 0
\end{aligned} \tag{9.18}$$

Comparing Equations 9.18 with the solution in Chapter 6, Equation 6.52, the current solution has additional terms with the factor ($c^2 - 3z^2$). For thin disks, the corresponding stresses are small and their resultant over the thickness of the disk is zero.

9.4 Force at a Point in an Infinite Solid

Consider a concentrated force P acting at the origin of cylindrical coordinates in an infinite solid. It is expected that some or all the induced stresses in the solid have singularities at the origin. For a stress function, we use the first expression in Equations 9.13, which satisfies the bi-harmonic equation in Equations 9.5.

$$\phi(r, z) = B(r^2 + z^2)^{1/2} \tag{a}$$

The induced stresses are obtained using Equations 9.5 as

$$\begin{aligned}
\sigma_r &= B[(1-2\upsilon)z(r^2+z^2)^{-3/2} - 3r^2z(r^2+z^2)^{-5/2}] \\
\sigma_\theta &= B(1-2\upsilon)z(r^2+z^2)^{-3/2} \\
\sigma_z &= -B[(1-2\upsilon)z(r^2+z^2)^{-3/2} + 3z^2(r^2+z^2)^{-5/2}] \\
\tau_{rz} &= -B[(1-2\upsilon)r(r^2+z^2)^{-3/2} + 3rz^2(r^2+z^2)^{-5/2}]
\end{aligned} \tag{9.19}$$

It is obvious that all the induced stresses are singular at the point of application of the force. In order to find the constant B, we consider the point of application of P to be the center of a small spherical cavity of radius r and use the stresses in Equations 9.19 to calculate the forces on the surface of the cavity.

Equilibrium of the forces on a ring-shaped element shown in Figure 9.4 indicates that the surface force in the z-direction is

$$F_z = -(\tau_{rz}\sin\eta + \sigma_{z\cos}\eta) \tag{b}$$

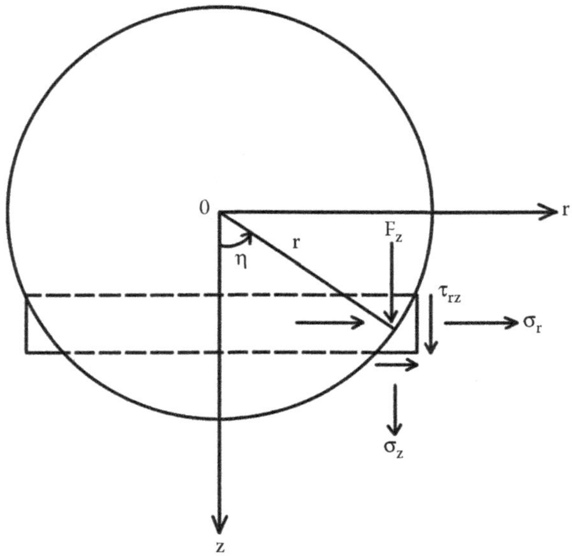

FIGURE 9.4
Spherical cavity around force P in an infinite solid.

Since $\sin\eta = r\,(r^2+z^2)^{-1/2}$ and $\cos\eta = z\,(r^2+z^2)^{-1/2}$, using Equations 9.19, F_z is

$$F_z = B[(1-2\upsilon)(r^2+z^2)^{-1} + 3z^2(r^2+z^2)^{-2}] \tag{c}$$

The resultant of these forces over the surface of the cavity gives in the z-direction

$$2\int_0^{\pi/2} F_z\sqrt{r^2+z^2}(2\pi r)d\eta = 8\pi(1-\upsilon)B = P \tag{d}$$

From symmetry, the resultant of the surface forces in the radial direction is zero. Equation (d) gives the magnitude of B as

$$B = \frac{P}{8\pi(1-\upsilon)} \tag{9.20}$$

Note that the force P is balanced by surface forces on the outer boundary of the cavity, spherical or otherwise, as required by equilibrium.

In particular, across the $z=0$ plane, Equations 9.19 and 9.20 give zero normal stresses and the shearing stresses are given as

$$\tau_{rz} = -\frac{(1-2\upsilon)}{8\pi(1-\upsilon)}\frac{P}{r^2} \tag{e}$$

These stresses are inversely proportional to the square of the distance r from the point of application of the force, that is they have singularity at $r=0$.

The solution of this problem was first given by Lord Kelvin [9.3].

EXAMPLE 9.2

Determine the stresses induced in a thick-walled spherical container subjected to internal and external pressures.

Solution

It is recalled that this problem is discussed in Chapter 7, Section 7.2. The induced stresses in the container are determined using spherical coordinates (Equations 7.41). An alternative method is to apply the 3D stresses generated by a force P at a point in an infinite elastic solid to determine the same stresses in Equations 7.41.

We begin the discussion by considering the problem of two equal and opposite forces, each of magnitude P, acting at a small distance d apart along the z-axis, as shown in Figure 9.5. The stresses induced by P at origin 0 in Figure 9.5 are given in Equations 9.19. The stresses induced by P at point 0* can be obtained from Equations 9.19 by replacing P by −P, and because distance d is infinitely small, any f(r, z) in Equations 9.19 need to be replaced by $-[f(r, z) + (\delta f/\delta z)d]$. Superposing the stresses induced by both forces and introducing new constant A = Bd, we have

$$\sigma_r = -A\frac{\partial}{\partial z}[(1-2\upsilon)z(r^2+z^2)^{-3/2} - 3r^2z(r^2+z^2)^{-5/2}]$$

$$\sigma_\theta = -A\frac{\partial}{\partial z}[(1-2\upsilon)z(r^2+z^2)^{-3/2}]$$

$$\sigma_z = A\frac{\partial}{\partial z}[(1-2\upsilon)z(r^2+z^2)^{-3/2} + 3z^3(r^2+z^2)^{-5/2}]$$

$$\tau_{rz} = A\frac{\partial}{\partial z}[(1-2\upsilon)r(r^2+z^2)^{-3/2} + 3rz^2(r^2+z^2)^{-5/2}]$$

Next, we consider the stress components σ_R and $\tau_{R\eta}$ acting at a point D in Figure 9.5 on an elemental area normal to a radius 0D in the rz-plane, of length R and making angle η

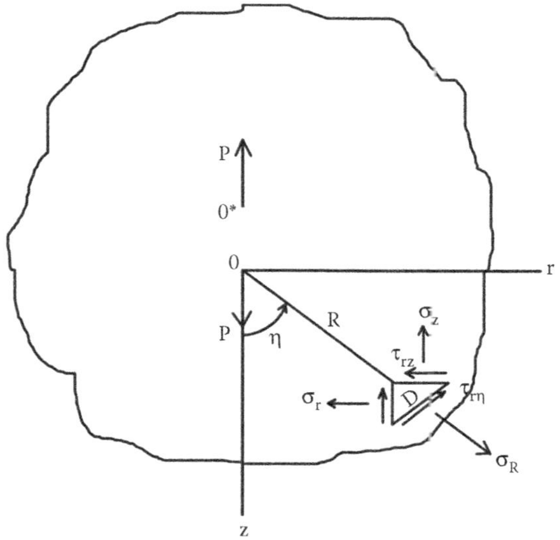

FIGURE 9.5
An infinite elastic body subjected to equal and opposite forces.

with the z-axis. Equilibrium of a small triangular area shown in Figure 9.5, we have

$$\begin{aligned}\sigma_R &= \sigma_r \sin^2\eta + \sigma_z \cos^2\eta + 2\tau_{rz}\sin\eta\cos\eta \\ \tau_{R\eta} &= (\sigma_r - \sigma_z)\sin\eta\cos\eta - \tau_{rz}(\sin^2\eta - \cos^2\eta)\end{aligned} \tag{a}$$

where

$$\begin{aligned}\sin\eta &= \frac{r}{(r^2+z^2)^{1/2}} = \frac{r}{R} \\ \cos\eta &= \frac{z}{(r^2+z^2)^{1/2}} = \frac{z}{R}\end{aligned}$$

In deriving Equations (a), we have neglected the small contribution of the stress component σ_θ acting on the sides of the element in the meridional section of the solid.

Inserting Equations (a) into 9.19, we obtain

$$\begin{aligned}\sigma_R &= -\frac{2(1+\upsilon)A}{R^3}\left[-\sin^2\eta + \frac{2(2-\upsilon)}{1+\upsilon}\cos^2\eta\right] \\ \tau_{R\eta} &= -\frac{2(1+\upsilon)A}{R^3}\sin\eta\cos\eta\end{aligned} \tag{b}$$

Note that these stresses are symmetrical with respect to the z-axis and with respect to the plane normal to z.

Center of compression: In addition to the system of double forces along the z-axis considered above, we consider an identical system of two forces acting along the r axis and another one along an axis perpendicular to the rz-plane. The combination of the three perpendicular systems of double forces is known as the center of compression. Because of symmetry, the resulting stress distribution of the three systems is symmetrical with the origin, and it follows that any sphere in the body with center at the origin is subjected only to uniformly distributed stress acting normal to its surface. The magnitude of this stress is obtained from the first equation in Equation (b). In order to calculate the stress at points on the circle in the rz-plane, the first equation of Equations (b) gives part of the normal stress due to the double force along z-axis. By interchanging $\sin\eta$ and $\cos\eta$, the normal stress round the circle by the double force along the r-axis is obtained. The normal stress due to the two forces along an axis normal to the rz-plane is obtained by substituting $\eta = \pi/2$ in the same equation.

By adding the actions of the three systems of double forces, the total normal stress acting on the surface of a sphere is found as follows:

$$\sigma_R = -\frac{4(1-\upsilon)A}{R^3} \tag{c}$$

This radial compressive stress is inversely proportional to the cube of the distance R and can be used to determine the stresses in a hollow sphere centered at the origin and subjected to internal and external pressures.

Consider a hollow sphere of internal and external radii a and b, respectively, subjected to internal pressure p_i and external pressure p_0. Superposing constant tension or compression on Equation (c), we can write the general expression for the radial stress in the form

$$\sigma_R = C + \frac{D}{R^2} \tag{d}$$

where C and D are arbitrary constants to be determined from the conditions on the inner and outer boundaries of the sphere, that is,

$$\frac{C}{a^3} + D = -p_i$$
$$\frac{C}{b^3} + D = -p_0$$

Solving for C and D, we have

$$C = \frac{a^3 b^3}{a^3 - b^3}(p_i - p_0)$$
$$D = \frac{b^3 p_0 - a^3 p_i}{a^3 - b^3} \tag{9.20a}$$

and

$$\sigma_R = \frac{b^3(R^3 - a^3)}{R^3(a^3 - b^3)} p_0 - \frac{a^3(b^3 - R^3)}{R^3(a^3 - b^3)} p_i$$

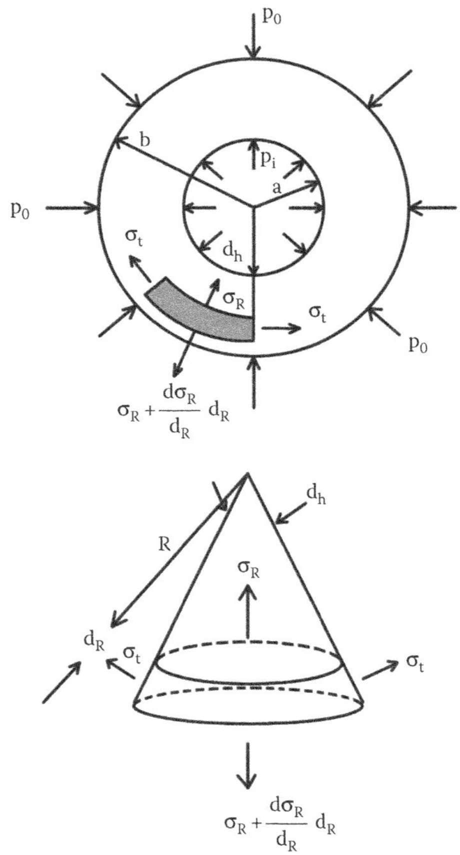

FIGURE 9.6
Example 9.2.

Also, the internal and external pressures produce tangential stress, its magnitude can be determined from equilibrium of an element cut out from the sphere by two concentric spherical surfaces of Radii R and R + dR and by a circular cone with a small angle $d\eta$ (Figure 9.6).

The equation of equilibrium is

$$\sigma_{tan}\frac{\pi R}{2}dR(d\eta)^2 = \frac{d\sigma_R}{dR}\frac{\pi R^2}{4}dR(d\eta)^2 + \sigma_R\frac{\pi R}{2}dR(d\eta)^2$$

which yields

$$\sigma_{tan} = \frac{d\sigma_R}{dR}\frac{R}{2} + \sigma_R \tag{e}$$

Using Equation 9.20a for the radial stress, Equation (e) gives

$$\sigma_{tan} = \frac{b^3(2R^3 + a^3)}{2R^3(a^3 - b^3)}p_0 - \frac{a^3(2R^3 + b^3)}{2R^3(a^3 - b^3)}p_i \tag{9.20b}$$

The stresses in Equations 9.20a and 9.20b are identical with σ_r and σ_θ in Equation 7.41.

9.5 Stress Concentration around a Spherical Cavity

Consider an elastic member containing a spherical cavity of radius a. The member is subjected to uniform tension of magnitude T (Figure 9.7). For a member without cavity, the normal and shearing stresses acting on a spherical surface are

$$\sigma_R = T\cos^2\eta, \quad \tau_{R\eta} = -T\sin\eta\cos\eta \tag{a}$$

In order to obtain the solution for a member containing spherical cavity of radius a, we need to superpose stress systems that nullify the stresses on the surface of the cavity and vanish at infinity. It appears that we need three stress systems to do the required task. The first system is the stresses in Equation (b) of Example 9.2. The second system is the stresses due to the double force along the z-axis, that is Equations 9.19, and the third system is stresses due to a center of compression.

The combined stresses acting on the spherical surface of radius a can be expressed as follows:

$$\begin{aligned} (\sigma_R)_1 &= -\frac{2(1+\upsilon)A}{a^3}\left(-1 + \frac{5-\upsilon}{1+\upsilon}\cos^2\eta\right) \\ (\tau_{R\eta})_1 &= -\frac{2(1+\upsilon)A}{a^3}\sin\eta\cos\eta \end{aligned} \tag{b}$$

and

$$\begin{aligned} (\sigma_R)_2 &= \frac{B}{a^3} \\ (\tau_{R\eta})_2 &= 0 \end{aligned} \tag{c}$$

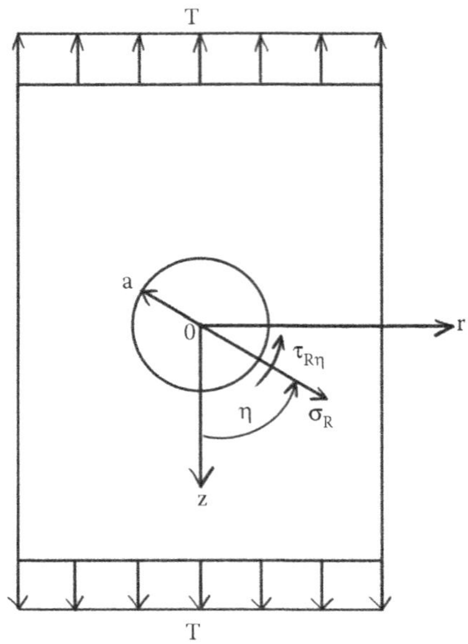

FIGURE 9.7
Member with shperical cavity subjected to uniform tension.

where A and B are constants to be adjusted subject to satisfying the required conditions. By adding stresses (b) and (c), it is clear that the stresses produced by T, Equation (a), cannot be made to vanish and that an additional stress system is required.

Taking a stress function from Equations 9.12

$$\phi(r, z) = Cz(r^2 + z^2)^{-3/2}$$

The corresponding stresses from Equations 9.5 are

$$\begin{aligned}
\sigma_r &= \frac{3C}{R^5}(-4 + 35 \sin^2 \eta \cos^2 \eta) \\
\sigma_z &= \frac{3C}{R^5}(3 - 30 \cos^2 \eta + 35 \cos^4 \eta) \\
\sigma_\theta &= \frac{3C}{R^5}(1 - 5 \cos^2 \eta) \\
\tau_{rz} &= \frac{15C}{R^5}(-3 \sin \eta \cos \eta + 7 \sin \eta \cos^2 \eta)
\end{aligned} \tag{d}$$

Using Equations (a) of Example 9.2, the stresses acting on a spherical surface of radius a are

$$\begin{aligned}
(\sigma_R)_3 &= \frac{12C}{a^5}(-1 + 3, \cos^2 \eta) \\
(\tau_{R\eta})_3 &= \frac{24C}{a^5} \sin \eta \cos \eta
\end{aligned} \tag{e}$$

Adding stress systems (b), (c), and (e), we find

$$\begin{aligned}\sigma_R &= \frac{2(1+\upsilon)A}{a^3} - \frac{2(5-\upsilon)A}{a^3}\cos^2\eta + \frac{B}{a^3} - \frac{12C}{a^5} + \frac{36C}{a^5}\cos^2\eta \\ \tau_{R\eta} &= -\frac{2(1+\upsilon)A}{a^3}\sin\eta\cos\eta + \frac{24C}{a^5}\sin\eta\cos\eta\end{aligned} \tag{f}$$

Next, we add Equations (a) and (f). The resulting stresses on the spherical surface of the cavity will vanish if the constants A, B, and C satisfy the following equations:

$$\begin{aligned}\frac{2(1+\upsilon)A}{a^3} + \frac{B}{a^3} - \frac{12C}{a^5} &= 0 \\ \frac{2(5-\upsilon)A}{a^3} - \frac{36C}{a^5} &= T \\ \frac{2(1+\upsilon)A}{a^3} - \frac{24C}{a^5} &= -T\end{aligned} \tag{g}$$

Solving Equations (g), we find

$$\begin{aligned}\frac{A}{a^3} &= \frac{5T}{2(7-5\upsilon)} \\ \frac{B}{a^5} &= \frac{(1-5\upsilon)T}{7-5\upsilon} \\ \frac{C}{a^5} &= \frac{T}{2(7-5\upsilon)}\end{aligned} \tag{h}$$

The complete stress at any point of the solid consists of Equations (a) and (d) above, Equations 9.19 due to double force system, and stress due to center of compression given by Equations (c) and (e) of Example 9.2.

In particular, the stresses acting on plane $z=0$, because of symmetry there are no shear stresses acting on this plane while the normal stress is computed as follows:

From Equation (d) above with $R=r$ and $\eta=\pi/2$, $\sigma_z = \dfrac{9C}{r^5} = \dfrac{9a^5T}{2(7-5\upsilon)r^5}$

From Equation 9.19 for $z=0$, $\sigma_z = \dfrac{(1-2\upsilon)A}{r^3} = \dfrac{5(1-2\upsilon)T}{2(7-5\upsilon)}\dfrac{a^3}{r^3}$ (i)

From Equation (e) in Example 9.2, $\sigma_z = (\sigma_{tan})_{z=0} = -\dfrac{B}{2r^3} = -\dfrac{(1-5\upsilon)T}{2(7-5\upsilon)}\dfrac{a^3}{r^3}$

The total normal stress on the plane $z=0$ is: T + the sum of the previous three expressions in Equations (f), that is,

$$\sigma_z = \left[1 + \frac{4-5\upsilon}{2(7-5\upsilon)}\frac{a^3}{r^3} + \frac{9}{2(7-5\upsilon)}\frac{a^5}{r^5}\right]T \tag{9.21}$$

At $r = a$, we have the following stress concentration factor:

$$\frac{(\sigma_z)_{max}}{T} = \frac{27 - 15\upsilon}{2(7 - 5\upsilon)} \tag{9.22}$$

For $\nu = 0.3$, the maximum stress is 45/22, that is, about twice as large as T and the stress concentration factor $= 45/22 = 2.05$.

When r is increased beyond a, the stress in Equation 9.21 rapidly approaches T.

Similarly, for points in the plane $z = 0$

$$(\sigma_\theta)_{z=0} = \frac{3C}{r^5} - \frac{(1 - 2\upsilon)A}{r^3} - \frac{B}{2r^3}$$

Inserting the values of the constants A, B, and C from Equation (h), we find that the tensile stress along the equator ($\eta = \pi/2$) of the cavity is

$$(\sigma_\theta)_{z=0,r=a} = \frac{(15\upsilon - 3)T}{2(7 - 5\upsilon)}$$

At the pole of the cavity ($\eta = 0$ or $\eta = \pi$), we have

$$\sigma_r = \sigma_\theta = \frac{2(1 - 2\upsilon)A}{a^3} - \frac{B}{2a^3} - \frac{12C}{a^5} = -\frac{3 + 15\upsilon}{2(7 - 5\upsilon)}T$$

indicating that the longitudinal tension T produces compression at this point.

For pure shear loading of the member, T in one direction and $-T$ in the perpendicular direction, it can be shown that the maximum shearing stress is

$$\tau_{max} = \frac{15(1 - \upsilon)}{7 - 5\upsilon}T \tag{9.23}$$

The solution of this problem is due to Southwell [9.4]. Further results and original references are given in Reference 5.3.

Stress concentration problems around holes in 2D solids are discussed in Section 7.3 of Chapter 7.

9.6 Concentrated Force on Boundary of a Semi-Infinite Solid

Consider a semi-infinite solid, boundary $z = 0$, subjected to a concentrated force P acting at the origin of cylindrical coordinates r, z in the z-direction, as shown in Figure 9.8. We seek to determine the stresses and displacements in the solid.

In Section 9.4, it is shown that when a force P acts in an infinite solid, the induced stresses are given by Equations 9.19, 9.20, and (e). In particular, making use of Equations 9.20 and (e), the shearing stresses on plane $z = 0$ are

$$\tau_{rz} = -\frac{(1 - 2\upsilon)B}{r^2} \tag{a}$$

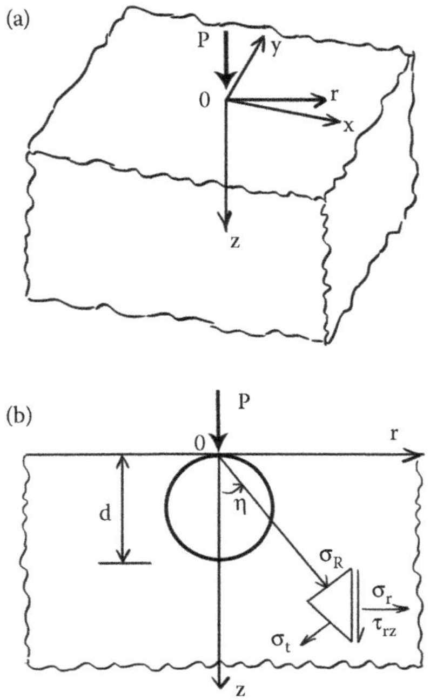

FIGURE 9.8
Force P acting on the boundary of a semi-infinite solid, (a) semifinite solid and (b) vertical section through point 0 elemental of area mn of diameter = d.

In order to arrive at the solution for a semi-infinite solid, we need to eliminate the stresses given in Equation (a). This can be done by using the stress distribution due to center of compression (Equation (c) in Example 9.2). In polar coordinates R, η, this stress distribution is

$$\begin{aligned}\sigma_R &= \frac{A}{R^3} \\ \sigma_{tan} &= \frac{d\sigma_R}{dR}\frac{R}{2} + \sigma_R = -\frac{A}{2R^3}\end{aligned} \tag{b}$$

where A is a constant. In cylindrical coordinates r, z, the stresses at the element shown in Figure 9.8 are

$$\begin{aligned}\sigma_r &= \sigma_R \sin^2\eta + \sigma_{tan}\cos^2\eta = A\left(r^2 - \frac{1}{2}z^2\right)(r^2 + z^2)^{-5/2} \\ \sigma_z &= \sigma_R \cos^2\eta + \sigma_{tan}\sin^2\eta = A\left(z^2 - \frac{1}{2}r^2\right)(r^2 + z^2)^{-5/2} \\ \tau_{rz} &= \frac{1}{2}(\sigma_R - \sigma_{tan})\sin 2\eta = \frac{3}{2}Arz(r^2 + z^2)^{-5/2} \\ \sigma_\theta &= \sigma_{tan} = -\frac{A}{2R^2} = -\frac{A}{2}(r^2 + z^2)^{-3/2}\end{aligned} \tag{9.24}$$

Next, consider that the centers of pressure are uniformly distributed along the z axis from $z = 0$ to $z = -0\,0$. It follows that upon superposition, the stress components in infinite solid, from Equations 9.24 with a new constant C, are

$$\sigma_r = C\int_z^{\infty}\left(r^2 - \frac{1}{2}z^2\right)(r^2+z^2)^{-5/2}dz = \frac{C}{2}\left[\frac{1}{r^2} - \frac{z(r^2+z^2)^{-\frac{1}{2}}}{r^2} - z(r^2+z^2)^{-3/2}\right]$$

$$\sigma_z = C\int_z^{\infty}\left(z^2 - \frac{1}{2}r^2\right)(r^2+z^2)^{-5/2}dz = \frac{C}{2}z(r^2+z^2)^{-3/2}$$

$$\sigma_\theta = -\frac{C}{2}\int_z^{\infty}(r^2+z^2)^{-3/2}dz = -\frac{C}{2}\left[\frac{1}{r^2} - \frac{z(r^2+z^2)^{-1/2}}{r^2}\right] \tag{9.25}$$

$$\tau_{rz} = \frac{3C}{2}\int_z^{\infty} rz(r^2+z^2)^{-5/2}dz = \frac{C}{2}r(r^2+z^2)^{-3/2}$$

On the $z = 0$ plane, we find that the normal stress σ_z is zero and the shearing stress is

$$(\tau_{rz})_{z=0} = \frac{C}{2r^2} \tag{c}$$

Upon equating Equations (a) and (c), we obtain the magnitude of C

$$C = 2(1 - 2\nu)B$$

which when inserted into Equations 9.25 and adding to Equations 9.19 we obtain

$$\sigma_r = B\left\{(1-2\upsilon)\left[\frac{1}{r^2} - \frac{z(r^2+z^2)^{-1/2}}{r^2}\right] - 3r^2z(r^2+z^2)^{-5/2}\right\}$$

$$\sigma_z = -3Bz^3(r^2+z^2)^{-5/2}$$

$$\sigma_\theta = B(1-2\upsilon)\left[-\frac{1}{r^2} + \frac{z(r^2+z^2)^{-1/2}}{r^2} + z(r^2+z^2)^{-5/2}\right] \tag{d}$$

$$\tau_{rz} = -3Brz^2(r^2+z^2)^{-5/2}$$

On $z = 0$, $\sigma_z = \tau_{rz} = 0$.

The next step is to find the constant B so that the forces distributed over a hemispherical surface with center at the origin are statically equivalent to the force P along the z axis. Consider the equilibrium of the element shown in Figure 9.8. The component in the z-direction of the forces on the hemispherical surface is

$$F_z = -(\tau_{rz}\sin\eta + \sigma_z\cos\eta) = 3Bz^2(r^2+z^2)^{-2}$$

and B is determined from the following relation:

$$P = 2\pi \int_0^{\pi/2} F_z r(r^2 + z^2)^{1/2} d\eta = 6\pi B \int_0^{\pi/2} \cos^2\eta \sin\eta d\eta = 2\pi B$$

$$B = \frac{P}{2\pi}$$

Inserting this value of B into Equations (d), we obtain the final solution for the stresses induced by a concentrated force P acting on the boundary of a semi-infinite solid (Figure 9.8).

$$\begin{aligned}
\sigma_r &= \frac{P}{2\pi}\left\{(1-2\upsilon)\left[\frac{1 - z(r^2+z^2)^{-1/2}}{r^2}\right] - 3r^2 z(r^2+z^2)^{-5/2}\right\} \\
\sigma_z &= -\frac{P}{2\pi} z^3 (r^2+z^2)^{-5/2} \\
\sigma_\theta &= \frac{P}{2\pi}(1-2\upsilon)\left[\frac{-1 + z(r^2+z^2)^{-1/2} + z(r^2+z^2)^{-5/2}}{r^2}\right] \\
\tau_{rz} &= -\frac{3P}{2\pi} r z^2 (r^2+z^2)^{-5/2}
\end{aligned} \tag{9.26}$$

This solution, which is the 3D analog of the 2D solution presented in Section 6.7.3 of Chapter 6, is due to Boussenesq [9.5].

Considering an elemental area mn perpendicular to the z-axis, the ratio of the normal and shearing stresses on this element, from Equations 9.26, is

$$\frac{\sigma_z}{\tau_{rz}} = \frac{z}{r} \tag{e}$$

It follows that the direction of the resultant stress passes though the origin 0. The magnitude of the resultant stress is

$$Re = \sqrt{\sigma_z^2 + \tau_{rz}^2} = \frac{3P}{2\pi}\frac{z^2}{(r^2+z^2)^2} = \frac{3P}{2\pi}\frac{\cos^2\eta}{(r^2+z^2)} \tag{9.27}$$

This stress is inversely proportional to the square of the distance from the point of application of the force P. For each point on the spherical surface of diameter d, tangent to the plane $z = 0$ at the origin, we have

$$r^2 + z^2 = d^2\cos^2\eta \tag{f}$$

Inserting Equation (f) into Equation 9.27, we conclude that for points of the sphere the resultant stress on horizontal planes is constant and equal to $3P/2\pi\, d^2$.

Next, the displacements produced in the semi-infinite solid by the load P, denoted u and w, are determined

$$u = r\varepsilon_\theta = \frac{r}{E}[\sigma_\theta - \upsilon(\sigma_r + \sigma_z)]$$

Inserting the stress components from Equations 9.26, the displacement u is

$$u = \frac{(1+\upsilon)(1-2\upsilon)P}{2\pi Er}\left[-1 + z(r^2+z^2)^{-1/2} + \frac{r^2 z(r^2+z^2)^{-3/2}}{1-2\upsilon}\right] \quad (9.28)$$

In order to determine the vertical displacement w, we start with

$$\frac{\partial w}{\partial z} = \varepsilon_z = \frac{1}{E}[\sigma_z - \upsilon(\sigma_r + \sigma_\theta)]$$

$$\frac{\partial w}{\partial r} = \gamma_{rz} - \frac{\partial u}{\partial z} = \frac{2(1+\upsilon)\tau_{rz}}{E} - \frac{\partial u}{\partial z}$$

Substituting for the shear stress τ_{rz} the fourth equation in Equations 9.26 and for from Equation 9.28, we obtain the following:

$$\frac{\partial w}{\partial z} = \frac{P}{2\pi E}\{3(1+\upsilon)r^2 z(r^2+z^2)^{-5/2} - [3+\upsilon(1-2\upsilon)]z(r^2+z^2)^{-3/2}\}$$

$$\frac{\partial w}{\partial r} = -\frac{P(1+\upsilon)}{2\pi E}[2(1-\upsilon)z^2(r^2+z^2)^{-3/2} + 2(1-\upsilon^2)(r^2+z^2)^{-5/2}]$$

Carrying out the integration and omitting arbitrary constants, the magnitude of w is

$$w = \frac{P}{2\pi E}[(1+\upsilon)z^2(r^2+z^2)^{-3/2} + 2(1-\upsilon^2)(r^2+z^2)^{-1/2}] \quad (9.29)$$

Across the boundary $z = 0$, the displacements are

$$(u)_{z=0} = -\frac{(1+\upsilon)(1-2\upsilon)P}{2\pi Er}$$

$$(w)_{z=0} = \frac{(1-\upsilon^2)P}{\pi Er} \quad (9.30)$$

Equations 9.30 show that (i) the product wr is constant at the boundary. Hence, after deformation, the radii drawn from the origin on the boundary surface are hyperbolas with asymptotes 0r and 0z, (ii) the displacements are singular at the origin, therefore the material becomes plastic and around the origin cut out by a hemispherical surface of small radius and P is replaced by statically equivalent forces distributed over the surface as required by equilibrium.

9.7 Uniform Load Distributed over a Part of the Boundary of a Semi-Infinite Solid

When a uniform load of intensity q is distributed over a given area on the surface $z = 0$ of a semi-infinite body, the solution for a force P can be used by the method of "*superposition*" to obtain the corresponding stresses and displacements in a semi-infinite solid due to distributed load on the boundary. Reference 5.3 gives detailed derivation of the solution for uniform load distributed over a circle of radius a and provides results for cases of uniform pressure distributed over the surface of a square with sides 2a, and over rectangles with various ratios $\beta = a/b$ of the sides.

9.7.1 Uniform Load Distributed over a Circular Area of Radius a

Consider a uniform load of intensity q distributed over a circle of radius a on the surface $z = 0$ of a semi-infinite elastic body. Using cylindrical coordinates r and z with origin located at the center of the circle, the induced displacements and stresses in the semi-infinite solid $z > 0$ can be determined [5.3].

Throughout the semi-infinite solid, the vertical displacement w is expressed in terms of complete elliptic integrals. For points inside the loaded region,

$$w = \frac{4(1-\upsilon^2)qa}{\pi E}\int_0^{\pi/2}\sqrt{1-(r^2/a^2)\sin^2\theta}\,d\theta \tag{9.31a}$$

Evaluating the integral [9.6], we obtain

$$w = \frac{(1-\upsilon^2)qa}{2E}\left[4 - \frac{r^2}{a^2} - \frac{3r^4}{16a^4} - \frac{5r^6}{256a^6} - \cdots\right], \quad \frac{r^2}{a^2} < 1$$

For points outside the loaded region, we have

$$w = \frac{4(1-\upsilon^2)qr}{\pi E}\left[\int_0^{\pi/2}\sqrt{1-(a^2/r^2)\sin^2\theta}\,d\theta - \left(1-\frac{a^2}{r^2}\right)\int_0^{\pi/2}\frac{d\theta}{\sqrt{1-(a^2/r^2)\sin^2\theta}}\right] \tag{9.31b}$$

The integrals in Equation 9.31b can be expressed in terms of infinite series containing a^2/r^2, same as for w inside the loaded region [9.6]. Maximum deflection occurs at the center of the loaded region obtained by setting $r = 0$ in Equation 9.31a

$$w_{max} = \frac{2(1-\upsilon^2)qa}{E} \tag{a}$$

The deflection at the boundary of the loaded circle is obtained by setting $r = a$ in Equation 9.31b

$$w_{r=a} = \frac{4(1-\upsilon^2)qa}{\pi E} \tag{b}$$

Note that $w_{max} = \pi/2\, w_{r=a}$. For a given value of q, the maximum deflection increases with the radius of the loaded area.

The stresses are

$$\sigma_z = q[-1 + z^3(a^2+z^2)^{-3/2}] \tag{9.32}$$

$$\sigma_r = \sigma_\theta = \frac{q}{2}\left[-(1+2\upsilon) + 2(1+\upsilon)z(a^2+z^2)^{-1/2} - z^3(a^2+z^2)^{-3/2}\right]$$

At the center of the loaded region, $z = 0$

$$\sigma_z = -q,\ \sigma_r = \sigma_\theta = -[(1+2\nu)/2]q. \text{ Thus, for } \nu = 0.3,\ \sigma_r = \sigma_\theta = -0.8\ q.$$

Also,

$$\frac{1}{2}(\sigma_\theta - \sigma_z) = \frac{q}{2}\left[\frac{1-2\upsilon}{2} + (1+\upsilon)z(a^2+z^2)^{-1/2} - \frac{3}{2}z^3(a^2+z^2)^{-3/2}\right] \tag{a}$$

And the maximum shearing stress at the origin, on planes at 45° to the z-axis, from Equation (a), $z = 0$ and $\nu = 0.3$, is 0.1 q. If yielding of the material depends on the maximum shearing stress (Tresca's criterion in Chapter 4), the origin is not the critical location. At any point on the z-axis, the maximum shearing stress is obtained by differentiating Equation (a) with respect to z. It is readily confirmed that the expression in Equation (a) becomes maximum when

$$\frac{z}{\sqrt{a^2+z^2}} = \frac{1}{3}\sqrt{2(1+\upsilon)} \tag{b}$$

which gives for z

$$z = \sqrt{\frac{2(1+\upsilon)}{7-2\upsilon}}a \tag{c}$$

Inserting Equations (c) in (a), the maximum shearing stress becomes

$$\tau_{max} = \frac{q}{2}\left[\frac{1-2\upsilon}{2} + \frac{2}{9}(1+\upsilon)\sqrt{2(1+\upsilon)}\right] \tag{d}$$

which gives for Poisson's ratio = 0.3, $z = 0.637a$, and $\tau_{max} = 0.33\ q$.

This indicates that the maximum shearing stress on the z-axis occurs at a depth of about 2/3 of the radius of the loaded circle and the magnitude of the maximum shearing stress is 1/3 of the applied loading.

The solution of this problem was given by Boussenesq [9.5]. See also References 9.7 through 9.9.

9.7.2 Uniform Pressure Distributed over Squares and Rectangles

For a square with sides = 2a subjected to uniform pressure q, the maximum deflection at the center of the square is [5.3]

$$w_{max} = \frac{8}{\pi}\ln\left(\sqrt{2}+1\right)\frac{qa}{E} = 2.24\frac{(1-\upsilon^2)}{E}qa \tag{9.33}$$

The deflection at the corners of the square is half that at the center. The average deflection is

$$w_{aver} = 1.90\frac{(1-\upsilon^2)qa}{E} \tag{9.34}$$

TABLE 9.1

Factors m in Equation 9.35

β	1	1.5	2	3	5	10	100
m	0.95	0.94	0.92	0.88	0.82	0.71	0.37

Similar calculations have been made for uniform pressure q over the surface of rectangles of various ratios, $\beta = a/b$. All the results including the circle can be put in the form [9.9]

$$w_{aver} = m\frac{(1-\upsilon^2)P}{E\sqrt{A}} \tag{9.35}$$

where m is a numerical factor depending on β, A is the magnitude of the area, and P is the total load, $P = qA$. For a circle, $m = 0.96$.

The values of m for rectangles are given in Table 9.1.

It is clear that for a given load P and area A, the deflections increase when the ratio of the perimeter of the loaded area to the area decreases.

Equation 9.35 is useful in discussing deflections of foundations of engineering structures. In order to maintain equal deflections of various portions of the structure, there must be certain relation between the shape and the magnitude of the loaded area.

9.8 Contact Problems in Three Dimensions

A rigid punch in the form of a solid of revolution is pressed against the boundary of an elastic semi-infinite solid. Neglecting friction between the solid and the punch, the punch will fit the elastic medium over the area between its apex and a circular area of radius a, as shown in Figure 9.9. It is required to determine the induced stresses and displacements in the solid.

The solution of a rigid circular cylindrical punch of radius a was first given by Boussenesq [9.5], other methods of solution are described in References 9.10 and 9.11. The vertical displacement w is constant over a circular base of radius a, and is given by

$$w = \frac{(1-\upsilon^2)P}{2aE} \tag{9.36}$$

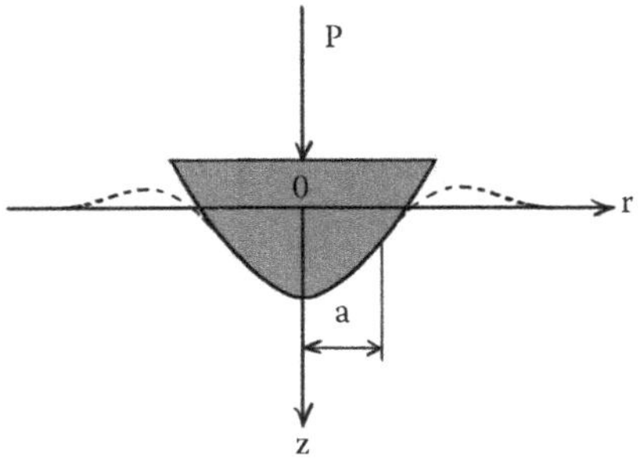

FIGURE 9.9
A rigid punch in contact with a semi-infinite elastic solid.

where P is the total load on the punch, $P = \pi a^2 q$, q is the pressure distribution beneath the punch given by

$$q = \frac{P}{2\pi a\sqrt{a^2 - r^2}} \tag{9.37}$$

The minimum pressure occurs at $r = 0$

$$q_{min} = \frac{P}{2\pi a^2} \tag{a}$$

which is half the average pressure, $q_{aver} = P/(\pi a^2)$.

At $r = a$, the pressure is infinite. In practical cases, however, the material will yield at the boundary of the circle below the punch and any yielding is localized to small distances from the boundary and does not affect the distribution of the pressure at points some distance from the base of the cylinder.

It is interesting to compare the magnitude of w in Equation 9.36 to the average value of w in Equation 9.31a, that is, to the integral

$$w_{av} = \frac{1}{\pi a^2}\int_0^a w 2\pi r\, dr \tag{b}$$

It turns out that they are approximately equal (see Problem P9.3).

9.8.1 Solution Depending on One Harmonic Function

For an elastic semi-infinite body subjected to either load or displacement at the boundary, which is free from shear stresses, the induced stresses and displacements depend on finding one harmonic function.

In rectangular coordinates, the equilibrium equations in term of displacements are

$$\begin{aligned}(1 - 2\upsilon)\nabla^2 u_x + \frac{\partial e}{\partial x} &= 0\\ (1 - 2\upsilon)\nabla^2 u_y + \frac{\partial e}{\partial y} &= 0\\ (1 - 2\upsilon)\nabla^2 u_z + \frac{\partial e}{\partial z} &= 0\end{aligned} \tag{9.38}$$

where ∇^2 is the usual Laplacian operator in rectangular coordinates and e is the dilatation.

The solution for a semi-infinite solid in terms of one harmonic function $\phi(x, y, z)$ is

$$\begin{aligned}2Gu_x &= (1 - 2\upsilon)\frac{\partial\phi}{\partial x} + z\frac{\partial^2\phi}{\partial x\partial z}\\ 2Gu_y &= (1 - 2\upsilon)\frac{\partial\phi}{\partial y} + z\frac{\partial^2\phi}{\partial y\partial z}\\ 2Gu_z &= -2(1 - \upsilon)\frac{\partial\phi}{\partial z} + z\frac{\partial^2\phi}{\partial z^2}\end{aligned} \tag{9.39a}$$

and the corresponding stresses related to the semi-infinite body are

$$\sigma_z = -\frac{\partial^2\phi}{\partial z^2} + z\frac{\partial^3\phi}{\partial z^3}$$
$$\tau_{zx} = z\frac{\partial^3\phi}{\partial x\partial z^2} \qquad (9.39b)$$
$$\tau_{zy} = z\frac{\partial^3\phi}{\partial y\partial z^2}$$

Similarly, in terms of cylindrical polar coordinates, r, θ, z, and denoting the displacements by u, v, and w, the displacements in terms of one harmonic function $\phi(r, \theta, z)$ are

$$2Gu = (1 - 2\upsilon)\frac{\partial\phi}{\partial r} + z\frac{\partial^2\phi}{\partial r\partial z}$$
$$2Gv = (1 - 2\upsilon)\frac{\partial\phi}{r\partial\theta} + \frac{z\partial^2\phi}{r\partial\theta\partial z} \qquad (9.40a)$$
$$2Gw = -2(1 - 2\upsilon)\frac{\partial\phi}{\partial z} + z\frac{\partial^2\phi}{\partial z^2}$$

and the corresponding stresses related to the $z = 0$ plane of the solid are

$$\sigma_z = -\frac{\partial^2\phi}{\partial z^2} + z\frac{\partial^3\phi}{\partial z^3}$$
$$\tau_{zr} = z\frac{\partial^3\phi}{\partial r\partial z^2} \qquad (9.40b)$$
$$\tau_{z\theta} = \frac{z}{r}\frac{\partial^3\phi}{\partial\theta\partial z^2}$$

The other stress components can be obtained in a straightforward manner. It is clear that the boundary of the body $z = 0$ is free from shear stresses and the normal displacements and stresses are given by

$$u_z = w = -\frac{2(1 - \upsilon^2)}{E}\frac{\partial\phi}{\partial z}$$
$$\sigma_z = -\frac{\partial^2\phi}{\partial z^2} \qquad (9.41)$$

Thus, once we find the harmonic function, which satisfies the prescribed boundary conditions the induced stresses and strains in the body are routinely determined. One example to illustrate the procedure is described below.

EXAMPLE 9.3

A rigid flat-ended cylinder of radius a is pressed a distance c against the horizontal boundary of an elastic semi-infinite body. Find the induced stresses and strains in the body, in particular, the magnitudes of the displacement and pressure below the cylinder.

Solution

Using the solution derived in Equations 9.40 and 9.41, the boundary conditions at $z = 0$ needed to determine the harmonic function $\phi(r, z)$ are

$$\begin{aligned} \frac{\partial \phi}{\partial z} &= -\frac{Ec}{2(1-\upsilon^2)}, \quad 0 < r \leq a \\ \frac{\partial^2 \phi}{\partial z^2} &= 0, \quad r > a \end{aligned} \tag{a}$$

The following function satisfies Laplace's equation in three dimensions:

$$\frac{\partial \phi}{\partial z} = \int_0^\infty \frac{A(s)}{s} e^{-sz} J_0(rs) ds, \quad z \geq 0 \tag{b}$$

where $J_0(rs)$ is Bessel's function of order zero and $A(s)$ is an arbitrary function of s such that the integral (b) exists and tends to zero as r approaches infinity through positive values of z. Application of boundary conditions (a) leads to the following dual integral equations:

$$\begin{aligned} &\int_0^\infty s^{-1} A(s) J_0(rs) ds = -\frac{Ec}{2(1-\upsilon^2)}, \quad 0 < r \leq a \\ &\int_0^\infty A(s) J_0(rs) ds = 0, \quad r > a \end{aligned} \tag{c}$$

A proper solution to Equations (c) is available by setting

$$A(s) = K \sin(as), \quad K = \text{constant} \tag{d}$$

Inserting expression (d) into the second integral in Equation (c) and making use of the result [9.12]

$$\int_0^\infty J_0(rs) \sin(as) ds = \begin{cases} (a^2 - r^2)^{-1/2} & 0 < r \leq a \\ 0 & r > a \end{cases}$$

It is readily confirmed that the second boundary condition is satisfied.

Inserting expression (d) into the first integral in Equation (c) and making use of the result [9.12]

$$\int_0^\infty s^{-1} J_0(rs) \sin(as) ds = \frac{\pi}{2}, \quad 0 \leq r < a$$

the first boundary condition in Equation (c) gives

$$K = -\frac{Ec}{(1-\upsilon^2)\pi} \tag{e}$$

Hence, the stress function in Equation (b) is determined and the induced stresses and strains in the body are routinely determined.

In particular, the normal stress beneath the cylinder is

$$\sigma_z = -\frac{\partial^2\phi}{\partial z^2} = \int_0^\infty K\sin(as)J_0(rs)ds = -\frac{Ec}{(1-\upsilon^2)\pi\sqrt{a^2-r^2}} \tag{f}$$

The total force exerted by the cylinder on the body is

$$P = -2\pi\int_0^a (\sigma_z)_{z=0} r dr = \frac{2Eca}{1-\upsilon^2} \tag{h}$$

which is identical with the result in Equation 9.36. Also, by combining Equations (f) and (h), the result in Equations 9.37 is recovered.

9.8.2 General Solution of 3D Contact Problems

Consider the pressure of a rigid punch on a semi-infinite elastic body when there is no friction between the punch and the body. A given load p(x, y) acts on the boundary $z = 0$ of the half-space. The general solution for the induced stresses and strains in the body consists of representing the displacements u, v, and w by four harmonic functions first introduced by Papkovitch and Neuber [9.13,9.14]

$$\begin{aligned} u &= \phi_1 - \frac{1}{4(1-\upsilon)}\frac{\partial}{\partial x}(\phi_0 + x\phi_1 + y\phi_2 + z\phi_3) \\ v &= \phi_2 - \frac{1}{4(1-\upsilon)}\frac{\partial}{\partial y}(\phi_0 + x\phi_1 + y\phi_2 + z\phi_3) \\ w &= \phi_3 - \frac{1}{4(1-\upsilon)}\frac{\partial}{\partial z}(\phi_0 + x\phi_1 + y\phi_2 + z\phi_3) \end{aligned} \tag{9.42}$$

On the boundary of the body, $z = 0$, we have the conditions

$$\tau_{xz} = 0, \quad \tau_{yz} = 0 \tag{a}$$

Using Equations 9.42, the following equations are reached:

$$\begin{aligned} \frac{\partial\phi_3}{\partial x} + \frac{2(1-\upsilon)}{1-2\upsilon}\frac{\partial\phi_1}{\partial z} - \frac{1}{1-2\upsilon}\frac{\partial}{\partial x}\left(x\frac{\partial\phi_1}{\partial z} + y\frac{\partial\phi_2}{\partial z} + \frac{\partial\phi_0}{\partial z}\right) &= 0 \\ \frac{\partial\phi_3}{\partial y} + \frac{2(1-\upsilon)}{1-2\upsilon}\frac{\partial\phi_2}{\partial z} - \frac{1}{1-2\upsilon}\frac{\partial}{\partial y}\left(x\frac{\partial\phi_1}{\partial z} + y\frac{\partial\phi_2}{\partial z} + \frac{\partial\phi_0}{\partial z}\right) &= 0 \end{aligned} \tag{b}$$

Equations (b) are satisfied if we assume the following relations hold when $z = 0$:

$$\begin{aligned} \frac{\partial\phi_1}{\partial z} &= -\frac{1-2\upsilon}{2(1-\upsilon)}\frac{\partial\phi_3}{\partial x} \\ \frac{\partial\phi_2}{\partial z} &= -\frac{1-2\upsilon}{2(1-\upsilon)}\frac{\partial\phi_3}{\partial y} \\ \frac{\partial\phi_0}{\partial z} &= \frac{1-2\upsilon}{2(1-\upsilon)}\left(x\frac{\partial\phi_3}{\partial x} + y\frac{\partial\phi_3}{\partial y}\right) \end{aligned} \tag{c}$$

Next, we make use of the fact that "two harmonic functions vanish at infinity, satisfy certain regularity conditions, and coincide on the boundary of a region, then they are identical throughout the region." In this way, the following relations hold throughout the half-space:

$$\begin{aligned}
\frac{\partial \phi_0}{\partial z} &= \frac{1-2\upsilon}{2(1-\upsilon)}\left(x\frac{\partial \phi_3}{\partial x} + y\frac{\partial \phi_3}{\partial y} + z\frac{\partial \phi}{\partial z}\right) \\
\frac{\partial \phi_1}{\partial z} &= -\frac{1-2\upsilon}{2(1-\upsilon)}\frac{\partial \phi_3}{\partial x} \\
\frac{\partial \phi_2}{\partial z} &= -\frac{1-2\upsilon}{2(1-\upsilon)}\frac{\partial \phi_3}{\partial y}
\end{aligned} \tag{9.43}$$

Introducing a new function

$$\psi = \frac{3-4\upsilon}{(1-\upsilon^2)}\phi_3 \tag{d}$$

It follows that

$$\begin{aligned}
\phi_1 &= \frac{4(1+\upsilon)(1-2\upsilon)}{3-4\upsilon}\int_z^\infty \frac{\partial \psi}{\partial x}dz \\
\phi_2 &= \frac{4(1+\upsilon)(1-2\upsilon)}{3-4\upsilon}\int_z^\infty \frac{\partial \psi}{\partial y}dz \\
\phi_0 &= -\frac{4(1+\upsilon)(1-2\upsilon)}{3-4\upsilon}\int_z^\infty \left(x\frac{\partial \psi}{\partial x} + y\frac{\partial \psi}{\partial y} + z\frac{\partial \psi}{\partial z}\right)dz
\end{aligned} \tag{9.44}$$

Inserting these expressions into Equations 9.42, using the appropriate rules for differentiation of definite integrals containing variable limits [9.2], the components of displacements become

$$\begin{aligned}
u &= (1-2\upsilon)\int_z^\infty \frac{\partial \psi}{\partial x}dz - z\frac{\partial \psi}{\partial x} \\
v &= (1-2\upsilon)\int_z^\infty \frac{\partial \psi}{\partial y}dz - z\frac{\partial \psi}{\partial y} \\
w &= 2(1-\upsilon)\psi - z\frac{\partial \psi}{\partial z}
\end{aligned} \tag{9.45a}$$

The corresponding stresses on the z-plane are

$$\tau_{xz} = -2Gz\frac{\partial^2 \psi}{\partial x \partial z}$$
$$\tau_{yz} = -2Gz\frac{\partial^2 \psi}{\partial y \partial z} \tag{9.45b}$$
$$\sigma_z = 2G\left(\frac{\partial \psi}{\partial z} - z\frac{\partial^2 \psi}{\partial z^2}\right)$$

The other components of stress can be routinely determined.

Assume that the normal pressure p(x, y) is distributed over a given area of the $z = 0$ plane, the boundary conditions are

$$\begin{aligned} \sigma_z &= -p(x, y), (x, y) \text{ within area } \Omega \\ \sigma_z &= 0, (x, y) \text{ outside area } \Omega \end{aligned} \tag{e}$$

In terms of the new function, condition (e) becomes

$$\begin{aligned} \frac{\partial \psi}{\partial z} &= -\frac{1+\upsilon}{E}p(x, y), (x, y) \text{ within area } \Omega \\ \frac{\partial \psi}{\partial z} &= 0, (x, y) \text{ outside area } \Omega \end{aligned} \tag{f}$$

From potential theory such a function can be represented by the integral

$$\psi(x, y) = -\frac{1+\upsilon}{E}\iint_{\Omega} \frac{p(\xi, \eta)\, d\xi\, d\eta}{\left[(x-\xi)^2 + (y-\eta)^2 + z^2\right]^{1/2}} \tag{9.46}$$

And the normal displacement assumes the form

$$w(x, y, 0) = -\frac{1-\upsilon^2}{E}\iint_{\Omega} \frac{p(\xi, \eta)\, d\xi\, d\eta}{\left[(x-\xi)^2 + (y-\eta)^2\right]^{1/2}} \tag{9.47}$$

Once the function defined in Equation 9.46 is found, the stresses and displacement in the body are obtained.

Reference 9.12 provides solutions of Laplace's equation, that is, harmonic functions, in a number of systems of curvilinear coordinates to solve problems of a rigid punch acting on the boundary of semi-infinite elastic solids. The coordinates include ellipsoidal, spheroidal, and spherical. Making use of such harmonic functions, the induced displacements and stresses caused by the action of rigid punches of elliptical cross section and punches which are solids of revolution are determined. Axisymmetric contact problems in which there is friction between the punch and the elastic solid are also discussed. The problem of an elliptical punch acting on the boundary of a half-space and its application in the design of buildings is also discussed in Reference 9.15.

The following general result has been established [9.11]. A rigid punch of elliptical cross section with semiaxes a and b is acting on the surface of a semi-infinite elastic body $z < 0$, there are no frictional forces between the punch and the body. Outside the area of contact between the punch and the body, the surface $z = 0$ is free from tractions.

If the surface bounding the base of the punch is described by the equation

$$z = P_n(x, y), \quad \text{where } P_n(x, y) \text{ is a polynominal of degree } n \tag{g}$$

then the pressured beneath the punch can be represented in the form

$$p(x, y) = \left(1 - \frac{x^2}{a^2} - \frac{y^2}{b^2}\right)^{-1/2} P_n^*(x, y) \tag{9.48}$$

where $P_n{}^*(x, y)$ is also a polynomial of degree n. Because the polynomial $P_n(x, y)$ in Equation (g) is initially known, Equation 9.48 provides a set of simultaneous equations, which determine the number of unknown coefficients contained in the polynomial $Pn^*(x, y)$.

EXAMPLE 9.4

Wedge-shaped punch with a flat smooth base acting on a semi-infinite solid.

Solution

Consider a wedge-shaped punch L exerting a pressure on a semi-infinite elastic solid, $z < 0$, as shown in Figure 9.10. The angle of the wedge is 2α. There is no friction between the base of the punch and the solid. When a vertical force P in the z-direction is applied to the punch, the punch is displaced vertically a prescribed amount c. It is assumed that the base of the punch remains parallel to the xy-plane. Using Equation 9.45, the problem of finding the induced displacements and stresses in the solid is reduced to finding a harmonic function, $\phi(x, y, z)$, such that it satisfies the following mixed boundary conditions on $z = 0$:

$$\begin{aligned} &\phi = c, \text{(constant)} : \text{on L} \\ &\frac{\partial \phi}{\partial z} = 0 : \text{outside L} \end{aligned} \tag{a}$$

Here, L is part of the $z = 0$ plane inside the wedge and c is the amount of vertical displacement of the wedge.

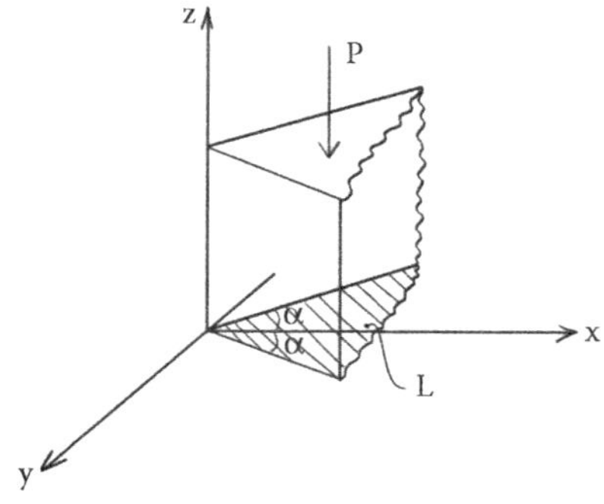

FIGURE 9.10
Wedge-shaped punch with plane base.

The resulting pressure under the punch is given by the following equation:

$$p(x, y) = -(\sigma_z)_{z=0} = \frac{E}{2(1-\upsilon^2)}\left(\frac{\partial\phi}{\partial z}\right)_{z=0} \tag{b}$$

In order to find the harmonic function in Equations (a) and (b), Galin [9.11] introduced spherical coordinates r, θ, θ* defined by the relations

$$x = r\cos\theta, \quad y = r\sin\theta\cos\theta^*, \quad z = r\sin\theta\cos\theta^* \tag{c}$$

In this coordinate system, Laplace equation assumes the form

$$\frac{\partial^2\phi}{\partial r^2} + \frac{2}{r}\frac{\partial\phi}{\partial r} + \frac{1}{r^2\sin\theta}\frac{\partial}{\partial\theta}\left(\sin\theta\frac{\partial\phi}{\partial\theta}\right) + \frac{1}{r^2\sin^2\theta}\frac{\partial^2\phi}{\partial\theta^{*2}} = 0 \tag{d}$$

Since the boundary conditions, Equations (a), are independent of r, the distance from the origin, then $\phi(x, y, z)$ is a spherical harmonic of degree zero and what is needed is a harmonic function containing θ and θ*, $\phi^*(\theta, \theta^*)$. The new function is governed by the equation

$$\sin\theta\frac{\partial}{\partial\theta}\left(\sin\theta\frac{\partial\phi^*}{\partial\theta}\right) + \frac{\partial^2\phi^*}{\partial\theta^{*2}} = 0 \tag{e}$$

$$\text{In terms of new variables: } \xi = \tan(\theta/2)\cos(2\theta^*), \quad \eta = \tan(\theta/2)\sin(2\theta^*) \tag{f}$$

The harmonic function satisfies the equation

$$\frac{\partial^2\psi}{\partial\xi^2} + \frac{\partial^2\psi}{\partial\eta^2} = 0 \tag{g}$$

Note that in the transformation (f), the sphere is transformed to the whole plane of the complex variable $\zeta = \xi + i\eta = e^{i2\theta}\tan(\theta/2)$.

While corresponding to the curve of intersection of the plane wedge L with the sphere, there is a right slit from the point $\xi = -\tan(\theta/2)$, $\eta = 0$ to the point $\xi = \tan(\theta/2)$, $\eta = 0$.

Galin [9.11] developed the solution and obtained the following formula for pressure under the rigid punch:

$$p(r, \theta) = k\frac{E}{2(1-\upsilon^2)}\left[\frac{1}{r(1+\cos\theta)}\right]\left[\tan^2(\alpha/2) - \tan^2(\theta/2)\right]^{-1/2} \tag{h}$$

where the angle θ is measured from the bisector of the angle at the vertex of the wedge and the factor k can be determined if the total pressure on a specified sector of the wedge is known.

Figure 9.11a and b shows the distribution of the pressure beneath two punches with wedge angles $2\alpha = 90°$ and $2\alpha = 45°$.

As α approaches zero, the pressure distribution in Equation (h) along the line normal to the bisector tends in the limit to the value found by solving the corresponding 2D problem. This can be confirmed by introducing the variable $s = \tan\theta$, so that along the straight line normal to the bisector at a distance H from the vertex of the wedge, $a = \tan\alpha$, $-a < s < a$, and Equation (h) gives in the limit

$$p_{\text{limit}} = k\frac{E(a^2 - s^2)^{-1/2}}{2(1-\upsilon^2)} \tag{k}$$

which is the contact pressure for the corresponding plane problem [9.11].

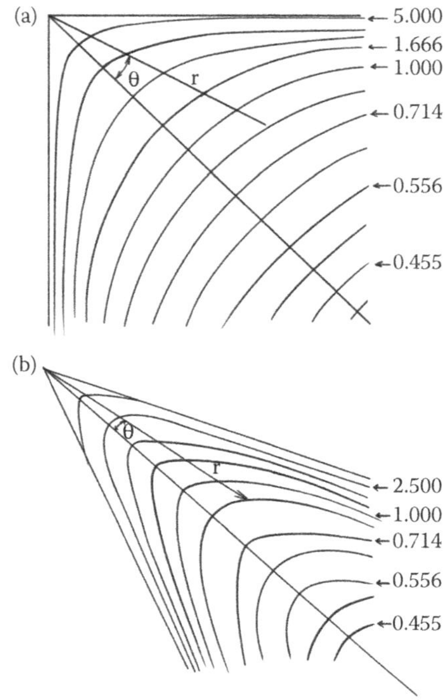

FIGURE 9.11
(a) Pressure distribution beneath a punch with angle $2\alpha = 90^\circ$, (b) pressure distribution beneath a punch with angle $2x = 45^\circ$.

9.8.3 Contact Problems for Solids Containing Cavities

A rigid, frictionless, cylindrical punch of radius a is pressed, under the actions of arbitrary forces and moments, against the surface of a body in the form of a half-space that contains concentric cylindrical hole of radius b, $b < a$, as shown in Figure 9.12. The contact area between the punch and the half-space is assumed to remain circular. It is required to determine the state of stress and deformation set up in the body. In particular, it is of interest to investigate the effects of the presence of the cavity on the distribution of stress, the deformed boundary of the body outside the punch, and the radial displacement along the cavity surface.

Bandyopadhayay and Kassir [9.16] provide a general formulation of the problem to account for nonaxisymmetric contact conditions. The formulation simplifies considerable in case of axially symmetric indentation in which the applied moments = 0.

Referring to the system of cylindrical coordinates shown in Figure 9.12 and denoting the displacements by (u, v, w) and the corresponding stresses by (σ_r, σ_θ, σ_z, τ_{rz}, $\tau_{r\theta}$, $\tau_{\theta z}$), the boundary conditions for the solid containing a stress-free hole can be expressed as follows:

$$
\begin{aligned}
&w(r,\theta) = \sum_{n=0}^{\infty} d_n(r)\cos n\theta; \quad z = 0,\ b \le r \le |\theta| \le \pi \\
&\sigma_z = 0; z = 0, r > a, |\theta| \le \pi \\
&\tau_{rz} = \tau_{\theta z} = 0; z = 0, r \ge b, |\theta| \le \pi \\
&\sigma_r = \tau_{rz} = \tau_{r\theta} = 0; z \ge 0, r = b, |\theta| \le \pi
\end{aligned}
\qquad (9.49)
$$

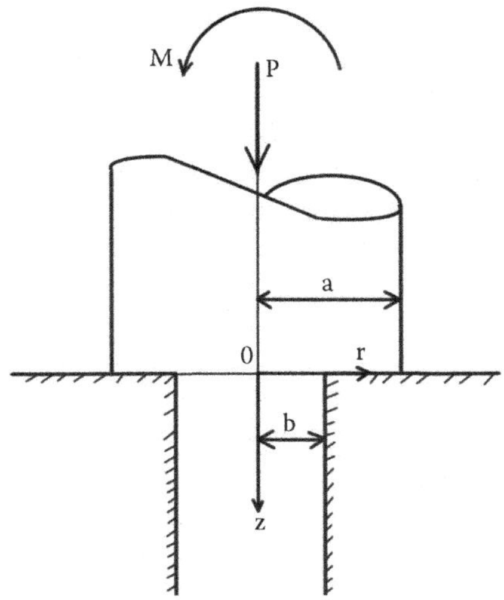

FIGURE 9.12
Rigid punch in contact with solid containing cylindrical cavity.

And all components of displacement and stress tend to zero as r and z tend to infinity.

The Fourier coefficients d_n, $n = 0, 1, 2, \ldots$ in the first of Equations 9.49 are related to the specified normal displacement by

$$d_0(r) = \frac{1}{\pi}\int_0^{\infty} w(r,\theta)d\theta$$
$$d_n(r) = \frac{2}{\pi}\int_0^{\infty} w(r,\theta)\cos(n\theta)d\theta, \quad n = 1, 2, \ldots \tag{9.50}$$

It is assumed that the normal displacement is generated in the contact region in an even function of the variable θ. A procedure similar to the one described here may be employed for odd variation in θ.

The normal force and moments required to accomplish the indentation are related to the contact stress by way of the relations

$$P = \int_0^a\int_0^{2\pi} \sigma_z(r,\theta,0) r\, dr\, d\theta$$
$$M_x = \int_b^a\int_0^{2\pi} y\sigma_z(r,\theta,0) r\, dr\, d\theta \tag{9.51}$$
$$M_y = \int_b^a\int_0^{2\pi} x\sigma_z(r,\theta,0) r\, dr\, d\theta$$

Note that by setting $n = 0$ in Equations 9.49 through 9.51 indicate that the solid is being pressed by means of axisymmetric forces and the resulting stress and deformation is independent of θ. In this case, the displacements (u, 0, w) and the nonvanishing stresses (σ_r, σ_θ, σ_z, τ_{rz}) are all functions of r and z only. When $n = 1,2, \ldots$, the resulting state of stress and deformation are functions of the three original coordinates and results in a cylindrical punch with inclined base under the action of moments.

In order to obtain the general solution, the displacement field at any point of the solid is represented in two parts:

1. One part is associated with a semi-infinite solid whose boundary is free from shear stresses. In this case, the displacements and stresses are represented by one harmonic function as indicated in Equations 9.39a and 9.39b.
2. The second part is a general solution of the equations of elasticity in which the displacements are represented by four harmonic functions, the well-known Papkovitch–Neuber potentials discussed in Equations 9.42.

Detailed derivations of the final solution as well as many pertinent results are available in Reference 9.16.

Here, we discuss the axisymmetric case in detail.

Axially symmetric case: When a rigid cylindrical punch is pressed against the half-space with axisymmetric forces P, $M_x = M_y = 0$, the resulting state of stress and deformation is independent of the variable θ.

The solution corresponding to part (i) from Equations 9.39a is

$$
\begin{aligned}
2Gu(r, z) &= (1 - 2\upsilon)\frac{\partial f}{\partial r} + z\frac{\partial^2 f}{\partial r \partial z} \\
2Gw(r, z) &= -2(1 - \upsilon)\frac{\partial f}{\partial z} + z\frac{\partial^2 f}{\partial z^2}
\end{aligned}
\tag{9.52a}
$$

in which f (r, z) is a space harmonic function given by

$$
f(r, z) = \int_0^\infty \frac{F(s)}{s} J_0(rs) e^{-sz}\, ds \tag{9.52b}
$$

where $J_0(rs)$ is a Bessel function of the first kind of order zero.

The Papkovitch–Neuber solution corresponding to part (ii) from Equations 9.42 is

$$
\begin{aligned}
2Gu(r, z) &= 4(1 - \upsilon)g_1 - \frac{\partial}{\partial r}(rg_1 - g_3) \\
2Gw(r, z) &= -\frac{\partial}{\partial z}(rg_1 + g_3)
\end{aligned}
\tag{9.53a}
$$

in which appropriate expressions for the potentials, g_n (r, z), $n = 1, 3$, are

$$
g_1 = \int_0^\infty A(s) K_1(rs) \cos(sz)\, ds
$$

$$g_2 = \int_0^\infty \frac{B(s)}{s} K_0(rs)\cos(sz)\,ds \qquad (9.53b)$$

Here, $K_n(rs)$, $n = 0, 1$ are modified Bessel functions of the second kind of order n.

In Equations 9.52b and 9.53b, F(s), A(s), and B(s) are integral transform parameters to be determined from the boundary conditions in Equations 9.49. Note that in the general solution a third potential function appears [9.16].

Combining the two-part solution (i) and (ii), we find

$$2Gu = \int_0^\infty \{(1 - 2\upsilon - sz)F(s)J'_0(rs)e^{-sz} + [(3 - 4\upsilon)K_1(rs) - rsK'_1]A(s) + K'_0(rs)B(s)\cos(sz)\}\,ds \qquad (9.54a)$$

$$2Gw = \int_0^\infty \{(2 - 2\upsilon + sz)F(s)J_0(rs)e^{-sz} + [K_0(rs)A(s) + rsK_1(rs)B(s)\sin(sz)]\}\,ds \qquad (9.54b)$$

$$\sigma_z = \int_0^\infty \{-(1 + sz)F(s)J_0(rs)e^{-sz} + [K_0(rs)A(s) + < rsK_1(rs) - 2\upsilon K_0(rs) > B(s)]\cos(sz)\}s\,ds \qquad (9.54c)$$

$$\sigma_r = \int_0^\infty \Bigg\{[(sz - 1)J_0(rs) + \frac{1}{rs}(1 - 2\upsilon - sz)J_1(rs)]F(s)e^{-sz}$$
$$- \frac{1}{rs}[< K_1(rs) + rsK_0(rs) > A(s) + < (4 - 4\upsilon) + r^2s^2 > K_1(rs)$$
$$+ < (3 - 2\upsilon)rsK_0(rs) > B(s)]\cos(sz)\Bigg\}s\,ds \qquad (9.54d)$$

$$\tau_{rz} = -\int_0^\infty \{szJ_1(rs)F(s)e^{-sz} + [K_1(rs)A(s) + < 2(1 - \upsilon)K_1(rs) + rsK_0(rs)B(s) >]\sin(sz)\}s\,ds \qquad (9.54e)$$

It is clear that $\tau_{rz} = 0$ at $z = 0$. The integral transform parameters F(s), A(s), and B(s) are determined by enforcing the boundary conditions stated in Equation 9.49. It turns out that F(s) is determined by way of two auxiliary functions $\phi(t)$ and $\psi(t)$ as follows:

$$F(s) = \frac{2}{\pi}\left[\int_0^\infty \phi(t)\cos(st)dt + \int_0^\infty \psi(t)\cos(st)dt\right] \qquad (9.55)$$

in which the auxiliary functions $\phi(t)$ and $\psi(t)$ are

$$\phi(t) = \frac{G}{1 - \upsilon}\frac{d}{dt}\int_b^r \frac{rd(r)dr}{\sqrt{t^2 - r^2}} \qquad (9.56a)$$

$$\psi(t) = \int_t^{\infty} \frac{rd^*(r)dr}{\sqrt{r^2 - t^2}}$$
$$d^*(r) = \int_0^{\infty} [K_0(rs)A(s) + < -2\upsilon K_0(rs) + rzK_1(rs) > B(s)]s\,ds \tag{9.56b}$$

Introducing the abbreviations

$$G(x) = x^2K_0^2(x) - (2 - 2\upsilon + x^2)K_1^2(x)$$
$$H(x) = x^2I_0(x)K_0(x) + (2 - 2\upsilon + x^2)I_1(x)K_1(x) \tag{9.57a}$$

and making use of the identities

$$K_{-(1/2)}(x) = K_{(1/2)}(x) = \left(\frac{\pi}{2x}\right)^{1/2} e^{-x}$$
$$K_{-(3/2)}(x) = K_{(3/2)}(x) = \left(\frac{\pi}{2x}\right)^{1/2} \left(1 + \frac{1}{x}\right) e^{-x} \tag{9.57b}$$

The values of A(s) and B(s) are determined in terms of the functions $\phi(t)$, $\psi(t)$, $G(x)$, and $H(x)$. They are

$$A(s) = -\frac{2}{\pi G(bs)} \left\{ \int_b^a [(3 - 2\upsilon)H(bs) + 4 - 4\upsilon + b^2x^2 - (2 - 2\upsilon) + H(bs)st] \right.$$
$$\left. \times\, \phi(t)e^{-st}dt + \int_a^{\infty} [(3 - 2\upsilon)H(bs) + 4 - 4\upsilon + b^2x^2 - (2 - 2\upsilon) + H(bs)]\psi(t)e^{-st}\,dt \right\} \tag{9.58a}$$

$$B(s) = \frac{2}{\pi G(bs)} \left\{ \int_b^a [1 + H(bs) - st]\phi(t)e^{-st}dt + \int_b^a [1 + H(bs) - st]\psi(t)e^{-st}dt \right\} \tag{9.58b}$$

Making use of Equations 9.56 through 9.58, it is found that the function $\psi(t)$ is governed by the equation

$$\psi(t) + \int_a^{\infty} L(\eta, t)\phi(\eta)\,d\eta = -\int_b^a \phi(\eta)L(\eta, t)\,d\eta \tag{9.59}$$

where

$$L(\eta, t) = \int_0^{\infty} \{3 - 2\upsilon + 2H(bs) - (st + s\eta)[1 + H(bs)] + s^2(b^2 + \eta t)\} \frac{e^{-(t+\eta)s}}{G(bs)}\,ds$$

Similarly, the shape of the deformed boundary of the body is found as

$$w(r,0) = \frac{2(1-\upsilon)}{\pi G}\left[\int_b^a \frac{\phi(t)dt}{\sqrt{r^2-t^2}} + \int_a^r \frac{\psi(t)dt}{\sqrt{r^2-t^2}}\right], \quad r > a \tag{9.60}$$

And the contact stress beneath the punch is computed from

$$\sigma_z(r,0) = \frac{2}{\pi}\left\{\frac{\psi(a)-\phi(a)}{\sqrt{a^2-r^2}} + \int_r^a \frac{\phi'(t)dt}{\sqrt{t^2-r^2}} + \int_a^\infty \frac{\psi'(t)dt}{\sqrt{t^2-r^2}}\right.$$

$$\left. + \frac{\pi}{2}\int_a^\infty [sK_0(rs)A(s) + < 2\upsilon sK'_1(rs) + \frac{1}{r}(2\upsilon + r^2s^2)K_1(rs) > B(s)]ds\right\} \tag{9.61}$$

$$b \leq r \leq a$$

For purposes of illustrating the use of the above formulas, consider the following example.

EXAMPLE 9.5

A flat-ended punch penetrating the solid with cylindrical hole to a specified constant depth $= d_0$. Find the response of the solid to axisymmetric loads.

Solution

With $d(r) = d_0$, Equation 9.56a gives

$$\phi(t) = \frac{Gd_0t}{(1-\upsilon)\sqrt{t^2-b^2}} \tag{9.62}$$

and Equation 9.60 leads to

$$w(r,0) = \frac{2}{\pi}\left[d_0\sin^{-1}\left(\frac{a^2-b^2}{r^2-b^2}\right)^{1/2} + \frac{1-\upsilon}{G}\int_a^r \frac{\psi(t)dt}{\sqrt{r^2-t^2}}\right], \quad r > a \tag{9.63}$$

The expression for $\phi(t)$ in Equation 9.62 is inserted in Equation 9.59 and the latter equation is solved numerically to yield the function ψ for values of $(b/a) = ½$ and $\nu = 0.3$. Making use of the values of ϕ and ψ,the normal displacement outside the punch Equation 9.63 and the contact stress Equation 9.61 are computed and displayed graphically in Figures 9.13 and 9.14, respectively.

For a solid without hole, that is, b approaches zero, it follows from the asymptotic expansions of Bessel functions that $K_0(bs) = \log(bs)$, $K_1(bs) = (1/bs)$, $I_0(bs) = 1$, $I_1(bs) = (bs/2)$, and Equation 9.57a indicates that G(bs) tends to infinity while H(bs) becomes finite. Accordingly, it follows from Equations 9.58 that A(s) and B(s) vanish and the solution reduces to the classical one, as shown in Example 9.3 and Reference 9.18. For the undisturbed solid, the variation of w/d_0 with r/b (where b now stands for half the radius of the punch) is shown in Figure 9.13.

In order to determine the influence of the cavity on the state of stress in the solid, the component, $\sigma_z(r, z)$, is computed. For this purpose, the expressions in Equations 9.55,

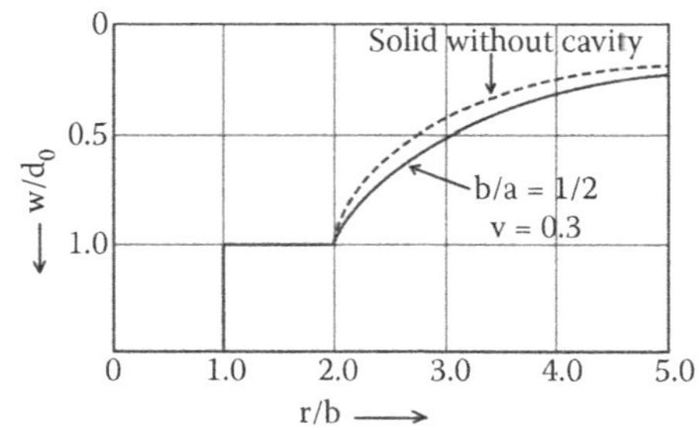

FIGURE 9.13
Normal displacement outside the punch (z = 0).

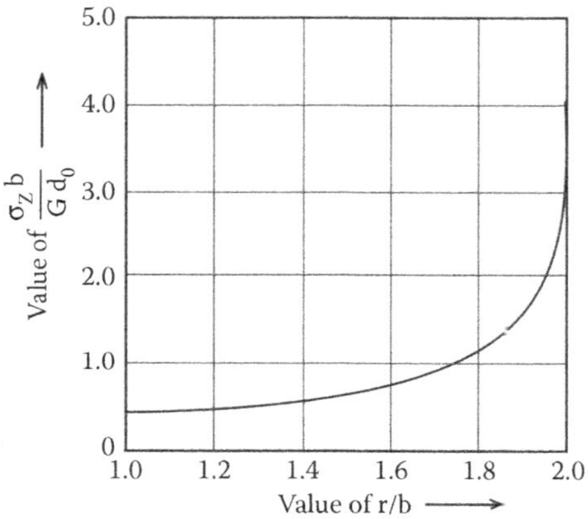

FIGURE 9.14
Contact stress between punch and solid.

9.56, and 9.58 are inserted in Equation 9.54c, and after some simplification, the later equation becomes

$$\sigma_z(r,z) = -\frac{4G}{\pi}\left\{\int_b^a \phi(t)[K_a(r,z,t) + K_b(r,z,t) + K_c(r,z,t)]\,dt + \int_a^\infty \psi(t)[K_a(r,z,t) + K_b(r,z,t) + K_c(r,z,t)]\,dt\right\} \tag{9.64}$$

where the following abbreviations have been adopted:

$$K_a(r,z,t) = \int_0^\infty s(1+sz)J_0(rs)e^{-sz}\cos(st)ds$$

$$K_b(r,z,t) = \int_0^\infty \frac{s}{G(bs)}\{[3H(bs) + 4 - 2\upsilon + b^2s^2]K_0(rs) - bs[1 + H(bs)]K_1(rs)\}$$
$$\times \cos(sz)e^{-st}dsK_c(r,z,t) \tag{9.65}$$

$$K_c(r,z,t) = -\int_0^\infty \frac{s^2}{G(bs)}\{[2 + H(bs)]K_0(rs) - bsK_1(rs)\}\cos(sz)e^{-st}ds$$

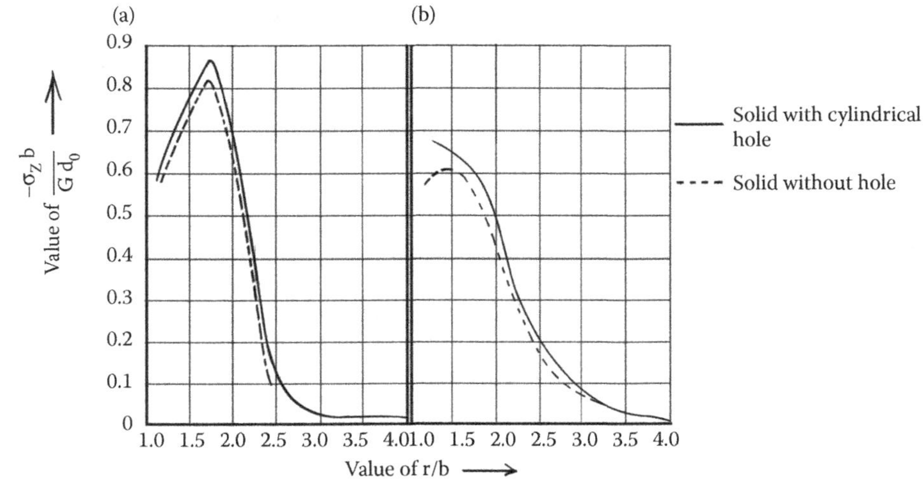

FIGURE 9.15
Stress component σ_Z (r,1/2), z = 1/2 (ν = 0.3, b/a = 1/2), σ_Z (r,1), z = 1.

Upper limits of the integrals in Equations 9.65 are established for different values of the variables r, z, and t, by employing Filon's modification of Simson rule [9.19]. Note that both integrals K_b and K_c are slowly convergent as r/b and t/b approach 1. The resulting values of these integrals are inserted into Equation 9.64 to compute the normal stress at depths z = ½ and z = 1. The results are shown in Figure 9.15. In order to assess the influence of the cavity on the magnification of stress in the solid, similar calculations are carried out for a solid without a hole and the results are shown dotted in the same figure. Here, again, in case of the solid without cavity, the parameter, b, stands for half the radius of the punch. It is apparent from these figures that the cavity has a moderate influence on the maximum stress in the solid.

Finally, Figure 9.16 shows the variation of the radial displacement u (r, z) in the solid at several locations below the surface. In particular, at depth z = 1/2, the magnitude of u, at the cavity surface, is about 75% of d_0, which is appreciable.

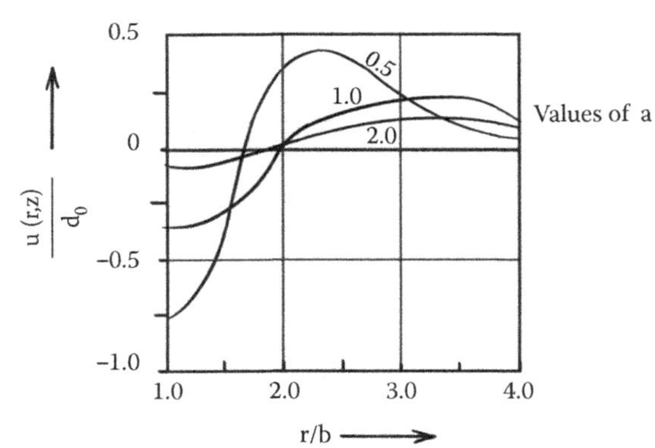

FIGURE 9.16
Varation of radial displacement below surface.

9.8.4 Contact between Two Spherical Bodies: Hertz Problem

The Hertz Problem [9.20] is concerned with determining the induced stresses and deformation when a loaded sphere is in contact with another sphere, a loaded sphere against a plane section, a loaded cylinder on a cylinder, etc. Such problems have many practical applications in structural and mechanical engineering.

Consider two elastic spherical bodies in contact and compressed by axial forces P, as shown in Figure 9.17. The radii of the spheres are R_1 and R_2. We neglect friction and consider smooth bodies, which exert normal pressures only.

When $P = 0$, the contact between the bodies is point 0 at the intersection of the x and y axes in the xy-plane. If the vertical distances from the horizontal tangent plane at 0 of points A and B located at a very small distance x from the y axis are y_1 and y_2 then approximately, $y_1 = x^2/(2R_1)$ and $y_2 = x^2/(2R_2)$. It follows that

$$y_1 + y_2 = \frac{x^2(R_1 + R_2)}{2R_1R_2} \tag{a}$$

If the contact is between a sphere and a horizontal plane, Figure 9.18a, then $R_2 =$ infinity and Equation (a) gives the distance between points A and B as $x^2/(2R_1)$. In case of contact between a sphere and a spherical seat, Figure 9.18b, then R_2 becomes $-R_2$ and the distance $y_1 - y_2$ is obtained from Equation (a).

When the spherical bodies are subjected to compressive axial forces, P, along the normal to the tangent plane at point 0, there will be local deformation over a small surface area with a circular boundary of radius a, called the *surface of contact*, shown shaded in Figure 9.17. The radii R_1 and R_2 are very large compared to a. Also, because of symmetry, the contact pressure q and the corresponding deformation are symmetrical with respect to the center 0 of the surface of contact.

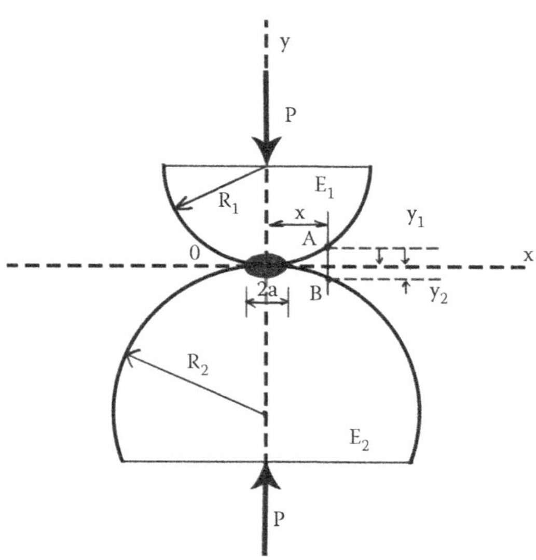

FIGURE 9.17
Two spherical bodies compressed by force P.

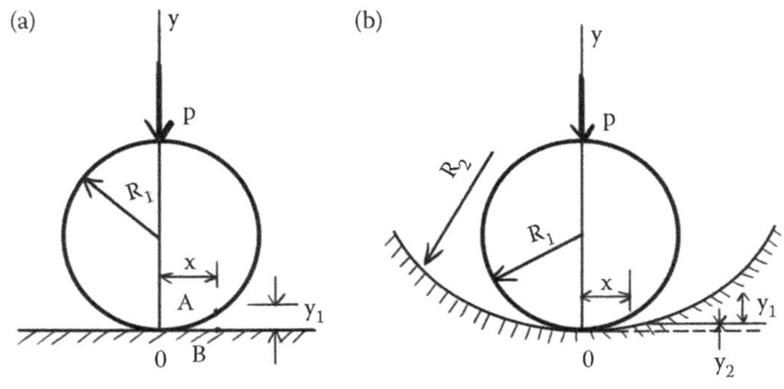

FIGURE 9.18
Contact between a sphere and (a) flat surface and (b) spherical seat.

The procedure and results of Section 9.7.1 can be used to determine the contact stress and deformation in the spherical bodies as outlined in Reference 9.20 (see also Reference 5.3).

It is shown in these references that the maximum pressure, q, at the contact surface is given by

$$q = 1.5\frac{P}{\pi a^2} \tag{9.66}$$

which is 1.5 times the average pressure over the surface of contact area. Note that the stress in Equation 9.66 represents the maximum principal stress at that point.

The corresponding radius, a, of the surface of contact is given by

$$a = 0.909\left[P\frac{(1-\upsilon_1^2)E_1 + (1-\upsilon_2^2)E_2}{E_1E_2}\frac{R_1R_2}{R_1+R_2}\right]^{1/3} \tag{9.67a}$$

The relative displacement of the centers of the two spherical bodies, d, due to deformation takes the form

$$d = 0.826\left[P^2\frac{(1-\upsilon_1^2)E_1 + (1-\upsilon_2^2)E_2}{(E_1E_2)^2}\frac{R_1+R_2}{R_1R_2}\right]^{1/3} \tag{9.67b}$$

If $\nu_1 = \nu_2 = 0.3$, Equations 9.67a and 9.67b become

$$a = 0.88\left[P\frac{(E_1+E_2)R_1R_2}{E_1E_2(R_1+R_2)}\right]^{1/3}$$
$$d = 0.775\left[P^2\left(\frac{E_1+E_2}{E_1E_2}\right)^2\frac{R_1+R_2}{R_1R_2}\right]^{1/3} \tag{9.68}$$

If $E_1 = E_2 = E$, Equations 9.68 yields

$$a = 1.109\left[\frac{P}{E}\frac{R_1R_2}{R_1 + R_2}\right]^{1/3}$$
$$d = 1.23\left[\left(\frac{P}{E}\right)^2\frac{R_1 + R_2}{R_1R_2}\right]^{1/3} \tag{9.69}$$

And for $R_1 = R_2 = R$, the last equation is reduced to

$$a = 0.698\left[\frac{PR}{E}\right]^{1/3}$$
$$d = 1.55\left[\frac{P^2}{E^2R}\right]^{1/3} \tag{9.70}$$

Next, we consider the following contact cases:
Sphere of radius R in contact with a plane surface of the same material (Figure 9.18a):
Here, $E_1 = E_2 = E$ and $R_1 = R$ while R_2 is infinity. Equations 9.68 give

$$a = 1.109\left[\frac{PR}{E}\right]^{1/3}$$
$$d = 1.23\left[\frac{P^2}{E^2R}\right]^{1/3} \tag{9.71a}$$

And Equations 9.66 and 9.71 give the following expression for the maximum pressure at the center of the surface contact:

$$q = 0.62\left[\frac{PE^2}{4R^2}\right]^{1/3} \tag{9.71b}$$

Sphere of radius R_1 in contact with a spherical seat of radius R_2 made of the same material (Figure 9.18b):
Here, R_2 is replaced by $-R_2$ and the modulus of elasticity in both spheres is E. Hence, Equation 9.68 gives

$$a = 1.109\left[\frac{P}{E}\left(\frac{R_1R_2}{R_2 - R_1}\right)\right]^{1/3}$$
$$d = 1.23\left[\left(\frac{P}{E}\right)^2\left(\frac{R_2 - R_1}{R_1R_2}\right)\right]^{1/3} \tag{9.72}$$

and Equations 9.66 and 9.71 yield

$$q = 0.392\left[PE^2\left(\frac{R_2 - R_1}{R_1R_2}\right)^2\right]^{1/3} \tag{9.73}$$

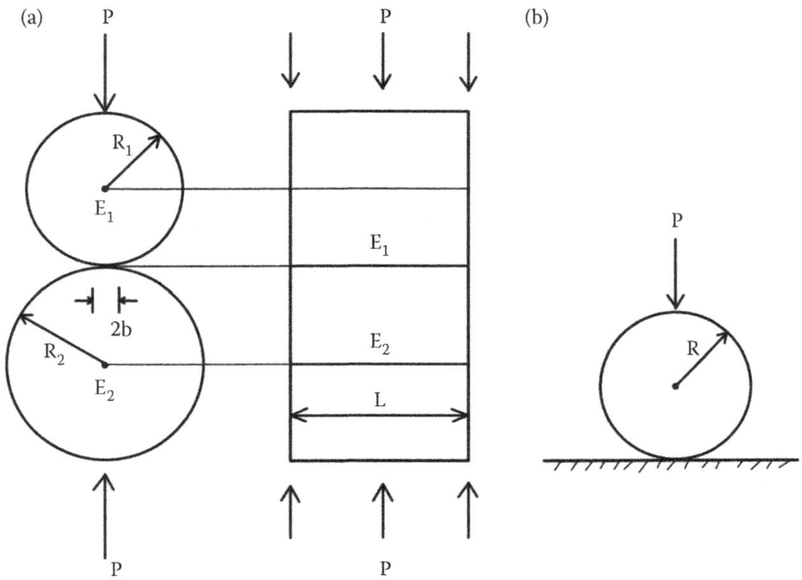

FIGURE 9.19
Contact between two parallel cylinders.

Contact between two parallel cylindrical rollers: Consider two cylinders of radii R_1, R_2, moduli of elasticity E_1, E_2, and Poisson's ratios ν_1 and ν_2, respectively, as shown in Figure 9.19. The contact area is a narrow rectangle of length L and width 2b.

The maximum contact pressure is given by

$$q = \frac{2}{\pi}\frac{P}{bL} \tag{9.74}$$

where

$$b = \left[\frac{4PR_1R_2}{\pi L(R_1 + R_2)}\left(\frac{1-\upsilon_1^2}{E_1} + \frac{1-\upsilon_2^2}{E_2}\right)\right]^{1/2} \tag{9.75}$$

If the cylinders are made of the same material, Equations 9.74 and 9.75 become

$$q = 0.419\left[\frac{PE(R_1 + R_2)}{L(R_1R_2)}\right]^{1/2} \tag{9.76}$$

and

$$b = 1.52\left[\frac{PR_1R_2}{EL(R_1 + R_2)}\right]^{1/2} \tag{9.77}$$

Also, if both have the same radius, R, then

$$q = 0.593(PE/LR)^{1/2} \quad \text{and} \quad b = 1.08(PR/E)$$

In the special case of contact between a cylinder of radius R and a flat surface, Figure 9.19b, both solids have the same modulus of elasticity, then we set $R_1 = R$ and $R_2 =$ infinity in the above equations to obtain

$$q = 0.419(PE/LR)^{1/2} \quad \text{and} \quad b = 1.52(PR/EL)^{1/2}$$

The following example illustrates numerical values of some of the formulas derived above.

EXAMPLE 9.6

A 100 mm diameter ball is pressed against a spherical seat of diameter 150 mm by a force $P = 12$ KN. The material of both bodies is steel with modulus of elasticity $E = 200$ GPa and Poisson's ratio = 0.25. Find

a. The radius of the contact surface, a, and the corresponding maximum contact pressure
b. Same as part (a) if the ball is pressed against a flat surface body with identical material
c. Same as above if the ball is pressed against an identical ball

Solution

Using Equation 9.67a, with $E_1 = E_2 = 200 \times 10^9\ \text{N/m}^2$ and $\nu_1 = \nu_2 = 0.25$, the radius of the contact surface is

$$a = 0.909\left[12 \times 10^3\text{N}\frac{2(0.9375)}{200 \times 10^9\text{N/m}^2}\frac{R_1R_2}{R_1 + R_2}\right]^{1/3} \tag{a}$$

Simplifying the numerical work, we obtain

$$a = 4.388\left[\frac{R_1R_2}{R_1 + R_2}\right]^{1/3}\frac{\text{m}}{10^3} \tag{b}$$

a. For spherical seat, we have $R_1 = 0.05$ m and $R_2 = -0.075$ m. Inserting these values in Equation (b) and performing the numerical work, we find $a = 2.33$ mm.
   The corresponding maximum contact pressure is computed from Equation 9.66 as

$$q = \frac{1.5 \times (12 \times 10^3\text{N})}{\pi(2.33 \times 10^{-3})^2\text{m}^2} = 1055 \times 10^6\frac{\text{N}}{\text{m}^2} = 1055\ \text{MPa}$$

b. For a flat surface, $R_1 = 0.05$ m and $R_2 =$ infinity. It follows that Equation (b) provides

$$a = 4.388[0.05]^{1/3}\ \text{m}/10^3 = 1.62\ \text{mm}$$

   and Equation 9.66 gives the corresponding maximum contact pressure as

$$q = 2183 \times 10^6\ \text{N/m}^2 = 2183\ \text{MPa}$$

c. For contact between identical spherical balls, $R_1 = R_2 = 0.05$ m, Equation (b) gives

$$a = 4.388[0.025]^{1/3}\ \text{m}/10^3 = 1.28\ \text{mm}$$

   and Equation 9.66 yields $q = 3497 \times 10^6\ \text{N/m}^2 = 3497$ MPa.

9.9 A Semi-Infinite Plane Cut or Crack in an Infinite Body

The purpose of this section is to determine the stresses and deformation in a large elastic structure containing a semi-infinite cut or crack, Figure 9.20. The structure is subjected to any system of external loads [9.20,9.21].

The equilibrium of an isotropic homogeneous elastic solid is governed by the Navier displacement equations, which, in the absence of body forces, appear in vectors notations as

$$\nabla^2 \mathbf{u} + \frac{1}{1-\upsilon}\nabla \cdot \nabla \mathbf{u} = 0 \tag{9.78}$$

where **u** is the displacement vector, υ designates Poisson's ratio of the material, and is the gradient operator. It is recalled that any crack problem in linear elasticity can be separated into two parts, the first part deals with the solid subjected to the specified external loads in the absence of the crack or cut, and the second part takes into account the crack with the applied surface tractions equal and opposite to those found at the prospective crack site in the first problem. Such procedure has been widely used to solve 2D and 3D crack problems.

9.9.1 Plane Crack Problem

Figure 9.20 shows the geometry of a semi-infinite crack in a large structure. In terms of rectangular coordinates, x, y, and z, the displacement components are denoted by u_x, u_y, and u_z, the crack edge is taken along the z-axis and the crack surfaces are described by $x < 0$, $y = +0$, $-00 < z < 00$. In terms of cylindrical polar coordinates, r, θ, z, the corresponding displacement components are denoted by u, v, and w, respectively, and the crack surfaces are described by $\theta = +\pi$, $0 < r < 00$.

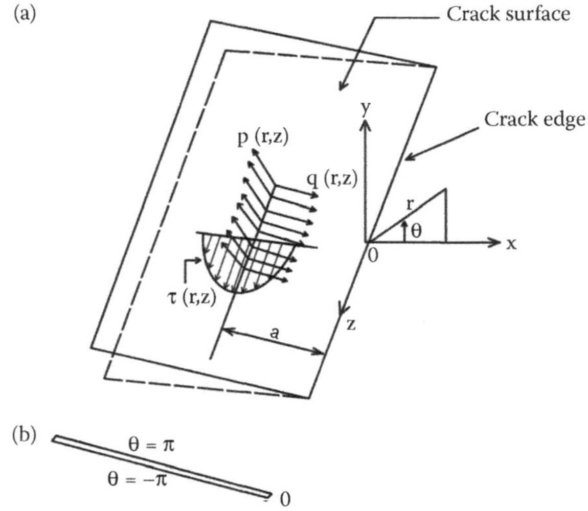

FIGURE 9.20
A semi-infinite plane cut (crack) in an infinite elastic body.

Basic equations: The general solution of this problem car. be expressed in terms of the Papkovich–Neuber harmonic functions [9.13,9.14]. In terms of x, y, and z coordinates, they are as follows:

$$u_x = 4(1-\upsilon)f_1 - \frac{\partial F}{\partial x}$$
$$u_y = 4(1-\upsilon)f_2 - \frac{\partial F}{\partial y} \tag{9.79}$$
$$u_z = 4(1-\upsilon)f_3 - \frac{\partial F}{\partial z}$$

where F is a linear function of the space harmonic functiors f_n (x, y, z), n = 0, 1, 2, and 3 given by

$$F = f_0 + xf_1 + yf_2 + zf_3 \tag{9.80}$$

The corresponding stress components on the y = 0 plane are

$$\frac{\sigma_y}{2G} = 2(1-\upsilon)\frac{\partial f_2}{\partial y} + 2\upsilon\left(\frac{\partial f_1}{\partial x} + \frac{\partial f_3}{\partial z}\right) - \frac{\partial^2 f_0}{\partial y^2} - x\frac{\partial^2 f_1}{\partial y^2} - y\frac{\partial^2 f_2}{\partial y^2} - z\frac{\partial^2 f_3}{\partial y^2} \tag{9.81a}$$

$$\frac{\tau_{yx}}{2G} = (1-2\upsilon)\left(\frac{\partial f_1}{\partial y} + \frac{\partial f_2}{\partial x}\right)\frac{\partial^2 f_0}{\partial x\partial y} - x\frac{\partial^2 f_1}{\partial x\partial y} - y\frac{\partial^2 f_2}{\partial x\partial y} - z\frac{\partial^2 f_3}{\partial x\partial y} \tag{9.81b}$$

$$\frac{\tau_{yz}}{2G} = (1-2\upsilon)\left(\frac{\partial f_2}{\partial z} + \frac{\partial f_3}{\partial y}\right) - \frac{\partial^2 f_0}{\partial y\partial z} - x\frac{\partial^2 f_1}{\partial y\partial z} - y\frac{\partial^2 f_2}{\partial y\partial z} - z\frac{\partial^2 f_3}{\partial y\partial \theta} \tag{9.81c}$$

where G is rigidity modulus of the material. The other stress components may be obtained in the same way.

Alternatively, the displacements u, v, and w, in the directions of the cylindrical coordinates, are given by

$$u(r,\theta,z) = 4(1-\upsilon)(f_1\cos\theta + f_2\sin\theta) - \frac{\partial F}{\partial r}$$
$$v(r,\theta,z) = 4(1-\upsilon)(f_2\cos\theta - f_1\sin\theta) - \frac{\partial F}{r\partial r} \tag{9.82}$$
$$w(r,\theta,z) = 4(1-\upsilon)f_3 - \frac{\partial F}{\partial z}$$

Here, F is a linear function of the space potentials, f_n (r, θ, z), n = 0, 1, 2, and 3, given by Equation 9.80.

The corresponding stresses are found as

$$\frac{\sigma_\theta}{2G} = -(1-2\upsilon)\left(\cos\theta\frac{\partial f_1}{\partial r} + \sin\theta\frac{\partial f_2}{\partial r}\right) + \frac{2(1-\upsilon)}{r}\left(\cos\theta\frac{\partial f_2}{\partial \theta} - \sin\theta\frac{\partial f_1}{\partial \theta}\right)$$
$$+ 2\upsilon\frac{\partial f_3}{\partial z} + \frac{\partial^2 f_0}{\partial r^2} + \frac{\partial^2 f_0}{\partial z^2} - \frac{1}{r}\left(\cos\theta\frac{\partial^2 f_1}{\partial \theta^2} + \sin\theta\frac{\partial^2 f_2}{\partial \theta^2}\right) + z\left(\frac{\partial^2 f_3}{\partial r^2} + \frac{\partial^2 f_3}{\partial z^2}\right) \tag{9.83a}$$

$$\frac{\tau_{\theta r}}{2G} = \frac{2(1-\upsilon)}{r}\left(\cos\theta\frac{\partial f_1}{\partial \theta} + \sin\theta\frac{\partial f_2}{\partial \theta}\right) + \frac{\partial H}{\partial r} \tag{9.83b}$$

$$\frac{\tau_{\theta z}}{2G} = \frac{2(1-\upsilon)}{r}\frac{\partial f_3}{\partial \theta} + \frac{\partial H}{\partial z} \tag{9.83c}$$

The function H (r, θ, z) is defined by the relation

$$H = (1-2\upsilon)(f_2 \cos\theta - f_1 \sin\theta) - \frac{\partial f_0}{r\partial\theta} - \cos\theta\frac{\partial f_1}{\partial\theta} - \sin\theta\frac{\partial f_2}{\partial\theta} - \frac{z\partial f_3}{r\partial\theta} \tag{9.84}$$

The unknown functions f_n (r, θ, z), n = 0, 1, 2, and 3 are harmonic functions in an infinite space with a cut occupying the space, $x < 0$, $y = 0$, $-00 < z < 00$. Since the displacement components must be bounded at infinity, the gradients of the functions f_n and their derivatives must be of the order $(r^2 + z^2)^{-1/2}$ and $(r^2 + z^2)^{-1}$, respectively, as $(r^2 + z^2)^{1/2}$ becomes infinity. Along the crack edge r = 0, the displacements must be finite and the stresses are expected to have singularities of the order $r^{-1/2}$. The self-equilibrating loads on the crack surfaces are always resolved into three pairs of load components along the x, y, and z coordinates. Each pair of the load components may be treated individually such that certain symmetry conditions prevail across the plane y = 0. In this way, the harmonic functions need only be determined in the half space, $-00 < x < 00$, $y > 0$ and $-00 < z < 00$ with the appropriate boundary conditions specified in the plane y = 0.

The stress intensity factors, k_1, k_2, and k_3, defined in Section 7.4 can be obtained from the appropriate stress components in the θ = 0 as follows:

$$\begin{aligned} k_1 &= \lim_{\delta x \to 0}(2x)^{1/2}\sigma_\theta(r, 0, z) \\ k_2 &= \lim_{\delta x \to 0}(2x)^{1/2}\tau_{\theta r}(r, 0, z) \\ k_3 &= \lim_{\delta x \to 0}(2x)^{1/2}\tau_{\theta z}(r, 0, z) \end{aligned} \tag{9.85}$$

9.9.2 Symmetric Normal Loads

Let the loads applied on the crack surfaces be such that the deformation is symmetric with respect to the variables y and z. On the upper surface of the crack $y = 0^+$ (or $\theta = \pi$), the applied surface conditions are

$$\sigma_y(x, 0^+, z) = p(r, z) \tag{9.86a}$$

$$\tau_{yx}(x, 0^+, z) = 0 \tag{9.86b}$$

$$\tau_{yz}(x, 0^+, z) = 0 \tag{9.86c}$$

where p(r, z), the normal pressure, is even in z, that is, p(r, −z) = p(r, z).

The boundary conditions on the plane y = 0 ahead of the crack are

$$u_y(x, 0, z) = \tau_{yx}(x, 0, z) = \tau_{yz}(x, 0, z) = 0 \tag{9.87}$$

For this problem, u_x, u_y, σ_y, τ_{yx} are even in z and u_z, τ_{zy} are odd in z.

Note that in cylindrical polar coordinates, the conditions (9.86) are replaced by

$$\sigma_q(r, \pi, z) = p(r, z), \tau_{qr}(r, \pi, z) = \tau_{\theta z}(r, \pi, z) = 0 \tag{9.88}$$

and at the same time for θ = 0, the displacement v = 0 and the shear stresses = 0.

Based on these considerations and making use of Equations 9.81 through 9.84, the solution for the symmetric normal loads may be expressed in terms of one harmonic function f(x, y, z)

$$f_0 = -(1-2\upsilon)\int_{\infty}^{y} f(x,y,z)\,dy \tag{9.89a}$$

$$f_2 = -f(x,y,z) \tag{9.89b}$$

With $f_1 = f_3 = 0$, Equations 9.86 may be inserted into the basic solution, Equations 9.79 through 9.84 to give the following conditions for the new function:

$$\begin{aligned} &f(r,0,z) = 0, \theta = 0 \\ &\frac{\partial f}{\partial \theta} = \frac{1}{2G} rp(r,z), \theta = \pi \end{aligned} \tag{9.90}$$

since $\partial/\partial y = -(1/r)\partial/\partial\theta$ at $\theta = \pi$.

The conditions in Equations 9.90 suggest that the function f(r, θ, z) can be found by applying a Fourier cosine transform on z and a Kantorovich–Lebedev transform on r [9.23,9.24].

Consider the representation

$$f(r,\theta,z) = \sqrt{\frac{2}{\pi}}\int_0^{\infty}\int_9^{\infty} A(s,t)\frac{\sinh(\theta t)}{t\cosh(\pi t)}K_{ir}(sr)\cos(sz)\,dsdr \tag{9.91}$$

where K_{ir} is the modified Bessel function of the second kind of imaginary order, which is also known as the Macdonald function [9.24]. The unknown function A(s, t) is determined such that Equation 9.90 is satisfied:

$$A(s,t) = \frac{t\sinh(\pi t)}{\pi^2 G}\int_0^{\infty} p_c(\xi,s)K_{it}(s\xi)\,d\xi \tag{9.92}$$

In which p_c stands for

$$p_c(r,s) = \sqrt{\frac{2}{\pi}}\int_0^{\infty} p(r,z)\cos(sz)\,ds \tag{9.93}$$

Once the normal load p(r, z) is specified, $p_c(r, s)$ can be obtained from Equation 9.93 and A(s, t) from Equation 9.92. This gives the function f(r, θ, z) in Equation 9.91 and the problem is basically solved.

The stress intensity factor k_1 may be found from σ_θ or σ_y on the plane $\theta = 0$.

Since $\partial/\partial y = (1/r)(\partial/\partial\theta)$ for $\theta = 0$, we have

$$\sigma_\theta = \sigma_y = -2G\frac{1}{r}\frac{\partial f}{\partial \theta}, \quad \theta = 0 \tag{9.94}$$

With the help of the identity [9.26]

$$\int_0^\infty t\tanh(\pi t)K_{it}(sr)K_{it}(s\xi)\,dt = \frac{\pi(r\xi)^{1/2}}{2(r+\xi)}e^{-s(r+\xi)} \tag{9.95}$$

the following k_1-formula is obtained

$$k_1 = -\frac{2}{\pi^{3/2}}\lim_{\delta x\to 0}\int_0^\infty e^{-sx}\cos(sz)ds\int_0^\infty \frac{\xi^{1/2}e^{-s\xi}}{x+\xi}p_c(\xi,s)\,d\xi \tag{9.96}$$

Assuming that the limiting process may be carried into the integrals, k_1 is further reduced to

$$k_1 = -\frac{2}{\pi^{3/2}}\int_0^\infty \cos(sz)ds\int_0^\infty \xi^{-1/2}e^{-s\xi}p_c(\xi,s)\,d\xi \tag{9.97}$$

Consider the special case of *a pair of equal and opposite concentrated normal forces P* applied to the crack surfaces at $r = a$, $\theta = +\pi$, $z = 0$, as shown in Figure 9.21.

The function $p(r, z)$ is

$$p(r,z) = -P\delta(z)\delta(r-a) \tag{9.98}$$

where $\delta(z)$ is the Dirac delta function. It follows immediately from Equation 9.93 that

$$p_c(r,s) = -\frac{P}{(2\pi)^{1/2}}\delta(r-a) \tag{9.99}$$

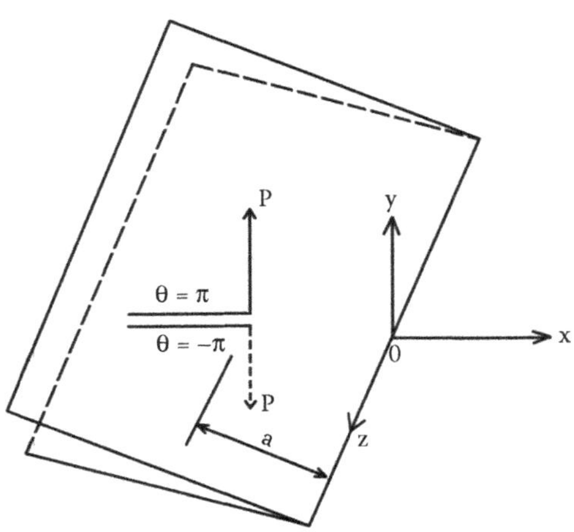

FIGURE 9.21
Concentrated normal forces P.

and hence, the stress intensity factor k_1 from Equation 9.96 is

$$k_1 = \frac{\sqrt{2}P}{\pi^2 a^{3/2}} \frac{1}{1+(z/a)^2} \tag{9.100}$$

Equation 9.100 is plotted in Figure 9.22. As expected, the stress intensity factor peaks in the region nearest to the point where the concentrated loads are applied and decreases in magnitude very rapidly as (z/a) increases. Equation 9.100 may be further used as Green's function to generate solutions of other problems that can be found in the appendix to this section.

The function f(r, θ, z) for the concentrated load problem can be expressed in terms of elementary functions in a form that can be conveniently used as Green's function to solve other problems. Let p_c (r, s) in Equation 9.92 be replaced by Equation 9.99 so that

$$A(s,t) = -\frac{Pt\sinh(\pi t)}{(2\pi)^{1/2}\pi^2 G} K_{it}(as) \tag{9.101}$$

which may be inserted in Equation 9.91 to give

$$f(r,\theta,z) = -\frac{P}{2G\pi^2(2ar)^{1/2}} \int_0^\infty \frac{\sinh(\theta t)\,dt}{\cosh(\pi t)} \int_0^\infty \frac{\sin(ts)\,ds}{[\cosh(s)-\cosh(\omega)]^{1/2}} \tag{9.102}$$

In deriving the above expression, use was made of the Fourier cosine transform [9.24]

$$\int_0^\infty K_{it}(sr)K_{it}(sa)\cos(sz)ds = \frac{\pi^2}{4(ar)^2\cosh(\pi t)} P_{it-1/2}[\cosh(\omega)]$$

$$= \frac{\pi}{2(ar)^{1/2}\sinh(\pi t)} \int_0^\infty \frac{\sin(st)ds}{\{2[\cosh(s)-\cosh(\omega)]\}^{1/2}} \tag{9.103}$$

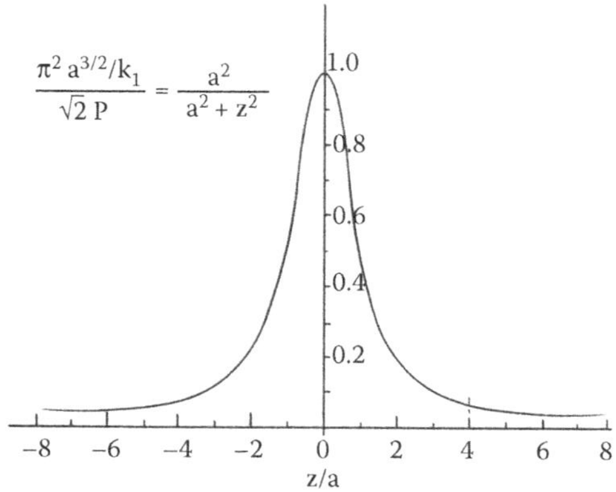

FIGURE 9.22
Variations of stress intensity factor k_1 with distance along the crack border.

In which, $P_{it-1/2}$ (x) is the Legendre coefficient of degree it −½ and

$$\cosh(\omega) = \frac{a^2 + r^2 + z^2}{2ar} \tag{9.104}$$

The order of integration in Equation 9.102 is interchanged so that the result [9.26]

$$\int_0 \frac{\sin h(\theta t)\sin(st)}{\cosh(\pi t)}dt = \frac{\sin(\theta/2)\sin h(s/2)}{\cos(\theta) + \cosh(s)}$$

can be used to render

$$f(r,\theta,z) = -\frac{P\sin(\theta/2)}{2G\pi^2(2ar)^{1/2}}\int_0 \frac{\sin h(s/2)ds}{(\cos\theta + \cosh s)(\cosh s - \cosh\omega)^{1/2}} \tag{9.105}$$

The integral in Equation 9.105 may be integrated by the substitution [9.21]

$$\cosh(s/2) = \cosh(u)\cosh(\omega/2)$$

and the final result for f(r, θ, z) is

$$f(r,\theta,z) = \frac{P}{2G\pi^2}\phi(r,\theta,z)$$

where (9.106)

$$\phi(r,\theta,z) = \frac{1}{\rho}\tan^{-1}\left[\frac{2(ar)^{1/2}}{\rho}\sin(\theta/2)\right]$$

where the quantity ρ is defined as

$$\rho = [(x+a)^2 + y^2 + z^2]^{1/2}$$

9.9.3 Antisymmetric Shear Loading Normal to Crack Edge

Assume that the crack surfaces are subjected to equal and opposite shears q(r, z) normal to the crack edge, as illustrated in Figure 9.20. The loading implies that the displacements u, w, and stress σ_θ must be odd in θ while v, $\tau_{\theta r}$, $\tau_{\theta z}$ even in θ. It suffices to consider the upper half space $y > 0$:

In the region $\theta = 0$, the surface of the half space is subjected to the conditions

$$u(r,0,z) = w(r,0,z) = \sigma_q(r,0,z) = 0 \tag{9.107a}$$

In the region $\theta = \pi$, we have the conditions

$$\sigma_\theta(r,\pi,z) = 0, \quad \tau_{\theta r}(r,\pi,z) = q(r,z), \quad \tau_{\theta z}(r,\pi,z) = 0 \tag{9.107b}$$

As the problem is symmetric in z then the function f_3(r, θ, z) in the basic Equations 9.79 through 9.81 may be set to zero. Inserting Equations 9.107a into the basic solution, we conclude that

$$f_0(r,0,z) = f_1(r,0,z) = 0, \quad (\partial f_2/\partial\theta)_{\theta=0} = 0 \tag{9.108}$$

While in the region $\theta = \pi$, Equations 9.107b when inserted in the basic solution provide the following requirements [9.22]:

$$\begin{aligned} &(1-2\upsilon)f_2 - \frac{\partial f_0}{\partial y} = -\frac{rq(r,z)}{4(1-\upsilon)G} \\ &\frac{\partial f_1}{\partial\theta} = -\frac{rq(r,z)}{4(1-\upsilon)G} \\ &\frac{\partial}{\partial y}\left[(1-2\upsilon]\int_{\infty}^{y}\frac{\partial f_0}{\partial y}dy - 2(1-\upsilon)f_2 + \frac{\partial f_0}{\partial y} + \frac{\partial f_1}{\partial\theta}\right] = 0 \end{aligned} \tag{9.109}$$

The integral and its limits in the third of Equations 9.109 are introduced to ensure that the quantity is bounded at infinity.

Equations 9.108 and 9.109 provide the necessary conditions for finding the three unknown potential functions f_0, f_1, and f_2.

Let f_0 (r, θ, z) be represented as the sum of two functions as follows:

$$f_0(r,\theta,z) = f_0^*(r,\theta,z) + f_0^{**}(r,\theta,z) \tag{9.110}$$

Such that

$$\begin{aligned} &f_0^*(r,0,z) = 0, \quad \theta = 0 \\ &\frac{\partial f_0^*}{\partial\theta} = -\frac{r^2q(r,z)}{4(1-\upsilon)G}, \quad \theta = \pi \end{aligned} \tag{9.111}$$

A relation between f_2 and f_0^{**} can be established by examining the above equations (more details are available in Reference 9.22) to arrive at

$$\begin{aligned} &\frac{\partial}{\partial y}\left[(1-2\upsilon)f_2 - \frac{\partial f_0^{**}}{\partial y}\right] = 0, \quad \theta = 0 \\ &(1-2\upsilon)f_2 - \frac{\partial f_0^{**}}{\partial y} = 0, \quad \theta = \pi \end{aligned} \tag{9.112}$$

Since the conditions represented by Equations 9.107 on f_0^* are the same as Equations 9.90 for the symmetric loading, it follows that the function f_0^* takes the same form as Equation 9.91 and is given by

$$f_0*(r,\theta,z) = \sqrt{\frac{2}{\pi}}\int_0^\infty\int_0^\infty \frac{\sinh(\theta t)}{t\cosh(\pi t)}A(s,t)K_{ir}(sr)\cos(sz)\,ds\,dt \tag{9.113}$$

The function (see Equation 9.92)

$$A(s,t) = -\frac{t\sinh(\pi t)}{2\pi^2(1-\upsilon)G}\int_0^\infty \xi q_c(\xi,s)K_{it}(s\xi)\,d\xi \tag{9.114}$$

is now related to the specified shear loads q(r, z) through the integral

$$q_c(\xi, s) = \sqrt{\frac{2}{\pi}} \int_0^\infty q(\xi, z) \cos(sz)\, dz \tag{9.115}$$

Examination of Equations 9.108 and 9.110 reveals that the potential $f_1(r, \theta, z)$ is governed by the following set of conditions:

$$\begin{aligned} &f_1(r, 0, z) = 0, \quad \theta = 0 \\ &\frac{\partial f_1}{\partial \theta} = -\frac{rq(r, z)}{4(1-\upsilon)G}, \quad \theta = \pi \end{aligned} \tag{9.116}$$

Although this set is similar to that in Equation 9.90, $f_1(r, \theta, z)$ cannot be written in the form of Equation 9.91 alone as this would lead to infinite displacements at the crack edge; an additional harmonic function odd in θ, which satisfies the requirements (9.112) and gives bounded displacements and stresses at infinity, is required [9.21]. The appropriate form is

$$f_1(r, \theta, z) = \sqrt{\frac{2}{\pi}} \int_0^\infty \int_0^\infty \frac{\sinh(\theta t)}{t \cosh(\theta \pi)} B(s, t) K_{it}(sr) \cos(sz)\, ds\, dt + C_1 g_1(r, \theta, z) \tag{9.117}$$

where C_1 is constant, g_1 is a singular solution of Laplace's equation in cylindrical coordinates [9.29], and B (s, t) is given by

$$B(s, t) = -\frac{t \sinh(\pi t)}{2\pi^2(1-\upsilon)G} \int_0^\infty q_c(\xi, s) K_{it}(s\xi)\, d\xi \tag{9.118}$$

The function g_1 (r, θ, z) is expressible in terms of an analytic function g(ζ) of the complex variable $\zeta = r + a + iz$, where

$$g_1(r, q, z) = r^{-1/2} \sin(q/2) \mathrm{Re}[g(z)] \tag{9.119}$$

Here, Re stands for "real part of."

The requirements on the nature of the displacements near the crack edge and Equation 9.112 suggest that we have

$$\begin{aligned} &(1-2\upsilon)f_2 - \frac{\partial f_0^{**}}{\partial y} = \frac{1}{2} C_2 g_2(r, \theta, z) \\ &g_2(r, \theta, z) = r^{-1/2} \sin(\theta/2) \mathrm{Re}[g(\varsigma)] \end{aligned} \tag{9.120}$$

The relation between the constants C_1 and C_2 needs to be determined based on the condition of finite displacements at $r = 0$, and the condition, $f_0^{**}(r, 0, z) = 0$, is used to determine the function g(ζ).

Consider the special case of $q(r \cdot z) =$ two equal and opposite concentrated forces Q at the location $r = a$, $\theta = +\pi$, $z = 0$, as illustrated in Figure 9.23. On the upper surface of

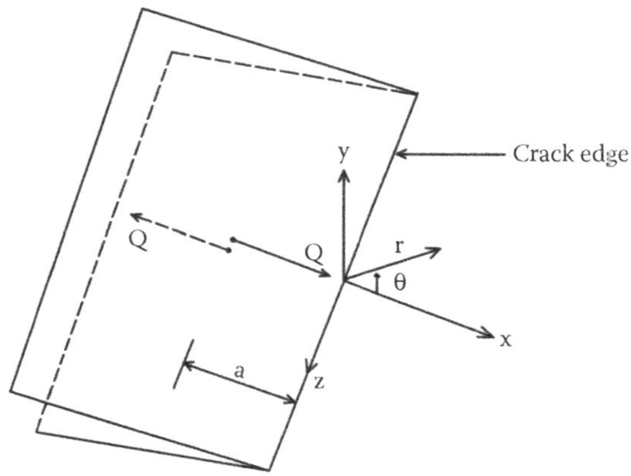

FIGURE 9.23
Concentrated shear forces Q normal to crack edge.

the crack, Q is directed in the positive x direction. The quantity q(r, z) in Equation 9.107b for the equivalent half space problem is

$$q(r, z) = -Qd(z)d(r - a)$$

and as in the symmetric normal loading case, the unknown potential functions may be expressed as

$$\begin{aligned} f_0^*(r, \theta, z) &= \frac{aQ}{4(1-\upsilon)G\pi^2}\phi(r, \theta, z) \\ f_1(r, \theta, z) &= \frac{Q}{4(1-\upsilon)G\pi^2}\phi(r, \theta, z) + C_1 g_1(r, \theta, z) \end{aligned} \tag{9.121}$$

where $\phi(r, \theta, z)$ is the elementary function given in Equation 9.106.

Detailed derivations of the displacements u, v, and w near the crack edge $r = 0$, the relations that determine the constants C_1 and C_2, and the function g (ζ) are available in Reference 9.22. This completes the steps needed to find the unknown functions f_0, f_1, and f_2.

The stress intensity factors, k_2 and k_3, can be extracted from the limits of the expressions for the shear stresses $\tau_{\theta r}$ (r, 0, z) and $\tau_{\theta z}$ (r, 0, z), respectively, as $r = 0$. They are

$$\begin{aligned} k_2 &= \frac{\sqrt{2}Q}{\pi^2 a^{3/2}} \frac{1}{1 + (z/a)^2}\left[1 + \left(\frac{2\upsilon}{2-\upsilon}\right)\frac{1-(z/a)^2}{1+(z/a)^2}\right] \\ k_3 &= \frac{\sqrt{2}Q}{\pi^2 a^{3/2}}\left(\frac{4\upsilon}{2-\upsilon}\right)\frac{z/a}{[1+(z/a)^2]^2} \end{aligned} \tag{9.122}$$

For values of Poisson's ratio $\nu = 0.3$, the variations of the stress intensity factors with distance z/a along the crack edge are shown in Figure 9.24. The following

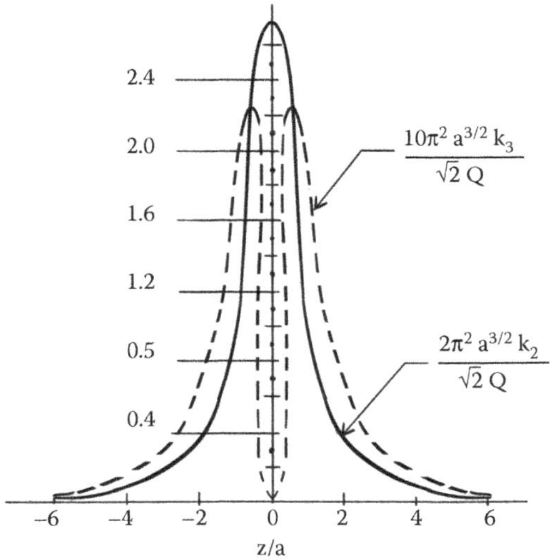

FIGURE 9.24
Variations of stress–intensity factor k_2 and k_3 with ration z/a.

quantities are plotted:

$$\frac{2\pi^2 a^{3/2} k_2}{\sqrt{2Q}} \text{ versus } \frac{z}{a}$$

$$\frac{10\pi^2 a^{3/2} k_3}{\sqrt{2Q}} \text{ versus } \frac{z}{a}$$

It appears that when the upper and lower crack surfaces are sheared normal to the crack edge, the k_2-factor is maximum at the point nearest to the applied shear Q and decays rapidly as z/a is increased. Due to the effect of Poisson's ratio, the material on the upper and lower crack surfaces tends to displace in the opposite directions parallel to the crack edge. This effect is evidenced by the variations of the k_3-factor, which attains two peak values at $z/a = +0.577$ and vanishes at $z = 0$, as illustrated by the dotted curve in Figure 9.24.

9.9.4 Antisymmetric Shear Loading Parallel to Crack-Edge

When equal and opposite shear tractions acting on crack surfaces are applied parallel to the crack edge, the material in that vicinity undergoes a tearing action. The displacements u, v, and w and stresses σ_θ and $\tau_{\theta r}$ are odd in z while w and $\tau_{\theta z}$ are even in z. Also, the conditions of symmetry with respect to the variable θ are such that u, w, and σ_θ are odd in θ while v, $\tau_{\theta r}$, and $\tau_{\theta z}$ are even in θ.

The problem is equivalent to specifying the following conditions on the surface $y = 0$ of the half-space $y > 0$:

On portion of the plane $\theta = 0$, it is necessary to have

$$u(r, 0, z) = w(r, 0, z) = s_q(r, 0, z) = 0 \tag{9.123a}$$

On the remaining portion $\theta = \pi$, we have

$$\begin{aligned} \tau_{\theta z}(r, \pi, z) &= \tau(r, z) \\ \tau_{\theta r}(r, \pi, z) &= 0 \end{aligned} \tag{9.123b}$$

where the applied shear stress $\tau(r, z)$ is a function of the variables r and z (Figure 9.20).

The solution of this problem requires use of all four harmonic functions f_n, $n = 0, 1, 2$, and 3 in the Papkovich–Neuber displacement representation of the equilibrium equation given by Equations 9.79 or 9.83 [9.29].

Using condition (9.123a), Equations 9.79 and 9.82 yield the following:

$$\begin{aligned} &f_0(r, 0, z) = f_1(r, 0, z) = f_3(r, 0, z) = 0 \\ &\left(\frac{\partial f_2}{\partial \theta}\right)_{\theta=0} = 0 \end{aligned} \tag{9.124a}$$

Similarly, condition (9.123b) gives the relations

$$\begin{aligned} &\frac{\partial f_3}{\partial \theta} = \frac{r\tau(r, z)}{4(1-\upsilon)G}, \quad \theta = \pi \\ &\frac{\partial f_3}{\partial \theta} = 0, \quad \theta = \pi \end{aligned} \tag{9.124b}$$

Equations 9.124a and 9.124b require that the function $f_3(r, \theta, z)$ satisfy the combinations

$$\begin{aligned} &f_3(r, 0, z) = 0, \quad \theta = 0 \\ &\frac{\partial f_3}{\partial \theta} = \frac{r\tau(r, z)}{4(1-\upsilon)G}, \quad \theta = \pi \end{aligned} \tag{9.125}$$

which are the same as condition (9.90) for the normal loading and conditions (9.107) for the shear loading normal to crack edge. Hence, the function $f_3(r, \theta, z)$ takes the form

$$f_3 = \left(\frac{2}{\pi}\right)^{1/2} \int_0^\infty \int_0^\infty A(s, t) \frac{\sinh(\theta t)}{\cosh(\pi t)} K_{it}(sr) \cos(sz)\, ds\, dt \tag{9.126}$$

and for the particular case of equal and opposite concentrated shear force R, shown in Figure 9.25, the function f_3 takes the form

$$f_3(r, \theta, z) = -\frac{R}{4(1-\upsilon)\pi^2 G} \frac{1}{\rho} \tan^{-1}\left(\frac{\sqrt{[2a(r-x)]}}{\rho}\right) \tag{9.127}$$

where ρ is the distance of any point in the region $y > 0$ from the point of application of R

$$\rho = [(x+a)^2 + y^2 + z^2]^{1/2}$$

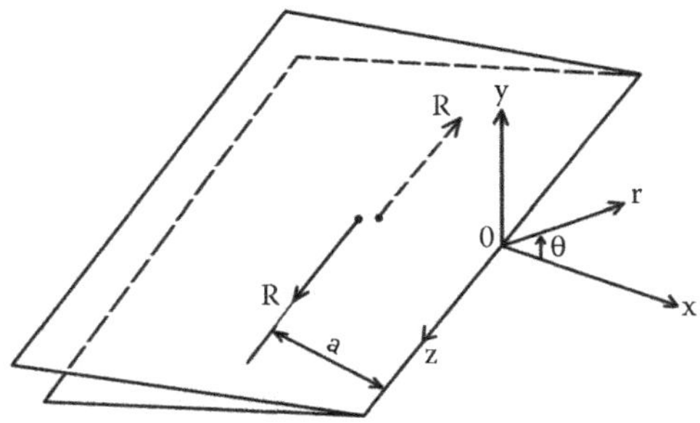

FIGURE 9.25
Concentrated shear forces R parallel to crack edge.

Using Equations 9.120a and 9.120b, the potential function $f_1(r, \theta, z)$ satisfies the "homogeneous" conditions:

$$f_1(r, 0, z) = 0, \quad \theta = 0$$
$$\frac{\partial f_1}{\partial \theta} = 0, \quad \theta = \pi \tag{9.128}$$

A proper solution for f_1 is a singular solution of Laplace equation in three dimensions, which satisfies Equations 9.128 and gives rise to well-defined displacements at infinity [9.28].

$$f_1(r, \theta, z) = C_1 \frac{\sin \theta/2}{\sqrt{r}} \mathrm{Im}[g(\varsigma)] \tag{9.129}$$

where C_1 is a constant and Im designates the imaginary part of an analytic function of the variable ζ defined as $\zeta = r + a + iz$, $i = (-1)^{1/2.}$

It is noted that the choice f_1 $(r, \theta, z) = 0$ in Reference 9.21 is inappropriate since it will not lead to a solution of the problem.

The remaining functions f_0 and f_2 can be shown to satisfy the conditions [9.22,9.29]

$$(1 - 2\upsilon)f_2 - \frac{\partial f_0}{\partial y} = 0, \quad \theta = \pi$$
$$\frac{\partial}{\partial y}\left[(1 - 2\upsilon)f_2 - \frac{\partial f_0}{\partial y}\right] = 0, \quad \theta = 0 \tag{9.130}$$

Equations 9.129 and 9.130 suggest the following relation [9.29]:

$$(1 - 2\upsilon)f_2 - \frac{\partial f_0}{\partial y} = \frac{A}{2} \frac{\cos \theta/2}{\sqrt{r}} \mathrm{Im}[g(\varsigma)] \tag{9.131}$$

where A is constant and the function $g(\zeta)$, defined in Equation 9.129, is given by

$$g(\varsigma) = \frac{1}{\varsigma} = \frac{1}{r + a + iz}, \quad i = \sqrt{-1} \tag{9.132}$$

The rest of the procedure of finding f_0 and f_2 and ensuring that the displacements u, v, and w become finite as r approaches 0 is described in References 9.22 and 9.29.

For the case of concentrated shears R in Figure 9.25, the shear stresses across the surface $\theta = 0$ are

$$\tau_{r\theta}(r, 0, z) = \frac{4\upsilon R\sqrt{a}}{(2-\upsilon)} \frac{(x+a)z}{(x)^{1/2}[(x+a)^2+z^2]^2} \tag{9.133a}$$

$$\tau_{\theta z}(r, 0, z) = -\frac{R\sqrt{a}}{(2-\upsilon)\pi^2} \frac{(2-3\upsilon)[(x+a)^2+(2+\upsilon)z^2]}{(x)^{1/2}[(x+a)^2+z^2]^2} \tag{9.133b}$$

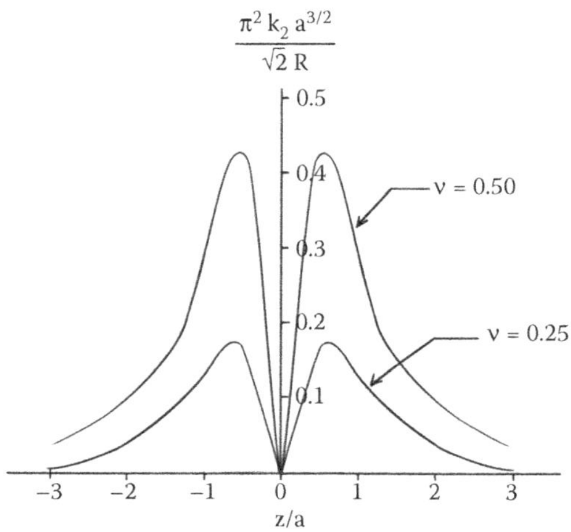

FIGURE 9.26
Variations of k_2 with z/a.

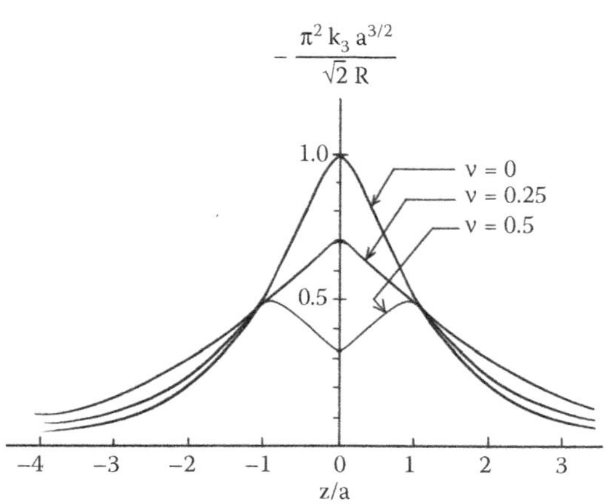

FIGURE 9.27
Variation of R_3 with z/a.

And the stress-intensity factors for the edge-sliding and tearing modes of crack extension are expressed in terms of the nondimensional parameter $z^* = z/a$ through the following relations:

$$k_2 = -\frac{\sqrt{2}R}{\pi^2 a^{3/2}(1 + z^{*2})}\left(\frac{\upsilon}{2 - \upsilon}\right)\frac{z^{*2}}{(1 + z^{*2})} \tag{9.134a}$$

$$k_3 = -\frac{\sqrt{2}R}{\pi^2 a^{3/2}(1 + z^{*2})}\left[1 - \frac{2\upsilon}{2 - \upsilon}\frac{1 - z^{*2}}{1 + z^{*2}}\right] \tag{9.134b}$$

The variations of these equations with $z^* = z/a$ are shown in Figures 9.26 and 9.27 for various values of Poisson's ratio.

EXAMPLE 9.7

Equal and opposite concentrated forces P, Q, and R acting at the point $x = -a$, $y = 0$, and $z = b$ of a semi-infinite crack in an infinite body as shown in Figure 9.28. Determine the stress intensity factors k_1, k_2, and k_3 in terms of the dimensionless variable $z_1 = (z - b)/a$.

Solution

a. Normal force P: In this case, Equation 9.98 gives

$$p(r, z) = -P\delta(z - b)\delta(r - a)$$

where $\delta(x)$ is the Dirac delta function of x. Equation 9.93 gives p_c and the first of Equations 9.85 yields

$$k_1 = \frac{\sqrt{2}P}{\pi^2 a^{3/2}}\left(\frac{1}{1 + z_1^2}\right)$$

b. Shear loads Q normal to crack edge: For this load,

$$q(r, z) = -Q\delta(z - b)\delta(r - a)$$

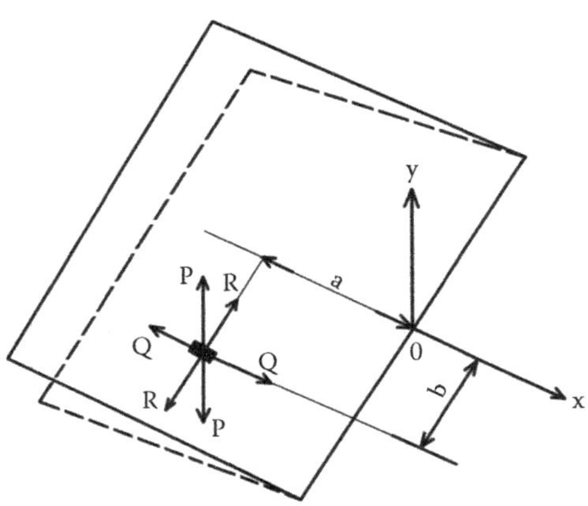

FIGURE 9.28
Example 9.7.

and the second and third of Equation 9.85 furnish the edge-sliding and edge-tearing modes of the stress-intensity factors

$$k_2 = \frac{\sqrt{2}Q}{\pi^2 a^{3/2}}\left(\frac{1}{1+z_1^2}\right)\left[1+\left(\frac{2\upsilon}{2-\upsilon}\right)\frac{1-z_1^2}{1+z_1^2}\right]$$

$$k_3 = \frac{\sqrt{2}Q}{\pi^2 a^{3/2}}\left(\frac{2\upsilon}{2-\upsilon}\right)\frac{1-z_1^2}{(1+z_1^2)^2}$$

c. Shear loads R parallel to crack edge: the load is given by

$$\tau(r,z) = -R\delta(z-b)\delta(r-a)$$

and the stress-intensity factors are

$$k_2 = \frac{\sqrt{2}R}{\pi^2 a^{3/2}}\left(\frac{2\upsilon}{2-\upsilon}\right)\left(\frac{1}{1+z_1^2}\right)$$

$$k_3 = \frac{\sqrt{2}R}{\pi^2 a^{3/2}}\left(\frac{1}{1+z_1^2}\right)\left[1-\left(\frac{2\upsilon}{2-\upsilon}\right)\frac{1-z_1^2}{1+z_1^2}\right]$$

9.10 Deformation of a Circular Cylinder

This section deals with the deformation of a circular cylinder subjected to forces acting on the lateral surface. The forces are symmetrically distributed with respect to the axis of the cylinder. The problem was first solved by Pochhammer [9.30]. Other references to earlier work are available in Reference 5.3. More recently, the method of Hankel transforms has been used by Sneddon [9.24] to solve such problems.

A suitable stress function that satisfies Equations 9.4, 9.5a, and 9.5b is

$$\phi(r,z) = \sin kz[C_0 J_0(ikr) + C_1(ikr)J_1(ikr)] \tag{9.135}$$

where J_0 and J_1 are the Bessel functions of orders zero and one, respectively, and k, C_0, and C_1 are constant. The argument of the Bessel functions is imaginary with $i = (-1)^{1/2}$. A proof of this result is next given in solved Example 9.8.

The associated stresses are obtained by inserting Equation 9.135 into 9.5a

$$\begin{aligned}\sigma_r &= \cos(kz)[C_0F_1(r) + C_1F_2(r)]\\ \tau_{rz} &= \sin(kz)[C_0F_3(r) + C_4F_4(r)]\end{aligned} \tag{a}$$

Here, F_j, j = 1, 2, 3, and 4 are certain functions of r containing the Bessel functions of imaginary argument ikr, which can be numerically calculated for any values of k and r.

Denoting the external radius of the cylinder by a, the forces applied on the surface of the cylinder are given by the stresses

$$\begin{aligned}\sigma_r &= \cos(kz)[C_0F_1(a) + C_1F_2(a)]\\ \tau_{rz} &= \sin(kz)[C_0F_3(a) + C_1F_4(a)]\end{aligned} \tag{b}$$

By suitable adjustments of the constants k, C_0, and C_1, various symmetrically distributed normal forces can discussed.

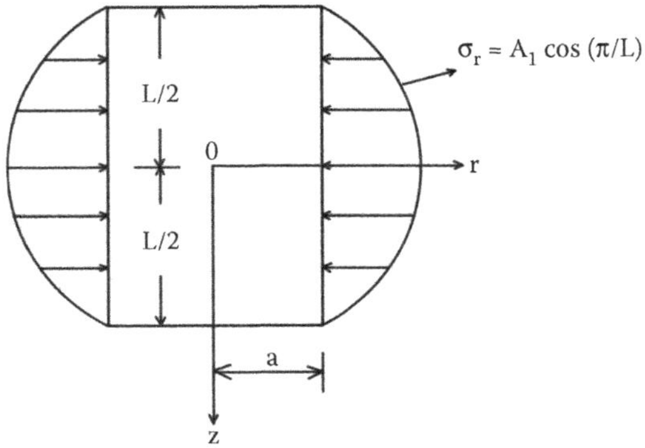

FIGURE 9.29
Circular cylinder subjected to normal pressure.

Denoting the length of the cylinder by L and taking $k = n\,\pi/L$, $n = 1, 2$, the constants C_0 and C_1, for the case when normal pressures $A_n \cos(n\pi/L)$ are acting on the surface, are determined by the relations

$$\begin{aligned} C_0F_1(a) + C_1F_2(a) &= A_n \\ C_0F_3(a) + C_1F_4(a) &= 0 \end{aligned} \qquad \text{(c)}$$

The case for $n = 1$ is shown in Figure 9.29.

In a similar manner, a solution can be obtained when shear stresses $B_n \sin(n\,\pi/L)$ act on the surface of the cylinder.

Hence, by superposition solutions of problems in which the normal pressures are represented by the series

$$A_1 \cos\frac{\pi}{L} + A_2 \cos\frac{2\pi}{L} + A_3 \frac{\cos 3\pi}{L} + \cdots \qquad \text{(d)}$$

and the shearing stresses by the series

$$B_1 \frac{\sin \pi}{L} + B_2 \frac{\sin 2\pi}{L} + B_3 \frac{\sin 3\pi}{L} + \cdots \qquad \text{(e)}$$

can be determined.

Besides the above solution, one can start with the stress function

$$\phi(r, z) = \cos(kz)[D_0J_0(ikr) + D_1ikrJ_1(ikr)] \qquad (9.136)$$

And suitably adjust the values of the constants k, D_0, and D_1 to arrive at the solution in which normal pressures on the cylinder are given in terms of sine series instead of the cosine series (d), and shearing pressures in term of cosine series instead of sine series (e).

By combining stress functions (9.135) and (9.136), we can obtain solutions of the cylinder when subjected to any axially symmetric loading normal to the lateral surface. Note that there will also arise forces distributed at the ends of the cylinder. By superposing simple tensile or compressive forces at the ends of the cylinder, the resultant can be arranged to be zero, and the effect on the stresses at distances away from the end can be neglected according to Saint-Venant's principle. Several examples of determining the deformation and stresses in circular cylinders subjected to axially symmetric loading have been discussed by Filon [9.31] and others [5.3].

EXAMPLE 9.8

Prove that Equation 9.135 is a valid stress function for obtaining the deformation of a circular cylinder subjected to axially symmetric lateral forces.

Solution

A solution of the harmonic equation

$$\nabla^2\phi(r,z) = \frac{\partial^2\phi}{\partial r^2} + \frac{\partial\phi}{r\partial r} + \frac{\partial^2\phi}{\partial z^2} = 0 \tag{a}$$

is sought in the form

$$\phi = f(r)\sin(kz) \tag{b}$$

where k is constant and f(r) is an arbitrary function. Inserting Equations (b) into (a) yields the following ordinary differential equation:

$$\frac{d^2f}{dr^2} + \frac{df}{rdr} - k^2f = 0 \tag{c}$$

A solution appropriate to the solid cylinder is obtained in a series form

$$f(r) = C_0 + C_1r^2 + C_2r^4 + C_3r^6 + C_4r^8 + \cdots \tag{d}$$

where C_j, $j = 0, 1, 2, \ldots$ are constants. Inserting Equations (d) into (c) the following relation between consecutive coefficients of the terms in the series is obtained:

$$(2n)^2C_n - k^2C_{n-1} = 0 \tag{e}$$

Hence, Equation (d) can be written as

$$f(r) = C_0\left[1 + \frac{k^2r^2}{2^2} + \frac{k^4r^4}{2^2(4^2)} + \frac{k^6r^6}{2^2(4^2)(6^2)} + \cdots\right] = C_0J_0(ikr) \tag{f}$$

in which the series between the brackets is the Bessel function of order zero and argument = ikr, $i = (-1)^{1/2}$, J_0 (ikr) and sometimes denoted by I_0 (kr). It follows from Equation (b) that

$$\phi_1(r,z) = C_0J_0(ikr)\sin(kz) \tag{g}$$

Upon differentiating Equation (g) with respect to ikr, it can be shown that a second solution of the harmonic equation is

$$\phi_2(r,z) = C_1(ikr)J_1(ikr)\sin(kz) \tag{h}$$

where $J_1(ikr)$ is the Bessel function of order 1 and argument ikr. Combining Equations (g) and (h), Equation 9.135 is obtained.

9.10.1 Pressure Band of Finite Length Acting on a Long Cylinder

When a short cylinder is shrunk onto a much longer one, the shrink-fit formula, Equation 7.28 derived in Section 7.1.3, is not accurate for finding the deformation in a long cylinder subjected to the pressure of a finite band. A more accurate approach is obtained by considering the cylinder shown in Figure 9.30. A long cylinder of radius a is subjected to uniform pressure in the band area ABCD of the surface of the cylinder.

The required solution is obtained by combining the two pressures acting on the cylinder, a pressure p/2 in the lower half of the cylinder and −p/2 in upper half. The length of the cylinder is assumed to be infinite and that of the band is b.

We begin by considering the stress function given in Equation 9.136. Writing $I_0(kr)$ for J_0 (ikr) and $-I_1$ (kr) for J_1 (ikr). Also, writing $b = \rho\, b_1$, ρ is a parameter to be determined later, then we have

$$\phi(r, z) = [\rho I_0(kr) - krI_1(kr)]b_1 \cos(kz) \tag{a}$$

If we consider k to take a wide range of values and b_1 to depend on k and increment in k, then we may write

$$b_1 = f(k)dk \tag{b}$$

where f(k) is an arbitrary function to be determined later. Inserting Equation (b) into (a) and summing all the resulting stress functions, we have the following stress function, which satisfies the governing Equations 9.5a and 9.5b:

$$\phi(r, z) = \int_0^\infty [\rho I_0(kr) - krI_1(kr)]f(k) \cos(kz)\, dk \tag{9.137}$$

The stresses are obtained using Equations 9.5a

$$\sigma_r = -\int_0^\infty \left[(1 - 2\upsilon - \rho)I_0(kr) + \left(kr + \frac{\rho}{kr}\right)I_1(kr)\right]k^2 f(k) \sin(kz)\, dk \tag{9.138a}$$

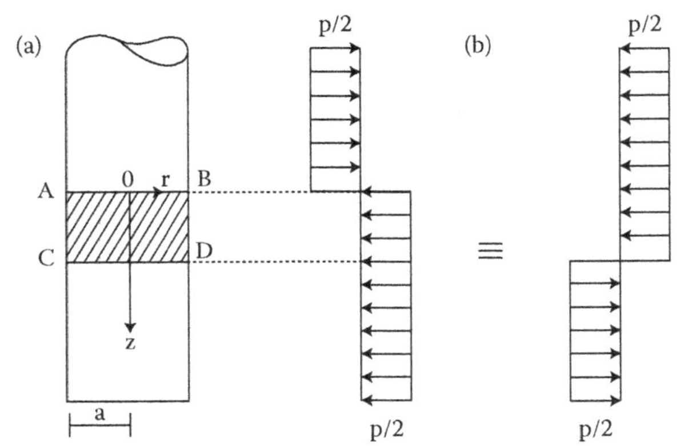

FIGURE 9.30
A long cylinder subjected to band pressure.

$$\tau_{rz} = \int_0^\infty [\rho k I'_0(kr) - k^2 r I'_1(kr) - k I_1(kr) - 2(1-\upsilon) I'_0(kr)] k^2 f(k) \cos(kz)\, dk \qquad (9.138b)$$

where the primes in Equation 9.134b indicate differentiation with respect to kr.

The associated boundary conditions are

i. $\tau_{rz} = 0$ at $r = a$

   Equation 9.134b provides an equation to give the value of the parameter ρ

$$\rho = 2(1-\upsilon) + ka\frac{I_0(ka)}{I_1(ka)} \qquad (c)$$

ii. For $r = a$, $z > 0$, $\sigma_r = p/2$ and if $z < 0$, $\sigma_r = -p/2$

   Making use of the result [9.25,9.30]

$$\frac{p}{\pi}\int_0^\infty \frac{\sin(kz)}{k} dk = \frac{p}{2} \text{ for } z > 0; \quad = 0 \text{ for } z = 0; \quad \text{and } -\frac{p}{2} \text{ for } z < 0$$

   Equation 9.134a furnishes the necessary equation to determine the function f(k)

$$-\left[(1 - 2\upsilon - \rho) I_0(ka) + \left(ka + \frac{\rho}{ka}\right) I_1(ka)\right] k^2 f(k) = -\frac{p}{\pi}\frac{1}{k} \qquad (d)$$

Having determined the unknown parameters ρ and f(k), the stresses are determined using Equations 9.5a and stress function (9.137). The procedure involves evaluation of several integrals similar to those in Equations 9.138a and 9.138b numerically. Curves for the variations of stresses and displacements for pressure bands of several widths are available in References 9.31 and 9.32, see also Reference 5.3.

The main conclusions are summarized below:

1. When the height of the band is equal to the radius of the cylinder, $b = a$, the tangential stress σ_θ at $r = a$ and middle of the band reaches a value about 10% higher than the applied pressure p.
2. The axial stress σ_z at the surface just outside the pressure band reaches a tensile value of about 45% of the applied pressure p.
3. The maximum shear stress τ_{rz} is equal to 31.8% of the applied pressure. It occurs at the edges of the band AB and CD just below the surface of the cylinder.
4. When the pressure is applied all over the curved surface of a cylinder of any length, we have compressive radial and tangential stresses equal to the applied pressure and $\sigma_z = \tau_{rz} = 0$.

Problems

9.1 A circular plate of radius a and thickness 2 c is subjected to a uniform radial stress of magnitude 10 MPa, as shown in Figure P9.1. The following stress function may be used to find the stresses in the plate:

$$\phi(r, z) = A(2z^3 - 3r^2z) + B(r^2z + z^3), \quad \text{A and B are constants.}$$

Show that the stress function is bi-harmonic and determine the stresses. Assume $\nu = 0.3$.

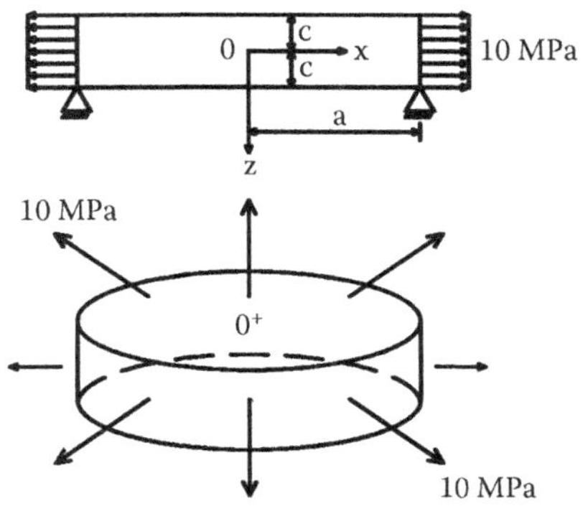

FIGURE P9.1

Ans: $\sigma_r = \sigma_\theta = 10$ MPa, $\sigma_z = \tau_{rz} = 0$.

9.2 Prove that a polynomial of the fourth degree derived from Equations 9.9 and 9.11 and mentioned in Equation (g) of Section 9.2.1 is a stress function. Determine the stresses generated in a circular plate of radius a and thickness 2 c. What is the state of deformation in the plate if the shear stress $\tau_{rz} = 0$? Assume $\nu = 0.3$.

Ans: $\tau_{rz} = \sigma_z = 0$, $\sigma_r = 28\,(1 + \nu)\,B_4\,z$, pure bending.

9.3 A force P = 1000 kip is acting at the origin of rectangular coordinates in the z-direction of a large elastic solid. Assume Poisson's constant = 0.3, find σ_θ at a point distant r = 10 in. from the origin in the (a) z = 0 plane; (b) z = 5 in. plane; (c) z = 10 in. plane.

Ans: 0, 81. 80.4 psi.

9.4 A 1 in. diameter sphere is pressed into a spherical seat of diameter = 2 in. by a force of 0.75 kip. The material of both sphere and seat is steel with E = 36 ksi and $\nu = 0.3$. Determine (a) the maximum contact pressure; (b) the radius of the contact area; (c) the relative displacement between the sphere and seat centers.

Ans: 6.2 kip/in.^{2}, 0.24 in., 0.12 in.

9.5 Repeat Problem 9.4 if the sphere is pressed into a flat surface made of the same material.

Ans: 6.14 kip/in.^{2}, 0.24 in., 0.107 in.

9.6 A 100 mm diameter ball is pressed into an identical ball by a force of 1000 N. Find the contact pressure σ_c. Assume E = 80 GPa.

Ans: 182 MPa.

9.7 Equation 9.100 in the text gives the formula for the opening mode stress-intensity factor k_1 for a pair of concentrated forces of magnitude P applied to the crack surfaces at r = a, $\theta = +\pi$, z = 0.

If the forces are applied to the crack surfaces at $r = a$, $\theta = +\pi$, $z = b$, as shown in Figure P9.7, find k_1

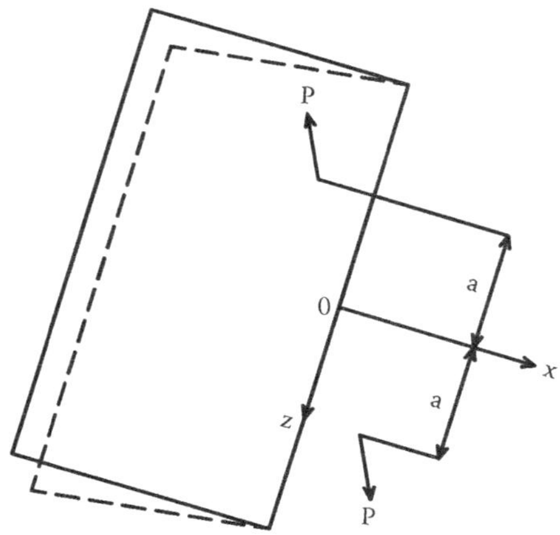

FIGURE P9.7

Ans:

$$k_1 = \frac{\sqrt{2}P}{\pi^2 a^{3/2}}\left(\frac{1}{1+z_1^2}\right)$$

$$z_1 = \frac{z-b}{a}$$

References

9.1. A. E. Love, *A Treatise on the Mathematical Theory of Elasticity*, 4th edition, Dover Publications, NY, 1927.
9.2. F. B. Hildebrand, *Advanced Calculus for Applications*, 2nd edition, Prentice-Hall, Englewood, New Jersey, 1962.
9.3. L. Kelvin, *Cambridge Dublin Math. J.*, 1848.
9.4. R. V. Southwell, *Phil. Magazine*, 1926.
9.5. J. Boussinesq, *Application des potentiels a l'etude de l'equilibreet du Movement des solids elasiques*, Gauthier-Villars, Paris, 1985.
9.6. E. Jahnke, F. Emde and F. Losch, *Tables of Higher Functions*, McGraw-Hill, New York, 1960.
9.7. H. Lamb, *Proc. London Math. Soc.*, vol. 34, p. 276, 1912.
9.8. A. E. H. Love, *Trans. R. Soc. Lond. A*, vol. 228, 1929.
9.9. F. Sehleiehar, *Bauingenieur*, vol. 7, 1926 and vol. 14, p. 242, 1923.
9.10. A. E. Green and W. Zerna, *Theoretical Elasticity*, Oxford University Press, London, 1954.
9.11. L. A. Galin, Contact problems in 3D, in *Contact Problems of the Theory of Elasticity (in Russian)*, English translation by H. Moss, edited by I. N. Sneddon, Department of Mathematics and Engineering Research, North Carolina State College, Rayleigh, NC, 1964, pp. 133–224.
9.12. I. N. Sneddon, *Mixed Boundary Value Problems in Potential Theory*, North-Holland Publishing Company, Amsterdam, 1966.
9.13. P. F. Papkovitch, *Comples rendus hebdomadaies des séances de l'academie des sciences*, Paris, vol. 195, pp. 513–515, 754–756, 1932.

9.14. H. Neuber, *Zeitschrift fur angewandte Mathematik und Mechanik*, vol. 14, p. 103, 1934 or his book: *Theory of Notch Stresses*, pp. 21–26, 1946.
9.15. H. Deresiewiez, The half-space under pressure distributed over an elliptical portion of its plane boundary, *J. Appl. Mech. ASME*, p. 111, 1960.
9.16. K. B. Bandyopadhyay and M. K. Kassir, Contact problems for solids containing cavities, *J. Eng. Mech. Div. ASCE*, vol. 104, No. EM6, pp. 1389–1402, 1978.
9.17. G. N. Watson, *Theory of Bessel Functions*, 2nd edition, Cambridge University Press, Cambridge, England, 1962.
9.18. J. W. Harding and I. N. Sneddon, The elastic stresses produced by the indentation of the plane surface of a semi-infinite solid by rigid punch, *Proc. Camb. Philos. Soc.*, vol. 41, p. 16, 1945.
9.19. C. J. Tranter, *Integral Transforms in Mathematical Physics*, Methuen and Co., London, 1951.
9.20. H. Hertz, *Gesammelte Werke*, vol. 1, Leipzig, p. 155, 1895.
9.21. Y. S. Uflyand, *Survey of Articles on the Application of Integral Transforms in the Theory of Elasticity*, North Carolina State University, 1965.
9.22. M. K. Kassir and G. C. Sih, *Three-Dimensional Crack Problems*, Chapter 5, Noordhoff International Publishing, Leyden, The Netherlands, 1975, pp. 134–160.
9.23. N. N. Lebedev, *Special Functions and Their Applications*, English translation by R. A. Silverman, Prentice-Hall Inc., Englewood Cliffs, NJ, 1965.
9.24. I. N. Sneddon, *The Use of Integral Transforms*, vols. 1 and 2, McGraw-Hill Inc., New York, 1964.
9.25. A. Erdelyi, Editor. *Tables of Integral Transforms*, vols. 1 and 2, McGraw-Hill Inc., New York, 1954.
9.26. I. S. Gradshteyn and I. M. Ryzhik, *Tables of Integrals, Series and Products*, Academic Press, New York, 1965.
9.27. O. D. Kellog, *Foundations of Potential Theory*, Dover Publications, New York, 1953.
9.28. H. Bateman, *Differential Equations*, London, 1926.
9.29. M. K. Kassir and G. C. Sih, Application of Papkovich–Neuber potentials to a crack problem, *Int. J. Solids Struct.*, vol. 9, pp. 643–654, 1974.
9.30. I. Pochhammer, *Crelle's J.*, vol. 81, 1876.
9.31. M. V. Barton, *J. Appl. Mech. ASME*, vol. 8, p. A-97, 1941.
9.32. A. W. Rankine, *J. Appl. Mech. ASME*, vol. 11, p. A-77, 1944.

10

Three-Dimensional Thermal Stresses

This chapter is a supplement to Chapter 8 and deals with analytical methods of determining 3D deformations and thermal stresses induced in bodies due to change in temperature. We begin by summarizing the basic equations required to determine the thermal effects.

10.1 General Equations

As mentioned in Chapter 8, the modified 3D Hooke's law in the presence of changes in temperature for elastic behavior becomes

$$\begin{aligned}\varepsilon_x - \alpha T &= \frac{1}{E}[\sigma_x - \upsilon(\sigma_y + \sigma_z)]\\ \varepsilon_y - \alpha T &= \frac{1}{E}[\sigma_y - \upsilon(\sigma_z + \sigma_x)]\\ \varepsilon_z - \alpha T &= \frac{1}{E}[\sigma_z - \upsilon(\sigma_x + \sigma_y)]\\ \gamma_{xy} = \tau_{xy}/G;\ \gamma_{yz} &= \tau_{yz}/G;\ \gamma_{zx} = \tau_{zx}/G\end{aligned} \tag{10.1}$$

where T(x, y, z) stands for change in temperature, α is the coefficient of thermal expansion, and E and G are the usual elastic moduli of elasticity. Changes in temperature of an isotropic elastic material do not affect the shear stresses.

Adding Equations 10.1 gives the dilatation, e, in terms of components of the first stress invariant I_1 as follows:

$$e = \frac{1}{E}(1 - 2\upsilon)I_1 + 3\alpha T \tag{10.2}$$

Solving for the stresses from Equations 10.1 and making use of Equations 10.2 and 4.9, we obtain

$$\begin{aligned}\sigma_x &= \lambda e + 2G\varepsilon_x - \frac{\alpha E T}{1 - 2\upsilon}\\ \sigma_y &= \lambda e + 2G\varepsilon_y - \frac{\alpha E T}{1 - 2\upsilon}\\ \sigma_z &= \lambda e + 2G\varepsilon_z - \frac{\alpha E T}{1 - 2\upsilon}\end{aligned} \tag{10.3}$$

Inserting relations (10.3) into the equilibrium equations (2.7) without body forces provides

$$\begin{aligned}
(\lambda + G)\frac{\partial e}{\partial x} + G\nabla^2 u - \frac{E\alpha}{1-2\upsilon}\frac{\partial T}{\partial x} &= 0 \\
(\lambda + G)\frac{\partial e}{\partial y} + G\nabla^2 V - \frac{E\alpha}{1-2\upsilon}\frac{\partial T}{\partial y} &= 0 \\
(\lambda + G)\frac{\partial e}{\partial z} + G\nabla^2 w - \frac{E\alpha}{1-2\upsilon}\frac{\partial T}{\partial z} &= 0
\end{aligned} \tag{10.4}$$

By comparing Equations 10.4 with 4.14, it is clear that the body forces in Equation 4.14, B_x, B_y, and B_z are replaced by the terms

$$-\frac{E\alpha}{1-2\upsilon}\left(\frac{\partial T}{\partial x}, \frac{\partial T}{\partial y}, \frac{\partial T}{\partial z}\right)$$

The associated boundary conditions, assuming there are no surface forces, are obtained using Equations 10.4 and 2.13 to provide

$$\begin{aligned}
\frac{\alpha ET}{1-2\upsilon}l &= \lambda e l + G\left(\frac{\partial u}{\partial x}l + \frac{\partial u}{\partial y}m + \frac{\partial u}{\partial z}n\right) + G\left(\frac{\partial u}{\partial x}l + \frac{\partial V}{\partial x}m + \frac{\partial w}{\partial x}n\right) \\
&\ldots\ldots\ldots\ldots\ldots \\
&\ldots\ldots\ldots\ldots\ldots
\end{aligned} \tag{10.5}$$

and two similar equations indicated by the dots above. Note that the terms $(\alpha ET)/(1-2\nu)$ replace the components of the surface forces, p_x, p_y, and p_x, Equation 2.13, in the regular stress problem.

Accordingly, the displacements u, v, and w produced by the temperature change, T(x, y, z), are the same as the displacements produced by the following body forces:

$$\begin{aligned}
B_x &= -\frac{\alpha E}{1-2\upsilon}\frac{\partial T}{\partial x} \\
B_y &= -\frac{\alpha E}{1-2\upsilon}\frac{\partial T}{\partial y} \\
B_z &= -\frac{\alpha E}{1-2\upsilon}\frac{\partial T}{\partial z}
\end{aligned} \tag{a}$$

and the following normal tensions

$$\frac{\alpha ET}{1-2\upsilon} \tag{b}$$

distributed over the surface.

In sum, the solution of Equation 10.4 which satisfies conditions (10.5) is found, giving the displacements u, v, and w, the shearing stresses from the last line in Equations 10.1 and the normal stresses from Equations 10.3. The normal stresses in Equation 10.3 consist of two components, one component derived from the strains and the other from a "hydrostatic" pressure $(\alpha ET)/(1-2\nu)$ at each point of the body. Hence, the total stress produced by nonuniform heating consists of superposing the hydrostatic pressure (b) on the stresses produced by body forces (a).

10.2 Thin Circular Disk (Plane Stress)

In this section, we discuss a different method of determining thermal stresses in thin circular disks from that used in Chapter 8. It is assumed that the applied temperature does not vary with the thickness. The induced displacement and stresses also do not vary over the thickness. The stresses σ_r and σ_θ satisfy the equilibrium equation

$$\frac{d\sigma_r}{dr} + \frac{\sigma_r - \sigma_\theta}{r} = 0 \quad \text{(a)}$$

Owing to symmetry of deformation, the shear stress $\tau_{r\theta}$ vanishes. The strains are due to thermal expansion and stresses in the disk.

$$\begin{aligned} \varepsilon_r - \alpha T &= \frac{1}{E}(\sigma_r - \upsilon\sigma_\theta) \\ \varepsilon_\theta - \alpha T &= \frac{1}{E}(\sigma_\theta - \upsilon\sigma_r) \end{aligned} \quad \text{(b)}$$

Solving for the stresses, we obtain

$$\begin{aligned} \sigma_r &= \frac{E}{1-\upsilon^2}[\varepsilon_r + \upsilon\varepsilon_\theta - (1+\upsilon)\alpha T] \\ \sigma_\theta &= \frac{E}{1-\upsilon^2}[\varepsilon_\theta + \upsilon\varepsilon_r - (1+\upsilon)\alpha T] \end{aligned} \quad \text{(c)}$$

If u(r) denotes the radial displacement, then

$$\varepsilon_r = \frac{du}{dr}, \quad \varepsilon_\theta = \frac{u}{r} \quad \text{(d)}$$

By inserting Equations (c) and (d) into Equation (a), simplifying the algebra, we arrive at the following second-order ordinary differential equation in u(r):

$$\frac{d^2u}{dr^2} + \frac{1}{r}\frac{du}{dr} - \frac{u}{r^2} = (1+\upsilon)\alpha r\frac{dT}{dr} \quad \text{(e)}$$

which can be written as follows:

$$\frac{d}{dr}\left[\frac{1}{r}\frac{d}{dr}(ru)\right] = (1+\upsilon)\alpha\frac{dT}{dr} \quad \text{(f)}$$

The integration of Equation (f) gives

$$u = (1+\upsilon)\alpha\int_a^r Trdr + C_1 r + \frac{C_2}{r} \quad \text{(g)}$$

Here, a in the lower limit of the integral is an arbitrary constant; it is zero for a solid disk and the magnitude of the inner radius for a hollow disk. The constants of integration are C_1 and C_2 to be determined from the boundary conditions.

To determine the stresses, relation (d) is inserted in Equation (g) to obtain the strains and the results are used in conjunction with Equations (c) to yield

$$\sigma_r = -\frac{\alpha E}{r^2}\int_a^r (Trdr) + \frac{E}{1-\upsilon^2}\left[C_1(1+\upsilon) - C_2(1-\upsilon)\frac{1}{r^2}\right]$$

$$\sigma_\theta = \frac{\alpha E}{r^2}\int_a^r (Trdr) - \alpha ET + \frac{E}{1-\upsilon^2}\left[C_1(1+\upsilon) + C_2(1-\upsilon)\frac{1}{r^2}\right] \tag{h}$$

For *a solid* disk of radius b, $C_2 = 0$ in order to eliminate undefined quantities at $r = 0$, a is taken to be zero, and the constant C_1 is found from the boundary condition $\sigma_r = 0$ giving

$$C_1 = (1-\upsilon)\frac{\alpha}{b^2}\int_0^b Trdr$$

It follows that

$$\sigma_r = \alpha E\left[\frac{1}{b^2}\int_0^b Trdr - \frac{1}{r^2}\int_0^r (Trdr)\right] \tag{10.6a}$$

and

$$\sigma_\theta = \alpha E\left[-T + \frac{1}{b^2}\int_0^b Trdr + \frac{1}{r^2}\int_0^r Trdr\right] \tag{10.6b}$$

Equations 10.6 are equivalent to those derived in Section 8.1 of Chapter 8.

Note that for a solid disk the radial stress at $r = 0$ is finite because of the limiting value as r approaches zero of the second integral in Equation 10.6a

$$\frac{1}{r^2}\int_0^r Trdr = \frac{1}{2}T_0$$

where T_0 is the temperature at the center of the disk [5.3].

10.3 Long Circular Cylinder (Plane Strain)

This section deals with determining the deformation and thermal stresses induced in a long circular cylinder (plain strain) with free ends when subjected to symmetrically distributed temperature T(r, z). Such a problem was first solved by Duhamel [10.1]. The stress–strain relations in cylindrical coordinates are

$$\varepsilon_r - \alpha T = \frac{1}{E}[\sigma_r - \upsilon(\sigma_\theta + \sigma_z)]$$
$$\varepsilon_\theta - \alpha T = \frac{1}{E}[\sigma_\theta - \upsilon(\sigma_z + \sigma_r)] \tag{a}$$
$$\varepsilon_z - \alpha T = \frac{1}{E}[\sigma_z - \upsilon(\sigma_r + \sigma_\theta)]$$

First, we assume that the axial displacement in the cylinder is zero, $w = 0$, and then modify the results to account for free ends. If $w = 0$, then the strain in the axial direction $= 0$ and the third equation in Equation (a) gives

$$\sigma_z = \upsilon(\sigma_r + \sigma_\theta) - E\alpha T \tag{b}$$

By inserting Equation (b) into the first two equations of Equation (a) and solving for the strains in the radial and tangential directions, the results are

$$\varepsilon_r - (1+\upsilon)\alpha T = \frac{1-\upsilon^2}{E}\left[\sigma_r - \frac{\upsilon}{1-\upsilon}\sigma_\theta\right]$$
$$\varepsilon_\theta - (1+\upsilon)\alpha T = \frac{1-\upsilon^2}{E}\left[\sigma_\theta - \frac{\upsilon}{1-\upsilon}\sigma_r\right] \tag{c}$$

Note that Equation (c) may be obtained from the corresponding equations for the plane stress case, Equations (b) in Section 10.2, by using the following analogy: In Equations (b) of Section 10.2, replace α by $(1+\nu)\,\alpha$, E by $E/(1-\nu^2)$, and ν by $\nu/(1-\nu)$.

The expressions for displacement u, and radial and tangential stresses for the present case can be written down from Equations (g) and (h) in Section 10.2 by using the just mentioned analogy between plane stress and plane strain. Hence, we have

$$u = \frac{1+\upsilon}{1-\upsilon}\frac{\alpha}{r}\int_a^r (Trdr) + C_1 r + \frac{C_2}{r}$$

$$\sigma_r = -\frac{\alpha E}{1-\upsilon}\frac{1}{r^2}\int_a^r Trdr + \frac{E}{1+\upsilon}\left[\frac{C_1}{1-2\upsilon} - \frac{C_2}{r^2}\right] \tag{d}$$

$$\sigma_\theta = \frac{\alpha E}{(1-\upsilon)r^2}\int_a^r Trdr - \frac{\alpha ET}{1-\upsilon} + \frac{E}{1+\upsilon}\left[\frac{C_1}{1-2\upsilon} + \frac{C_2}{r^2}\right]$$

and Equation (b) furnishes

$$\sigma_z = -\frac{E\alpha T}{1-\upsilon} + \frac{2\upsilon EC_1}{(1+\upsilon)(1-2\upsilon)} \tag{e}$$

In order to maintain $w = 0$ in the cylinder, normal forces distributed at the ends of the cylinder according to Equation (e) need to be applied. Accordingly, we add a uniform axial stress $\sigma_z = C_3$, a constant. By Saint-Venant's principle, such stress gives rise to local effects at the ends.

The remaining stresses in the radial and tangential directions are still given by Equation (d). However, the expression for the displacement u needs a term equal to $-C_3 r (\nu/E)$.

10.3.1 Solid Cylinder

For a solid cylinder, the constant C_2 in Equation (d) must vanish to eliminate the undefined quantities and the lower limit in the integrals for u and the radial and tangential stresses are taken as $a = 0$. The radius of the solid cylinder is $r = b$.

The constant C_1 is determined from the condition that the surface of the cylinder is free from any stress, that is from the boundary condition

$$(\sigma_r)_{r=b} = 0$$

Using Equation (d) with $r = b$, $a = 0$, and $C_2 = 0$, the constant C_1 is found as

$$C_1 = (1 + \upsilon)(1 - 2)\frac{\alpha}{(1 - \upsilon)b^2}\int_0^b Trdr \tag{a}$$

The resultant of the axial stress C_3 is $C_3 (\pi b^2)$ and the resultant of the axial stress in Equation (e) is given as follows:

$$\int_0^b \sigma_z 2\pi rdr = -\frac{2\pi\alpha E}{1 - \upsilon}\int_0^b Trdr + \frac{2\upsilon EC_1}{(1 + \upsilon)(1 - 2\upsilon)} \tag{b}$$

Here, the magnitude of constant C_1 is given in Equation (a).

Since the resultant of axial forces, C_3 and that of the axial stress must vanish, the constant C_3 is found to be given by

$$C_3\pi b^2 = \frac{2\pi\alpha E}{1 - \upsilon}\int_0^b Trdr - \frac{2\upsilon EC_1}{(1 + \upsilon)(1 - 2\upsilon)}\pi b^2 \tag{c}$$

where the constant C_1 is given in Equation (a).

We consider two end conditions of the cylinder and determine in each case the resulting displacement and stresses.

1. *Zero axial strain $\varepsilon_z - 0$*: The displacement u is obtained by combining its values from Equations (d) of Section 10.3 and Equation (a) of Section 10.3.1 to give

$$u = \frac{1 + \upsilon}{1 - \upsilon}\alpha\left[(1 - 2\upsilon)\frac{r}{b^2}\int_0^b Trdr + \frac{1}{r}\int_0^b Trdr\right] \tag{10.7}$$

The radial and tangential stresses are obtained using Equation (d) in Section 10.3, in which C_1 is given in Equation (a) of Section 10.3.1, and $C_2 = 0$. The result is

$$\sigma_r = \frac{\alpha E}{1-\upsilon}\left[\frac{1}{b^2}\int_0^b Trdr - \frac{1}{r^2}\int_0^r Trdr\right]$$
$$\sigma_\theta = \frac{\alpha E}{1-\upsilon}\left[\frac{1}{b^2}\int_0^b Trdr + \frac{1}{r^2}\int_0^r Trdr - T\right] \tag{10.8}$$

The axial stress is obtained from Equation (e) in Section 10.3, C_1 from Equation (a) in Section 10.3.1, and $C_2 = 0$

$$\sigma_z = \frac{\alpha E}{1-\upsilon}\left[\frac{2\upsilon}{b^2}\int_0^b Trdr - T\right] \tag{10.9}$$

2. *Zero axial force $F_z = 0$*: In this case, the radial and tangential stresses are identical to the previous case, that is, Equation 10.8. However, displacement u and axial stress σ_z have different values. For u, we need to add the term $-\nu C_3 (r/E)$ in which C_3 is given in Equation (c) and C_1 from Equation (a). The result is

$$u = \frac{1+\upsilon}{1-\upsilon}\alpha\left[\frac{1-3\upsilon}{1+\upsilon}\frac{r}{b^2}\int_0^b Trdr + \frac{1}{r}\int_0^r Trdr\right] \tag{10.10}$$

For the axial stress, a similar procedure leads to the following expression:

$$\sigma_z = \frac{\alpha E}{1-\upsilon}\left[\frac{2}{b^2}\int_0^b Trdr - T\right] \tag{10.11}$$

EXAMPLE 10.1

The lateral surface of a long cylinder, radius b and temperature T_0 at time zero, is suddenly cooled to temperature T_1, $T_1 < T_0$. Find the stresses induced in the cylinder.

Solution

The induced radial, tangential, and axial stresses are due to heat conduction in the cylinder, cooler heat passes from the surface to other parts of the cylinder. Solutions of the governing equations of heat conduction in solid materials are discussed in detail in many books, for example, References 10.2 through 10.4.

It is found that the distribution of temperature at any instant t is governed by

$$T(r, z) = (T_0 - T_1)\sum_{n=1}^{\infty} A_n J_0\left(\beta_n \frac{r}{b}\right)e^{-p_n t} \tag{a}$$

where J_0 denotes the Bessel function of order zero, β_n, n = 1, 2, 3,... are roots of the equation $J_0(\beta_n) = 0$. The coefficients A_n of the infinite series (a) are given as

$$A_n = \frac{2}{\beta_n J_1(\beta_n)}$$

Here, J_1 is the Bessel function of order one. The constants p_n are given by the equation

$$p_n = \frac{k\beta_n^2}{c\rho b^2} \tag{b}$$

Here, k is the thermal conductivity, c is the specific, and ρ is the density of the material.

Substituting the series (a) into Equations 10.8 and 10.9 and making use of the facts that References 10.5, 10.6 provide

$$\int_0^r J_0\left(\beta_n \frac{r}{b}\right) r dr = \frac{br}{\beta_n} J_1\left(\beta_n \frac{r}{b}\right)$$

and [10.7]

$$\int_0^b J_0\left(\beta_n \frac{r}{b}\right) r dr = \frac{1}{\beta_n} J_1(\beta_n)$$

The stresses in the cylinder are given by

$$\begin{aligned}
\sigma_r &= \frac{2\alpha E(T_0 - T_1)}{1 - \upsilon} \sum_{n=1}^{\infty} \frac{e^{-p_n t}}{\beta_n^2}\left[1 - \frac{b}{r}\frac{J_1\left(\beta_n \frac{r}{b}\right)}{J_1(\beta_n)}\right] \\
\sigma_\theta &= \frac{2\alpha E(T_0 - T_1)}{1 - \upsilon} \sum_{n=1}^{\infty} \frac{e^{-p_n t}}{\beta_n^2}\left[1 + \frac{bJ_1\left(\beta_n \frac{r}{b}\right)}{rJ_1(\beta_n)} - \frac{\beta_n J_0\left(\beta_n \frac{r}{b}\right)}{J_1(\beta_n)}\right] \\
\sigma_z &= \frac{2\alpha E(T_0 - T_1)}{1 - \upsilon} \sum_{n=1}^{\infty} \frac{e^{-p_n t}}{\beta_n^2}\left[2 - \frac{\beta_n J_0\left(\beta_n \frac{r}{b}\right)}{J_1(\beta_n)}\right]
\end{aligned} \tag{c}$$

The three expressions in Equation (c) represent the stresses in the cylinder. A number of numerical examples are available in Reference 10.7.

10.3.2 Cylinder with a Concentric Central Hole

A long cylinder of outer radius b contains a concentric circular hole of radius a, a < b, which is subjected to temperature T(r, z). It is required to find the thermal stresses induced in the cylinder. In this case, the boundary conditions required to determine the constants C_1 and C_2 appearing in Equations (d) and (e) of Section 10.3 are: $\sigma_r = 0$ at r = a and r = b. The results are

$$\begin{aligned}
C_1 &= \frac{1+\upsilon}{1-\upsilon}(1-2\upsilon)\frac{\alpha}{b^2 - a^2}\int_a^b Tr dr \\
C_2 &= \frac{1+\upsilon}{1-\upsilon}\frac{a^2}{b^2 - a^2}\alpha\int_a^b Tr dr
\end{aligned} \tag{a}$$

Inserting Equations (a) into the second and third of Equation (d) in Section 10.3, the radial and tangential stresses in the present case are as follows:

$$\sigma_r = \frac{\alpha E}{1-\upsilon}\frac{1}{r^2}\left[\frac{r^2 - a^2}{b^2 - a^2}\int_a^b Trdr - \int_a^r Trdr\right] \tag{10.12}$$

$$\sigma_\theta = \frac{\alpha E}{1-\upsilon}\frac{1}{r^2}\left[\frac{r^2 + a^2}{b^2 - a^2}\int_a^b Trdr + \int_a^r Trdr - Tr^2\right] \tag{10.13}$$

Inserting Equations (a) into (e) of Section 10.3 and adding from Equation (c) of Section 10.3.1, the axial stress C_3 required to make the resultant axial force zero, the axial stress in the present case is

$$\sigma_z = \frac{\alpha E}{1-\upsilon}\left[\frac{2}{b^2 - a^2}\int_a^b Trdr - T\right] \tag{10.14}$$

The following example illustrates the use of the above equations; see also Reference 10.8 for further details.

EXAMPLE 10.2

A cylinder of radius b contains a concentric hole of radius a, $a < b$, which is subjected to steady-state heat flow such that the temperature at the inner surface is T_i and at the outside surface is T_O, $T_O < T_i$. Determine the induced thermal stresses in the cylinder.

Solution

As discussed in Chapter 8, Section 8.2.1, the steady-state temperature distribution in solid materials is governed by the Laplace equation. In this case, the temperature at any distance r from the center of the cylinder is given by

$$T(r) = (T_i - T_O)\frac{\text{Log}\frac{b}{r}}{\text{Log}\frac{b}{a}} \tag{a}$$

By inserting Equation (a) into Equations 10.12 through 10.14 and evaluating the resulting integrals, the induced stresses are determined

$$\sigma_r = \frac{\alpha E(T_i - T_O)}{2(1-\upsilon)\text{Log}(b/a)}\left[-\text{Log}\frac{b}{r} - \frac{a^2}{b^2 - a^2}\left(1 - \frac{b^2}{r^2}\right)\text{Log}\frac{b}{a}\right] \tag{b}$$

$$\sigma_\theta = \frac{\alpha E(T_i - T_O)}{2(1-\upsilon)\text{Log}(b/a)}\left[1 - \text{Log}\frac{b}{r} - \frac{a^2}{b^2 - a^2}\left(1 + \frac{b^2}{r^2}\right)\text{Log}\frac{b}{r}\right] \tag{c}$$

$$\sigma_z = \frac{\alpha E(T_i - T_O)}{2(1-\upsilon)\text{Log}(b/a)}\left[1 - 2\text{Log}\frac{b}{r} - \frac{2a^2}{b^2 - a^2}\text{Log}\frac{b}{a}\right] \tag{d}$$

Note that if T_i is positive, the radial stress is compressive at all points of the cylinder and takes the value zero at the inner and outer surfaces, at $r = a$ and $r = b$.

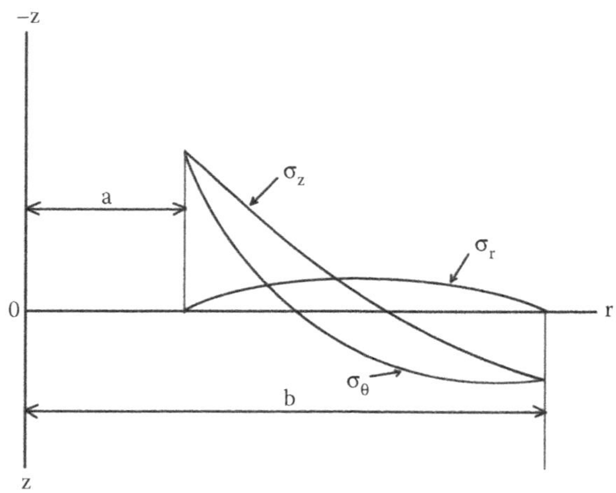

FIGURE 10.1
Thermal stress in thick-walled cylinder with $a/b = 0.3$.

The tangential and axial stresses have their largest numerical values at the inner and outer surfaces of the cylinder. For example, at $r = a$

$$(\sigma_\theta)_{r=a} = (\sigma_z)_{r=a} = \frac{\alpha E(T_i - T_0)}{2(1-\upsilon)\mathrm{Log}(b/a)}\left[1 - \frac{2b^2}{b^2 - a^2}\mathrm{Log}\frac{b}{a}\right] \tag{e}$$

while at $r = b$

$$(\sigma_\theta)_{r=b} = (\sigma_z)_{r=b} = \frac{\alpha E(T_i - T_0)}{2(1-\upsilon)\mathrm{Log}(b/a)}\left[1 - \frac{2a^2}{b^2 - a^2}\mathrm{Log}\frac{b}{a}\right] \tag{f}$$

Charts for calculating numerical values of the stresses are available in Reference 10.9. For the particular case of a cylinder with thickness $a/b = 0.3$, the distribution of thermal stresses over the wall thickness is shown in Figure 10.1, Reference 5.3.

10.4 Thermal Stresses in a Sphere

We consider a sphere subjected to a temperature change symmetrical with respect to the center, that is, the temperature is a function of r, the radial distance only, T(r). Owing to the symmetry there are three nonvanishing stresses, a radial stress σ_r, and two tangential components of stress σ_θ.The corresponding equation of equilibrium is

$$\frac{d\sigma_r}{dr} + \frac{2}{r}(\sigma_r - \sigma_\theta) = 0 \tag{a}$$

The stress–strain relations are

$$\begin{aligned} \varepsilon_r - \alpha T &= \frac{1}{E}[\sigma_r - 2\upsilon\sigma_\theta] \\ \varepsilon_\theta - \alpha T &= \frac{1}{E}[(1-\upsilon)\sigma_\theta - \upsilon\sigma_r] \end{aligned} \tag{b}$$

Solving for the stresses from Equations (b), we have

$$\sigma_r = \frac{E}{(1+\upsilon)(1-2\upsilon)}[(1-\upsilon)\varepsilon_r + 2\upsilon\varepsilon_\theta - (1+\upsilon)\alpha T]$$
$$\sigma_\theta = \frac{E}{(1+\upsilon)(1-2\upsilon)}[\varepsilon_\theta + \upsilon\varepsilon_r - (1+\upsilon)\alpha T] \tag{c}$$

If u(r) denotes the radial displacement, the strains are

$$\varepsilon_r = \frac{du}{dr}, \quad \varepsilon_\theta = \frac{u}{r} \tag{d}$$

By inserting relation (d) into Equations (c) and using Equation (a), we obtain the following differential equation in terms of u(r):

$$\frac{d^2u}{dr^2} + \frac{2}{r}\frac{du}{dr} - \frac{2u}{r^2} = \frac{1+\upsilon}{1-\upsilon}\alpha\frac{dT}{dr}$$

or (e)

$$\frac{d}{dr}\left[\frac{1}{r^2}\frac{d}{dr}(r^2u)\right] = \frac{1+\upsilon}{1-\upsilon}\alpha\frac{dT}{dr}$$

Solving Equation (e), we have

$$u = \frac{1+\upsilon}{1-\upsilon}\alpha\frac{1}{r^2}\int_a^r Tr^2dr + C_1r + \frac{C_2}{r^2} \tag{f}$$

where a is any convenient lower limit of the integral and C_1 and C_2 are constants to be found from the boundary conditions.

By using Equations (f) and (d) to determine the strains and inserting the results in Equations (c), the following stresses are obtained:

$$\sigma_r = -\frac{2\alpha E}{(1-\upsilon)r^3}\int_a^r Tr^2dr + \frac{EC_1}{1-2\upsilon} - \frac{2EC_2}{(1+\upsilon)r^3}$$
$$\sigma_\theta = \frac{\alpha E}{(1-\upsilon)r^3}\int_a^r Tr^2dr + \frac{EC_1}{1-2\upsilon} + \frac{EC_2}{(1+\upsilon)r^3} - \frac{\alpha ET}{1-\upsilon} \tag{g}$$

The constants C_1 and C_2 are determined from the boundary conditions and the lower limit a is taken as zero for a solid sphere or the inner radius of a hollow sphere.

10.4.1 Solid Sphere

For a solid sphere, C_2 must vanish in order to eliminate infinite displacement and stresses. Also, the lower limit of the integral may be chosen as $a = 0$, and the stresses in Equations (g) need to be finite at $r = 0$. The stresses are finite at $r = 0$ because

$$(\mathrm{Lim})_{r\to 0}\frac{1}{r^3}\int_0^r Tr^2dr = \frac{T_0}{3}$$

where T_0 is the temperature at the center, and similarly, it can be shown that $u = 0$ at $r = 0$.

The constant C_1 is found from the condition that the outer surface of the sphere, $r = b$, is free from stress. Equation (g) provides

$$C_1 = \frac{2(1-2\upsilon)\alpha E}{1-\upsilon}\frac{1}{b^3}\int_0^b Tr^2 dr \tag{a}$$

and the stresses are

$$\begin{aligned}\sigma_r &= \frac{2\alpha E}{1-\upsilon}\left[\frac{1}{b^3}\int_0^b Tr^2 dr - \frac{1}{r^3}\int_0^r Tr^2 dr\right]\\ \sigma_\theta &= \frac{\alpha E}{1-\upsilon}\left[\frac{2}{b^3}\int_0^b Tr^2 dr + \frac{1}{r^3}\int_0^r Tr^2 dr - T\right]\end{aligned} \tag{10.15}$$

Note that the volume of a sphere of radius r, $r < b$, is equal to $(4\pi r^3)/3$, and the mean temperature for such a sphere is

$$\left[4\pi\int_0^r Tr^2 dr\right]/[(4\pi r^3)/3] = \frac{3}{r^3}\int_0^r Tr^2 dr$$

Hence, the radial stress σ_r in Equation 10.15 is proportional to the difference between the mean temperature of the whole sphere and the mean temperature of a sphere of radius r. Once T(r) is known, the stresses can be determined.

10.4.2 Sphere with a Spherical Hole at the Center

Consider a sphere of outer radius b containing a spherical hole of radius a. The constants C_1 and C_2 in Equation (g) above are determined from the conditions that $\sigma_r = 0$ at $r = a$ and $r = b$. By finding the constants and inserting them in Equations (g), the following stresses are obtained:

$$\begin{aligned}\sigma_r &= \frac{2\alpha E}{1-\upsilon}\left[\frac{r^3-a^3}{b^3-a^3}\frac{1}{r^3}\int_a^b Tr^2 dr - \frac{1}{r^3}\int_a^r Tr^2 dr\right]\\ \sigma_\theta &= \frac{2\alpha E}{1-\upsilon}\left[\frac{2r^3+a^3}{2(b^3-a^3)}\frac{1}{r^3}\int_a^b Tr^2 dr + \frac{1}{2r^3}\int_a^r Tr^2 dr - \frac{1}{2}T\right]\end{aligned} \tag{10.16}$$

Once the temperature T(r) is known, the displacement and stresses are obtained.

EXAMPLE 10.3

Consider a sphere, outer radius b, containing a central hole of radius a. The sphere is subjected to a steady-state heat flow in which the temperature at the inner surface is T_i and at the outer surface is zero. Find the induced stress components.

Solution

The steady-state temperature at any distance r from the center is

$$T(r) = \frac{T_i a}{b - a}\left(\frac{b}{r} - 1\right) \tag{a}$$

By inserting Equation (a) into Equations 10.16 and evaluating the integrals, the stresses are

$$\sigma_r = \frac{\alpha E T_i}{1 - \upsilon}\frac{ab}{b^2 - a^2}\left[a + b - \frac{1}{r}(b^2 + ab + a^2) + \frac{a^2 b^2}{r^3}\right]$$

$$\sigma_\theta = \frac{\alpha E T_i}{1 - \upsilon}\frac{ab}{b^2 - a^2}\left[a + b - \frac{1}{2r}(b^2 + ab + a^2) - \frac{a^2 b^2}{2r^3}\right] \tag{b}$$

The radial stress σ_r is zero at $r = a$ and $r = b$, and it has a turning point (maximum or minimum value) when r is given by

$$r^2 = \frac{3a^2 b^2}{a^2 + ab + b^2} \tag{c}$$

When $T_i > 0$, the tangential stress σ_θ increases as r increases, and its values at the inner and outer surfaces, $r = a$, $r = b$, are as follows:

$$r = a: \quad \sigma_\theta = -\frac{\alpha E T_i}{2(1 - \upsilon)}\frac{b(b - a)(a + 2b)}{b^3 - a^3} \tag{d}$$

$$r = b: \quad \sigma_\theta = \frac{\alpha E T_i}{2(1 - \upsilon)}\frac{a(b - a)(2a + b)}{b^3 - a^3} \tag{e}$$

For a spherical of small thickness, we write $b = a + x$, where x is a small quantity and higher powers in x can be neglected; then for $r = a$

$$\sigma_\theta = -\frac{\alpha E T_i}{2(1 - \upsilon)}\left[1 + \frac{2}{3}x\right]$$

and for $r = b$

$$\sigma_\theta = \frac{\alpha E T_i}{2(1 - \upsilon)}\left[1 - \frac{2}{3}x\right]$$

(see Section 8.3.1 of Chapter 8 for a different derivation of this problem)

10.5 Thermoelastic Displacement Potential

A displacement potential exists which forms a particular solution of the 3D thermal stress problem governed by Equations 10.4. Any such particular solution will reduce Equations 10.4 to an ordinary problem of surface forces.

Such a displacement potential is obtained by writing the components of the displacement, u, v, and w, in the following form:

$$u = \frac{\partial \psi}{\partial x}, \quad V = \frac{\partial \psi}{\partial y}, \quad w = \frac{\partial \psi}{\partial z} \tag{a}$$

where Ψ, the thermos-elastic displacement potential, is a function of x, y, and z, and also of time t if the temperature varies with time. By inserting Equation (a) in Equations 10.4 and making use of the elastic constants given in Equations 4.6 and 4.9 of Chapter 4, Equations 10.4 become

$$\begin{aligned}
\frac{\partial e}{\partial x} + (1-2\upsilon)\nabla^2 u &= 2(1+\upsilon)\alpha\frac{\partial T}{\partial x} \\
\frac{\partial e}{\partial y} + (1-2\upsilon)\nabla^2 V &= 2(1+\upsilon)\alpha\frac{\partial T}{\partial y} \\
\frac{\partial e}{\partial z} + (1-2\upsilon)\nabla^2 w &= 2(1+\upsilon)\alpha\frac{\partial T}{\partial z}
\end{aligned} \tag{10.17}$$

Here, e is the dilatation, $e = (\partial u/\partial x) + (\partial V/\partial y) + (\partial w/\partial z)$ since

$$e = \nabla^2 \psi \tag{b}$$

By inserting Equation (b) in Equations 10.17, the following equations are obtained:

$$\begin{aligned}
(1-\upsilon)\frac{\partial}{\partial x}\nabla^2\psi &= (1+\upsilon)\alpha\frac{\partial T}{\partial x} \\
(1-\upsilon)\frac{\partial}{\partial y}\nabla^2\psi &= (1+\upsilon)\alpha\frac{\partial T}{\partial y} \\
(1-\upsilon)\frac{\partial}{\partial z}\nabla^2\psi &= (1+\upsilon)\alpha\frac{\partial T}{\partial z}
\end{aligned} \tag{10.18}$$

Equations 10.18 are satisfied if Ψ is the solution of the potential equation

$$\nabla^2\psi = \frac{1+\upsilon}{1-\upsilon}\alpha T \tag{10.19}$$

A solution of Equation 10.19 can be written as the gravitational potential of a body of density

$$-\frac{(1+\upsilon)\alpha T}{4\pi(1-\upsilon)} \tag{c}$$

Such a solution assumes the form [10.10]

$$\psi = -\frac{(1+\upsilon)\alpha}{4\pi(1-\upsilon)}\iiint_V T(\xi,\eta,\zeta)\frac{d\xi\,d\eta\,d\zeta}{R} \tag{10.20}$$

where T is the temperature at a point in an element of volume $d\xi\,d\eta\,d\zeta$, R is the distance between this point and the point x, y, z, and V is the volume of the element. Equations 10.17 through 10.20 provide the displacement and stresses in an infinite solid at temperature zero except for a heated (cooled) region of volume V.

References 10.11–10.13 provide solutions for uniformly hot regions in the form of an ellipsoid of revolution and semi-infinite circular cylinder.

For the ellipsoidal region, the maximum stress occurs normal to the surface of the ellipsoid at the two points of the sharpest curvature at the ends of the major axis and has the value

$$\frac{\alpha ET}{1-\upsilon}$$

For a spherical region at the same temperature, the maximum value is two-thirds as great [5.3].

When T is independent of z, there are two cases to be considered: a plane strain case and a thin case in which the temperature is constant through the thickness.

10.5.1 Plane Strain Case

In this case, T is independent of z and $w = 0$. Also, displacements u and V are independent of z and Equation 10.19 becomes

$$\frac{\partial^2\psi}{\partial x^2} + \frac{\partial^2\psi}{\partial y^2} = \frac{1+\upsilon}{1-\upsilon}\alpha T \tag{10.21}$$

A particular solution of Equation 10.21 is the following *logarithmic* potential:

$$\psi(\xi,\eta) = \frac{1}{2\pi}\frac{1+\upsilon}{1-\upsilon}\alpha \iint_A T(\xi,\eta)\mathrm{Log}(R)\, d\xi\, d\eta \tag{d}$$

$$\text{where} \quad R = [(x-\xi)^2 + (y-\eta)^2]^{1/2}$$

10.5.2 Plane Stress (Thin Plate)

In this case, there is no variation of temperature through the thickness of the element. The stresses σ_z, τ_{xz}, and τ_{yz} vanish, and u, v, σ_x, σ_y, τ_{xy} are independent of z. The stress–strain relations are obtained by solving the following equations:

$$\begin{aligned} \varepsilon_x - \alpha T &= \frac{1}{E}(\sigma_x - \upsilon\sigma_y) \\ \varepsilon_y - \alpha T &= \frac{1}{E}(\sigma_y - \upsilon\sigma_x) \end{aligned} \tag{a}$$

giving

$$\begin{aligned} \sigma_x &= \frac{E}{1-\upsilon^2}\left[\frac{\partial u}{\partial x} + \upsilon\frac{\partial V}{\partial y} - (1+\upsilon)\alpha T\right] \\ \sigma_y &= \frac{E}{1-\upsilon^2}\left[\frac{\partial V}{\partial y} + \upsilon\frac{\partial u}{\partial x} - (1+\upsilon)\alpha T\right] \\ \tau_{xy} &= \frac{E}{2(1+\upsilon)}\left(\frac{\partial u}{\partial y} + \frac{\partial V}{\partial x}\right) \end{aligned} \tag{b}$$

By inserting Equations (b) in the equilibrium equations for the stresses without body forces, Equations 2.6 in Chapter 2, and simplifying the algebra, we find (see problem P 10.6 for hints)

$$\begin{aligned} \frac{\partial}{\partial x}\left(\frac{\partial u}{\partial x}+\frac{\partial V}{\partial y}\right)+\frac{1-\upsilon}{1+\upsilon}\nabla^2 u &= 2\alpha\frac{\partial T}{\partial x} \\ \frac{\partial}{\partial y}\left(\frac{\partial V}{\partial y}+\frac{\partial u}{\partial x}\right)+\frac{1-\upsilon}{1+\upsilon}\nabla^2 V &= 2\alpha\frac{\partial T}{\partial y} \end{aligned} \tag{c}$$

Equations (c) are satisfied by writing

$$\begin{aligned} u(x,y) &= \frac{\partial\psi(x,y)}{\partial x} \\ V(x,y) &= \frac{\partial\psi(x,y)}{\partial y} \end{aligned} \tag{d}$$

By inserting Equation (d) into (c), we obtain

$$\frac{\partial^2\psi}{\partial x^2}+\frac{\partial^2\psi}{\partial y^2} = (1+\upsilon)\alpha T(x,y) \tag{10.22}$$

By comparing Equation 10.22 with 10.21, it is clear that the solution is the same as the logarithmic potential in Equation (d) in Section 10.5.1 with the factor $(1-\nu)$ in the denominator omitted.

EXAMPLE 10.4

Determine the induced deformation and stresses in an infinite thin plate maintained at zero temperature except a rectangular region ABCD of sides 2a by 2b in which the temperature is maintained at a uniform T(x, y), Figure 10.2.

Solution

A complete solution of this problem is available in Reference 10.11. The required logarithmic potential is

$$\psi = \frac{1}{2\pi}(1+\upsilon)\alpha T\int_{-a}^{a}\int_{-b}^{b}\mathrm{Log}\left[(x-\xi)^2+(y-\eta)^2\right]d\xi\, d\eta \tag{a}$$

The displacements u and V are obtained by differentiating the logarithmic potential, Equation (a) above, as indicated in Equation (a) of Section 10.5.

The stresses at points outside the hot rectangular area such as point P in Figure 10.2 can be reduced to

$$\begin{aligned} \sigma_x &= \frac{E\alpha T}{2\pi}(\theta_1-\theta_2) \\ \tau_{xy} &= \frac{E\alpha T}{4\pi}\mathrm{Log}\frac{r_1 r_3}{r_2 r_4} \end{aligned} \tag{b}$$

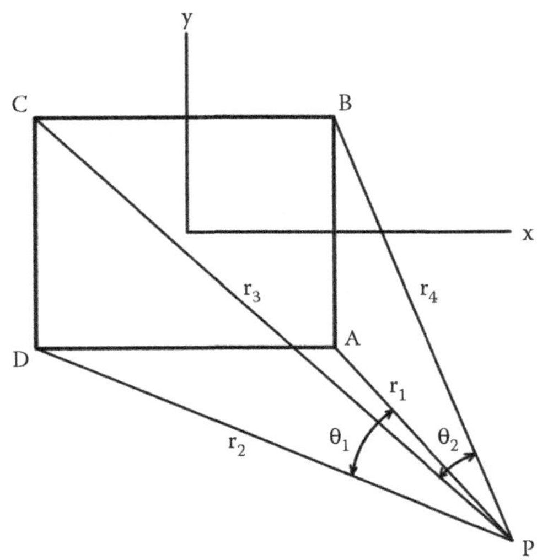

FIGURE 10.2
Rectangular heated region (Example 10.4).

where the angles θ_1, θ_2 and distances r_1, r_2, r_3, and r_4 are shown in Figure 10.2. The angles at P are θ_1 between lines PA and PD and θ_2 between lines PB and PC.

The stress σ_y turns out to be the same as σ_x above if the two angles are replaced by the angles subtended at P by the other two sides AB and CD of the rectangle.

The magnitude of σ_x just below line AD to the left of point A is

$$\frac{E\alpha T}{2\pi}\left(2\pi - \tan^{-1}\frac{a}{b}\right) \tag{c}$$

It becomes maximum for a rectangle which is infinitely long in the y direction and has the magnitude (1/2) αET.

The magnitude of shear stress τ_{xy} becomes infinity at a corner of the heated rectangle as expected because of the presence of a sharp corner.

Examples of other heated regions in an infinite plate: The problem of a heated ellipse $x^2/a^2 + y^2/b^2 = 1$ in an infinite thin plate at zero temperature is solved in Reference 10.11.

The magnitude of stress σ_y near the end of the major axis just outside the ellipse is

$$\frac{E\alpha T}{1 + (b/a)}$$

which approaches $E\alpha T$ for a very slender ellipse.

The magnitude of stress σ_x just beyond the end of the minor axis is

$$\frac{E\alpha T}{1 + (a/b)}$$

which approaches zero for a very slender ellipse.

For a heated circular region in an infinite plate, $a = b$, the above values become (1/2) $E\alpha T$.

10.6 Structural Members Containing Flaws or Cracks

When a temperature field is disturbed by the presence of flaws or cracks in a structural member, there is a high elevation of the temperature gradient accompanied by very large thermal stress at the crack edge. Thermal disturbances of this kind may in some cases cause crack propagation resulting in severe damage to the member.

Neglecting the influences of both inertia and coupling between the temperature and strain fields, the problem of determining the induced deformation and thermal stresses separates into two distinct boundary value problems to be solved consecutively:

1. The first is a potential problem in the theory of steady-state heat flow conduction which requires finding the temperature T(x, y, z) at every point of the member using solutions of the Laplace equation in three dimensions

$$\nabla^2 T(x, y, z) = 0 \tag{a}$$

A convenient alternative way of finding the temperature is to introduce a thermoelastic potential Ω(x, y, z) such that

$$\frac{\partial \Omega}{\partial z} = \frac{(1+\upsilon)}{2(1-\upsilon)} \alpha T \tag{b}$$

and then using potential theory to find the heat flow potential function.

2. The second boundary value problem involves finding the induced deformation and stresses by solving the following displacement equations of static equilibrium:

$$(1-2\upsilon)\nabla^2 u + \nabla\nabla \cdot u = 2(1+\upsilon)\alpha \nabla T \tag{c}$$

and then using the corresponding stress–displacement relation to find the stresses, References 10.14 through 10.16.

In Equation (c), u denotes the displacement vector, $u = u_x i + u_y j + u_z k$, ∇ and ∇^2 are the gradient and Laplacian operators, and ν,α are Poisson's ratio and the coefficient of linear expansion of the solid whose mechanical and thermal properties are assumed to be elastic, isotropic, and homogeneous.

The shapes of internal flaws or cracks considered include circular, elliptical, and semi-infinite planes of discontinuities. An external circular crack or plane of discontinuity is also considered. For each crack surface, we can prescribe either the temperature (symmetrical problem) or temperature gradient (skew-symmetric problem). In each case, the local thermal stresses at crack edges are determined and formulas for the associated stress-intensity factors (Chapter 7, Section 7.47) are derived.

10.6.1 Circular Crack

The geometry of internal circular crack in an infinite solid is of practical importance since it represents an idealization of the shape of internal flaws or imperfections that are present in many engineering materials. In terms of cylindrical coordinates r, θ, z, the crack is located in the $z = 0$ plane of an infinite homogeneous and isotropic solid, centered at the origin of coordinates, and has a radius equal to a, Figure 10.3.

The steady-state temperature field, Equation (a) of the previous article, is governed by the equation

$$\frac{\partial^2 T}{\partial r^2} + \frac{1}{r}\frac{\partial T}{\partial r} + \frac{\partial^2 T}{\partial z^2} = 0 \tag{10.23}$$

whose solution may be represented by

$$T(r, z) = \int_0^\infty sA(s)J_0(rs)e^{-sz}\, ds \tag{10.24}$$

where $J_0(rs)$ is the Bessel function of order zero and A(s) is a parameter to be determined from the boundary conditions.

Prescribed temperature: Consider a circular of radius a whose surfaces are heated and maintained at a certain temperature, $T_0(r)$, Figure 10.3. The region outside of the crack on the plane $z = 0$ is insulated and free of stress. These boundary conditions on the crack plane are expressed as follows:

$$\begin{aligned} &T(r, 0) = T_0(r), \quad 0 \le r \le a \\ &\left(\frac{\partial T}{\partial z}\right)_{z=0} = 0, \quad r \ge a \end{aligned} \tag{10.25a}$$

and

$$\begin{aligned} &\tau_{rz}(r, 0) = 0, \quad r \ge 0 \\ &\sigma_z(r, 0) = 0, \quad 0 \le r \le a \\ &u_z(r, 0) = 0, \quad r \ge a \end{aligned} \tag{10.25b}$$

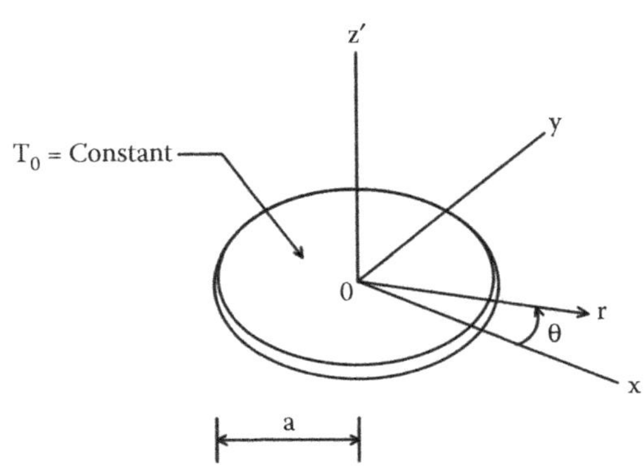

FIGURE 10.3
Circular crack subjected to constant temperature.

A solution which satisfies equilibrium and boundary Equations 10.25 can be expressed in terms of two potential functions, Reference 10.14, as follows:

$$u_r = (1 - 2\upsilon)\frac{\partial f}{\partial r} + z\frac{\partial F}{\partial r} + \int_z^\infty \frac{\partial \Omega}{\partial r} dz$$

$$u_z = -2(1 - \upsilon)\frac{\partial f}{\partial x} + z\frac{\partial F}{\partial z} \tag{10.26a}$$

The z-components of the stress tensor corresponding to the displacements in Equations 10.26a are

$$\sigma_z = 2G\left(-\frac{\partial F}{\partial z} + \frac{\partial^2 F}{\partial z^2}\right)$$

$$\tau_{rz} = 2Gz\frac{\partial^2 F}{\partial r \partial z} \tag{10.26b}$$

In Equations 10.26a and 10.26b, F (r, z) stands for

$$F(r, z) = \Omega + \frac{\partial f}{\partial z} \tag{10.26c}$$

where the potential Ω is related to the temperature T(x, y, z) by Equation (b) in the previous article.

It follows that the thermal and mechanical boundary conditions in Equations 10.25a and 10.25b may be expressed in terms of the potential function f(r, z) as follows:

$$\frac{\partial^2 f}{\partial z^2} = -\frac{\alpha}{2}\left(\frac{1 + \upsilon}{1 - \upsilon}\right)T_0(r), \quad z = 0, 0 \leq r \leq a$$

$$\frac{\partial f}{\partial z} = 0, \quad z = 0, r \geq a \tag{10.27}$$

Using dual integral equations to solve for the function f(r, z) from Equation 10.27, the displacements and stresses in the solid are obtained and the thermal stress-intensity factor at the edge of the crack is determined. Without going into the details, the result is [10.16]

$$k_1 = -\frac{\alpha E}{\pi(1 - \upsilon)a^{1/2}}\int_0^a T_0(r)\frac{r dr}{(a^2 - r^2)^{1/2}} \tag{10.28}$$

If $T_0(r)$ is constant, say T_0, the entire crack surfaces are uniformly covered with constant temperature different from that of the surrounding material in the infinite solid, and the stress-intensity factors are

$$k_1 = \frac{\alpha E T_0 a^{1/2}}{\pi(1 - \upsilon)}$$

$$k_2 = k_3 = 0 \tag{10.29}$$

Temperature gradient: Alternatively, the temperature gradient or the heat flux on the crack surfaces can be specified. There are two cases to consider: case (i) involves specifying the temperature gradient on crack surfaces and zero temperature gradient on the plane $z = 0$ outside the crack region, and case (ii) considers specifying the temperature gradient on crack surfaces and temperature on the outside region of the $z = 0$ plane.

Case (i)—The temperature gradients on the plane $z = 0$ are

$$\left(\frac{\partial T}{\partial z}\right)_{z=0} = Q(r), \quad 0 \leq r \leq a$$
$$\left(\frac{\partial T}{\partial z}\right)_{z=0} = 0, \quad r \geq a \tag{10.30}$$

The temperature field is determined from Equation 10.24 where the parameter A(s) is

$$sA(s) = \int_0^a rQ(r)J_0(rs)\, ds \tag{10.31}$$

Using Equation 10.24 to determine the temperature field throughout the solid, it follows by the same procedure used to derive the formula given in Equation 10.28 that in this case

$$k_1 = \frac{\alpha E}{2(1-\upsilon)a^{1/2}} \int_0^a rQ(r)\, dr \tag{10.32}$$

For the constant temperature gradient, Q_0, we have k_1 from above and $k_2 = k_3 = 0$.

$$k_1 = \frac{\alpha E Q_0 a^{3/2}}{4(1-\upsilon)}$$
$$k_2 = k_3 = 0$$

Case (ii)—On the plane $z = 0$, we have the boundary conditions

$$\left(\frac{\partial T}{\partial z}\right)_{z=0} = Q(r), \quad 0 \leq r \leq a$$
$$T(r, 0) = 0, \quad r \geq a \tag{10.33}$$

The conditions of skew symmetry across the $z = 0$ plane require that

$$\sigma_z(r, 0) = 0, \quad r \geq a \tag{10.34a}$$

and the stress free conditions on the crack surfaces are

$$\tau_{rz}(r, 0) = 0, \quad 0 \leq r \leq a$$
$$u_r(r, 0) = 0, \quad r \geq a \tag{10.34b}$$

The appropriate displacement field for this problem is given by [10.15]

$$\begin{aligned} u_r &= 2(1-\upsilon)\frac{\partial g}{\partial r} + z\frac{\partial G}{\partial r} \\ u_z &= \Omega - (1-2\upsilon)\frac{\partial g}{\partial z} + z\frac{\partial G}{\partial z} \end{aligned} \tag{10.35a}$$

where g(r) is a harmonic function and

$$G = \Omega + \frac{\partial g}{\partial z} \tag{10.35b}$$

The stresses on the z-plane are

$$\begin{aligned} \sigma_z &= 2\mu z\frac{\partial^2 G}{\partial z^2} \\ \tau_{rz} &= 2\mu\frac{\partial}{\partial r}\left(G + z\frac{\partial G}{\partial z}\right) \end{aligned} \tag{10.35c}$$

In Equation (10.35c), μ denotes the shear modulus of the solid. The boundary conditions (10.34b) when expressed in terms of g(r, z) become

$$\begin{aligned} &\frac{\partial^2 g}{\partial r \partial z} = -\frac{\partial \Omega}{\partial r}, \quad z = 0, r \le a \\ &\frac{\partial g}{\partial r} = 0, \quad z = 0, r \ge a \end{aligned} \tag{10.36}$$

Making use of Equations 10.24, 10.33, and (b) in the previous article, the following relation is obtained:

$$\left(\frac{\partial \Omega}{\partial r}\right)_{z=0} = \frac{2}{\alpha}\left(\frac{1+\upsilon}{1-\upsilon}\right)\frac{1}{r}\int_0^r rQ(r)dr, \quad r \le a \tag{10.37}$$

The function g(r) in Equation 10.36 can be found by following the procedure used in Reference 10.17, and the stress-intensity factor for this case becomes

$$k_2 = -\frac{\alpha E}{\pi(1-\upsilon)a^{3/2}}\int_0^a rQ(r)\left(a^2 - r^2\right)^{1/2} dr \tag{10.38}$$

For $Q(r) = Q_0$, constant, Equation 10.38 gives

$$\begin{aligned} k_2 &= \frac{\alpha E Q_0 a^{3/2}}{3(1-\upsilon)\pi} \\ k_1 &= k_3 = 0 \end{aligned} \tag{10.39}$$

which provides the result when a uniform heat flow of constant gradient Q_0 in an infinite elastic solid is disturbed by an insulated penny-shaped crack of radius a. The crack is located

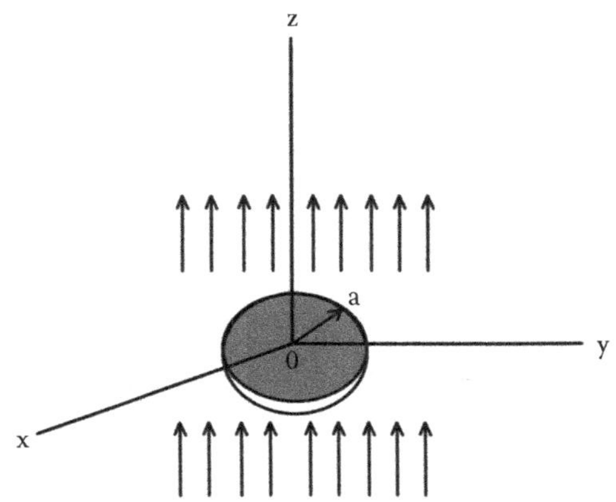

FIGURE 10.4
Uniform heat flow disturbed by an insulated circular crack.

normal to the direction of the heat flow, Figure 10.4. This problem was first solved in Reference 10.18.

Next, we present results of a few other cases concerning circular cracks and the references where full solutions can be found.

Two parallel circular cracks: Constant temperatures T_0 are applied to the surfaces of two circular cracks of equal radii a spaced at a distance 2b apart, Figure 10.5. The cracks are embedded in an otherwise isothermal solid free of mechanical tractions. For small values of the ratio a/b, the k-factors can be approximated by [10.19]

$$
\begin{aligned}
k_1 &= \frac{\alpha E T_0}{\pi(1-\upsilon)}\left[1 + 0.318\left(\frac{a}{b}\right) + 0.101\left(\frac{a}{b}\right)^2 - 0.339\left(\frac{a}{b}\right)^3 + \cdots\right] \\
k_2 &= \frac{2\alpha E T_0}{\pi^2(1-\upsilon)}\left[-1 + 0.3\left(\frac{a}{b}\right) + 0.232\left(\frac{a}{b}\right)^2 - 0.021\left(\frac{a}{b}\right)^3 + \cdots\right] \\
k_3 &= 0
\end{aligned}
\tag{10.40}
$$

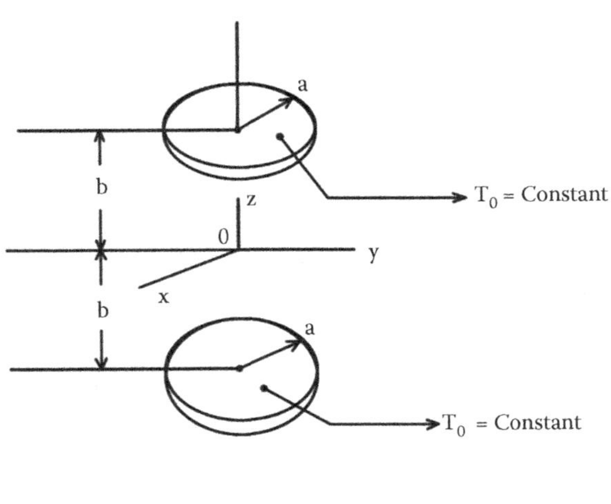

FIGURE 10.5
Two parallel circular cracks subjected to constant temperature.

Heat flow around two parallel insulated cracks: A heat flow with a constant gradient Q_0 passes by two parallel penny-shaped cracks of radius a. The cracks are spaced at a distance 2b apart and insulated such that heat cannot travel through them, Figure 10.6. The approximate solution for small values of the ratio a/b gives the following stress-intensity factors [10.19]

$$\begin{aligned} k_1 &= \frac{\alpha E Q_0 a}{3\pi^2(1-\upsilon)}\left(\frac{a}{b}\right)^2\left[-1+\left(\frac{a}{b}\right)^2+\cdots\right] \\ k_2 &= \frac{\alpha E Q_0 a}{3\pi(1-\upsilon)}\left[-1-0.318\left(\frac{a}{b}\right)+0.122\left(\frac{a}{b}\right)^3+\cdots\right] \\ k_3 &= 0 \end{aligned} \tag{10.41}$$

Another interesting problem consists of two semi-infinite elastic solids of different thermal and mechanical properties, joined together along a common border in the plane $z=0$ and subjected to temperature changes. The plane $z=0$ contains a flaw or imperfection which can be modeled by a circular crack. Next, the thermal stresses induced in the dissimilar materials are discussed.

Thermal stresses in dissimilar media: When two different materials are bonded together, changes in temperature will cause the materials to expand or contract depending on their coefficients of thermal expansion. High elevation of thermal stresses is produced near cracks or the surface of discontinuities around which the materials cannot expand or contract

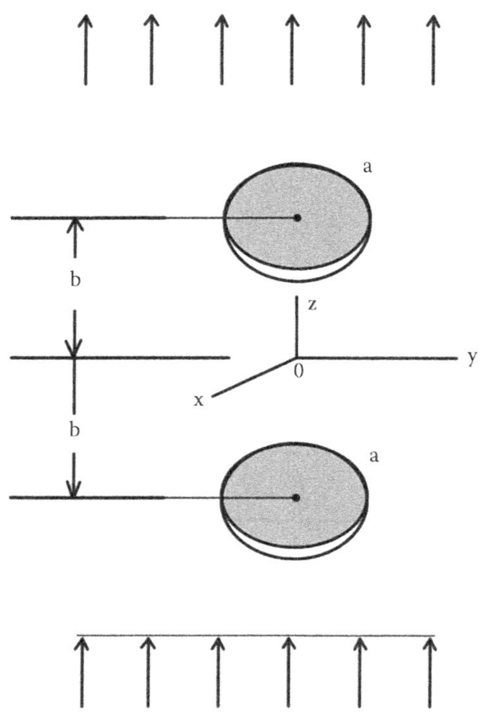

FIGURE 10.6
Heat flow around two parallel insulated circular cracks.

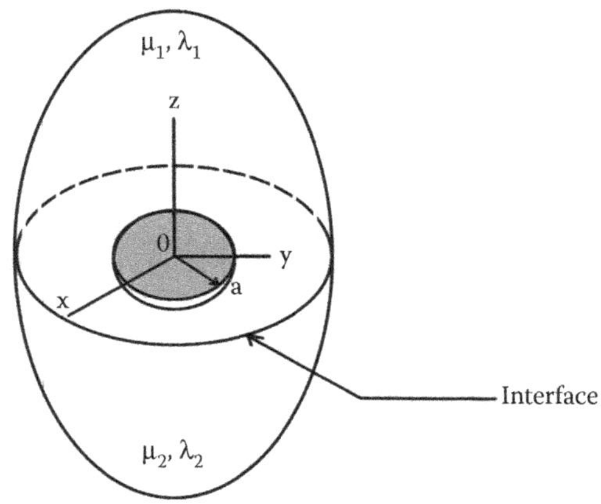

FIGURE 10.7
An interface circular crack between two solids.

freely. Problems of this type can be formulated and solved in the same way as those involving mechanical loadings [10.15, 10.20].

The geometry of the thermal stress problem is shown in Figure 10.7 where a uniform heat flow passes through the dissimilar media containing a penny-shaped crack at the interface. In addition to specifying the mechanical properties λ_j, μ_j, $j = 1, 2$, of the two bonded materials, it is also necessary to know the thermal conductivity κ_1 and coefficient of linear thermal expansion α_1 of the material in region $z > 0$ and κ_2, α_2 for region $z < 0$. Quantities such as temperatures, temperature gradients, displacements, and stresses are assumed to be continuous across the interface outside the crack region where the two materials are bonded.

Let a constant temperature gradient of magnitude Q_0 be applied in the z-direction. In the absence of the crack, the temperature field will be undisturbed and given by

$$Q_0 z \ \text{ for } z > 0 \quad \text{and} \quad (k_1/k_2)Q_0 z \ \text{ for } z < 0$$

The total temperature field with the crack present may be written as

$$[T(r, z)]_{z=0} = Q_0 z + T_1(r, z), \quad z > 0 \tag{a}$$

$$[T(r, z)]_{z=0} = (\kappa_1/\kappa_2)Q_0 z + T_2(r, z), \quad z < 0$$

in which T_j (r, z), $j = 1, 2$, represent the undisturbed temperatures due to the interface crack. If the crack surfaces are insulated, then T_j, $j = 1, 2$, which are governed by Equation 10.23 subject to the boundary conditions

$$\begin{aligned} &\frac{\partial T_1}{\partial z} = -Q_0; \ \frac{\partial T_2}{\partial z} = -\frac{\kappa_1}{\kappa_2} Q_0; \quad \text{for } z = 0 \text{ and}, 0 \le r \le a \\ &\kappa_1 \frac{\partial T_1}{\partial z} = \kappa_2 \frac{\partial T_2}{\partial z}; T_1 = T_2; \quad \text{for } z = 0 \text{ and}, r > a \end{aligned} \tag{b}$$

and the requirements that all the quantities must remain finite at infinity are to be satisfied.

It is not difficult to show that the required temperature field is given by

$$T_1(r,z) = \left(\frac{2}{\pi}\right)^{1/2} Q_0 a^{1/2} \int_a^{\infty} s^{-1/2} J_{3/2}(as) J_0(rs) e^{-sz}\, ds$$

$$T_2(r,z) = -\left(\frac{2}{\pi}\right)^{\frac{1}{2}} \left(\frac{\kappa_1}{\kappa_2}\right) Q_0 a^{1/2} \int_0^{\infty} s^{-1/2} J_{3/2} J_0(rs) e^{sz}\, ds \tag{10.42}$$

where J_n is the Bessel function of the first kind of order n.

The induced displacements and stresses in the dissimilar media can be found with the aid of the thermos-elastic potential introduced in Equation (b) of Section 10.6 and the use of harmonic functions as outlined in Reference 10.15.

The local stress field is always characterized by both the stress-intensity factors k_1 and k_2. The thermal stresses possess the same square singularity as in the case of mechanical loading. They also exhibit oscillations of the form sin (or cos) ε log r where ε is a bi-material constant.

The stress-intensity factors k_1 and k_2 are obtained in the limit as r approaches the boundary of the circular a. They are [10.20]

$$k_1 = \frac{4(\mu_2)^2}{(\pi a)^{\frac{1}{2}}(\lambda_2 + 3\mu_2)} \operatorname{Im}\left\{ [d + 2a^2 c(1 + 2i\varepsilon - 2\varepsilon^2)] \frac{\Gamma(1 + i\varepsilon)}{\Gamma\left(\frac{1}{2} + i\varepsilon\right)} \right\}$$

$$k_2 = \frac{4(\mu_2)^2}{(\pi a)^2(\lambda_2 + 3\mu_2)} \operatorname{Re}\left\{ [d + 2a^2 c(1 + 2i\varepsilon)] \frac{\Gamma(1 + i\varepsilon)}{\Gamma\left(\frac{1}{2} + i\varepsilon\right)} \right\} \tag{10.43}$$

where the constants c, d, and ε are as follows:

$$c = -\frac{(1 + \upsilon_2)(3 - 4\upsilon_2)\alpha_2\kappa_1}{8(1 - \upsilon_2)(1 - 2\upsilon_2)\kappa_2} Q_0$$

$$d = -\frac{2}{3}(1 - 2\varepsilon^2)ca^2 + \frac{4\gamma\psi}{3(4\gamma^2 - 1)\varepsilon}$$

$$\varepsilon = \frac{1}{2\pi} \operatorname{Ln}\left[\frac{\mu_1 + \mu_2(3 - 4\upsilon_1)}{\mu_2 + \mu_1(3 - 4\upsilon_2)}\right]$$

and

$$\psi = Q_0 \left[\frac{a^2(3 - 4\upsilon_2)}{4\pi}\right] \left[\frac{(2\mu_1)(1 + \upsilon_1)(1 - \upsilon_2)\alpha_1\kappa_2 - (1 + \upsilon_2)[\mu_1 + (1 - 2\upsilon_1)\mu_2]\alpha_2\kappa_1}{(1 - \upsilon_2)(1 - 2\upsilon_2)[(1 - 2\upsilon_2)\mu_1 - (1 - 2\upsilon_1)\mu_2]\kappa_2}\right]$$

In the homogeneous case, $\alpha_1 = \alpha_2$, etc., Equations 10.43 reduces to

$$k_1 = 0$$

$$k_2 = \frac{E\alpha Q_0}{3(1 - \upsilon)\pi} a^{3/2}$$

as in Equation 10.39.

The numerical values of the stress-intensity factors in Equation 10.43 are computed for the special case of steel bonded to copper. Using the necessary thermos-elastic constants, the results are

$$k_1 = 61.36\,Q_0 a^{1/2}$$

$$k_2 = 1906.8\,Q_0 a^{1/2}$$

If the steel in the region $z > 0$ is bonded to glass, then the results become

$$k_1 = 0.116\,Q_0 a^{1/2}$$

$$k_2 = -1.187\,Q_0 a^{1/2}$$

10.6.2 External Circular Crack

This section deals with the steady-state axisymmetric thermos-elastic problem concerning two semi-infinite solids joined over a circular region; for the purpose of mathematical simplicity, we assume that the radius of the circular region is unity. The unconnected portion of the solids may be regarded as an external circular crack occupying the region is $1 < r < \infty$ of the $z = 0$ plane [10.21].

The applicable axisymmetric equations of thermoelasticity are those used previously in Section 10.6.1. If the displacements and stresses induced by the temperature T are of the symmetric pattern, then

$$u_r, \sigma_r, \sigma_q, \sigma_z \text{ even in } z \quad \text{and} \quad u_z, \tau_{rz} \text{ are odd in } z$$

The corresponding displacements u_r and u_z are expressed in terms of harmonic functions as indicated by Equations 10.26.

In the antisymmetric problem where

$$u_r, \sigma_r, \sigma_q, \sigma_z \text{ are odd} \quad \text{and} \quad u_z,\ \tau_{rz} \text{ are even in } z$$

Equations 10.35 are used to express the induced displacements and stresses.

Considerations of the evenness and oddness of the displacements and stresses stated above, together with the prescribed thermal conditions on the crack surfaces, reduce the crack problem to one of an elastic half-space subjected to mixed boundary conditions on the plane $z = 0$. Owing to symmetry, the plane $z = 0$ must be free from the shearing stress τ_{rz} and u_z (r, 0) must vanish inside the circular region $r < 1$.

The thermal and elastic boundary conditions on the $z = 0$ plane are

$$\frac{\partial T}{\partial z} = 0, \quad 0 \le r < 1$$

$$T = T(r), \quad r > 1$$

and (10.44)

$$u_z = 0, \quad 0 \le r < 1$$

$$\sigma_z = 0, \quad r > 1$$

$$\tau_{rz} = 0, \quad 0 \le r < \infty$$

Steady-state temperature distribution: For a semi-infinite solid $z > 0$ that is free from disturbance at infinity, the solution for the temperature in Equation 10.44 can be obtained by writing

$$T(r,z) = \int_0^\infty A(s)e^{-sz}J_0(rs)\,ds, \quad z \geq 0 \tag{10.45}$$

where $A(s)$ is a function of the parameter s to be determined from the boundary condition and J_0 (rs) is the Bessel function of the first kind. Writing $T(r) = T_0\, g(r)$, T_0 is constant, the function $g(r)$ is bounded at infinity, and the integral

$$\int_1^\infty g(r)\,dr \tag{a}$$

is to be absolutely convergent.

The boundary conditions for the temperature, $T(r)$, in Equations 10.44 lead to the following dual integral equations:

$$\begin{aligned} &\int_0^\infty sA(s)J_0(rs)ds = 0, \quad 0 \leq r < 1 \\ &\int_0^\infty A(s)J_0(rs)ds = T_0 g(r), \quad r > 1 \end{aligned} \tag{b}$$

which determines the unknown function $A(s)$. The solution of Equations (b) is available in the open literature; see, for example, Reference 10.22.

$$A(s) = -T_0\left(\frac{2s}{\pi}\right)^{1/2}\int_1^\infty t^{1/2}J_1(st)dt\left[\frac{d}{dt}\int_t^\infty \frac{rg(r)dr}{\sqrt{(r^2 - t^2)}}\right] \tag{c}$$

Introducing the following function:

$$\phi(t) = \frac{2T_0}{\pi}\int_t^\infty \frac{rg(t)\,dr}{\sqrt{r^2 - t^2}} \tag{d}$$

Equation (c) is reduced to

$$A(s) = -\int_1^\infty \frac{d\phi}{dt}\sin(st)\,d \tag{e}$$

Equation (e) may be inserted into Equation 10.45 to give the temperature distribution at any point of the solid. However, for the purpose of setting up the mechanical boundary conditions in the subsequent work, it suffices to compute the temperature on the $z = 0$ plane.

It may be shown that

$$T(r,0) = -\int_{\max(1,r)}^{\infty} \frac{d\phi}{dt}\frac{dt}{\sqrt{t^2 - r^2}} \tag{10.46}$$

in which the derivative of ϕ can be calculated from Equation (d) once $g(r)$ is given. Two examples of interest are cited:

1. Consider the problem of heating up the face of an external circular crack over a ring whose inner and outer radii are unity and a, respectively. In this case, $g(r)$ takes the form $g(r) = H(a - r)$, where $H(r)$ represents the Heaviside step function defined as

$$H(a - r) = [{}^{1,a>r}_{0,a<r}], \quad r > 1 \tag{f}$$

A straightforward calculation gives

$$\phi(t) = \frac{2T_0}{\pi}\sqrt{a^2 - t^2}H(a - t), \quad t < a \tag{g}$$

and hence $T(r, 0)$ may be found from Equation 10.46. The result is

$$T(r,0) = \frac{2T_0}{\pi}\int_1^a \frac{tdt}{\sqrt{[(a^2 - t^2)(t^2 - r^2)]}} = \frac{2T_0}{\pi}\sin^{-1}\left(\frac{a^2 - 1}{a^2 - r^2}\right)^{1/2}, \quad 0 \le r < 1 \tag{10.47a}$$

and the condition $T(r, 0) = T_0$ for $r > 1$ is obviously satisfied.

2. If the temperature variation on the crack faces is such that $g(r) = r^{-n}$, $n > 1$, $r > 1$, then Equation 10.44 yields

$$\phi(t) = \frac{T_0}{\sqrt{\pi}}\frac{\Gamma\left(\frac{n}{2} - \frac{1}{2}\right)}{\Gamma\left(\frac{n}{2}\right)}t^{1-n}, \quad n > 1 \tag{h}$$

where $\Gamma(n)$ is the Gamma function. Putting Equation (h) into 10.46 and carrying out the integration, $T(r,0)$ is obtained

$$T(r,0) = \frac{T_0(n - 1)\Gamma\left(\frac{n - 1}{2}\right)}{2\sqrt{\pi}\Gamma\frac{n}{2}}r^{-n}B_{r^2}\left(\frac{n}{2}, \frac{1}{2}\right), \quad 0 \le r < 1, n > 1 \tag{10.47b}$$

Here, $B_x(m, n)$ is the incomplete Beta function defined by

$$B_x(m,n) = \int_0^x y^{m-1}(1 - y)^{n-1}\,dy; \quad \mathrm{Re}[m] > 0; \mathrm{Re}[n] > 0$$

Note that the complete Beta function, B(m, n), may be related to the Gamma functions as

$$B(m,n) = \int_0^1 y^{m-1}(1-y)^{n-1}dy = \frac{\Gamma(m)\Gamma(n)}{\Gamma(m+n)}$$

when $n = 2$, 3, etc., the incomplete Beta function in Equation 10.47b reduces to elementary functions:

a. $n = 2$

$$T(r,0) = \frac{T_0}{r^2}[1-(1-r^2)^{1/2}]; \quad 0 \le r < 1$$

b. $n = 3$

$$T(r,0) = \frac{2T_0}{\pi r^3}[\sin^{-1}r - r(1-r^2)^{1/2}]; \quad 0 \le r < 1$$

Similar expressions of T(r, 0) for other values of n may also be deduced. For $r > 1$, the prescribed temperature distributions of $T(r, 0) = T_0 r^{-n}$ can be easily recovered from Equation 10.45.

Temperature distributions of other types of thermal boundary conditions can be worked out in a similar manner.

Displacements, thermal stresses, and stress-intensity factors: The evaluation of the displacements, thermal stresses, and stress-intensity factors involves a considerable amount of work. It is given in detail by Kassir and Sih [10.21]. The final results for the stress-intensity factors are as follows:

a. *Step function*: Constant temperature T_0 over a ring of the external circular crack with inner and outer radii, unity and a, respectively

$$k_1 = -\frac{E\alpha T_0}{(1-\upsilon)\pi}\sqrt{(a^2-1)} \tag{10.48}$$

b. *Radial decay*: In this case, the temperature varies with $T_0 r^{-n}$, $n > 1$, $r > 1$

$$k_1 = -\frac{E\alpha T_0 \Gamma(n-1)}{2(1-\upsilon)\sqrt{(\pi)\Gamma\left(\frac{n}{2}\right)}}; \quad n > 1 \tag{10.49}$$

The stress-intensity factors given in Equations 10.48 and 10.49 can be associated with the forces which motivate and produce crack extension owing to thermal disturbances. The critical values of k_1 for a particular material can usually be measured experimentally. Moreover, if the material undergoes plastic yielding at the crack border, where the thermal stresses are exceedingly high, there will be a localized zone of plasticity surrounding the periphery of the crack. The size of this plastic zone for an external penny-shaped crack will be estimated in the next section.

Thermal plastic zone size: An ideal elastic–plastic model for the plane extension problem of a straight crack in a thin sheet has been proposed by Dugdale [10.23]. This simple model is adopted here to estimate the extent of plastic yielding at the edge of an external circular crack. The material near the crack is assumed to flow after yielding at a constant tensile stress σ_0 and the plastic zone is confined to a thin layer of width ω around the uncracked portion of the plane $z = 0$. The parameter ω is determined from the finiteness condition of σ_z at the leading edge of the plastic zone.

Mathematically, the solid may be assumed to deform elastically under the action of thermal loading together with a mechanical compressive stress, $-\sigma_0$, distributed over the surface of a ring of inner radius $r = 1$ and outer radius $r = 1 + \omega$. For this problem, the stress σ_z is obtained by superposing the thermal and mechanical stresses. By requiring σ_z to be bounded at $r = 1$, the extent of the plastic zone ω can be estimated.

The details of the procedure are available in References 10.21, 10.24. Formulas for the sizes of the thermal plastic zones for the two previously mentioned examples are worked out. The results are

1. Step function: If β denotes the width of the region heated by the constant temperature T_0, the width of the plastic zone ω is given by

$$\omega = -1 + [1 + \beta(\beta + 2)\gamma^2]^{1/2}$$

where

$$\gamma = +\frac{E\alpha T_0}{2(1 - \upsilon)\sigma_0} \tag{10.50}$$

Note that the quantity γ in Equation 10.50 may be interpreted as the ratio of the applied thermal stress to the yield stress of the material σ_0.

2. For $g(r) = r^{-n}$, $n > 1$, the plastic zone size ω is found to be

$$\omega = -1 + \left\{1 + \frac{\pi}{4}\left[\gamma \frac{\Gamma\left(\frac{n-1}{2}\right)}{\Gamma\left(\frac{n}{2}\right)}\right]^2\right\}^{1/2} \tag{10.51}$$

Variations of γ for different values of n and plots of ω versus γ are available in References 10.22, 10.25.

10.6.3 Elliptical Crack

This section deals with the induced deformation and stresses in an infinite elastic solid weakened by an elliptical plane of discontinuity or crack whose surfaces are exposed to changes in temperature and/or temperature gradient [10.14, 10.25].

Figure 10.8 shows an elliptical crack with major and minor axes a and b, respectively, located in the $z = 0$ plane, where the origin of the x, y, z coordinates coincides with the center of the crack. Points on the crack surfaces are described by the relations:

$$\frac{x^2}{a^2} + \frac{y^2}{b^2} \leq 1, \quad z = \pm 0 \tag{a}$$

while points on the crack boundary are described by the parametric equations

$$x = a\cos\phi, \quad y = b\sin\phi \tag{b}$$

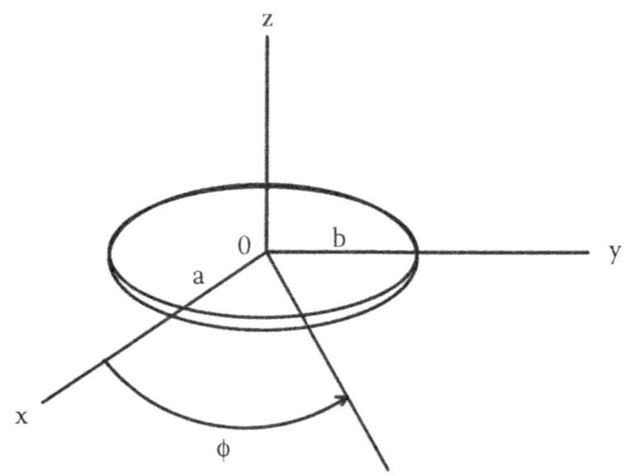

FIGURE 10.8
A schematic of an elliptical crack geometry.

where the parametric angle is denoted by ϕ. In order to solve boundary value problems for such geometry, it is expedient to use ellipsoidal coordinates (ξ, η, ζ) which are related to rectangular coordinates (x, y, z) by the relations

$$
\begin{aligned}
a^2(a^2 - b^2)x^2 &= (a^2 + \xi^2)(a^2 + \eta^2)(a^2 + \zeta^2) \\
b^2(b^2 - a^2)y^2 &= (b^2 + \xi^2)(b^2 + \eta^2)(b^2 + \zeta^2) \\
a^2b^2z^2 &= \xi\eta\zeta
\end{aligned}
\tag{10.52a}
$$

where

$$
-a^2 < \zeta < -b^2 < \eta < 0 < \xi < \infty \tag{10.52b}
$$

In the plane $z = 0$, the inside of the ellipse is given by $\xi = 0$ and the outside by $\eta = 0$. Points on the periphery of the ellipse are identified by $\xi = \eta = 0$.

For changes in temperature of the solid, the steady-state temperature field T(x, y, z) is governed by Equations (a) and (b) of Section 10.6. Once the temperature field is known, the induced components of the displacement vector $\mathbf{u} = u\,\mathbf{i} + v\,\mathbf{j} + w\,\mathbf{k}$ can be obtained from the equations of equilibrium of the solid and the corresponding stress–displacement relations, Equation (c) in Section 10.6.

There are two stress systems to be considered in the deformed body: the first deals with the application of surface temperatures that are the same on both sides of the crack (symmetric system), and the second considers the case where the temperature on the upper surface, $z = 0^+$, are equal and opposite on the lower surface, $z = 0^-$ (skew-symmetric system).

In the symmetric system, the stresses and displacements induced by thermal changes will depend on the variable z as follows:

$$
u, v, \sigma_x, \sigma_y, \sigma_z : \text{even in } z \tag{c}
$$

$$
w, \tau_{xz}, \tau_{yz} : \text{odd in } z \tag{d}
$$

These functions are required to be continuous outside the crack region on the plane $z = 0$, vanish at infinity, and the odd ones in Equation (d) must be zero on the plane $z = 0$.

Furthermore, the crack surfaces are free from applied stress, then $\sigma_z = 0$, and in view of symmetry, the conditions $\tau_{xz} = \tau_{yz} = 0$ must hold everywhere on the plane $z = 0$.

While in the skew-symmetric system, we have

$$u, v, \sigma_x, \sigma_y, \sigma_z : \text{odd in } z \tag{e}$$

$$w, \tau_{xz}, \tau_{yz} : \text{even in } z \tag{f}$$

these functions are also required to be continuous across the $z = 0$ plane with the exception of the crack region, vanish at infinity, and the odd ones in Equation (e) must vanish on the plane of continuity. On the crack surfaces, the shear stresses vanish, and the normal stress $\sigma_z = 0$ on the entire $z = 0$ plane. Next, we consider the thermos-elastic problems.

10.6.3.1 Symmetrical Case: Constant Temperature on Crack Surface

Let the temperature distribution T(x, y, z) in the solid be an even function of z vanishing at infinity. The boundary conditions of interest are

$$\begin{aligned} T &= -T_0, \xi = 0 \\ \frac{\partial T}{\partial z} &= 0, \eta = 0 \end{aligned} \tag{10.53}$$

where T_0 is constant. The solution of this problem is well known since T(x, y, z) is equivalent to the velocity potential of a perfect fluid passing through an elliptic aperture of a rigid partition (see, e.g., Reference 10.26, p. 150, Equation 10.1). Hence, it can be shown that

$$T(x, y, z) = -\frac{aT_0}{2K(k)}\int_{\xi}^{\infty}\frac{ds}{\sqrt{Q(s)}} = -T_0\frac{u}{K(k)} \tag{10.54}$$

where $Q(s) = s\,(a^2 + s)\,(b^2 + s)$ and K(k) is the complete elliptic integral of the first kind with argument $k^2 = 1 - (b/a)^2$.

The variable u is associated with the Jacobian elliptic functions, sn u, cn u, …, and should not be confused with the x-component of the displacement vector u = (u, v, w). The relationship between u and the ellipsoidal coordinate ξ is

$$\xi = a^2\left(\frac{\text{cn}u}{\text{sn}u}\right)^2 = a^2(\text{sn}^{-2}u - 1) \tag{g}$$

Note that because Equation 10.54 gives $T = -T_0$ for $\xi = 0$ and

$$T = -\frac{T_0 u}{K(k)}, \eta = 0$$

it follows that the initial boundary conditions for T are recovered.

Differentiating Equation 10.54, we obtain

$$\frac{\partial T}{\partial z} = \frac{T_0}{bK(k)}\frac{[\eta(a^2 + \xi)(b^2 + \xi)]^{1/2}}{(\xi - \eta)(\xi - \zeta)} \tag{10.55}$$

$$\text{Since } \frac{\partial T}{\partial z} = 0, \eta = 0, \quad \text{and} \quad \frac{\partial T}{\partial z} = \frac{T_0}{bK(k)}\left(1 - \frac{x^2}{a^2} - \frac{y^2}{b^2}\right)^{-1/2}, \xi = 0$$

it follows that Equation 10.55 also recovers the initial boundary conditions.

By inserting Equation 10.54 into Equation (b) of Section 10.6, we obtain

$$\frac{\partial\Omega}{\partial z} = -\left(\frac{1+\upsilon}{1-\upsilon}\right)\frac{a\alpha T_0}{4K(k)}\int_{\xi}^{\infty}\frac{ds}{\sqrt{Q(s)}} \tag{10.56}$$

Using Equations 10.26a, 10.26b, and 10.56, the induced displacements and stresses in the solid can be obtained. The applicable boundary conditions are

$$\sigma_z = 0, \xi = 0$$
$$w = 0, \eta = 0$$

Equations 10.26a and 10.26b require the following:

$$\frac{\partial F}{\partial z} = 0, \xi = 0$$
$$\frac{\partial f}{\partial z} = 0, \eta = 0$$

where F(r, z) is defined in Equation 10.26c. An appropriate form of the function f(x, y, z) is

$$f(x, y, z) = \frac{A}{2}\int_{\xi}^{\infty}\left(\frac{x^2}{a^2+s}+\frac{y^2}{b^2+s}+\frac{z^2}{s}-1\right)\frac{ds}{\sqrt{Q(s)}} \tag{10.57a}$$

in which the constant A is given by

$$A = \left(\frac{1+\upsilon}{1-\upsilon}\right)\frac{ab^2\alpha T_0}{4E(k)} \tag{10.57b}$$

The remaining steps required to determine the displacements and stresses in the solid are described in Reference 10.14.

The knowledge of thermal stresses in a small region ahead of the crack border is essential in the investigation of the stability behavior of the crack. The onset of rapid crack extension in structural members occurs when the magnitude of the crack border stress field or simply the stress-intensity factor reaches a critical value. This value will in general depend on the properties of the material such as the shear modulus, Poisson's ratio, and the coefficient of thermal expansion. It follows that a knowledge of the stress-intensity factors k_j, $j = 1, 2, 3$, where each one of the k-factors refers to a particular mode of crack surface displacement.

The "opening mode" of crack extension, governed by k_1, occurs in the symmetrical problem. For finding k_1, it suffices to calculate σ_z in the plane $z = 0$ outside the elliptical crack region. This can be done by using Equations 10.57a, b and 10.26b.

In the neighborhood of the crack border, the ellipsoidal coordinates ξ and η have the limits

$$\xi = 2abr(a^2\sin^2\phi + b^2\cos^2\phi)^{-1/2}$$
$$\zeta = -(a^2\sin^2\phi + b^2\cos^2\phi)$$

which are valid for $\eta = 0$ and $r < 1$. Substituting the above in the expression for σ_z, which is available in Reference 10.14, and expanding for small values of r (or ξ) results in

$$\sigma_z = \left(\frac{1+\upsilon}{1-\upsilon}\right)\frac{G\alpha T_0}{E(k)}\sqrt{\frac{b}{a}}(a^2\sin^2\phi + b^2\cos^2\phi)^{1/2}\frac{1}{\sqrt{2r}} + O(r^{1/2}) \tag{10.58}$$

The stress-intensity factor can now be extracted from Equation 10.58 as

$$k_1 = \left(\frac{1+\upsilon}{1-\upsilon}\right)\frac{G\alpha T_0}{E(k)}\sqrt{\frac{b}{a}(a^2\sin^2\phi + b^2\cos^2\phi)}, \quad a > b \tag{10.59}$$

The variation of k_1 with the parametric angle ϕ is plotted in Figure 10.9 for $\nu = 1/3$ and different values of the ratio a/b. A glance at the curves shows that k_1 is always greatest at $\phi = 90°$, that is, the intersection of the crack boundary with the minor axis of the ellipse. The crack propagation, if it occurs, would first take place at the point (0, b, 0) and tend to produce a penny-shaped crack.

The remaining strength of the member whose continuity is now interrupted by an approximate circular crack can be estimated from Equations 10.58 and 10.59 by setting $a = b$, and $\xi = r^2 - a^2$. Hence,

$$\begin{aligned} \sigma_z &= 2G\left(\frac{1+\upsilon}{1-\upsilon}\right)\frac{\alpha T_0}{\pi}\left(\frac{r^2}{a^2} - 1\right)^{-1/2}, \quad r > 1 \\ k_1 &= \left(\frac{1+\upsilon}{1-\upsilon}\right)\frac{2G\alpha T_0}{\pi}\sqrt{a} \end{aligned} \tag{10.60}$$

The derivation of the stress-intensity factors k_2 and k_3, corresponding, respectively, to the "edge-sliding" and "tearing" modes of fracture, requires the expansion of formulas for the shear stresses τ_{xz} and τ_{yz} near the crack border. This is available in References 10.14, 10.15.

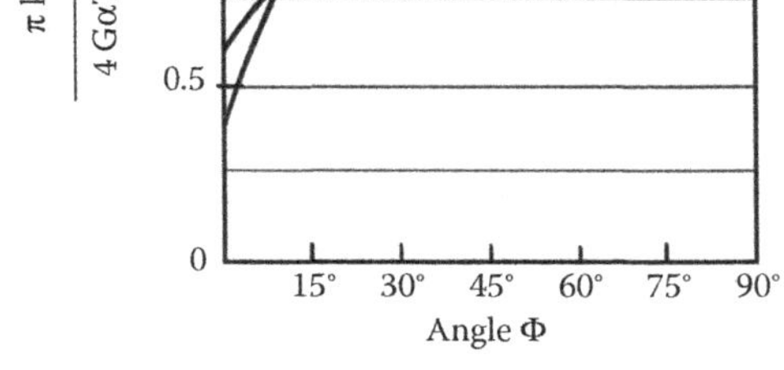

FIGURE 10.9
Variation of k_1 with Φ for an elliptical crack.

10.6.3.2 Skew-Symmetric Case: Insulated Elliptical Crack

The problem of a uniform steady-state heat flow disturbed by an insulated elliptical crack is a skew-symmetric problem and is equivalent to assigning a uniform temperature gradient, say Q_0, to the crack surfaces. The temperature T(x, y, z) is an odd function of z and

$$\tau_{xz} = \tau_{yz} = 0, \xi = 0; u = v = 0, \eta = 0$$

The boundary conditions are

$$\begin{aligned} \frac{\partial T}{\partial z} &= -Q_0, \xi = 0 \\ T &= 0, \eta = 0 \end{aligned} \tag{10.61}$$

Aside from a multiplying constant, T(x, y, z) is identical with the velocity potential for axial low past an elliptic disk in an infinite fluid (Reference 10.26, p. 144, Equation 10.8). From this hydrodynamical analogy, it is found that

$$T(x, y, z) = \frac{ab^2 Q_0}{2E(k)} \cdot z \int_{\xi}^{\infty} \frac{ds}{s\sqrt{Q(s)}} = \frac{Q_0}{E(k)} \left[\frac{snu \cdot dnu}{cnu} - E(u) \right] \cdot z \tag{10.62}$$

where E(k) is the complete elliptical of the second kind, and

$$E(u) = \int_0^u dn^2 t \cdot dt$$

Formulas for the corresponding displacements, stresses, and stress-intensity factors are available in References 10.14, 10.15.

In this problem, there exists a combination of "edge-sliding" and "tearing" movements of the crack surfaces as both k_2 and k_3 occur simultaneously. Hence, the shape to which the crack would grow will depend upon a function of k_2 and k_3 and is no longer *a priori* evident as in the case of the opening mode of the previous case.

The formulas for k_2 and k_3 are

$$\begin{aligned} k_2 &= -\frac{4G}{\sqrt{(ab)}} \frac{(B\cos^2\phi + C\sin^2\phi)}{(a^2\sin^2\phi + b^2\cos^2\phi)^{1/2}} \\ k_3 &= \frac{4G(1-\upsilon)}{\sqrt{(ab)}} \frac{\left(\frac{a}{b}B - \frac{b}{a}C\right)}{(a^2\sin^2\phi + b^2\cos^2\phi)^{1/2}} \end{aligned} \tag{10.63}$$

where the following abbreviations for constants B and C have been used:

$$B = \frac{m_1 q_2 - m_2 q_1}{p_1 q_2 - p_2 q_1}$$

$$C = \frac{m_1 p_2 - m_2 p_1}{q_1 p_2 - q_2 p_1}$$

in which the constants m_1 and m_2 are

$$m_1 = \left(\frac{1+\upsilon}{1-\upsilon}\right) \cdot \frac{a\alpha Q_0}{4k^2}\left[\frac{K(k)}{E(k)} - 1\right]\left(\frac{b}{a}\right)^2$$

$$m_2 = \left(\frac{1+\upsilon}{1-\upsilon}\right) \cdot \frac{a\alpha Q_0}{4k^2}\left[1 - \left(\frac{b}{a}\right)^2 \frac{K(k)}{E(k)}\right]$$

The constants p_1 and p_2 are

$$p_1 = -\frac{E(k)}{b^2} + \frac{1}{a^2 - b^2}\left\{E(k) - K(k) + \upsilon\left[\left(\frac{a}{b}\right)^2 E(k) - K(k) - \frac{(a^2+b^2)K(k) - 2a^2E(k)}{a^2 - b^2}\right]\right\}$$

$$p_2 = \frac{\upsilon}{a^2 - b^2}\left[K(k) - \left(\frac{a}{b}\right)^2 E(k) + \frac{(a^2+b^2)K(k) - 2a^2E(k)}{a^2 - b^2}\right]$$

and the constants q_1 and q_2 are

$$q_1 = \frac{\upsilon}{a^2 - b^2}\left[E(k) - K(k) + \frac{(a^2+b^2)K(k) - 2a^2E(k)}{a^2 - b^2}\right]$$

$$q_2 = -\frac{E(k)}{b^2} + \frac{1}{a^2 - b^2}\left\{K(k) - \left(\frac{a}{b}\right)^2 E(k) + \upsilon\left[K(k) - E(k) \frac{(a^2+b^2)K(k) - 2a^2E(k)}{a^2 - b^2}\right]\right\}$$

It is seen that the solution of this problem is a complicated problem of the material constants and the geometry of the problem.

For $\nu = 1/3$, the values of k_2 and k_3 in Equation 10.63 are plotted against the angle ϕ as shown in Figures 10.10 and 10.11. Note that the variation of k_2 with the angle ϕ is similar to that of k_1 in Figure 10.9. Inspection of Figure 10.10 reveals the interesting fact that the four points at which the crack border intersects the major and minor axes are under the action of "edge-sliding" type of displacement, since k_3 vanishes at those places. The maximum values of k_3 shift as the ellipticity of the crack geometry is changed. This is clearly illustrated in Figure 10.11.

In the degenerate case of decreasing ellipticity, that is, when a approaches b, we have

$$B = C = -\left(\frac{1+\upsilon}{1-\upsilon}\right)\frac{a^3\alpha Q_0}{6\pi}$$

and Equation 10.63 yields

$$k_2 = \left(\frac{1+\upsilon}{1-\upsilon}\right)\frac{2G\alpha Q_0}{3\pi} a^{3/2} \qquad (10.64)$$

$$k_3 = 0$$

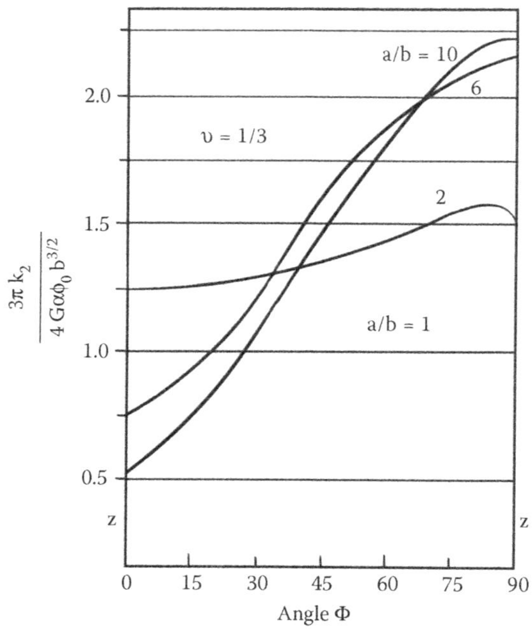

FIGURE 10.10
Variation of k_2 with Φ for an elliptical crack.

and the "tearing mode" disappears completely. This result is in agreement with that obtained by Florence and Goodier for the circular crack; Reference 10.18.

So far in this section, we have considered an elliptical crack whose surfaces are opened out by constant temperature T_0 and temperature gradient Q_0 (an insulated crack disturbing a uniform heat flow).

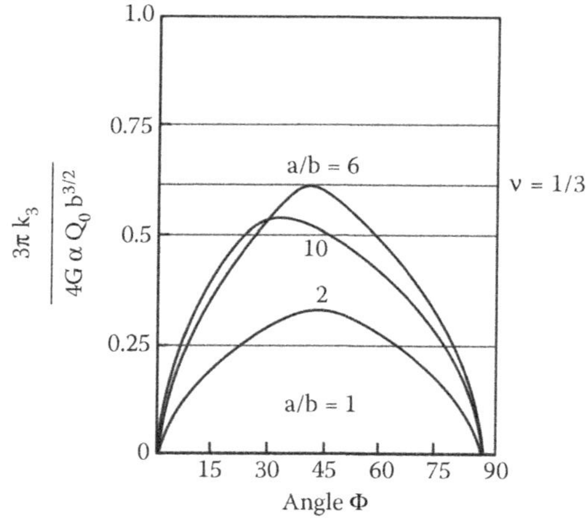

FIGURE 10.11
Variation of k_3 with Φ for an elliptical crack.

Analytical solutions also exist when the prescribed temperature and temperature gradient are arbitrary polynomials in x and y [10.25]. Essentially, the method of solution is one of undetermined coefficients. A combination of harmonic functions is sought which reduces, on the crack surface, to a polynomial in x and y of the same degree as the prescribed boundary conditions. On matching coefficients of similar terms in x, x^2, y, y^2, xy, etc., sufficient algebraic equations are found to determine uniquely the harmonic functions in question. Such problems are discussed next.

10.6.3.3 Symmetrical Case: Polynomial Temperature Specified on Crack Surface

Consider a semi-infinite space $z > 0$ with the following thermal and mechanical boundary conditions on the $z = 0$ plane:

$$T = t(x, y), \xi = 0$$
$$\frac{\partial T}{\partial z} = 0, \eta = 0 \qquad \text{(a)}$$

and

$$\sigma_z = 0, \xi = 0$$
$$w = 0, \eta = 0$$

where t(x, y) is a polynomial in x and y. A harmonic function which is suitable to satisfy the above boundary conditions is

$$V^{(n)} = A_n \int_{\xi}^{\infty} \left[1 - \frac{x^2}{a^2 + s} - \frac{y^2}{b^2 + s} - \frac{z^2}{s}\right]^n \frac{ds}{[Q(s)]^{n+1}}, (n = 0, 1, 2, \ldots) \qquad (10.65)$$

$$Q(s) = s(a^2 + s)(b^2 + s)$$

where A_n, $n = 0, 1, 2, \ldots$ are arbitrary constants.

On the crack surfaces, $\xi = 0$, Equation 10.65 yields a polynomial of degree n in x^2 and y^2. The partial derivatives of Equation 10.65 with respect to x and y are also harmonic functions and generate on the crack surface arbitrary polynomials in x and y. Thus, a suitable solution for the temperature field in the solid is obtained by writing

$$T(x, y, z) = \sum A \frac{\partial^m V^{(n)}}{\partial x^i \partial y^{m-i}} \qquad (10.66)$$

In particular, if t(x, y) is a quadratic polynomial in x and y (without the constant term), that is,

$$t(x, y) = t_1 x + t_2 y + t_3 x^2 + t_4 y^2 + t_5 xy \qquad (10.67)$$

then a suitable expression for T in Equation 10.66 may be chosen as

$$T = A_0 V^{(0)} + A_1 V^{(1)} + A_2 \frac{\partial V^{(1)}}{\partial x} + A_3 \frac{\partial V^{(1)}}{\partial y} + A_4 \frac{\partial^2 V^{(2)}}{\partial x^2} + A_5 \frac{\partial^2 V^{(2)}}{\partial x \partial y} \qquad (10.68)$$

where A_j, $j = 0, 1, 2, 3, 4$, and 5 are arbitrary constants.

Making use of the abbreviations

$$J_{mn} = \int_{\xi}^{\infty} \frac{ds}{(a^2+s)^m (b^2+s)^n [Q(s)]^{1/2}}$$

and for points inside the crack

$$I_{mn} = \int_{0}^{\infty} \frac{ds}{(a^2+s)^m (b^2+s)^n [Q(s)]^{1/2}}$$

the temperature field in Equation 10.68 becomes

$$\begin{aligned} T(x,y,z) = {} & (A_0 + A_1)J_{00} - 4A_4J_{10} - 2A_2J_{10}x - 2A_2J_{01}y + 8A_5J_{11}xy \\ & - (A_1J_{10} - 12A_4J_{20})x^2 - (A_1J_{01} - 4A_4J_{11})y^2 \\ & - A_1z^2 \int_{\xi}^{\infty} \frac{ds}{s[Q(s)]^{1/2}} + 4A_4z^2 \int_{\xi}^{\infty} \frac{ds}{s(a^2+s)[Q(s)]^{1/2}} \end{aligned} \tag{10.69}$$

Letting $\xi = 0$ and matching coefficients of equal terms between Equation 10.69 and 10.67, a system of simultaneous equations is obtained whose solution yields the following values for the constants A_j ($j = 0, 1, 2, \ldots, 5$):

$$\begin{aligned} A_0 &= \frac{4}{\Delta_1 I_{00}}[(I_{00}I_{11} - I_{10}I_{01})t_3 + (I_{10}^2 - 3I_{00}I_{20})t_4] \\ A_1 &= \frac{4}{\Delta_1}[-I_{11}t_3 + 3I_{20}t_4] \\ A_2 &= -\frac{t_1}{2I_{10}} \\ A_3 &= -\frac{t_2}{2I_{01}} \\ A_4 &= \frac{1}{\Delta_1}(-I_{01}t_3 + I_1t_4) \\ A_5 &= \frac{t_5}{8I_{11}} \end{aligned} \tag{10.70}$$

where the following abbreviation is used:

$$\Delta_1 = 4(I_{10}I_{11} - 3I_{01}I_{20})$$

and the integrals J_{mn} and I_{mn} introduced above are evaluated in Reference 10.25 as follows:

$$\begin{aligned} J_{mn} &= 2a^{-(2m+2n+1)} \int_0^a (\text{sn}t)^{2m}(\text{sd}t)^{2n}\,dt \\ I_{mn} &= 2a^{-(2m+2n+1)} \int_0^{K(k)} (\text{sn}t)^{2m}(\text{sd}t)^{2n}\,dt \end{aligned} \tag{b}$$

where K(k) is the complete elliptic integral of the first kind and $k^2 = 1 - (b/a)^2$. Values of the above integrals are available in the handbook of integrals; Reference [10.27].

The temperature gradient may be easily calculated from Equation 10.69. In particular, for $\xi = 0$

$$\frac{\partial T}{\partial z} = \frac{2Z^{-1/2}}{ab}\left[-A_0 - A_1 + \frac{4A_4}{a^2} + \frac{2A_2}{a^2}x + \frac{2A_3}{b^2}y - \frac{8A_5}{ab}xy + \left(A_1 - \frac{12A_4}{a^2}\right)\frac{x^2}{a^2} + \left(A_1 - \frac{4A_4}{b^2}\right)\frac{y^2}{b^2}\right] - \frac{2Z^{-1/2}}{ab}\left(A_1 - \frac{4A_4}{a^2}\right) \quad (10.71)$$

where

$$z = 1 - \frac{x^2}{a^2} - \frac{y^2}{b^2}$$

Boundary conditions (a) in conjunction with the solution given in Equations 10.26a and 10.26b written in terms of x, y, z coordinates require finding an ellipsoidal harmonic function f(x, y, z) such that

$$\begin{aligned} \frac{\partial^2 f}{\partial z^2} &= -\frac{\alpha}{2}\left(\frac{1+\upsilon}{1-\upsilon}\right)(t_1 x + t_2 y + t_3 x^2 + t_4 y^2 + t_5 xy), \xi = 0 \\ \frac{\partial f}{\partial z} &= 0, \eta = 0 \end{aligned} \quad (10.72)$$

Such a potential function may be sought in the form

$$f(x, y, z) = B_0 V^{(1)} + B_1 V^{(2)} + B_2 \frac{\partial V^{(2)}}{\partial x} + B_3 \frac{\partial V^{(2)}}{\partial y} + B_4 \frac{\partial^2 V^{(3)}}{\partial x^2} + B_5 \frac{\partial^2 V^{(3)}}{\partial x \partial y} \quad (10.73)$$

where B_j, $j = 0, 1, 2, \ldots, 5$ are arbitrary constants. The second of Equation 10.72 is assured and the first gives

$$\begin{aligned} B_0 &= \frac{3\alpha}{2\Delta_2}\left(\frac{1+\upsilon}{1-\upsilon}\right)\{[3(I_{12} + I_{21}) - (I_{11} + 3I_{20})(I_{11} + 3I_{02})/(I_{10} + I_{01})]t_3 \\ &\quad + [(I_{11} + 3I_{20})^2/(I_{10} + I_{20}) - 3(I_{21} + 5I_{30})]t_4\} \\ B_1 &= -\frac{9\alpha}{4\Delta_2}\left(\frac{1+\upsilon}{1-\upsilon}\right)[(I_{12} + I_{21})t_3 - (I_{21} + 5I_{30})]t_4 \\ B_2 &= \frac{\alpha}{16(I_{11} + 3I_{20})}\left(\frac{1+\upsilon}{1-\upsilon}\right)t_1 \\ B_3 &= \frac{\alpha}{16(I_{11} + 3I_{02})}\left(\frac{1+\upsilon}{1-\upsilon}\right)t_2 \\ B_4 &= \frac{\alpha}{8\Delta_2}\left(\frac{1+\upsilon}{1-\upsilon}\right)[-(I_{11} + 3I_{02})t_3 + (I_{11} + 3I_{20})t_4] \\ B_5 &= -\frac{\alpha}{288(I_{12} + I_{21})}\left(\frac{1+\upsilon}{1-\upsilon}\right)t_5 \end{aligned} \quad (10.74)$$

where

$$\Delta_2 = 18[I_{11}(5I_{30} - I_{12}) - 3I_{20}(I_{12} + I_{21}) + 3I_{02}(5I_{30} + I_{21})]$$

The opening-mode stress-intensity factor, which controls the stability behavior of the crack, may be obtained by calculating the stress σ_z near the vicinity of the crack, $\xi = 0$ and $\eta = 0$.

It may be shown that

$$\sigma_z = \frac{k_1}{\sqrt{(2r)}} + \cdots O(r^0)$$

where k_1 is given by

$$k_1 = \frac{8G(b^2\cos^2\phi + a^2\sin^2\phi)^{1/4}}{(ab)^{3/2}}\left[B_0 - 4\frac{B_2}{a}\cos\phi - 4\frac{B_3}{b}\sin\phi \right.$$
$$\left. + 24\frac{B_4}{a^2}\cos^2\phi + 24\frac{B_5}{ab}\sin\phi\cos\phi\right] \qquad (10.75)$$

and ϕ is the parametric angle of the ellipse. Additional information is available in Reference 10.27.

10.6.3.4 Skew-Symmetric Case: Polynomial Temperature Gradient Specified on Crack Surface

Owing to the antisymmetry of the stress system, the solution is obtained by considering a half-space with the following thermal and mechanical boundary conditions on $z = 0$:

$$\begin{aligned} &\frac{\partial T}{\partial z} = Q(x, y), \xi = 0 \\ &T = 0, \eta = 0 \\ &\tau_{xz} = \tau_{yz} = 0, \xi = 0 \\ &u = v = 0, \eta = 0 \end{aligned} \qquad (10.76)$$

where $Q(x, y)$ is a polynomial in x and y. We will consider the following quadratic polynomial to illustrate the procedure:

$$Q(x, y) = Q_0 + Q_1x + Q_2y + Q_3x^2 + Q_4y^2 + Q_5xy \qquad (a)$$

where Q_j, $j = 0, 1, 2, \ldots, 5$ are constants.

The associated thermoelastic potential $\phi(x, y, z)$ is required to satisfy the conditions

$$\begin{aligned} &\frac{\partial^2\phi}{\partial z^2} = \frac{\alpha}{2}\left(\frac{1+\upsilon}{1-\upsilon}\right)[Q_0 + Q_1x + Q_2y + Q_3x^2 + Q_4y^2 + Q_5xy], \xi = 0 \\ &\frac{\partial\phi}{\partial z} = 0, \eta = 0 \end{aligned} \qquad (10.77)$$

A suitable expression for ϕ in terms of the harmonic function $V(x, y, z)$ in Equation 10.65 is

$$\phi = D_0V^{(1)} + D_1V^{(2)} + D_2\frac{\partial V^{(2)}}{\partial x} + D_3\frac{\partial V^{(2)}}{\partial y} + D_4\frac{\partial^2 V^{(3)}}{\partial x^2} + D_5\frac{\partial^2 V^{(3)}}{\partial x\partial y} \qquad (10.78)$$

in which $D_j, j = 0, 1, 2, \ldots, 5$, are constants related to the constants $B_j, j = 0, 1, 2, \ldots, 5$, in Equations 10.74 as follows:

$$D_0 = \left(\frac{1+\upsilon}{1-\upsilon}\right)\frac{\alpha ab^2 Q_0}{8E(k)} - B_0$$
$$D_j = -B_j, j = 1, 2, \ldots, 5 \tag{b}$$

Here, E(k) is the complete elliptic integral of the second kind with argument $k^2 = 1 - (b^2/a^2)$, and B_j, $j = 0, 1, 2, \ldots, 5$ are given in Equation 10.74 but in this case with the t_j's replaced by the Q_j's.

To complete the solution, boundary conditions (10.76) require finding two new harmonic functions, g(x, y, z) and h(x, y, z), such that on the crack surface for $\xi = 0$

$$(1-\upsilon)\frac{\partial^2 g}{\partial y^2} + \frac{\partial^2 g}{\partial x^2} + \upsilon\frac{\partial^2 h}{\partial x \partial y} = -\frac{\partial \phi}{\partial x}$$
$$(1-\upsilon)\frac{\partial^2 h}{\partial x^2} + \frac{\partial^2 h}{\partial y^2} + \upsilon\frac{\partial^2 g}{\partial x \partial y} = -\frac{\partial \phi}{\partial y} \tag{10.79a}$$

and for $\eta = 0$

$$\frac{\partial g}{\partial z} = \frac{\partial h}{\partial z} = 0 \tag{10.79b}$$

In order to find functions g and h in Equations 10.79, one approach is to divide the procedure into four steps in order to simplify the algebra resulting from using Equation 10.78.

Step (1): keeping Q_0 only, Q_j, $j = 1, 2, 3, 4, 5 = 0$
Equation 10.78 gives $\xi = 0$

$$\frac{\partial \phi}{\partial x} = -\frac{(1+\upsilon)ab^2\alpha Q_0}{4(1-\upsilon)E(k)} I_{10}x$$
$$\frac{\partial \phi}{\partial y} = -\frac{(1+\upsilon)ab^2\alpha Q_0}{4(1-\upsilon)E(k)} I_{01}y \tag{c}$$

It is easily shown that the appropriate solution for the potential function g and h is

$$g(x, y, z) = c_1\frac{\partial V^{(2)}}{\partial x}$$
$$h(x, y, z) = d_1\frac{\partial V^{(2)}}{\partial y} \tag{d}$$

where c_1 and d_1 are constants determined in terms of Q_0 when Equation (d) is inserted in Equations 10.79 and (c) to give

$$c_1 = \frac{(1+\upsilon)ab^2\alpha Q_0}{32(1-\upsilon)E(k)\Delta_3}[(1-\upsilon)I_{11}I_{10} + 3I_{10}I_{02} - \upsilon I_{01}I_{11}]$$
$$d_1 = \frac{(1+\upsilon)ab^2\alpha Q_0}{32(1-\upsilon)E(k)\Delta_3}[(1-\upsilon)I_{11}I_{01} + 3I_{01}I_{20} - \upsilon I_{10}I_{11}] \tag{e}$$

in which the following abbreviation is used:

$$\Delta_3 = (1 - 2\upsilon)I_{11}^2 + 3(1 - \upsilon)I_{11}(I_{20} + I_{02}) + 9I_{20}I_{02}$$

In the plane $z = 0$, a system of orthogonal coordinates (n, t, z) which traverses the periphery of the crack is introduced to determine the tangential stresses

$$\tau_{nz} = \frac{k_2}{\sqrt{(2r)}} + O(r^0)$$
$$\tau_{tz} = \frac{k_3}{\sqrt{(2r)}} + O(r^0)$$

in which the stress-intensity factors for edge-sliding and tearing modes, k_2 and k_3, respectively, are found to be given by

$$k_2 = -\frac{32G(c_1b^2\cos^2\phi + d_1a^2\sin^2\phi)}{(ab)^{3/2}(b^2\cos^2\phi + a^2\sin^2\phi)^{1/4}}$$
$$k_3 = \frac{32G(1-\upsilon)(c_1 - d_1)\sin\phi\cos\phi}{(ab)^{3/2}(b^2\cos^2\phi + a^2\sin^2\phi)^{1/4}} \quad (10.80)$$

where c_1 and d_1 are given in Equation (e). Note that, in the degenerate case of $a = b$, we have

$$c_1 = d_1 = \frac{(1+\upsilon)Q_0a^5}{24(1-\upsilon)\pi}$$

and the values pertaining to the penny-shaped crack are recovered [10.18].

Step (2): Using Q_1 and Q_2, all other Q_j's $= 0$
Here, the thermos-elastic potential

$$\phi = -\left(\frac{1+\upsilon}{1-\upsilon}\right)\frac{\alpha}{16}\left(\frac{Q_1}{I_{11}+3I_{20}}\frac{\partial V^{(2)}}{\partial x} + \frac{Q_2}{I_{11}+3I_{02}}\frac{\partial V^{(2)}}{\partial y}\right) \quad (10.81)$$

which yields $\xi = 0$

$$\frac{\partial\phi}{\partial x} = -\left(\frac{1+\upsilon}{1-\upsilon}\right)\frac{\alpha}{4}\left[\frac{Q_1}{I_{11}+3I_{20}}(-I_{10} + 3I_{20}x^2 + I_{11}y^2) - \frac{2Q_2I_{11}xy}{I_{11}+3I_{02}}\right]$$
$$\frac{\partial\phi}{\partial y} = -\left(\frac{1+\upsilon}{1-\upsilon}\right)\frac{\alpha}{4}\left[\frac{2Q_1I_{11}xy}{I_{11}+3I_{20}} + \frac{Q_2}{I_{11}+3I_{02}}(-I_{01} + I_{11}x^2 + 3I_{02}y^2)\right]$$

Suitable potential functions g and h are as follows:

$$\begin{pmatrix} g \\ h \end{pmatrix} = \begin{pmatrix} c_2 \\ d_2 \end{pmatrix}V^{(1)} + \begin{pmatrix} c_3 \\ d_3 \end{pmatrix}V^{(2)} + \begin{pmatrix} c_4 \\ d_4 \end{pmatrix}\frac{\partial^2 V^{(3)}}{\partial x^2} + \begin{pmatrix} c_5 \\ d_5 \end{pmatrix}\frac{\partial^2 V^{(3)}}{\partial x \partial y} \quad (10.82)$$

which satisfy Equation 10.79b. In order to satisfy Equation 10.79a and find the values of the eight constants, c_j and d_j, $j = 1, 2, 3,$ and 4, so that the solution is complete, it is best to solve the resulting system of simultaneous equations numerically for a given material of the solid and various values of the ratio (a/b).

For $\nu = 0.3$ and $(a/b) = 1, 2, 4, 6$, and 10, the numerical result is shown in the table below for $Q_1 = Q_2 = Q^*$ and the constants should be multiplied by $\alpha\, Q^* \times 10^{-3}$.

	Ratio (a/b)				
Constants	**1**	**2**	**4**	**6**	**10**
C_2/a^2	11.59	5.09	0.745	0.209	0.036
C_3/a^2	24.63	−2.003	−0.377	−0.119	−0.022
C_4/a^2	−0.821	−0.296	−0.074	−0.022	−0.004
C_5/a^2	−0.822	−0.823	0.002	0.000	0.000
D_2/a^2	11.59	11.591	0.053	0.015	0.002
D_3/a^2	24.63	26.63	0.126	0.024	0.003
D_4/a^2	−0.821	−0.821	−0.02	0.001	0.000
D_5/a^2	−0.833	−0.823	−0.012	−0.003	−0.001

Once the harmonic functions have been determined, the displacements and stresses in the solid can be determined. The stress-intensity factors are

$$\begin{aligned} k_2 &= \frac{8G}{(ab)^{3/2}(b^2\cos^2\phi + a^2\sin^2\phi)^{1/4}}\left[b\cos\phi\left(c_2 + c_4\frac{24\cos^2\phi}{a^2} + c_5\frac{12\sin 2\phi}{ab}\right)\right. \\ &\quad \left. + a\sin\phi\left(d_2 + d_4\frac{24\cos^2\phi}{b^2} + d_5\frac{12\sin 2\phi}{ab}\right)\right] \\ k_3 &= \frac{8G(1-\upsilon)}{(ab)^{3/2}(b^2\cos^2\phi + a^2\sin^2\phi)^{1/4}}\left[b\cos\phi\left(d_2 + d_4\frac{24\cos^2\phi}{a^2} + d_5\frac{12\sin 2\phi}{ab}\right)\right. \\ &\quad \left. + a\sin\phi\left(c_2 + c_4\frac{24\cos^2\phi}{a^2} + c_5\frac{12\sin 2\phi}{ab}\right)\right] \end{aligned} \tag{10.83}$$

Step (3): Using Q_3 and Q_4 only. Neglecting the remaining Q's
The thermoelastic potential is

$$\phi = D_0V^{(1)} + D_1V^{(2)} + D_4\frac{\partial^2 V^{(2)}}{\partial x^2}$$

which gives $\xi = 0$

$$\begin{aligned} \frac{\partial\phi}{\partial x} &= -[(D_0 + 2D_1)I_{10} - 36D_4I_{20}]x + 4(D_1I_{20} - 30D_4I_{30})x^3 \\ &\quad + 4(D_1I_{11} - 18D_4I_{21})xy^2 \\ \frac{\partial\phi}{\partial y} &= -2[(D_0 + 2D_1)I_{01} - 12D_4I_{11}]y + 4(D_1I_{02} - 6D_4I_{12})y^3 \\ &\quad + 4(D_1I_{11} - 18D_4I_{21})x^2y \end{aligned}$$

and suitable functions g and h are

$$g = c_6 \frac{\partial V^{(2)}}{\partial x} + c_7 \frac{\partial V^{(3)}}{\partial x} + c_8 \frac{\partial^3 V^{(4)}}{\partial x^3}$$

$$h = d_6 \frac{\partial V^{(2)}}{\partial y} + d_7 \frac{\partial V^{(3)}}{\partial y} + d_8 \frac{\partial^3 V^{(4)}}{\partial y^3}$$

where the constants ($c_6, \ldots, d_8$) are found by solving a system of six simultaneous algebraic equations. As in the previous case, numerical procedure is used.

For $\nu = 0.3$, $Q_3 = Q_4 = Q^*$ and $a/b = 1, 2, 4, 6$, and 10, the following numerical values were obtained, and each entry in the table is to be multiplied by $\alpha Q^* \times 10^{-3}$:

Constants	**1**	**2**	**4**	**6**	**10**
c_6/a^7	3.583	−0.125	−0.133	−0.055	−0.013
c_7/a^7	−1.194	0.402	0.137	0.053	0.012
c_8/a^9	0.000	0.018	0.006	0.002	0.001
d_6/a^7	3.583	0.204	0.064	0.019	0.004
d_7/a^7	−1.194	−0.028	−0.019	−0.006	−0.001
d_8/a^9	0.001	0.001	0.001	0.001	0.001

The stress intensity factors k_2 and k_3 are

$$k_2 = -\frac{32G}{(ab)^{5/2}(b^2\cos^2\phi + a^2\sin^2\phi)^{1/4}}\left[b^2\cos^2\phi\left(c_6 + c_7\frac{48\cos^2\phi}{a^2}\right)\right.$$
$$\left. + a^2\sin^2\phi\left(d_6 + d_8\frac{48\sin^2\phi}{b^2}\right)\right] \tag{10.84}$$
$$k_3 = \frac{32G(1-\upsilon)\sin\phi\cos\phi}{(ab)^{3/2}(b^2\cos^2\phi + a^2\sin^2\phi)^{1/4}}\left[c_6 + c_8\frac{48\cos^2\phi}{a^2} - d_6 - d_8\frac{48\sin^2\phi}{b^2}\right]$$

Step (4): Q_5

In this case, we have

$$\phi = \frac{(1+\upsilon)\alpha Q_5}{288(1-\upsilon)(I_{12}+I_{21})}\frac{\partial^2 V^{(3)}}{\partial x \partial y}$$

which gives

$$\begin{bmatrix} \dfrac{\partial\phi}{\partial x} \\ \dfrac{\partial\phi}{\partial y} \end{bmatrix} = \frac{(1+\upsilon)\alpha Q_5}{24(1-\upsilon)(I_{12}+I_{21})}\begin{bmatrix} I_{11}y - I_{12}y^3 - 3I_{21}x^2y \\ I_{11}x - I_{21}x^3 - 3I_{12}xy^2 \end{bmatrix}$$

Functions g and h are sought in the form

$$g = c_9 \frac{\partial V^{(2)}}{\partial y} + c_{10} \frac{\partial V^{(3)}}{\partial y} + c_{11} \frac{\partial^3 V^{(4)}}{\partial y^3}$$

$$h = d_9 \frac{\partial V^{(2)}}{\partial x} + d_{10} \frac{\partial V^{(3)}}{\partial x} + d_{11} \frac{\partial^3 V^{(4)}}{\partial x^3}$$

A system of six simultaneous equations is used to determine the six constants, $c_9 \ldots d_{11}$. Numerical values of the constants are obtained for $\nu = 0.3$ and various ratios (a/b) as indicated in the table below (each entry is to be multiplied by $\alpha\, Q_5 \times 10^{-3}$).

	Ratio (a/b)				
Constants	**1**	**2**	**4**	**6**	**10**
c_9/a^7	1.987	−0.515	−0.014	−0.003	−0.001
c_{10}/a^7	−3.284	0.123	−0.002	−0.002	0.001
c_{11}/a^9	−0.059	0.001	0.001	0.001	0.001
d_9/a^7	1.987	1.647	1.494	0.635	0.217
d_{10}/a^7	−3.284	−0.952	−0.025	−0.005	−0.007
d_{11}/a^9	−0.059	−0.025	0.001	−0.001	0.000

In this case, the stress-intensity factors are

$$k_2 = -\frac{16G\sin 2\phi}{(ab)^{3/2}(b^2\cos^2\phi + a^2\sin^2\phi)^{1/4}}\left[c_9 + d_9 + 48\left(c_{11}\frac{\sin^2\phi}{b^2} + d_{11}\frac{\cos^2\phi}{a^2}\right)\right]$$

$$k_3 = \frac{32G(1-\upsilon)}{(ab)^{3/2}(b^2\cos^2\phi + a^2\sin^2\phi)^{1/4}}\left[a^2\sin^2\phi\left(c_9 + c_{11}\frac{48\sin^2\phi}{b^2}\right)\right. \qquad (10.85)$$

$$\left. - b^2\cos^2\phi\left(d_9 + d_{11}\frac{48\cos^2\phi}{a^2}\right)\right]$$

10.6.4 Semi-Infinite Plane Crack

This section deals with the 3D stress distribution that arises when a semi-infinite plane crack in a large body is opened out by the application of thermal changes to arbitrary regions of its surface [10.28,10.29]. Figure 10.12 shows the geometry of the crack.

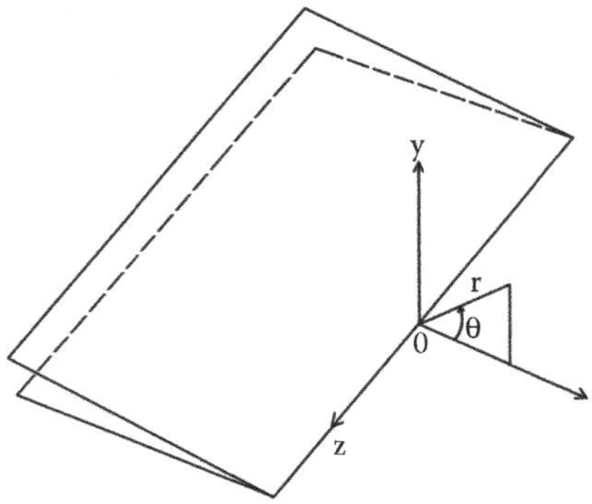

FIGURE 10.12
A semi-infinite plane crack in an infinite solid.

In a system of rectangular coordinates x, y, and z, the crack edge coincides with the z-axis and the crack surfaces are described by $y = 0^+$, $x < 0$, $-00 < z < 00$. The local coordinates are denoted by r and θ. When the crack surfaces are exposed to changes in temperature, thermal stresses of high intensity arise near the crack edge. Neglecting the influence of both inertia and coupling, the thermal stress problem involves the solution of two distinct boundary value problems. The first is a potential problem in the classical theory of heat conduction which requires finding the temperature distribution T(r, θ, z) from the Laplace equation for the steady state

$$\frac{\partial^2 T}{\partial r^2} + \frac{1}{r}\frac{\partial T}{\partial r} + \frac{1}{r^2}\frac{\partial^2 T}{\partial \theta^2} + \frac{\partial^2 T}{\partial z^2} = 0 \tag{10.86}$$

The second is concerned with the determination of the induced displacements and stresses from the Navier equations of static equilibrium and the Duhamel–Neumann stress displacement relations.

Prescribed temperature: When the surfaces of the crack are subjected to the application of prescribed temperature over a certain region of exposure, the temperature field is determined subject to the boundary conditions

$$\begin{aligned} &T(r, \pi, z) = T_0(r, z), y = 0 \\ &\frac{\partial T}{\partial y} = 0, y = 0, \theta = 0 \end{aligned} \tag{10.87}$$

where $T_0(r, z)$ is a prescribed function of r and z. Representing the temperature by the double integral

$$T(r, \theta, z) = \left(\frac{2}{\pi}\right)^{1/2} \int_0^\infty \int_0^\infty A(s, t) \frac{\cosh(\theta t)}{\cosh(\pi t)} K_{it}(sr) \cos(sz)\, ds\, dt \tag{10.88}$$

in which K_{it} is the modified Bessel function of the second kind of imaginary order (also known as the Macdonald function). The equation in Equation 10.87 is automatically satisfied and the first equation implies that

$$A(s, t) = \frac{2}{\pi^2} t \sinh(\pi t) \int_0^\infty \frac{1}{r} T_c(r, s) K_{it}(rd) dr \tag{a}$$

where T_c is the cosine transform of the specified temperature

$$T_c(r, s) = \left(\frac{2}{\pi}\right)^{1/2} \int_0^\infty T_0(r, z) \cos(sz) dz \tag{b}$$

provided that $T_0(r, z) = 0$. In deriving Equations (a) and (b), the inversion formulas for Fourier and Kontorovich–Lebedev transforms have been used [10.29]. When $T_0(0, z) = 0$, Lebedev's modified inversion formula [10.30] can be used to the integral in

Equation (a) to give

$$A(s,t) = \frac{2}{\pi}\left\{T_c(0,s) + \left(\frac{2}{\pi^5}\right)^{1/2} t\sinh(\pi t)\int_0^\infty\int_0^\infty [T_0(r,z) - T_0(0,z)e^{-sr}]\frac{1}{r}K_{it}(sr)\cos(sz)dr\,dz\right\} \tag{10.89}$$

As an example, if a constant temperature of magnitude T_0 is concentrated at the point $r = a$, $\theta = \pi$, $z = 0$, then Equation 10.89 can be evaluated to yield

$$A(s,t) = \left(\frac{2}{\pi^3}\right)^{1/2} T_0\left[1 + \frac{2t}{a\pi}\sinh(\pi t)(1 - e^{-as})K_{it}(as)\right] \tag{10.90}$$

Inserting Equation 10.90 into Equation 10.88, the variation of the temperature throughout the solid is determined. In particular, across the surface $\theta = 0$, the temperature is

$$T(r,0,z) = \frac{T_0}{2\pi^2}\frac{(ar)^{3/2}}{(r^2 + a^2 + z^2)^2} \tag{c}$$

Induced displacements and stresses: Introducing a thermos-elastic potential Ω (x, y, z) related to the temperature by the relation

$$\frac{\partial\Omega}{\partial y} = \frac{(1+\upsilon)\alpha}{2(1-\upsilon)}T(x,y,z) \tag{d}$$

a solution which satisfies the Navier equations of static equilibrium and automatically clears the $y = 0$ plane from shear stresses admits the displacement **u** (u, v, w) representation [10.28]

$$u = \int_\infty^y \frac{\partial}{\partial x}[(1-2\upsilon)f + \Omega]\,dy + y\frac{\partial F}{\partial x}$$

$$V = -2(1-\upsilon)f + y\frac{\partial F}{\partial y} \tag{10.91a}$$

$$w = \int_\infty^y \frac{\partial}{\partial z}[(1-2\upsilon)f + \Omega]\,dy + y\frac{\partial F}{\partial z}$$

where f(x, y, z) is a space harmonic function and $F = f + \Omega$. Some of the shear stresses are

$$\sigma_y = 2G\left[-\frac{\partial F}{\partial y} + y\frac{\partial^2 F}{\partial y^2}\right]$$

$$\tau_{yx} = 2Gy\frac{\partial^2 F}{\partial x\partial y} \tag{10.91b}$$

$$\tau_{yz} = 2Gy\frac{\partial^2 F}{\partial y\partial z}$$

The crack plane boundary conditions are

$$\begin{aligned} \sigma_y(x, 0, z) &= 0, \theta = \pi \\ V(x, 0, z) &= 0, \theta = 0 \end{aligned} \tag{e}$$

The symmetry requirements are

$$\tau_{yx}(x, 0, z) = \tau_{yz}(x, 0, z) = 0, \theta = 0, \theta = \pi \tag{f}$$

and the regularity conditions that the displacements and stresses must vanish at infinity. Moreover, near the crack border, r approaches 0, the displacements must be finite, and the stresses are expected to have the usual square root singularity, $r^{-1/2}$.

In this case, the function f(x, y, z) is determined subject to the requirements on $y = 0$

$$\begin{aligned} \frac{\partial f}{\partial y} &= -\frac{(1+\upsilon)\alpha}{2(1-\upsilon)} T_0(r, z), y = 0, \theta = \pi \\ f &= 0, y = 0, \theta = 0 \end{aligned} \tag{10.92}$$

An appropriate expression for f(r, θ, z) which satisfies Equation 10.92 is

$$f(r, \theta, z) = \left(\frac{2}{\pi}\right)^{1/2} \int_0^{\infty} B(s, t) \frac{\sinh(\theta t)}{t \cosh(\pi t)} K_{it}(sr) \cos(sz) ds\, dt \tag{10.93}$$

with

$$B(s, t) = \frac{(1+\upsilon)\alpha}{2(1-\upsilon)\pi^2} t \sinh(\pi t) \int_0^{\infty} T_c(r, s) K_{it}(rs) dr$$

where T_c is given in Equation (b). Making use of the identity, $\frac{\partial}{\partial} = r\frac{\partial}{\partial \theta}, \theta = 0$, the normal stress σ_z in the region $\theta = 0$ is computed from the relation

$$\sigma_z(r, 0, z) = 2G\left[r\frac{\partial f}{\partial \theta} + \frac{t(1+\upsilon)\alpha}{2(1-\upsilon)} T(r, 0, z)\right] \tag{g}$$

By inserting the appropriate expressions from Equations (c) and (10.93), using the identity [10.31]

$$\int_0^{\infty} t \tanh(\pi t) K_{it}(sr) K_{it}(s\eta) dt = \frac{\left(\frac{\pi}{2}\right)(r\eta)^{1/2} e^{-s(r+\eta)}}{r + \mu} \tag{h}$$

and expanding near the crack border, Equation (g) can be put in the standard form to reveal the stress-intensity factor for the specified temperature as

$$k_1 = -\frac{E\alpha}{(1-\upsilon)\pi^{3/2}}\left[{}_z\mathrm{Lim}_0 \int_0^{\infty} e^{-sz} \cos(sz) ds \int_0^{\infty} \frac{\eta^{1/2} e^{-s\eta} T_c(\eta, s) d\eta}{z + \eta}\right] \tag{10.94}$$

For the case of the concentrated temperature T_0, Equation 10.94 gives

$$k_1 = \frac{E\alpha T_0}{\sqrt{2}(1-\upsilon)\pi^2 a^{3/2}}\left[\frac{1}{1+z_0^2}\right] \tag{10.95}$$

where $z_0 = z/a$.

The variation of k_1 with z/a is shown by the solid line in Figure 10.13. Formula (10.95) is similar to that in Equation 9.100 which corresponds to a concentrated force P acting at the same location of the semi-infinite crack surface as T_0 (both have the same variation with z/a).

Note that the expression in Equation 10.95 may be used as Green's function to generate solutions for any temperature prescribed over a given portion of the crack surfaces.

Prescribed heat flux: Suppose that the crack is opened out by the application of a prescribed heat flux Q(x, z) to its surfaces such that the deformation is symmetrical with respect to the cylindrical coordinates r, θ, z. The temperature field satisfies the following boundary conditions on the y = 0 plane:

$$\begin{aligned}\frac{\partial T}{\partial y} &= Q(x,z), y = 0, \theta = \pi \\ \frac{\partial T}{\partial y} &= 0, y = 0, \theta = 0\end{aligned} \tag{a}$$

Equations (a) indicate that the temperature field is equivalent to a Neumann problem for a half-space whose solution is known [10.31]. In the case of a heat flux Q_0 concentrated at the point r = a, θ = π, z = 0, the temperature at any point of the solid is determined as

$$\begin{aligned}T(r,\theta,z) &= \frac{Q_0}{2\pi\rho} \\ \rho &= [(x+a)^2 + y^2 + z^2]^{1/2}\end{aligned} \tag{10.96}$$

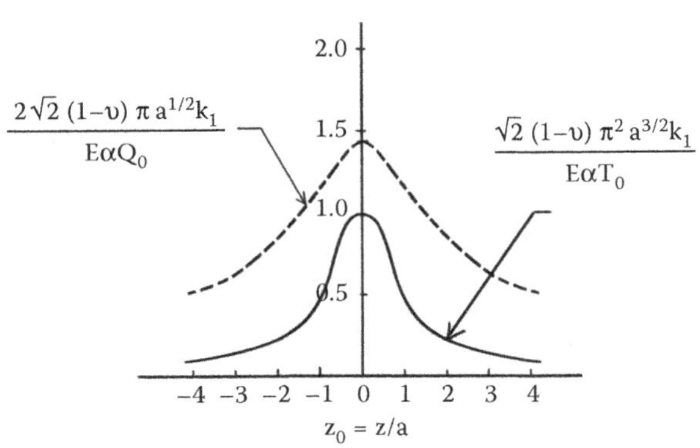

FIGURE 10.13
Variations of k_3 along crack border for a semi-infinite crack.

The induced displacements and stresses are obtained using Equations 10.91 subject to the boundary conditions

$$\frac{\partial f}{\partial y} = -\frac{\partial \Omega}{\partial y}, y = 0, \theta = \pi$$
$$f = 0, y = 0, \theta = 0 \tag{10.97}$$

The solution for f(r, θ, z) can be obtained by applying the methods of Fourier and Kontorovich–Lebedev integral transforms as was done above for the prescribed temperature case. However, in the present case, it is possible to adopt an alternative method of attack involving the use of the following singular solution of the Laplace equation in cylindrical coordinates [10.32]

$$\frac{1}{\sqrt{r}} e^{\pm i\theta/2} g(\varsigma)$$

where

$$\varsigma = r + a = iz \tag{a}$$

in which g is an arbitrary function.

Utilizing the identity, $\partial^2/\partial y^2 = -(\partial^2/\partial r^2 + \partial^2/\partial z^2), \theta = 0$, the boundary conditions on function $F = f + \Omega$ are

$$\frac{\partial F}{\partial y} = 0, \theta = \pi$$
$$\frac{\partial^2 F}{\partial y^2} = 0, \theta = 0 \tag{b}$$

Equation (b), coupled with the knowledge that the deformation is symmetric in the variables θ and z, suggests the relation

$$\frac{\partial F}{\partial y} = \frac{\cos\frac{\theta}{2}}{\sqrt{r}} \mathrm{Re}[g(\varsigma)] \tag{c}$$

The physical significance of this solution is immediate from the fact that the characteristic square root singularity, $r^{-1/2}$, has been retained in the expression for the stresses near the crack border. The displacements, however, are finite as r approaches 0.

Making use of Equation (d) in the previous section, Equation (c) yields

$$\frac{\partial f}{\partial y} = -\frac{(1+\upsilon)\alpha Q_0}{4(1-\upsilon)\pi\rho} + \frac{\cos\frac{\theta}{2}}{\sqrt{r}} \mathrm{Re}[g(\varsigma)] \tag{d}$$

The function g(ζ) can be found explicitly by imposing the second boundary condition in Equation 10.97. However, since the integral of the term 1/ρ diverges on using the infinite limit of integration, instead of the second boundary condition in Equation 10.97, the following equivalent requirement is applied (note that an additive constant term in the expression for the potential f does not influence the stress distribution in the solid):

$$\frac{\partial f}{\partial z} = 0, y = 0, \theta = 0 \tag{e}$$

By differentiating Equation (d) with respect to z, then integrating over y between the limits y and infinity, the following relation is reached:

$$\frac{\partial f}{\partial z} = -\frac{(1+\upsilon)\alpha Q_0 z}{4(1-\upsilon)\pi\rho(\rho+y)} - \mathrm{Im}\cdot\left[\int_{\infty}^{y}\frac{\cos(\theta/2)}{\sqrt{r}}g'(\varsigma)dy\right] \tag{f}$$

where

$$g'(\xi) = \frac{d}{d\varsigma}[g(\varsigma)]$$

This is equivalent to

$$\frac{\partial f}{\partial z} = -\frac{(1+\upsilon)\alpha Q_0 z}{4(1-\upsilon)\rho(\rho+y)} - \frac{1}{\sqrt{r}}\mathrm{Im}.\int_{\infty}^{\varsigma}\frac{g'(t)dt}{\sqrt{t-\varsigma_0}} \tag{g}$$

where

$$\varsigma_0 = [\varsigma]_{\theta=0} = x + a + iz$$

By imposing condition (e) on (g), we conclude that the function $g(\zeta)$ satisfies the Abel-type integral equation

$$\int_{\varsigma_0}^{\infty}\frac{g'(t)dt}{\sqrt{t-\varsigma_0}} = -\frac{\sqrt{2}(1+\upsilon)\alpha Q_0}{4(1-\upsilon)\pi\varsigma_0} \tag{h}$$

which yields [10.27]

$$g(\varsigma) = \frac{\sqrt{2}(1+\upsilon)\alpha Q_0}{4(1-\upsilon)\pi\sqrt{\varsigma}} \tag{i}$$

By inserting Equation (i) into (d), performing an integration with respect to the variable y between the limits θ and y, and noting the identity

$$\mathrm{Re}\int_{\theta}^{y}\frac{\cos\theta/2}{\sqrt{r}}\frac{dy}{\sqrt{r+a+iz}} = \frac{1}{\sqrt{2}}\mathrm{Log}\frac{\left[\sqrt{\varsigma}+\sqrt{2r}\sin(\theta/2)\right]^2}{\sqrt{\rho^2-y^2}} \tag{j}$$

we conclude that the expression for the potential f is

$$f(r,\theta,z) = -\frac{(1+\upsilon)\alpha Q_0}{2(1-\upsilon)\pi}\mathrm{Log}\frac{\sqrt{\rho+y}}{\left[\sqrt{\varsigma}+\sqrt{\varsigma-\varsigma_0}\right]} \tag{10.98}$$

and the solution is basically complete. The physical quantities of interest may be readily obtained from Equations 10.91 when the appropriate expressions for the harmonic functions in Equations 10.96, d in the previous section, and (10.98). In particular, the normal stress across the surface $\theta = 0$ is computed as

$$\sigma_z(x, 0, z) = -\frac{E\alpha Q_0}{2(1-\upsilon)\pi}\frac{\mathrm{Re}(x+a+iz)^{-1/2}}{\sqrt{2x}} \tag{10.99}$$

which may be expressed in the standard form

$$\sigma_z(x, 0, z) = \frac{k_1}{\sqrt{2x}} + O(x^0)$$

and the stress-intensity factor for the opening mode of crack extension is given by the relation

$$k_1 = -\frac{E\alpha Q_0}{2(1-\upsilon)\pi\sqrt{2a}}\left[\frac{1+(1+z_0)^{1/2}}{1+z_0^2}\right]^{1/2} \tag{10.100}$$

where $z_0 = z/a$.

The variations of k_1 with z/a are shown in Figure 10.13 dotted line.

Note that, for the case of a heat flux applied at the points $(-a, +0, b)$ instead of $(-a, +, 0)$, the expression in Equation 10.100 still holds provided that z_0 is replaced by $(z_0 - (b/a))$.

Insulated semi-infinite crack: When a heat flow with a constant thermal gradient is diverted around a sharp edge of an insulated crack or flaw, there is a local intensification of the temperature gradient accompanied by singular thermal stress which may cause crack propagation resulting in serious damage to structural members. Since the critical value of the intensity of the local stress field can be associated with the fracture toughness of the material, it follows that by knowing the stress-intensity factors as functions of the temperature gradients, material properties, and flaw size, it is possible to predict a critical temperature gradient which will not result in failure of the component. By confining attention to 3D cracks or planes of discontinuity, the local intensity of the thermal stress has been determined for the circular crack [10.18] and the more general elliptical crack [10.14]. In case of the elliptical crack, both factors k_2 and k_3, associated with the edge-sliding and tearing modes of fracture, are found to be operating, while for the axially symmetric crack, the local stress field is controlled by k_2 only.

Temperature field: The temperature field at every point of the solid is obtained from a solution to Equation 10.86 in the region $y > 0$ subject to satisfying the boundary conditions

$$\begin{aligned} \frac{\partial T}{\partial y} &= -Q(x, z), y = 0, \theta = \pi \\ T &= 0, y = 0, \theta = 0 \end{aligned} \tag{10.101}$$

where Q(x, z) is the specified temperature gradient. The desired expression for the temperature gradient is constructed by using a Fourier cosine transform in the variable z and a Kontorovich–Lebedev transform in the variable r. Toward this end, let

$$T = \left(\frac{2}{\pi}\right)^{1/2}\int_0^\infty\int_0^\infty A(s, t)\frac{\sinh(\theta t)}{t\cosh(\pi t)}K_{it}(sr)\cos(sz)\,ds\,dt \tag{a}$$

where $K_{it}(sr)$ is the modified Bessel function of the second kind of imaginary order and A(s, t) is an arbitrary function such that the second condition in Equation (a) is satisfied. Differentiation T with respect to y and inserting it in the first of Equation 10.101, A(s, t) is determined as

$$A(s,t) = \frac{2^{3/2}}{\pi^{5/2}} t \sinh(\pi t) \int_0^\infty Q_c(r,s) K_{it}(sr)\, dr \tag{b}$$

in which

$$Q_c(r,s) = \int_0^\infty Q(r,z)\cos(sz) dz \tag{c}$$

Suppose that at the point $r = a$, $\theta = \pi$, $z = 0$ of the crack surface there is a concentrated temperature gradient of magnitude Q_0 (constant), then Equations (b) and (c) provide

$$A(s,t) = \frac{\sqrt{2}}{\pi^{5/2}} Q_0 t \sinh(\pi t) K_{it}(as) \tag{d}$$

By inserting Equation (d) into Equation (a) and performing the integration, the temperature field is obtained as follows:

$$T(r,\theta,z) = \frac{Q_0}{\pi^2 \rho} \tan^{-1}\left(\frac{\sqrt{2a(r-x)}}{\rho}\right)$$
$$\rho = [(x+a)^2 + y^2 + z^2]^{1/2} \tag{10.102}$$

Note that ρ represents the distance between any point in the region $y > 0$ and the point $(r = a,\ \theta = \pi,\ z = 0)$.

Thermal stresses: Denoting the projections of the displacement vector **u** along the directions of cylindrical coordinates by (u_r, u_θ, u_z), the Papkovich–Neuber potential representation of the displacement field is

$$\begin{aligned}
u_r(r,\theta,z) &= 4(1-\upsilon)(f_1\cos\theta + f_2\sin\theta) - \frac{\partial F}{\partial r} + \sin\theta \frac{\partial}{\partial r}(r\Omega) \\
u_\theta(r,\theta,z) &= 4(1-\upsilon)(f_2\cos\theta - f_1\sin\theta) - \frac{1}{r}\frac{\partial F}{\partial \theta} + \frac{\partial}{\partial \theta}(\sin\theta\,\Omega) \\
u_z(r,\theta,z) &= 4(1-\upsilon)f_3 - \frac{\partial F}{\partial z} + r\sin\theta \frac{\partial \Omega}{\partial z} \\
F &= f_0 + (r\cos\theta)f_1 + (r\sin\theta)f_2 + zf_3
\end{aligned} \tag{10.103}$$

where f_n, $n = 0, 1, 2, 3$ are harmonic functions and $\Omega(r, \theta, z)$ is determined from a knowledge of the temperature field through the relation

$$\frac{\partial \Omega}{\partial y} = \frac{(1+\upsilon)\alpha}{2(1-\upsilon)} T(r,\theta,z) \tag{10.104}$$

The corresponding stress field $(\sigma_\theta, \tau_{\theta r}, \tau_{\theta z})$ is readily obtained by using Equations 10.103, 10.104, and the Duhamel–Neumann stress–displacement relations in linear thermos-elasticity [10.35,10.36].

By virtue of the fact that the induced deformation is skew-symmetric with respect to the variable θ, the quantities u_r, u_z, σ_θ are odd in the variable θ while the quantities u_θ, $\tau_{\theta r}$, and $\tau_{\theta z}$ are even in the same variable. This circumstance suggests that the problem can be formulated for the upper half-space $y > 0$ of the solid with appropriate boundary conditions specified in the region $y = 0$ ($\theta = \pi$, $\theta = 0$).

The continuity of the solid outside the crack region in the $y = 0$ plane implies that

$$\begin{aligned} u_r(r, 0, z) &= 0 \\ u_z(r, 0, z) &= 0 \\ \sigma_\theta(r, 0, z) &= 0 \end{aligned} \tag{10.105a}$$

Since the crack surface is free from mechanical loading, the following conditions must be satisfied in the plane $y = 0$, $\theta = \pi$:

$$\begin{aligned} \sigma_\theta(r, \pi, z) &= 0 \\ \tau_{\theta r}(r, \pi, z) &= 0 \\ \tau_{\theta z}(r, \pi, z) &= 0 \end{aligned} \tag{10.105b}$$

In addition, conditions (10.105) must be accompanied by the regularity requirements at infinity, namely, the displacements and stresses must vanish at infinity. Also, near the crack border, r approaches 0, the displacements must be finite, and the stresses are expected to have the usual square root singularity $r^{-1/2}$.

Making use of Equations 10.103 and the associated stresses, the boundary conditions in Equations 10.105b are found to be satisfied by the requirement

$$\begin{aligned} f_0(r, 0, z) &= 0 \\ f_1(r, 0, z) &= 0 \\ f_3(r, 0, z) &= 0 \\ \frac{\partial f_2}{\partial \theta}(r, 0, z) &= 0 \end{aligned} \tag{10.106}$$

In the region $\theta = \pi$, it can be shown that the following relations must hold [10.34]:

$$\begin{aligned} \frac{\partial f_1}{\partial y} &= -\frac{1}{2(1-\upsilon)}\frac{\partial \Omega}{\partial x}, \quad \theta = \pi \\ \frac{\partial f_3}{\partial y} &= -\frac{1}{2(1-\upsilon)}\frac{\partial \Omega}{\partial z}, \quad \theta = \pi \end{aligned} \tag{10.107a}$$

and

$$\begin{aligned} &(1-2\upsilon)f_2 - \frac{\partial f_0}{\partial y} - \frac{\partial f_1}{\partial \theta} + y\frac{\partial f_3}{\partial z} - z\frac{\partial f_3}{\partial y} = 0, \quad \theta = \pi \\ &2(1-\upsilon)\frac{\partial f_2}{\partial y} - (1-2\upsilon)\frac{\partial f_1}{\partial x} + 2\upsilon\frac{\partial f_3}{\partial z} - \frac{\partial^2 f_0}{\partial y^2} - \frac{\partial^2 f_1}{\partial \theta \partial y} - z\frac{\partial^2 f_3}{\partial y^2} = 0, \quad \theta = \pi \end{aligned} \tag{10.107b}$$

The mixed relations consisting of the second and third equations in Equations 10.106 and 10.107a provide the necessary information for determining f_1 and f_3. In Equations 10.107a, the derivatives of the thermos-elastic potentials are determined using Equation 10.104 and knowledge of the temperature field in the solid. In order to determine the remaining potentials f_0 and f_2, the particular case of a concentrated flux at a given point of the crack surface is used to illustrate the procedure.

Constant heat flux: Consider a constant heat flux Q_0 situated at the point $(a, \pi, 0)$. Equations 10.102 and 10.104 yield

$$\frac{\partial\Omega}{\partial x}+i\frac{\partial\Omega}{\partial z}=\frac{(1+\upsilon)\alpha Q_0}{2(1-\upsilon)\pi^2(x+a-iz)}\left[\left(\frac{2a}{x-a-iz}\right)^{1/3} x\mathrm{Ln}\left(\frac{\sqrt{r+x}+\sqrt{x-a-iz}}{\sqrt{r+a=iz}}\right)\right.$$
$$\left.-\frac{y}{\rho}\tan^{-1}\frac{\sqrt{2a(r-x)}}{\rho}\right] \qquad (10.108)$$

Suitable expressions for f_1 and f_3 are as follows:

$$f_1(r,\theta,z)=-\frac{1}{2(1-\upsilon)}\left[y\frac{\partial\Omega}{\partial x}-(x+a)\frac{\partial\Omega}{\partial y}\right]+C_1\frac{\sin\frac{\theta}{2}}{\sqrt{r}}\mathrm{Re}\cdot[g(\varsigma)]$$
$$f_3(r,\theta,z)=-\frac{1}{2(1-\upsilon)}\left[y\frac{\partial\Omega}{\partial z}-z\frac{\partial\Omega}{\partial y}\right] \qquad (10.109a)$$

where C_1 is a constant introduced for convenience and Re. designates the real part of an analytic function g of the variable ζ defined by

$$\varsigma=r+a+ix,\quad i=\sqrt{-1} \qquad (10.109b)$$

Note that the term containing C_1 is a singular solution of the Laplace equation which satisfies the required conditions of f_1 and conforms to the requirements at infinity.

The next step in the analysis is to derive another set of relations which governs the remaining potentials f_0 and f_2. In order to achieve this, it is expedient to add to the first of Equation 10.107b the term, $y(\partial f_3/\partial z)$, which vanishes in the region $\theta=\pi$. It is not difficult to prove that the resulting relation yields the following equation:

$$\frac{\partial}{\partial y}\left[(1-2\upsilon)f_2-\frac{\partial f_0}{\partial y}-\frac{\partial f_1}{\partial\theta}+y\frac{\partial f_3}{\partial z}-z\frac{\partial f_3}{\partial y}\right]=0,\quad \theta=0 \qquad (10.110)$$

A glance at the quantity inside the bracket in Equation 10.110 reveals that it is a harmonic function.

It follows that the first of Equations 10.107b and 10.110 suggest the following relation involving the singular solution introduced in Equation 10.108:

$$(1-2\upsilon)f_2-\frac{\partial f_0}{\partial y}-\frac{\partial f_1}{\partial\theta}+y\frac{\partial f_3}{\partial z}-z\frac{\partial f_3}{\partial y}=C_2\frac{\cos(\theta/2)}{\sqrt{r}}\mathrm{Re}\cdot[g(\varsigma)] \qquad (10.111)$$

where C_2 is constant. A second relation between f_0 and f_2 may be derived from Equations 10.106 and the second equation of Equation 10.107b, where the result is [10.34].

$$2(1-\upsilon)f_2 - \frac{\partial f_0}{\partial y} = \frac{\partial f_1}{\partial \theta} + z\frac{\partial f_3}{\partial y} - y\frac{\partial f_3}{\partial z} + (1-2\upsilon)\int_{\infty}^{y}\left(\frac{\partial f_1}{\partial x} + \frac{\partial f_3}{\partial z}\right)dy, \quad y \geq 0 \tag{10.112}$$

Solving Equations 10.111 and 10.112 simultaneously and using Equation 10.109a results in

$$\begin{aligned} f_2 &= [(1-2\upsilon)C_1 - C_2]\frac{\cos(\theta/2)}{\sqrt{r}}\mathrm{Re}\cdot[g(\xi)] \\ &\quad - (1-2\upsilon)\int_{\infty}^{y}\frac{\sin(\theta/2)}{\sqrt{r}}\mathrm{Re}\cdot[g'(\varsigma)]dy + \frac{(1-2\upsilon)}{2(1-\upsilon)}\left[\Omega + (x+a)\frac{\partial \Omega}{\partial x} + y\frac{\partial \Omega}{\partial y} + z\frac{\partial \Omega}{\partial z}\right]. \end{aligned} \tag{10.113a}$$

$$\begin{aligned} \frac{\partial f_0}{\partial y} &= \left[\left(\frac{1}{2} - 4\upsilon + 4r^2\right)C_1 - 2(1-\upsilon)C_2\right]\frac{\cos(\theta/2)}{\sqrt{r}}\mathrm{Re}\cdot[g(\varsigma)] \\ &\quad - (1-2\upsilon)^2 C_1\int_{\infty}^{y}\frac{\sin(\theta/2)}{\sqrt{r}}\mathrm{Re}\cdot[g'(\varsigma)]dy + \frac{1}{2(1-\upsilon)}x\left\{\left[(1-2\upsilon)^2 + y\frac{\partial}{\partial y}\right]\right. \\ &\quad \times\left[\Omega + (x+a)\frac{\partial \Omega}{\partial x} + y\frac{\partial \Omega}{\partial y} + z\frac{\partial \Omega}{\partial z}\right] + x\left[\frac{\partial \Omega}{\partial x} + y\frac{\partial^2 \Omega}{\partial x \partial y} - (x+a)\frac{\partial^2 \Omega}{\partial y^2}\right] \\ &\quad \left. + z\left[\frac{\partial \Omega}{\partial z} + y\frac{\partial^2 \Omega}{\partial y \partial z} - z\frac{\partial^2 \Omega}{\partial y^2}\right]\right\} \end{aligned} \tag{10.113b}$$

where g' is the derivative of function g with respect to the argument, and Ω is given in Equation 10.108.

The remaining conditions to be satisfied are the first of Equation 10.106 and the requirement that the displacements are finite at $r = 0$. These conditions are used to determine the function $g(\zeta)$ and the constants C_1 and C_2. The procedure is shown in detail in Reference 10.34. The results are, for the function g we obtain

$$g(\zeta) = \mathrm{Ln}\zeta \tag{10.114a}$$

By expanding the expressions for displacements u_r and u_θ near the crack border, we obtain

$$\begin{aligned} C_1 &= -\frac{\upsilon(1+\upsilon)\alpha Q_0\sqrt{a}}{4(2-\upsilon)(1-\upsilon)^2\pi^2} \\ C_2 &= -\frac{\upsilon(1+\upsilon)\alpha Q_0\sqrt{a}}{2(2-\upsilon)(1-\upsilon)\pi^2} \end{aligned} \tag{10.114b}$$

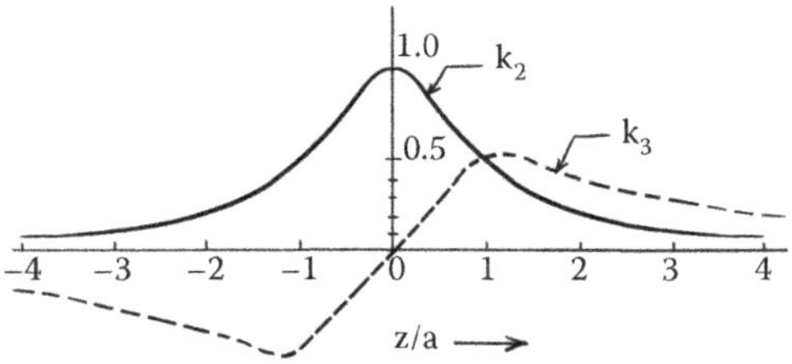

FIGURE 10.14
Variations of k_2 and k_3 along crack border of semi-infinite crack.

The shear stresses across the surface $\theta = 0$ are found as follows:

$$\tau_{\theta r}(r, 0, x) = \frac{E\alpha Q_0 \sqrt{a}(x + a)}{(2 - \upsilon)\pi^2 \sqrt{x}[(x + a)^2 + z^2]} + \cdots$$
$$\tau_{\theta z}(r, 0, x) = \frac{E\alpha Q_0 \sqrt{az}}{(2 - \upsilon)\pi^2 \sqrt{x}[(x + a)^2 + z^2]} + \cdots \quad (10.115a)$$

where E is Young's modulus of the material and the nonsingular terms have been neglected.

Equations 10.115 may be expressed in the standard form

$$\tau_{\theta r}(r, 0, z) = \frac{k_2}{\sqrt{2x}} + O(x^0)$$
$$\tau_{\theta z}(r, 0, z) = \frac{k_3}{\sqrt{2x}} + O(x^0) \quad (10.115b)$$

in which the stress-intensity factors, k_2 and k_3, associated with the edge-sliding and tearing modes of crack extensions, respectively, are given in terms of the nondimensional parameter $z_0 = z/a$ by

$$k_2 = \frac{\sqrt{2}E\alpha Q_0}{(2 - \upsilon)\pi^2 \sqrt{a}} \left(\frac{1}{1 + z_0^2} \right)$$
$$k_3 = \frac{\sqrt{2}E\alpha Q_0}{(2 - \upsilon)\pi^2 \sqrt{a}} \left(\frac{z_0}{1 + z_0^2} \right) \quad (10.116)$$

Figure 10.14 shows the variations of the stress-intensity factors in Equation 10.116 with the nondimensional distance z_0.

Problems

10.1 Determine σ_r and σ_θ induced in a long solid cylinder subjected to a change in temperature of magnitude T°F assuming elastic behavior, $\varepsilon_z = 0$ and $\nu = 1/3$. Explain any similarity of the results.

Ans:

$$C \quad \frac{\sigma_r}{E} = \frac{3}{2}\varepsilon_r + \frac{3}{4}\varepsilon_\theta - 3\alpha T, \quad \frac{\sigma_\theta}{E} = \frac{3}{2}\varepsilon_\theta + \frac{3}{4}\varepsilon_r - 3\alpha T$$

, Radial and tangential stresses are independent of z

10.2 A long annular cylinder, inner radius = a and outer radius = b, is subjected to steady-state heat flow such that the temperature on the inner surface is T_i and on the outer surface is zero. If the thickness of the cylinder is small in comparison with b, find expressions of the tangential stresses at r = a and r = b in terms of m where b/a = 1 + m, m is small.

Ans:

$$\frac{\alpha E T_i}{2(1-\upsilon)}\left[-\left(1+\frac{m}{3}\right);\left(1-\frac{m}{3}\right)\right]$$

10.3 Show that the results for the tangential stress in Problem (10.2) also hold for the axial stress σ_z.

10.4 Examine Equations 10.12 through 10.14 related to an annular cylinder subjected to temperature T(r, z) when the internal radius, a, is zero. Compare the results with the corresponding equations for solid cylinder, Equations 10.8 and 10.9. What are the reasons for any discrepancy?

Ans: Equation 10.9 is not recovered, Equation 10.11 is recovered due to end condition.

10.5 Show that by using Equations 10.17 where the derivatives of the dilatation are given by

$$\frac{\partial e}{\partial x}=\frac{(1+\upsilon)\alpha}{(1-\upsilon)}\frac{\partial T}{\partial x}$$

$$\frac{\partial e}{\partial y}=\frac{(1+\upsilon)\alpha}{(1-\upsilon)}\frac{\partial T}{\partial y}$$

$$\frac{\partial e}{\partial Z}=\frac{(1+\upsilon)\alpha}{(1-\upsilon)}\frac{\partial T}{\partial z}$$

Equations 10.18 can be confirmed.

10.6 An infinite elastic solid contains a circular crack of radius = a in the z = 0 plane.

a. Find the stress-intensity factor k_1 when a uniform temperature T_0 is kept over a concentric circular area of radius b, b < a.

b. Same as part (a) if the constant temperature is maintained over a ring area $\pi(a^2-b^2)$ outside the concentric circle of radius b.

Ans:

$$k_1=\frac{\alpha E T_0 a^{1/2}}{\pi(1-\upsilon)}\left\{1-\left[1-(b/a)^2\right]^{1/2}\right\}$$

10.7 Find the stress-intensity factor k_1 when a constant temperature gradient Q_0 is maintained over a circular area of radius b on the surface of a circular crack of radius a, a > b, in the z = 0 plane of an infinite solid.

$$k_1=\frac{\alpha E Q_0 a^{3/2}}{4(1-\upsilon)}$$

$$k_2=k_3=0$$

10.8 Repeat Problem 10.8 if a circular ring of inner radius b and outer radius a in the circular crack is heated by a uniform temperature gradient Q_0.

Ans:

$$k_1 = \frac{\alpha E Q_0 (a^2 - b^2)}{4(1 - \upsilon) a^{1/2}}$$

$$k_2 = k_3 = 0$$

10.9 An external circular crack is subjected to constant temperature $T_0 = 20°C$ over a ring area bounded by inner and outer radii 1 and 3 cm, respectively. Find the extension of the plastic zone in the bonded area using Dugdale's model. Assume $E = 200$ GPa, $\sigma_0 = 250$ MPa, $\alpha = 11.7 \times 10^{-6}/°C$, and $\nu = 0.3$.

Ans: 0.13 cm.

10.10 Repeat the last problem if the heated region on the crack surfaces is given by $T(r) = T_0 r^{-n}$, $n = 3$.

Ans: 0.01 cm.

References

10.1. J. M. C. Duhammel, *Memoires par Divers Savants*, vol. 5, Paris, 1838.
10.2. T. M. MacRobert, *Spherical Harmonics*, Pergamon Press Ltd., New York, 1967.
10.3. R. V. Churchill, *Fourier Series and Boundary Value Problems*, McGraw-Hill Book Co., New York, 1963.
10.4. W. E. Byerly, *Fourier Series and Spherical Harmonics*, 1902.
10.5. E. Jahnke, F. Emde and F. Losch, *Tables of Higher Functions*, McGraw-Hill Book Co., New York, 1950.
10.6. I. S. Gradshteyn and I. M. Ryzhik, *Tables of Integrals, Series and Products*, Academic Press, New York and London, 1965.
10.7. C. H. Lees, *Proc. Roy. Soc.* (London), vol. 101, p. 411, 1922.
10.8. R. Lorenz, *Ver. Deutch. Ing.*, vol. 51, p. 343, 1907.
10.9. L. Barker, *Engineering*, vol. 124, p. 443, 1927.
10.10. W. D. MacMillan, *Theory of Potentials*, New York, 1930.
10.11. J. N. Goodier, *Phil. Magazine*, vol. 23, p. 1007, 1937.
10.12. N. O. Myklestand, *J. Appl. Mech., ASME*, p. A-131, 1942.
10.13. R. H. Edwards, *J. Appl. Mech., ASME*, vol. 18, pp. 13–30, 1951.
10.14. M. K. Kassir and G. C. Sih, Three-dimensional thermo-elastic problems of planes of discontinuities or cracks in solids, edited by W.A. Shaw, *Developments in the Theoretical and Applied Mechanics*, vol. 3, pp. 117–146, Pergamon Press, 1967.
10.15. M. K. Kassir and G. C. Sih, *Three-Dimensional Crack Problems*, Noordhoff International Publishing, Lyden, The Netherlands, 1975.
10.16. Z. Olesiak and I. N. Sneddon, The distribution of thermal stress in an infinite elastic solid containing a penny-shaped crack, *Arch. Ration. Mech. Anal.*, vol. 4, pp. 238–254, 1959.
10.17. M. K. Kassir, Thermal crack propagation, *J. Basic Eng., ASME* paper No. 71 – Met/ - N, 1972.
10.18. A. L. Florence and J. N. Goodier, The linear thermos-elastic problem of uniform heat flow disturbed by a penny-shaped crack, *Int. J. Eng. Sci.*, vol. 1, pp. 533–540, 1963.
10.19. M. K. Kassir and A. M. Bergman, Thermal stresses in a solid containing parallel circular cracks, *Appl. Sci. Res.*, vol. 25, pp. 262–280, 1971.
10.20. A. M. Bergman and M. K. Kassir, Thermal fracture of bonded dissimilar media containing a penny-shaped crack, *Int. J. Fract. Mech.*, vol. 10, No. 1, pp. 87–98, 1974.

10.21. M. K. Kassir and G. C. Sih, Thermal stresses in a solid weakened by an external circular crack, *Int. J. Solids Struct.*, vol. 5, pp. 351–367, 1969.
10.22. M. Lowengrub and I. N. Sneddon, The solution of a pair of dual integral equations, *Proc. Glasgow Math. Assoc.*, vol. 6, pp. 14–18, 1963.
10.23. D. S. Dugdale, Yielding of steel sheets containing slits, *J. Mech. Phys. Solids*, vol. 8, pp. 100–104, 1964.
10.24. M. K. Kassir, Size of thermal plastic zones around external cracks, *Int. J. Fract. Mech.*, vol. 5, No. 3, pp. 167–177, 1969.
10.25. M. K. Kassir, On the distribution of thermal stresses around an elliptical crack in an infinite solid, *Int. J. Eng. Sci.*, vol. 7, pp. 769–784, 1969.
10.26. E. T. Whittaker and G. N. Watson, *Modern Analysis*, 4th edition, Cambridge University Press, England, pp. 548–552, 1962.
10.27. P. F. Byrd and M. D. Friedman, *Handbook of Elliptic Integrals for Engineers and Physicists*, Springer Verlag, 1954.
10.28. M. K. Kassir, Thermal stresses in an elastic solid containing a plane crack, *Int. J. Eng. Sci.*, vol. 13, pp. 703–711, 1975.
10.29. I. N. Sneddon, *The Use of Integral Transforms*, McGraw-Hill Book Co., New York, 1972.
10.30. N. N. Lebedev, I. P. Skalskaya and Y. S. Uflyand, *Problems of Mathematical Physics*, English translation by R. A. Silverman, Prentice-Hall, Englewood Cliffs, NJ, 1965.
10.31. O. D. Kellog, *Foundations of Potential Theory*, Dover Publications, New York, 1953.
10.32. H. Bateman, *Differential Equations*, 1926.
10.33. W. Nowacki, *Thermoelasticity*, Addison-Wesley Publishing Co., Reading, MA, 1962.
10.34. M. K. Kassir, Stress-intensity factors for an insulated half-plane crack, *J. Appl. Mech. ASME*, Paper no. 76-APM-17, 1975.
10.35. A. Erdelyi (editor), *Tables of Integral Transforms*, vol. 2, McGraw-Hill Book Co., New York, 1954.
10.36. M. K. Kassir and G. C. Sih, Application of Papkovich–Neuber potentials to a crack problem, *Int. J. Solids Struct.*, vol. 9, pp. 643–654, 1973.

11

Solutions to Selected Problems

Chapter 2

P2.7: We consider a plane through the origin of a system of rectangular coordinates x, y, and z whose normal is line 0A in the element of the solid.

The direction cosines of 0A are (l, m, n) and the projection of 0A on the xy-plane is line 0B.

A unit vector along 0A is denoted by $\mathbf{e_A} = l\mathbf{i} + m\mathbf{j} + n\mathbf{k}$ and a unit vector along 0B is denoted by

$$\mathbf{e_B} = \cos 40^\circ \mathbf{i} + \cos 50^\circ \mathbf{j} = 0.766\mathbf{i} + 0.643\mathbf{j}$$

Since $\mathbf{e_A}.\ \mathbf{e_B} = \cos 45^\circ = 0.7071$, we have

$$0.766\,l + 0.643\,m = 0.7071$$

Since $l^2 + m^2 + n^2 = 1$, solving for the direction cosines l, m, and n, we obtain

$$l = 0.555,\ m = 0.439, \quad \text{and} \quad n = 0.707$$

Equation 2.13 yields $p_x = 0.099$ MPa, $p_y = 3.609$ MPa, $p_z = -2.908$ MPa, and the normal stress $\sigma_n = lp_x + mp_y + np_z = -0.42$ MPa.

Equation 2.16 provides the tangential stress $\tau_n = 4.617$ MPa; it follows that the resultant of the normal and tangential stresses is $R = 4.636$ MPa.

Equation 2.17 gives $l_s = 0.072$, $m_s = 0.822$, $n_s = -0.566$, and using Equation 2.19, the direction cosines of the resultant stress R are

$$l = \cos\theta_x = p_x/R_n = 0.714$$

$$m = \cos\theta_y = p_y/R_n = 0.778$$

$$n = \cos\theta_z = p_z/R_n = -0.627$$

P2.24: Given that $\sigma_1 > \sigma_2 = \sigma_3$ and $m^2 + n^2 = 1 - l^2$, Equation 2.14 in terms of principal stresses provides

$$\sigma_n = l^2\sigma_1 + m^2\sigma_2 + n^2\sigma_3 = l^2\sigma_1 + (m^2 + n^2)\sigma_2 = \sigma_2 + l^2(\sigma_1 - \sigma_2) \tag{1}$$

Equation 2.16 gives

$$\sigma_n^2 + \tau_n^2 = p_x^2 + p_y^2 + p_z^2 = l^2\sigma_1^2 + m^2\sigma_2^2 + n^2\sigma_3^2 = \sigma_2^2 + l^2(\sigma_1 - \sigma_2)(\sigma_1 + \sigma_2) \tag{2}$$

The next step is to eliminate σ_1 from Equations (1) and (2) above and end up with the following quadratic equation for σ_2 containing terms with l^2 and normal and shearing stresses only:

$$(1-l^2)\sigma_2^2 - 2(1-l^2)\sigma_n\sigma_2 + (1-l^2)\sigma_n^2 - l^2\tau_n^2 = 0 \tag{3}$$

The solution of Equation (3) is

$$\sigma_2 = \frac{1}{2(1-l^2)}\left\{2(1-l^2)\sigma_n \pm \left[4(1-l^2)^2\sigma_n^2 - 4(1-l^2) < (1-l^2)\sigma_n^2 - l^2\tau_n^2 >\right]^{1/2}\right\} \tag{4}$$

Simplifying Equation (4), we find that

$$\sigma_2 = \sigma_n - \frac{l\tau_n}{\sqrt{1-l^2}}$$

and Equation (1) yields

$$\sigma_1 = \frac{1}{l^2}\left[\sigma_n - (1-l^2)\sigma_2\right] = \sigma_n + \frac{\sqrt{1-l^2}\,\tau_n}{l}$$

as required.

P2.33: Equation 2.46b gives the octahedral shearing stress as $(2/3\,J_2)^{1/2}$, where J_2 is the second stress deviator invariant defined in Equation 2.42, $J_2 = R$.

The principal stresses are $\sigma_1 > \sigma_2 > \sigma_3$. Hence, $\tau_{max} = (1/2)\,(\sigma_1 - \sigma_3)$.

Using Equation 2.31, we have

$$\tau_{max} = \left(\frac{J_2}{3}\right)^{1/2}\left[\cos\theta - \cos(\theta + 120°)\right] = \frac{1}{2}\sqrt{J_2}\left[\sqrt{3}\cos\theta + \sin\theta\right]$$

It follows that

$$\frac{\tau_{oct}}{\tau_{max}} = y(\theta) = \frac{2\sqrt{2/3}}{\sqrt{3}\cos\theta + \sin\theta}$$

The turning points of $y(\theta)$ are found by setting the first derivative of y with respect to θ to be zero. This gives $\tan\theta = 1/(3)^{1/2}$ or $\theta = 30°$. Hence,

$$\frac{\tau_{oct}}{\tau_{max}} = y = 0.8165 = \frac{\sqrt{6}}{3}$$

For $\theta = 60°$, $y = 0.9428 = (8)^{1/2}/3$. Hence, we have

$$\frac{\sqrt{6}}{3} \le \frac{\tau_{oct}}{\tau_{max}} \le \frac{\sqrt{8}}{3}$$

Chapter 3

P3.6: Point B moves to B′ and point C to C′. Angle C′0B = Angle B0B′ = β.
Angle 0C′A = Angle AC′B′ = Angle 0AC′ = Angle AC′B′ = γ
Angle C0C′ = α. It follows that

$$2\beta + \alpha = \frac{\pi}{2}, \quad 2\beta + 2\gamma = \pi$$

$$\beta = \frac{\pi}{4} - \frac{\alpha}{2}, \quad \beta + \gamma = \frac{\pi}{2}$$

$$< 0DA = \frac{\pi}{2}$$

Denoting length 0D = DB′ = e and AD = DC′ = f, then we have $\sin\beta = f/L$, $\cos\beta = e/L$, and

$$\frac{0B'}{AC'} = \frac{2e}{2f} = \frac{\cos\beta}{\sin\beta} = \frac{\cos\left(\frac{\pi}{4} - \frac{\alpha}{2}\right)}{\sin\left(\frac{\pi}{4} - \frac{\alpha}{2}\right)} = \frac{\left(\cos\frac{\alpha}{2} + \sin\frac{\alpha}{2}\right)}{\left(\cos\frac{\alpha}{2} - \sin\frac{\alpha}{2}\right)} = \frac{1 + \tan\alpha/2}{1 - \tan\alpha/2}$$

It follows that

$$\log\frac{0B'}{AC'} = \log\left(\frac{1 + \tan\alpha/2}{1 - \tan\alpha/2}\right) = 2\tan^{-1}\frac{\alpha}{2}$$

P3.7: Points A and B in Figure P3.7 move to A* and B*.
Consider point D located at coordinates (x L, h), the point move to point D* located at coordinates (a x L, b h).

i. Since 0D = 0D* or $x^2 L^2 + h^2 = a^2 x^2 L^2 + b^2$, solving for $1/x$, we have

$$\frac{1}{x} = \frac{L}{h}\sqrt{\frac{a^2 - 1}{1 - b^2}}$$

Let angle A0D = η_0 and angle A0D* = η, then

$$\tan\eta_0 = \frac{h}{xL} = \sqrt{\frac{a^2 - 1}{1 - b^2}}$$

and

$$\tan\eta = \frac{bh}{axL} = \frac{b}{a}\sqrt{\frac{a^2 - 1}{1 - b^2}} = \frac{b}{a}\tan\eta_0$$

ii. Assume thickness of plate = t, initial volume = Lht, and final volume = aLbht.

So, If ab = 1, there is no change in volume and the deformation is equivalent to that of pure torsion.

P3.9: Equation of the initial plate is $x^2 + y^2 = a^2$, and the polar coordinates of an arbitrary point in the plate are $x = r\cos\theta$, $y = r\sin\theta$.

For the deformed plate, the equation of the ellipse in polar coordinates is

$$\frac{r^2\cos^2\theta}{(1+\beta)^2} + r^2\sin^2\theta = a^2$$

$$r^2\cos^2\theta(1+\beta)^{-2} + r^2\sin^2\theta = a^2$$

$$r^2\cos^2\theta(1-2\beta+\cdots) + r^2\sin^2\theta = a^2$$

$$r = [a^2(1-2\beta\cos^2\theta)^{-1}]^{1/2} = a(1+\beta\cos^2\theta)$$

Now distance $PP' = r - a = \beta a\cos^2$ while $QQ' = \beta a\cos^2(\theta + d\theta)$. The shear strain between the radial and tangential directions is

$$\gamma_{rt} = \underbrace{\text{Lim}\frac{QQ' - PP'}{PQ}}_{Q\to P} - = \underbrace{\text{Lim}\frac{\beta a\cos^2(\theta+d\theta) - \beta a\cos^2\theta}{a\,d\theta}}_{d\theta\to 0}$$

Simplifying the algebra and noting that in the limit, $\cos d\theta$ becomes 1 while $\sin d\theta = 0$, the shear strain is

$$\gamma_{rt} = -2\beta\sin\theta\cos\theta = -2\beta\frac{xy}{a^2}$$

The minus indicates that the angles become larger.

Chapter 4

P4.8: A valid solution must satisfy the equations of compatibility of strains, Equation 3.13 and equations of equilibrium, Equations 2.6. In this case, we have

$$\varepsilon_x = \frac{\partial u}{\partial x} = 3c_1x^2 + c_2y^2, \quad \varepsilon_y = \frac{\partial V}{\partial y} = c_3x^2$$

$$\gamma_{xy} = \frac{\partial u}{\partial y} + \frac{\partial V}{\partial x} = 2(c_2 + c_3)xy$$

The derivatives of the strains which appear in Equation 3.13 are

$$\frac{\partial^2\varepsilon_x}{\partial y^2} = 2c_2, \quad \frac{\partial^2\varepsilon_y}{\partial x^2} = 2c_3, \quad \frac{\partial^2\gamma_{xy}}{\partial x\partial y} = 2(c_2 + c_3)$$

Hence, Equation 3.13 is satisfied for all values of the constants.

Next, we have to check Equation 2.6. The stresses are derived as follows:

$$\sigma_x = \frac{E}{1-\upsilon^2}(\varepsilon_x + \upsilon\varepsilon_y) = \frac{E}{1-\upsilon^2}(3c_1x^2 + c_2y^2 + \upsilon c_3x^2)$$

$$\sigma_y = \frac{E}{1-\upsilon^2}(\varepsilon_y + \upsilon\varepsilon_x) = \frac{E}{1-\upsilon^2}(c_3x^2 + 3\upsilon c_1x^2 + \upsilon c_2y^2)$$

$$\tau_{xy} = \frac{E}{2(1+\upsilon)}\gamma_{xy} = \frac{E}{1+\upsilon}(c_2 + c_3)$$

Inserting these expressions in Equation 2.6, we obtain

$$6c_1 + (1 - \upsilon)(c_2 + c_3) = 0$$
$$c_2 + c_3 + \upsilon(c_2 - c_3) = 0$$

Hence, Equation 2.6 is not satisfied and the solution is not valid.

P4.12: Let $d\varepsilon'$ denote the increment of lateral strain in the tensile test. This increment consists of an elastic and plastic component as follows:

$$d\varepsilon' = \left(\frac{1}{2} - \upsilon\right)d\varepsilon^e - \frac{1}{2}d\varepsilon$$

The lateral contraction is given by

$$\eta = -\frac{d\varepsilon'}{d\varepsilon} = \frac{1}{2} - \left(\frac{1}{2} - \upsilon\right)\frac{d\varepsilon^e}{d\varepsilon}$$

We have $E_t = d\sigma/d\varepsilon$ from Equation 4.29 and since $d\varepsilon^e = d\sigma/E$, it follows that

$$\eta = \frac{1}{2} - \left(\frac{1}{2} - \upsilon\right)\frac{1}{E}\frac{d\sigma}{d\varepsilon} = \frac{1}{2} - \left(\frac{1}{2} - \upsilon\right)\frac{E_t}{E}$$

P4.14: Equation 4.41b defines the instability condition as: $d\sigma/d\varepsilon = \sigma$. The Ramberg–Osgood relation is

$$\varepsilon = \frac{\sigma}{E} + 0.3\sigma_0\left(\frac{\sigma}{\sigma_0}\right)^n$$

Differentiating the above relation with respect to the strain, noting that $\varepsilon_0 = \sigma/\sigma_0$, we obtain

$$1 = \frac{1}{E}\frac{d\sigma}{d\varepsilon} + 0.3\sigma_0 n\left(\frac{\sigma}{\sigma_0}\right)^{n-1}\frac{1}{\sigma_0}\frac{d\sigma}{d\varepsilon}$$

$$1 = \left[\frac{1}{E} + 0.3\varepsilon_0\frac{n}{\sigma}\left(\frac{\sigma}{\sigma_0}\right)^n\right]\frac{d\sigma}{d\varepsilon}$$

$$\sigma = \left[\frac{\sigma}{E} + 0.3\varepsilon_0 n\left(\frac{\sigma}{\sigma_0}\right)^n\right]\frac{d\sigma}{d\varepsilon}$$

and the stress at instability is obtained as

$$1 = \frac{\sigma}{E} + 0.3\varepsilon_0 n\left(\frac{\sigma}{\sigma_0}\right)^n$$

Neglecting the elastic strain (small quantity), the instability condition is

$$\varepsilon = \varepsilon_0(0.3\varepsilon_0 n)^{-(1/n)} + 0.3\varepsilon_0\left[(0.3\varepsilon_0 n)^{-(1/n)}\right]^n = (0.3n)^{-(1/n)}\varepsilon_0^{1-(1/n)} + \frac{1}{n}$$

Hence,

$$\varepsilon = \frac{1}{n} + \frac{(\varepsilon_0)^{1-(1/n)}}{(0.3n)^{(1/n)}}$$

while for the power law, $\sigma = C\,\varepsilon^{1/n}$, $C =$ constant, and the instability condition is $\varepsilon = 1/n$.

For $n = 6$ and $\varepsilon_0 = 0.002$, the power law gives $\varepsilon = 0.125$ and the Ramberg–Osgood relation gives

$$\varepsilon = \frac{1}{6} + \frac{(0.002)^{5/6}}{(1.8)^{1/6}} = \frac{1}{6} + 0.0051085$$

The % error is 0.51.

P4.17: The stress–strain relation $\sigma = A\,[1 - m\,e^{-n\varepsilon}]$ for $n = 0$ and $n =$ infinity yield constant values, $A\,(1 - m)$ and A, respectively. If one plots the relation for other values of n, using $e = 2.7186$, the behavior of the material is basically that of a rigid-plastic behavior.

At the onset of instability, we have $d\sigma/d\varepsilon = \sigma$, or

$$A\,m\,n\,e^{-n\varepsilon} = A(1 - n\,e^{-n\varepsilon})$$

Simplifying, we get

$$e^{-n\varepsilon} = \frac{1}{m(1+m)}$$

$$-n\varepsilon = \mathrm{Ln}\frac{1}{m(1+m)}$$

$$\varepsilon = \frac{1}{n}[\mathrm{Ln}(1+m)m]$$

and the true stress at instability is given as

$$me^{-n\varepsilon} = \frac{1}{1+n}$$

$$\frac{\sigma}{A} = 1 - \frac{1}{1+n}$$

$$\sigma = \frac{nA}{1+n}$$

P4.28: Since the length of the vertical member (member 2) is L, the length of the inclined member (members 1 and 3) is $L/\cos\alpha$. The horizontal and vertical displacements of joint D are denoted by u and v, respectively. Refer to solved Example 4.8 in the text.

Equilibrium of the truss gives

$$\sigma_1\cos\alpha + \sigma_2 + \sigma_3\cos\alpha = \frac{P}{A}$$

$$\sigma_1 - \sigma_3 = \frac{Q}{A\sin\alpha}$$

The strains and displacements of joint D are as follows:

$$\varepsilon_1 = \frac{u}{L}\sin\alpha\cos\alpha + \frac{V}{L}\cos^2\alpha$$

$$\varepsilon_2 = \frac{V}{L}$$

$$\varepsilon_3 = -\frac{u}{L}\sin\alpha\cos\alpha + \frac{V}{L}\cos^2\alpha$$

and

$$\varepsilon_1 = \varepsilon_3 = \varepsilon_2\cos^2\alpha = \frac{V}{L}\cos^2\alpha$$

The elastic solution yields

$$\sigma_1 = \frac{\cos^2\alpha}{1 + 2\cos^3\alpha}\frac{P}{A} + \frac{Q}{2A\sin\alpha}$$

$$\sigma_2 = \frac{1}{1 + 2\cos^3\alpha}\frac{P}{A}$$

$$\sigma_3 = \frac{\cos^2\alpha}{1 + 2\cos^3\alpha} - \frac{Q}{2A\sin\alpha}$$

a. Magnitude of P at collapse is $P_p = (1 + 2\cos\alpha)\sigma_0$, and the stresses, strains, and displacements are as follows:

$$\sigma_1 = \sigma_2 = \sigma_3 = \sigma_0$$

$$\varepsilon_1 = \varepsilon_3 = \varepsilon_0, \quad \text{and} \quad \varepsilon_2 = \frac{\varepsilon_0}{\cos^2\alpha}$$

$$u = 0, \quad \text{and} \quad \frac{V}{L} = \frac{\varepsilon_0}{\cos^2\alpha}$$

Next, increase load Q. Since the change in the vertical displacement is zero, there are no changes in the stress and strain of the middle member.

Since $\sigma_1 = 0$, increasing Q causes σ_3 to decrease. Both stresses in bars 1 and 2 do not change. So load P must decrease as Q increases, and we have

$$\frac{\Delta P}{A} = -\frac{\cos\alpha}{\sin\alpha}\frac{\Delta Q}{A}$$

The truss fails when the three bars yield in reverse loading, that is, at

$$\frac{\Delta Q}{A} = (2\sin\alpha)\sigma_0$$

$$\frac{\Delta P}{A} = -\frac{\cos\alpha}{\sin\alpha}(2\sin\alpha)\sigma_0 = -(2\cos\alpha)\sigma_0$$

The failure loads are

$$\frac{P}{A} = (2\cos\alpha)\sigma_0$$

$$\frac{Q}{A} = (2\sin\alpha)\sigma_0$$

Using the basic equations, we arrive at

$$\sigma_1 = \sigma_0, \sigma_3 = -\sigma_0, \quad \text{and} \quad \sigma_2 = (2\cos\alpha)\sigma_0$$

$$\varepsilon_1 = \frac{u}{L}\sin\alpha\cos\alpha + \frac{V}{L}\cos^2\alpha$$

$$\varepsilon_3 = -\frac{u}{L}\sin\alpha\cos\alpha + \frac{V}{L}\cos^2\alpha$$

Hence

$$\frac{V}{L} - \frac{\varepsilon_0}{4\cos^4\alpha}$$

$$\frac{u}{L}\sin\alpha\cos\alpha = \frac{\varepsilon_0}{2\cos^2\alpha} + \varepsilon_0 - \frac{\varepsilon_0}{4\cos^2\alpha}$$

$$\frac{u}{L} = \frac{\varepsilon_0(4\cos^2\alpha + 1)}{4\sin\alpha\cos^3\alpha}$$

Chapter 5

P5.4: For a solid bar of radius a and no work-hardening material, Section 5.2.1 gives

$$T_p = \frac{2\pi}{3}\tau_0 a^3$$

$$\theta_e = \frac{\gamma_0}{a} = \frac{\tau_0}{aG}$$

$$\frac{a}{c} = \frac{\theta}{\theta_e}$$

For work-hardened material with $G_p/G = 0.3$, $a/c = \theta/\theta_e$, Equation 5.12b provides

$$(1 + 0.1)T = \frac{\pi\tau_0 a^3}{6}\left[4 - \frac{1}{(\theta/\vartheta_e)^3} + 0.3\frac{\theta}{\theta_e}\right]$$

Replacing T by T_p and solving for $x = \theta/\theta_e$, we obtain the following equation:

$$0/3x - 1/x^3 = 0.4$$

Solving for x by trial and error, we find

$$\theta/\theta_e = 1.86$$

P5.10: Using the coordinate system shown in Figure P5.10, we find the equations of the sides of the triangle as follows:

$$\text{SideBC} : x - \frac{a}{2\sqrt{3}} = 0$$

$$\text{SideAB} : y = \frac{x}{\sqrt{3}} + \frac{a}{3}$$

$$\text{SideAC} : y = \frac{x}{\sqrt{3}} - \frac{a}{3}$$

The proper torsion stress function is obtained by multiplying together the above equations

$$\phi(x, y) = k\left[\frac{a^2}{9} - (x^2 + y^2) - \frac{2x}{\sqrt{3}a}(x^2 - 3y^2)\right]$$

where k is constant. Differentiating the stress function with respect to x and y results in confirming that Equation 5.22 is satisfied provided that $k = G\theta/2$.

The stresses are obtained using Equation 5.21

$$\tau_{xz} = \frac{\partial\phi}{\partial y} = G\theta y\left(-1 + 4\sqrt{3}\frac{x}{\sqrt{a}}\right)$$

$$\tau_{yz} = -\frac{\partial\phi}{\partial x} = G\theta\left(x + \sqrt{3}\frac{x^2}{a} - \sqrt{3}\frac{y^2}{a}\right)$$

At the midpoint of side BC, $y = 0$ and $x = a/(2)(3)^{1/2}$, so

$$\tau_{xz} = 0$$

$$\tau_{yz} = \frac{\sqrt{3}}{4}G\theta a$$

as in Equation (5.33), (g).

P5.14: Upon differentiation of the given function, we find that

$$\nabla^2\phi = \frac{\partial^2\phi}{\partial r^2} + \frac{1}{r}\frac{\partial\phi}{\partial r} + \frac{1}{r^2}\frac{\partial^2\phi}{\partial\theta^2} = -2G\theta^*$$

for all values of the constants A_0 and A_3.

Next, we find the values of the constants by requiring the stress function $\phi(r,\theta)$ to vanish at points B and D of the boundary of Figure P5.14. Now Tan < BAC − Tan < 0AC = 1.732 a − 0.577 a = 1.155 a = 0B. Hence, the stress function vanishes at point B.

Also, 0D = 2a − 0A = 2a − 1.155 a = 0.845 a, so for $\theta = 0$, $\phi = 0$ at point D, and for $\theta = 60^\circ$, $\phi = 0$ at point B. Hence, we obtain

$$\phi = 0 = -\frac{G\theta^*}{2}(0.845a)^2 + A_0 + A_3(0.845)^3$$

$$\phi = 0 = -\frac{G\theta^*}{2}(1.155a)^2 + A_0 - A_3(1.155a)$$

Solving for A_0 and A_3

$$A_0 = 0.444G\theta^*a^2, \quad \text{and} \quad A_3 = -0.145G\theta^*a^2$$

Chapter 6

P6.9: In order for the stress function to be bi-harmonic, it must satisfy the equation

$$\nabla^4\phi = \frac{\partial^4\phi}{\partial x^4} + 2\frac{\partial^4\phi}{\partial x^2\partial y^2} + \frac{\partial^4\phi}{\partial y^4} = 0$$

Performing the necessary differentiation, we obtain

$$\nabla^4\phi = 8C + 24Dy + 24G + 120Hy$$

Hence, $G = -(1/3)\,C$ and $H = -D/5$.

The corresponding stresses are

$$\sigma_x = \frac{\partial^2\phi}{\partial y^2} = 2Cx^2 + 6Dx^2y + 2E + 6Fy - 4Cy^2 - 4Dy^3$$

$$\sigma_y = \frac{\partial^2\phi}{\partial x^2} = 2A + 2By + 2Cy^2 + 2Dy^3$$

$$\tau_{xy} = -\frac{\partial^2\phi}{\partial x\partial y} = -2Bx - 4Cxy - 6Dxy^2$$

Six boundary conditions are needed to determine the six unknown constants. On the lower surface, $y = -d$, we have $\sigma_y = \tau_{xy} = 0$, while on the upper surface, $y = d$, $\sigma_y = 0$, and $\tau_{xy} = \tau\ x/L$. These boundary conditions give $A = \tau d/4L$, $B = \tau/8L$, $C = -\tau/(8dL)$, and $D = -\tau/(8d^2L)$. The remaining constants E and F are determined from the conditions on the fixed edge of the cantilever, $x = 0$, $-d < y < d$. The resultant horizontal force and moment at the support edge must balance the applied shearing stress on the upper surface. Assuming the thickness of the cantilever = t, we have

$$\iint_A (\sigma_x)_{x=0} dA = \int_0^L \tau\frac{x}{L}t\,dx = \frac{\tau t L}{2}$$

$$\iint_A (\sigma_x)_{x=0} y\,dA = \frac{\tau t L}{2} d$$

Evaluating the integrals (it is simpler to evaluate line integrals in the variable y between limits $-d$ and d), we find $E = \tau L/8d - \tau d/12L$ and $F = -\tau/10L + \tau L/4d^2$.

The solution is imperfect because the boundary conditions for σ_x and τ_{xy} at the free edge $x = L$ are not satisfied.

P6.12: Differentiating the tangential strain with respect to r yields

$$\frac{\partial\varepsilon_\theta}{\partial r} = \frac{1}{r}\frac{\partial u}{\partial r} - \frac{u}{r^2} = \frac{1}{r}(\varepsilon_r - \varepsilon_\theta) \tag{1}$$

Since $\varepsilon_z = 0$, giving $\sigma_z = \nu(\sigma_r + \sigma_\theta)$, and inserting in Hooke's law, we obtain

$$\varepsilon_r = \frac{1+\upsilon}{E}[(1-\upsilon)\sigma_r - \upsilon\sigma_\theta]$$

$$\varepsilon_\theta = \frac{1+\upsilon}{E}[(1-\upsilon)\sigma_\theta - \upsilon\sigma_r] \tag{2}$$

$$\varepsilon_r - \varepsilon_\theta = \frac{1+\upsilon}{E}(\sigma_r - \sigma_\theta)$$

Using the third equation of Equation (2) in (1), we have

$$\frac{\partial\varepsilon_\theta}{\partial r} = \frac{1+\upsilon}{E}\left(\frac{\sigma_r - \sigma_\theta}{r}\right) \tag{3}$$

Differentiating the second equation of Equation (2) with respect to r yields

$$\frac{\partial \varepsilon_\theta}{\partial r} = \frac{1+\upsilon}{E}\left[(1-\upsilon)\frac{\partial \sigma_\theta}{\partial r} - \upsilon\frac{\partial \sigma_r}{\partial r}\right] \tag{4}$$

Equating Equation (3) to (4) and simplifying gives the result

$$\upsilon r\frac{\partial \sigma_r}{\partial r} - (1-\upsilon)r\frac{\partial \sigma_\theta}{\partial r} + \sigma_r - \sigma_\theta = 0$$

P6.13: As mentioned in the text, the constants A and B appearing in the elastic solution of curved beams may be determined by using boundary condition stated in Equation (d) of the text, which is

$$\iint_A \sigma_\theta r dA = \int_a^b \sigma_\theta r(1)dr = -M$$

However, a simpler boundary condition exists (which in fact is used in the text) and is proven below. The equilibrium of the beam is governed by the equation

$$r\frac{d\sigma_r}{dr} = \sigma_\theta - \sigma_r$$

$$r\frac{d\sigma_r}{dr} + 2\sigma_r = \sigma_\theta + \sigma_r$$

$$r^2\frac{d\sigma_r}{dr} + 2r\sigma_r = r(\sigma_\theta + \sigma_r)$$

$$\frac{d}{dr}(r^2\sigma_r) = r(\sigma_\theta + \sigma_r)$$

Integrating with respect to r between the limits a and b and observing that σ_r vanishes at both boundaries, we obtain

$$\int_a^b \sigma_\theta r\,dr = -\int_a^b \sigma_r r\,dr$$

Since

$$-M = \int_a^b \sigma_\theta r\,dr = \frac{1}{2}\int_a^b \sigma_\theta r\,dr + \left(-\frac{1}{2}\right)\int_a^b \sigma_r r\,dr$$

it follows that

$$M = \frac{1}{2}\int_a^b (\sigma_r - \sigma_\theta) r\,dr$$

P6.31:

i. Solid disc: Extent of the plastic zone is given by Equation 6.56 as

$$\rho\omega_c^2 b^2 = \frac{3\sigma_0}{1 + 0.25\left(1 - \dfrac{c^2}{b^2}\right)^2} \tag{1}$$

The angular velocity needed to initiate yield is given in Equation 6.54

$$\rho\omega_e^2 b^2 = 2.4\sigma_0 \tag{2}$$

The critical speed of the disc is given in Equation 6.58

$$\rho\omega_u^2 b^2 = 3\sigma_0 \tag{3}$$

Equating the average speed from Equations (2) and (3) to that in Equation (1), we have the following equation to determine the extent of the plastic zone c.

$$1 + 0.25\left(1 - \frac{c^2}{b^2}\right)^2 = 1.11$$

which gives $c = 0.57\ b$.

ii. Annular disc: In this case, the velocity needed to initiate yield is obtained from Equation 6.62

$$\rho\omega_e^2 b^2 = 1.174\sigma_0 \tag{4}$$

and Equation 6.65 provides the critical velocity

$$\rho\omega_u^2 b^2 = 2.077 \tag{5}$$

The average angular speed obtained using Equations (4) and (5) is equated to Equation 6.64 to provide the necessary equation to find extent of the plastic zone. After simplification, the resulting equation is as follows:

$$2 - \frac{b}{3c} - \frac{c}{3b} = 0.531\left[2.5 + \frac{1}{27}\left(\frac{c}{b} - \frac{b}{c}\right) - \frac{c^2}{b^2} + 0.5\frac{c^4}{b^4}\right]$$

The magnitude of c is determined using a computer program or trial and error as

$$C = 0.72b$$

Chapter 7

P7.2: Equilibrium of the cylinder is given by Equation 7.1. If σ_θ is constant, say B, then

$$r\frac{d\sigma_r}{dr} + \sigma_r = B$$

$$\frac{d}{dr}(r\sigma_r) = B$$

The solution is $r\sigma_r = Br + A$, $A = \text{constant}$, so $\sigma_r = A/r + B$.

At $r = a$, $\sigma_r = -p$, and at $r = b$, $\sigma_r = 0$. This gives $A = -p(ab/T)$ and $B = p(a/T)$ where $T = b - a$.

The stresses are

$$\sigma_r = -\frac{pa}{T}\left(\frac{b}{r} - 1\right)$$

$$\sigma_\theta = \frac{pa}{T}$$

At any thickness t, we have $r = a + t$, and

$$\sigma_r = -\frac{pa}{T}\left(\frac{T - t}{a + t}\right)$$

P7.5: Since yield occurs according to Tresca's criterion, then in Equation 7.10 if a is replaced by c, the equation becomes

$$q = \frac{\sigma_0}{2}\left(1 - \frac{c^2}{b^2}\right)$$

Equation 7.1 and Tresca's criterion give

$$\sigma_r = -\sigma_0 \text{ Ln } r + A$$

where $A = \text{constant}$ determined from the condition that at $r = c$, $\sigma_r = -q$, giving $A = q - \sigma_0$ Ln c and the stresses become

$$\sigma_r = -q - \sigma_0 \text{Ln}c = -\frac{\sigma_0}{2}\left(1 - \frac{c^2}{b^2}\right) + \sigma_0 \text{Ln}\frac{r}{c}$$

$$\sigma_\theta = \sigma_r + \sigma_0 = \frac{\sigma_0}{2}\left(1 + \frac{c^2}{b^2}\right) + \sigma_0 \text{Ln}\frac{r}{c}$$

At $r = a$, $\sigma_r = -p$, which gives $p = q - \text{Ln } a/c$ as required.

Equation 7.9b for a closed cylinder gives

$$u = \frac{(1 + \upsilon)pa^2 r}{E(b^2 - a^2)}\left[\frac{1 - 2\upsilon}{1 + \upsilon} + \frac{b^2}{r^2}\right]$$

At $a = c$, $p = q$, and $r = c$, we obtain

$$\frac{u_b}{b} = (2 - \upsilon)\frac{qc^2}{Eb(b^2 - c^2)}$$

Since Poisson's ratio in the plastic region is equal to ½ and $q = \sigma_0/2(1 - c^2/b^2)$, it follows that

$$\frac{u_b}{b} = \frac{3\sigma_0}{4E}\left(\frac{c^2}{b^2}\right)$$

$$\frac{4Eu_b}{3b\sigma_0} = \frac{c^2}{b^2}$$

Furthermore, since

$$a \leq c \leq b, \quad \frac{a^2}{b^2} \leq \frac{c^2}{b^2} \leq 1$$

$$\frac{a^2}{b^2} \leq \frac{4Eu_b}{3b\sigma_0} \leq 1$$

P7.13: Equation 7.10 gives $p_2 = 125\ (1 - 0.25) = 93.75$ MPa.
Now $P > p_e$, extent of the plastic zone is determined by Equation 7.19a as

$$150 = 125[1 - c^2/0.09 + 2\ln(c/0.15)]$$

or

$$1 - c^2/0.09 + 2\ln(c/0.15) = 1.2$$

This equation may be solved by simple programing or trial and error. The result is

$$C = 0.215\,\text{m}$$

P7.15: For a thick-walled spherical shell, Equations 7.44 and 7.49 give

$$\frac{p}{p_e} = \frac{1 - c^3/b^3 + 3\text{Ln}c/a}{1 - a^3/b^3}$$

For $0 < c/a < 2$, we have

$$\begin{aligned}\ln c/a &= (c/a - 1) - (1/2)(c/a - 1)^2 + (1/3)(c/a - 1)^3 - (1/4)(c/a - 1)^4 + \cdots \\ &= -11/6 + 3c/a - 3/2(c^2/a^2) + 1/3c^3/a^3\end{aligned}$$

$$3\ln c/a = -11/2 + 9c/a - 9/2c^2/a^2 + c^3/a^3 - \cdots$$

$$p/p_e < (c/a)^3$$

$$\left(\frac{p}{p_e}\right)^{\frac{1}{3}} < \frac{c}{a}$$

$$a\left(\frac{p}{p_e}\right)^{\frac{1}{3}} < c$$

P7.18: We have $\sigma_x = -p$ and $\sigma_z = 0$ (thin plate with fixed upper and lower sides). Hence $\varepsilon_y = 0$. Hooke's law gives

$$\varepsilon_y = \frac{1}{E}[\sigma_y - \upsilon(\sigma_x + \sigma_z)] = 0$$

$$\sigma_y = -\upsilon p$$

The maximum value of σ_θ $\alpha\tau\rho = \alpha$ is obtained from Equation (c) in Section 7.3.1 of the text:

$$\begin{aligned}\sigma_\theta &= -p(1 - 2\cos 2\theta) + (-\upsilon p)[1 - 2\cos(\pi + 2\theta)] \\ &= -(1 + \upsilon)p + 2(1 - \upsilon)p\cos 2\theta\end{aligned}$$

Hence, at $\theta = 0$, $\sigma_\theta = 0.1$ p, and at $\theta = \pi/2$, $\sigma_\theta = (-3 + \nu)$, it follows that the

$$\text{S.C.F.} = [(-3 + 0.3)p]/(-p) = 2.7$$

Chapter 8

P8.7: $T = T_0 + T_1 r$, that is, $90 = 20 + T_1$ (0.1 m), $T_1 = 700°C/m$

The disc is rotating with angular velocity ω RPM during the change in temperature.

For elastic behavior, we can combine the stresses. Thus, combining Equations 8.13 and 6.52, the stresses are

$$\sigma_r = \frac{E\alpha T_1}{3}(a - r) + \frac{3 + \upsilon}{8}\rho\omega^2(a^2 - r^2)$$

$$\sigma_\theta = \frac{E\alpha T_1}{3}(a - 2r) + \frac{\rho\omega^2}{8}[(3 + \upsilon)a^2 - (1 + 3\upsilon)r^2]$$

$$\sigma_z = 0$$

Maximum stresses occur at $r = 0$

$$(\sigma_r)_{max} = (\sigma_\theta)_{max} = \frac{E\alpha T_1 a}{3} + \frac{3 + \upsilon}{8}\rho\omega^2 a^2$$

If the disc remains elastic, then $\sigma_\theta - \sigma_z = \sigma_0$, that is,

$$\frac{E\alpha T_1 a}{3} + \frac{3 + \upsilon}{8}\rho\omega^2 a^2 = \sigma_0$$

Inserting the given numerical values, we obtain

$$\omega^2 a^2 = 0.0269 \times 10^6$$

$$\omega a = 1640\ \text{rad/s}$$

Since 1 RPM $= 2\pi/60 = 0.1047$ rad/s

$$\omega a = 15.664\ \text{RPM}$$

P8.13: For steady-state heat flow, Equation 8.27 gives $T(r) + C_1 \ln r + C_2$, that is

$$200°C = C_1 \ln 0.2 + C_2$$
$$100°C = C_1 \ln 0.3 + C_2$$

Solving for C_1, we obtain $C_1 = -247°C$.
Equation 8.35 gives

$$\sigma_r = A + \frac{B}{r^2} - \frac{205 \times 10^9 \times 11 \times 10^{-6}}{2 \times 0.7} = 1.611 \times 10^6$$

At $r = 0.3$ m, $\sigma_r = 0$, and $T(r) = 100°C$, and at $r = 0.2$ m, $\sigma_r = 0$, and $T(r) = 200°C$.
Inserting these conditions in the previous equation and solving for the constants A and B, we find

$$A = 0.32 \times 10^8 \quad \text{and} \quad B = -11.6 \times 10^6$$

The tangential stress from Equation 8.35 is

$$\sigma_\theta = A + \frac{B}{r^2} - \frac{E\alpha T(r)}{2(1-\upsilon)} - \frac{E\alpha C_1}{2(1-\upsilon^2)}$$

Inserting the numerical values determined earlier, we have

$(\sigma_\theta)_{r=a} = -182\ \text{MN/m}^2$
$(\sigma_\theta)_{r=b} = 140\ \text{MN/m}^2$

P8.15: Using Equations (a) and (b) in Section 8.2 of the text in conjunction with Equation 8.22, the radial displacement of a thick-walled cylinder subjected to heat flow T(r) is given as

$$r\frac{du}{dr} + u = -2\upsilon r\varepsilon_z + 2(1+\upsilon)\alpha rT + \frac{1+\upsilon}{E}(1-2\upsilon)(\sigma_r + \sigma_\theta)$$

The equation governing the equilibrium of the cylinder is

$$r\frac{d\sigma_r}{dr} = \sigma_\theta - \sigma_r$$

Using the above two equations, we have

$$\frac{d}{dr}(ru) = -2\upsilon r\varepsilon_z + 2(1+\upsilon)\alpha rT + \frac{1}{E}(1+\upsilon)(1-2\upsilon)\frac{d}{dr}(r^2\sigma_r)$$

Integration of this equation with respect to r yields

$$ru = -\upsilon r^2\varepsilon_z + (1+\upsilon)\alpha r^2 T + \frac{1}{E}(1+\upsilon)(1-2\upsilon)r^2\sigma_r + D \qquad (1)$$

where D is constant. By dividing Equation (1) by r^2 and modifying the constant D by adding other constants for reasons to be seen later, Equation (1) is equivalent to

$$\frac{u}{r} = -\upsilon\varepsilon_z + (1+\upsilon)\alpha T + (1+\upsilon)(1-2\upsilon)\frac{\sigma_r}{E} + (1-\upsilon^2)\left[\frac{D}{r^2} - \frac{\beta}{\ln(b/a)}\right] \qquad (2)$$

The constant D is determined from the condition that the hoop strain u/r at r = b is identical to that Equation 8.40 in the text. At r = b, $\sigma_{r=0}$, and

$$\frac{u}{r} = -\upsilon\varepsilon_z + (1+\upsilon)\alpha T + \frac{1-\upsilon^2}{E}\sigma_\theta$$

$$\sigma_\theta = \left(\frac{\sigma_0}{E} - \frac{\beta}{\ln b/a}\right)\frac{c^2}{b^2} + \frac{\beta}{\ln b/a}$$

It follows that

$$(1-\upsilon^2)\left[\frac{D}{b^2} + \frac{\beta}{\ln b/a}\right] = (1-\upsilon^2)\left[\left(\frac{\sigma_0}{E} - \frac{\beta}{\ln b/a}\right)\frac{c^2}{b^2} + \frac{\beta}{\ln b/a}\right]$$

$$D = \left(\frac{\sigma_0}{E} - \frac{\beta}{\ln b/a}\right)c^2$$

Hence,

$$\frac{u}{r} = -\upsilon\varepsilon_z + (1+\upsilon)\alpha T(r) + (1+\upsilon)(1-2\upsilon)\frac{\sigma_r}{E} + (1-\upsilon^2)\left[\frac{\sigma_0 c^2}{Er^2} - \frac{\beta}{\ln b/a}\left(\frac{c^2}{r^2} - 1\right)\right]$$

as required.

Chapter 9

P9.2: For a polynomial of the fourth degree, we use the terms from Equations 9.9 and 9.11

$$\phi = A(8z^4 - 24r^2z^2 + 3r^4) + B(2z^4 + r^2z^2 - r^4)$$

and determine the following derivatives:

$$\nabla^2\phi = \frac{\partial^2\phi}{\partial r^2} + \frac{1}{r}\frac{\partial\phi}{\partial r} + \frac{\partial^2\phi}{\partial r^2} = B(28z^2 - 14r^2)$$

$$\nabla^4\phi = \nabla^2\nabla^2\phi = 0$$

Hence, $\phi(r, z)$ is a stress function. The stresses are determined using Equations 9.5a

$$\sigma_r = \frac{\partial}{\partial z}\left(\upsilon\nabla^2\phi - \frac{\partial^2\phi}{\partial r^2}\right) = 96Az + 4B(14\upsilon - 11)z$$

$$\sigma_z = \frac{\partial}{\partial z}\left[(2-\upsilon)\nabla^2\phi - \frac{\partial^2\phi}{\partial z^2}\right] = -192Az + 8B(8 - 2\upsilon)z$$

$$\tau_{rz} = \frac{\partial}{\partial r}\left[(1-\upsilon)\nabla^2\phi - \frac{\partial^2\phi}{\partial z^2}\right] = 96Ar - 4(8 - 7\upsilon)Br$$

If $\tau_{rz} = 0$, then $A = (1/24)\,(8 - 7\,\nu)\,B$. We obtain

$$\tau_{rz} = \sigma_z = 0$$
$$\sigma_r = 28(1+\upsilon)Bz = 36.4Bz$$

which represents pure bending of the circular plate

P9.6: The contact pressure is obtained from Equation 9.66 as

$$\sigma_c = q = 1.5\frac{P}{\pi a^2}$$

where $P = 1000$ N and a is given in Equation 9.68 with $r_1 = r_2 = 0.05$ m and $E_1 = E_2 = 8 \times 10^9$ N/m^2. It follows that

$$a = 0.88\left[\frac{pr}{E}\right]^{\frac{1}{3}} = 0.88\left[\frac{1000(0.05)}{8 \times 10^9}\right]^{\frac{1}{3}} = 0.00162\,\text{m}$$

$$\sigma_c = \frac{1.5 \times 1000}{\pi(0.00162)^2} = 182\,\text{MPa}$$

P9.7: When the forces P are applied at points $r = a$, $\theta = +\pi$, $\theta = -\pi$, and $z = 0$ of the crack surfaces, the opening mode stress-intensity factor is given as

$$k_1 = \frac{\sqrt{2}P}{\pi^2 a^{\frac{3}{2}}}\frac{1}{1 + \left(\frac{z}{a}\right)^2}$$

If the forces are at $r = a$, $\theta = \pi$, $\theta = -\pi$, and $z = b$ of the crack surfaces, then

$$k_1 = \frac{\sqrt{2}P}{\pi^2 a^{\frac{3}{2}}}\frac{1}{1 + \left(\frac{z-b}{a}\right)^2} = \frac{\sqrt{2}P}{\pi^2 a^{\frac{3}{2}}}\frac{1}{1 + z_1^2}$$

$$z_1 = \frac{z - b}{a}$$

Chapter 10

P10.2: The tangential stresses at $r = a$ and $r = b$ are given in Equations (e) and (f) of solved Example 10.2, Section 10.3.2 of the text with $T_0 = 0$. Since the thickness of the cylinder is small in comparison with the outer radius b, let $b/a = 1 + m$ where m is a small number.

We have $\text{Log}\ b/a = \text{Log}\ (1 + m) = m - m^2/2 + m^3/3 - \cdots$

Equation (e) gives

$$(\sigma_\theta)_{r=a} = \frac{\alpha ET_i}{(2 - \upsilon)}\left[\frac{1}{\log(1+m)} - \frac{2}{1 - \dfrac{1}{(1+m)^2}}\right] = \frac{\alpha ET_i}{(2 - \upsilon)}\left[\frac{1}{1 - \dfrac{m^2}{2} + \dfrac{m^3}{3} - ..} - \frac{2(1+m)^2}{2m + m^2}\right]$$

Simplifying the algebra and neglecting terms of order m^2 and higher, we arrive at

$$(\sigma_\theta)_{r=a} - \frac{\alpha ET_i}{2(1-\upsilon)}\left(1 + \frac{m}{3}\right)$$

Similarly, Equation (f) yields

$$(\sigma_\theta)_{r=b} = \frac{\alpha ET_i}{2(1 - \upsilon)}\left[\frac{1}{\log(1+m)} - \frac{2}{(1+m)^2 - 1}\right] = \frac{\alpha ET_i}{2(1 - \upsilon)}\left[\frac{1}{m - \dfrac{m^2}{2} + \dfrac{m^3}{3}} - \cdots - \frac{2}{2m + m^2}\right]$$

$$= \frac{\alpha ET_i}{2(1-\upsilon)}\left(1 - \frac{m}{3}\right)$$

P10.3: Equation (d) of solved Example 10.2 gives

$$(\sigma_z)_{r=a} = \frac{\alpha ET_i}{2(1-\upsilon)\log\dfrac{b}{a}}\left[1 - \log\frac{b}{a}\left(2 + \frac{2a^2}{b^2 - a^2}\right)\right]$$

$$= \frac{\alpha ET_i}{2(1-\upsilon)}\left[\frac{1}{\log(1+m)} - \frac{2}{1 - \dfrac{1}{(1+m)^2}}\right]$$

which is identical to the expression of (σ_θ) at $r = a$ in the previous problem.

Similarly,

$$(\sigma_z)_{r=b} = \frac{\alpha ET_i}{2(1-\upsilon)\log\dfrac{b}{a}}\left[1 - \frac{2a^2}{b^2 - a^2}\log\frac{b}{a}\right] = \frac{\alpha ET_i}{2(1-\upsilon)}\left[\frac{1}{\log(1+m)} - \frac{2}{(1+m)^2 - 1}\right]$$

which is identical to (σ_θ) at $r = b$ in the previous problem.

P10.4: When $a = 0$ is inserted in Equations 10.12 through 10.14, the resulting expressions for the radial and tangential stresses in an annular cylinder are identical to the corresponding expressions for the solid cylinder in Equation 10.8. However, for $r = a = 0$, the expression for the axial stress σ_z in the annular cylinder is not identical with Equation 10.9 for the solid cylinder (it is identical with the expression in Equation 10.11).

Apparently, the end conditions of the cylinder create the discrepancy. Note that for the solid cylinder, Equation 10.9 corresponds to zero axial strain, $\varepsilon_z = 0$, while Equation 10.11 corresponds to zero axial force, $F_z = 0$.

P10.10: In the present application, Equation 10.51 provides the extent of the plastic zone at the edge of an external circular crack as

$$\omega = -1 + \left\{1 + \frac{\pi}{4}\left[\gamma \frac{\Gamma\left(\frac{n-1}{2}\right)}{\Gamma\left(\frac{n}{2}\right)}\right]^2\right\}^{\frac{1}{2}}$$

where γ is given in Equation 10.50

$$\gamma = \frac{E\alpha T_0}{2(1-\upsilon)\sigma_0}$$

For $n = 3$, ω becomes

$$\omega = -1 + \left\{1 + \frac{\pi}{4}\left[\gamma \frac{\Gamma(1)}{\Gamma(1.5)}\right]^2\right\}^{\frac{1}{2}}$$

The definition of the gamma function is

$$\Gamma(x) = \int_0^\infty t^{x-1} e^{-t}\, dt$$

$$\text{So, } \Gamma(1) = \int_0^\infty e^{-t} dt = 1, \Gamma(1.5) = 0.8862$$

Also,

$$\gamma = \frac{200 \times 10^9 \times 11.7 \times 10^{-6} \times 20}{2(0.7)(250 \times 10^6)} = 0.1337143$$

$$\gamma^2 = 0.01788$$

and

$$\omega = -1 + \left\{1 + \frac{\pi}{4}\left[0.01788\left(\frac{1}{0.8862}\right)^2\right]\right\}^{\frac{1}{2}} = 0.01 \text{ cm}$$

Index

D

E

R

S

U

V

W

Y